WITHDRAWN

FOREST RESOURCES POLICY

PROCESS, PARTICIPANTS, AND PROGRAMS

McGraw-Hill Series in Forest Resources

Avery and Burkhart: Forest Measurements
Brown and Davis: Forest Fire: Control and Use
Dana and Fairfax: Forest and Range Policy
Daniel, Helms, and Baker: Principles of Silviculture
Davis and Johnson: Forest Management
Dykstra: Mathematical Programming for Natural Resource Management
Ellefson: Forest Resources Policy: Process, Participants, and Programs
Harlow, Harrar, Hardin, and White: Textbook of Dendrology
Knight and Heikkenen: Principles of Forest Entomology
Laarman and Sedjo: Global Forestry: Issues for Six Billion People
Panshin and De Zeeuw: Textbook of Wood Technology
Sharpe, Hendee, and Sharpe: Introduction to Forestry
Sinclair: Forest Products Marketing
Stoddart, Smith, and Box: Range Management

Walter Mulford was *Consulting Editor* of this series from its inception in 1931 until January 1, 1952.

Henry J. Vaux was *Consulting Editor* of this series from January 1, 1952 until July 1, 1976.

Paul V. Ellefson, *University of Minnesota* is currently our Consulting Editor.

FOREST RESOURCES POLICY

PROCESS, PARTICIPANTS, AND PROGRAMS

Paul V. Ellefson

Professor of Forest Economics and Policy
Department of Forest Resources
University of Minnesota

McGRAW-HILL, INC.

New York St. Louis San Francisco Auckland Bogotá Caracas
Lisbon London Madrid Mexico Milan Montreal
New Delhi Paris San Juan Singapore Sydney Tokyo Toronto

This book was set in Times Roman by Arcata Graphics/Kingsport.
The editors were Anne C. Duffy and Margery Luhrs;
the production supervisor was Friederich W. Schulte.
The cover was designed by David Romanoff.
R. R. Donnelley & Sons Company was printer and binder.

FOREST RESOURCES POLICY
Process, Participants, and Programs

1 2 3 4 5 6 7 8 9 0 DOC DOC 9 0 9 8 7 6 5 4 3 2 1

ISBN 0-07-019415-7

Library of Congress Cataloging-in-Publication Data

Ellefson, Paul V.
Forest resources policy: process, participants, and programs
Paul V. Ellefson.
p. cm.—(McGraw-Hill series in forest resources)
Includes bibliographical references and indexes.
ISBN 0-07-019415-7
1. Forest policy—United States. 2. Forest management—United States. I. Title. II. Series.
SD565.E44 1992
333.75'0973—dc20 91–24517

Photo Credits
Part I and Part III photographs appear courtesy of USDA-Forest Service.

Part II photograph appears courtesy of Minnesota State Senate.

Part IV, Part V, and Part VI photographs appear courtesy of Minnesota Agricultural Experiment Station.

ABOUT THE AUTHOR

PAUL V. ELLEFSON is a professor of forest economics and policy at the University of Minnesota's Department of Forest Resources. His academic training was at Michigan State University and the Universities of Michigan, Minnesota and New Mexico.

His professional experiences have included U.S. Forest Service professional forester in California; U.S. Forest Service research forest economist in Pennsylvania and West Virginia; resource economist in the Michigan Department of Natural Resource's Office of Planning Services; and Director of Forest Policy Programs for the Society of American Foresters (SAF) in Washington, D.C.

Ellefson has served in a number of advisory capacities, including chair of the SAF Task Force on Federal Lands in Alaska; vice-chair of the Minnesota Environmental Quality Board's Advisory Committee on Power Plant Siting and Transmission Line Location; chair of the Subject Group on Forest Policy, Law and Administration of the International Union of Forest Research Organizations; member of the Advisory Committee to the Minnesota Legislature's Joint Select Committee on Forestry; and chair of the SAF National Committee on Forest Policy.

He has been a National Science Foundation Visiting Scientist to the University of Illinois and a lecturer at Auburn University's Mosley Distinguished Environmental Lecture Program. He has served as a consultant to the U.S. Congress Office of Technology Assessment (Advisory Panel on Technologies for Planning Public Natural Resource Programs) and to the Food and Agricultural Organization of the United Nations. Focusing on forest management and research, he has traveled extensively in foreign countries, including Japan, France, Ireland, Finland, Scotland, Norway, Sweden, England, Italy, Greece, Yugoslavia, Australia, New Zealand, Austria, Switzerland, and Germany.

Ellefson received an American Forestry Association award for outstanding service to conservation, and a resolution of appreciation from the Minnesota State Senate for service to forestry. He is a recipient of a Norwegian Marshall Fund Award and is a Fellow of the Society of American Foresters. He is also a consulting editor to McGraw-Hill Publishers (forestry series), and he serves as an associate editor of *Evaluation Review*. He has published over 100 articles in various journals and books. He is the author of U.S. *Wood-Based Industry: Industrial Organization and Performance* (Praeger) and the editor of *Forest Resource Economics and Policy Research: Strategic Directions for the Future* (Westview).

To
Peggy, Jennifer, and Bonnie
and
Jewel V. and Dagney H. Ellefson

CONTENTS

LIST OF FIGURES xix
LIST OF TABLES xxi
PREFACE xxiii

PART I INTRODUCTION 1

1 Presence and Importance of Forests 3

PHYSICAL PRESENCE 3
SOCIAL AND ECONOMIC IMPORTANCE 6
Wood Products 6
Watersheds 6
Fish and Wildlife 7
Recreation 7
Vegetative Support 8
Minerals 8
POLITICAL AND INSTITUTIONAL PROMINENCE 8
Public Acknowledgment 8
Environmental Issue Linkages 9
Legislative Recognition 9
Government-Established Agencies 10
Financial Investments 10
HISTORICAL SIGNIFICANCE 11
National Treasure 11
Exploitation Challenged 11
Deliberate Management 12
Environmental Concerns 12
REFERENCES 13

2 **Forest Resource Policies** 14
DEFINITIONS 15
CHARACTERISTICS 17
DEVELOPMENT 20
Democratic Principles 20
Group Competition 20
Powerful Elites 21
Policy Process 21
Decision-Making Tenets 27
REFERENCES 28

PART II POLICY PROCESS 29

3 **Policy Agenda: Issues to Government** 31
ISSUES AND AGENDAS 31
Issues 31
Agendas 32
AGENDA-SETTING PROCESSES 33
Functional Activity 34
Power and Influence Activity 35
Opportunity Activity 35
Strategic Activity 38
Summary 47
DISPOSITION OF ISSUES 48
Diversion, Deferment, and Displacement 48
Issues and Demands Ignored 49
Agenda Status Obstructed 51
ORGANIZATIONAL AND CITIZEN INVOLVEMENT 54
Governmental Participants 54
Nongovernmental Participants 55
FORESTRY PROFESSIONALS AND AGENDA SETTING 56
CONTEMPORARY FOREST RESOURCE AGENDAS 58
Congressional 59
Federal Forestry Agency 59
State Forestry Agency 60
Agency Policy Analysis Unit 60
Councils and International Agencies 61
Interest Groups 61
Wood-Based Industry 62
AGENDA SETTING IN THE POLICY PROCESS 63
REFERENCES 65

4 **Policy Formulation: Action in Government** 67
FORMS AND POLITICAL CONTEXT 68

IDEAS FOR POLICY OPTIONS 71
Creativity and Imagination 71
Knowledgeable Authorities 73
Deductive Reasoning 73
Institutional Sources 73
COMMUNITIES OF FORMULATORS 75
FORMULATION IN THE POLICY PROCESS 77
REFERENCES 79

5 Policy Selection: Action in Government 81

SELECTION IN CONTEXT 82
Programmed versus Unconventional Selection 82
Selection as a Nonevent 83
Priorities, Scope, and Timing 84
Leverage, Information, and Personality 86
Group versus Individual Decisions 88
SELECTION MODELS AND CONCEPTS 91
Rational Comprehensive 92
Rational Incremental 93
Mixed Scanning 99
Organized Anarchy 100
Summary 103
BARGAINING AND NEGOTIATION 103
Cooperation and Mutual Adjustment 103
Strategies for Securing Agreement 107
Tactics for Securing Agreement 112
Mediation 112
CRITERIA GUIDING CHOICE 114
Technical and Ecological Criteria 115
Efficiency and Effectiveness Criteria 116
Equity and Ethical Criteria 116
Values and Ideologies 118
Procedural Criteria 120
SELECTION IN THE POLICY PROCESS 122
REFERENCES 123

6 Legitimizing Policy Choices: Action in Government 126

STYLES AND FORMS 126
POLITICAL FEASIBILITY 131
Organizing Information 133
Strategies and Tactics 135
LEGITIMIZING IN THE POLICY PROCESS 136
REFERENCES 138

7 Implementing Policies and Programs: Government to Issues **139**

INTERPRETATION, ORGANIZATION, AND APPLICATION 140
Interpretation 140
Organization 141
Application 141
Guidelines 141
SUCCESSFUL IMPLEMENTATION 142
Clarity of Intent 143
Cause-and-Effect Linkage 144
Support of Intent 145
Sufficient Resources 145
Administrative Simplicity 146
Administrative Compliance 147
SIGNIFICANT PARTICIPANTS 148
General Public 148
Government Agencies 148
Legislatures 149
Judicial Systems 150
Special-Interest Groups 151
Policy Beneficiaries 151
IMPLEMENTATION IN THE POLICY PROCESS 151
REFERENCES 153

8 Evaluating Policies and Programs: Programs to Government **154**

CONCEPT AND PROCESS 155
Concept 155
Process 155
Scope 156
PURPOSE AND INTENTION 156
Accountability 156
Control 157
Fostering Change 157
Clarification 157
Ritual and Symbolic 158
Legal Requirement 158
Obstacles 159
OUTCOME AND EFFECT 160
Redefining an Issue 160
Additional Policies Needed 160
APPROACHES TO EVALUATION 161
Informal Evaluation 161

Formal Evaluation 161
Performance Standards 163
Methods and Techniques 165
CLIENT-EVALUATOR RELATIONSHIPS 168
Clients and Administrators 168
Evaluators and Analysts 170
OBSTACLES TO EVALUATION 174
Conflicting Expectations 174
Official Resistance 175
Mistrust of Results 175
Administrative Disparities 176
Operational Problems 176
COMMUNITIES OF EVALUATORS 178
Legislatures, Commissions, and Citizens 178
Operating Staff and Specialized Staff 182
EVALUATION IN THE POLICY PROCESS 185
REFERENCES 186

9 Terminating Polices and Programs 189

CONCEPT AND PURPOSE 190
Issues 190
Degrees 190
Purpose 191
TYPES AND LEVELS 192
Agency or Organization 192
Policies 192
Programs 193
Partial Termination 193
STRATEGIC AND TACTICAL DESIGNS 194
Termination and Policy Development 194
Positive Attitude 195
Gradual Steps 195
Budgetary Growth Periods 195
Leadership Changes 196
Administrative Design 196
Financial Incentives 197
Zero-Based Budgeting and Sunset Laws 197
Specific Tactics 197
OBSTACLES AND DETERRENTS 199
Moral and Intellectual Resistance 199
Ideology of Permanence 200
Political Opposition 200
Legal and Procedural Deterrents 201

TERMINATION IN THE POLICY PROCESS 201
REFERENCES 203

PART III PARTICIPANTS 205

10 Legislative Systems and Processes 207

FUNCTIONS AND RESPONSIBILITIES OF LEGISLATURES 207
Constituent Representation 208
Information and Education 208
Enactment of Law 209
Oversight and Review 214
ORGANIZATION AND PARTICIPANTS 217
Characteristics of Chambers 218
Leadership and Rules of Procedure 218
Legislators 219
Leadership 227
Committees 235
Staff 235
Support Agencies 237
LEGISLATIVE-BUREAUCRACY INTERACTION 238
LEGISLATIVE CHALLENGES 240
Unresponsiveness 240
Parochialism 241
Lack of Innovation 241
Cumbersome Processes 241
Minority Power 242
REFERENCES 242

11 Judicial Systems and Processes 244

OCCURRENCE OF LEGAL ACTIONS 245
LAW AND LEGAL PROCEEDINGS 248
Categories 249
Adversary Process 250
Forms of Illegal Conduct 251
Virtues of Courts 251
ORGANIZATION AND PROCESS 252
Structure 252
Jurisdiction 253
Trial Courts 254
Appellate Courts 255
Participants in Judicial Processes 256

POLICY MAKING 258
Characteristics 258
Appropriate Scope 260
JUDICIAL CHALLENGES 260
REFERENCES 263

12 Bureaucratic Systems and Processes 264

CHARACTER AND ORGANIZATION 265
Structure 265
Procedure 265
Organization 266
MANAGEMENT AND DECISION MAKING 269
Bureaucratic Power 270
Power Struggles 270
Policy Decisions 270
Leadership 272
Legitimizing Policies 272
BUREAUCRATIC LANDSCAPE 274
Federal Agencies 275
State Agencies 282
Agencies Worldwide 286
United Nations' FAO Forestry Department 287
BUREAUCRATIC CHALLENGES 289
Destructive Ideologies 289
Confusion of Interests 290
Ineffective Policy Selection 290
Political versus Technical Skills 291
Accountability, Advocacy, and Information Sharing 291
REFERENCES 292

13 Interest-Group Systems and Processes 294

FUNCTION AND PURPOSE 295
Conflicting Views 296
Policies through Conflict 296
Variety of Focal Points 297
UNIVERSE AND CHARACTER 298
Growth in Numbers 298
Categories of Groups 298
GOVERNANCE AND ORGANIZATION 307
Leadership 307
Administrative Structure 308
Governing Boards, Officers and Staffs 308
Government Regulation 309

POWER AND INFLUENCE 310
Decision to Influence 311
Strategies and Tactics 316
Measures of Success 322
INTEREST GROUP CHALLENGES 325
REFERENCES 328

14 The General Public, Political Parties, and the Mass Media 330

GENERAL PUBLIC 330
POLITICAL PARTIES 332
MASS MEDIA 333
REFERENCES 336

PART IV POLICIES AND PROGRAMS 337

15 Planning and Budgeting Use and Management of Forests 339

STRATEGIC PROGRAM PLANNING: THE RPA 339
Strategic Planning 339
Resource Assessment 340
Resource Program 341
Presidential Policy Statement 345
Statements of Intent and Accomplishment 345
Design versus Reality 345
Planning Principles 352
LAND USE AND MANAGEMENT PLANNING: THE NFMA 353
Planning Process 354
Administrative Procedures 355
Management Actions 356
Standards and Guidelines 357
Clearwater National Forest: An Example 357
Design versus Reality 359
FISCAL AND BUDGETARY PLANNING 360
Character of Budget Process 361
Budget Development Strategies 365
Budgetary Challenges 366
REFERENCES 367

16 Private Forestry Program Initiatives 370

NONINDUSTRIAL PRIVATE FORESTS 370
Extent of Ownership 370

Benefits Produced 371
Deterrents to Management 373
Policy and Program Options 373
INDUSTRIAL PRIVATE FORESTS 375
Extent of Ownership 375
Timber Production 376
Timberland Management Policies 377
Ownership Disadvantages 378
Professional Forestry Staff 379
Public Interest in Industrial Forests 379
REFERENCES 379

17 Information and Service Program Initiatives 381

DIRECT TECHNICAL ASSISTANCE 383
Public Sponsorship 383
Private Sponsorship 388
Technical Assistance Challenges 389
EXTENSION OF INFORMATION 390
Educational Activities 390
Federal, State, and Local Cooperation 391
Program Planning and Funding 391
Extension Challenges 392
INDIRECT TECHNICAL ASSISTANCE 392
Pest Management 393
Fire Protection 393
Research 393
REFERENCES 395

18 Fiscal and Tax Program Initiatives 397

FISCAL INCENTIVES 398
Forestry Incentives Program 398
Stewardship Incentives Program 400
Agricultural Conservation Program 401
Conservation Reserve Program 402
State Programs 403
Fiscal Incentive Challenges 404
TAX INITIATIVES 404
Income Taxes 405
Property Taxes 407
Tax Program Challenges 409
REFERENCES 409

19 Government Regulatory Initiatives **412**

LAND-USE REGULATION 413
- Federal Regulations 413
- State Regulations 414

POLLUTANT REGULATION 415
- Pesticides 415
- Water Pollutants 415
- Air Pollutants 416

FOREST PRACTICES REGULATION 417
- Administrative Environment 418
- Regulatory Landscape 421
- Costs and Accomplishments 427
- Forest Practice Regulation Challenges 431

REFERENCES 432

20 Government Forest Ownership **434**

EXTENT AND CHARACTER 435

PRODUCTS AND SERVICES 436

POLICIES AND PROGRAMS 437
- USDA-Forest Service Policies 437
- USDI-Bureau of Land Management 439
- USDI-National Park Service Policies 441
- USDI-Fish and Wildlife Service Policies 441
- Other Federal Policies 442
- State and County Policies 442
- Government Ownership Challenges 443

REFERENCES 444

PART V SUMMARY AND OBSERVATIONS **445**

FOREST SCIENCE IN THE POLICY PROCESS 448

ETHICS AND POLICY DEVELOPMENT 452

INDUSTRIAL-SECTOR POLICY DEVELOPMENT 458

RESEARCH ON POLICY DEVELOPMENT 460

REFERENCES 470

PART VI GLOSSARY AND SELECTED FEDERAL LAWS **473**
- Glossary 475
- Selected Major Federal Laws Addressing the Use and Management of Forest Resources 487

INDEX 495

LIST OF FIGURES

2-1 Policy Development and Implementation Process, by Policy Event and Policy Product 22
3-1. Sequence of Events Involved in Institutional Agenda Setting Undertaken as a Strategic Activity 38
3-2. Strategies for Preventing Institutional Agenda Status for an Issue, by Orientation and Directness of Strategy 52
5-1. Rational Incrementalism as a Sequence of Events and Activities 94
5-2. A Bargaining Relationship Example Involving Government and Private Owners of Forestland 105
8-1. Research Designs for Policy and Program Evaluation 166
10-1. The Sequence of Events Involved in the Enactment of a Law 210
10-2. The Sequence of Events Involved in Legislative Voting Decisions of Legislators 225
11-1. Factors Influencing the Role and Decisions of Judges in a Judicial System 257
12-1. Administrative Organization of the USDA-Forest Service, 1989 277
12-2. Administrative Organization of the U.S. Environmental Protection Agency, 1988 281
12-3. Administrative Organization of Public Forestry Organizations in Georgia, Minnesota, and Oregon, 1989 284
13-1. Interest Group Competition Model of Policy Development 297
13-2. Society of American Foresters Procedures for Developing National Positions on Forest Resource Issues, 1988 315
19-1. Forest Practices Application-Notification Form Required of Forest Landowners Intending to Harvest Timber in the State of Washington, 1990 426
S-1. Possible Responses to Ethical and Value Conflicts 457

LIST OF TABLES

1-1. Forest and Woodland Area in the World, by Region, 1985–1987 4
1-2. Forestland Area in the United States, by Region, 1987 5
2-1. Forest and Related Resource Policy Statements 17
5-1. Hypothetical Net Social Benefits Produced by Forest Resource Policies, by Policy Option and Society Beneficiary 118
8-1. Types of Policy Analysts by Political and Technical Behavior 171
10-1. Importance of Various Actors on Legislative Voting Decisions of Members of the U.S. House of Representatives 223
10-2. State Legislative Policy Committee (Forestry and Natural Resources) Staff Communication with Organizations Active in State Forest Policy Matters, by Type of Organization and Frequency of Communication, 1990 228
10-3. State Legislative Policy Committee (Forestry and Natural Resources) Staff Assessment of the Adequacy of Forestry and Related Information Available for Legislative Use, by Type of Information and Adequacy Rating, 1990 229
11-1. U.S. Federal Court Case Decisions, by Forestry Subject Area and by Type of Court, 1975–1988 246
12-1. Selected Federal Agencies Responsible for the Use and Management of Forest and Related Resources, by Agency Title, Number of Employees, and Annual Budget, 1988 274
13-1. Organized Interest Groups Having Interests in Forest and Related Resources, 1987–1989 300
13-2. Strategies and Tactics Used by Organized Interest Groups to Influence Development of Public Policies, 1982 317
13-3. Financial Contributions to Candidates for Federal Office made by PACs with Interests in Forest and Related Resources, 1983–1984 321
15-1. Selected Findings and Opportunities Identified by the 1989 RPA Assessment of U.S. Forest and Rangeland Resources 342
15-2. Recommended Renewable Resources Program (1990–2040) and Long-Term Strategy for the National Forest System 346
15-3. USDA-Forest Service State and Private Forestry Accomplishments in 1987 Compared to RPA-Recommended Program Levels for 1987 350
15-4. National Forest Land-Management Planning Process 355

15-5. Development and Execution of a Federal Budget, by Stage, Timing, and Government Participant 361
16-1. Timberland Area in the United States, by Region and Ownership, 1987 372
17-1. State-Administered Forestry Programs Focused on Major Private Forestry Activities, by Program Type and Region, 1985 386
17-2. Federal-State Government Cooperative Private Forest-Management Assistance Activities, by Region, 1989 388
18-1. Funding and Accomplishments of the Forestry Incentives Program and the Agricultural Conservation Program, 1977–1990 399
19-1. Major Components of a State's Private Forest Practices Regulatory System 419
19-2. Administrative and Enforcement Expenditures for State Forest Practice Programs, by Selected States and State Agencies, 1984 428
19-3. Public- and Private-Sector Costs of Administering and Complying with State Forest Practices Programs, by Selected States, 1984 429
S-1. Code of Ethics: American Society for Public Administration, 1985 455
S-2. Code of Ethics: Society of American Foresters, 1989. 456

PREFACE

The study of forest resource policy has been an intergal part of the formal training of forestry professionals since the turn of the century. There has been an implicit recognition that forestry is practiced within the context of various political and administrative environments. In fact, some of the most exciting forestry history describes the excursions of noted forestry professionals into the halls of legislatures, the chambers of courtrooms, and the inner workings of large bureaucracies. Over time, forest policy has been presented to students in a number of fashions. A historical framework was especially popular prior to the 1950s; through the 1960s, attention focused on case examples of how specific laws were enacted and implemented. In recent years, university schools of public affairs have recognized the need to instill in students a better appreciation of the political, administrative, and institutional forces which repeatedly merge to produce an often incredible variety of public policies and programs. The fundamental structure of this book is based on: the *process* by which forest resource policies are developed, implemented, and disposed of; the *participants* who trigger and energize the process; and examples of major forest resource *policy and program* initiatives which have resulted from the interaction of process and participants.

Approaching the study of forest resource policy from the perspective of process, participants, and programs has a number of advantages. For example, students avoid the danger of focusing their energies on a single event in the policy process (such as implementation or evaluation) with the expectation that doing so will result in a socially desirable policy. Development of effective forest policies requires understanding of a number of political events ranging from agenda setting and formulation to policy selection and termination. Moreover, since the policy process is not unique to any one organization, students' understanding of key process elements can often be transferred to any organization that becomes engaged in the process. To be sure, policies and programs are seldom the product of a single political institution; many organizations (including legislatures, bureaucracies, courts, and interest groups) can and do become involved in the development and implementation of policies. By broadening their recognition and understanding of the organizational landscape within which policy development occurs, students become better equipped to avoid the perils of focusing on one institution (for example, a legislature) at the expense of others that may be more active and more influential in the politics of developing forest resource policies. Then too, by focusing on policies and programs in the context of process and participants, students become more appreciative of the essence (and especially the shortcomings) of policies which have been produced by the interaction of the two.

The material presented here is designed for upper-level undergraduate and graduate-level courses. The subject matter can be used as a core around which instructors can foster additional student insight via examination of contemporary forest policy literature and discussion of student and instructor experiences with policy development. The descriptions of policies and programs presented here are only highlights of very complex activities. The literature devoted to the specifics of contemporary forest resource policies is especially rich; it should be used liberally to complement the central features introduced here. Since public forest resource policies have been developed, challenged, and redeveloped at a heightened pace in recent years, the often-changing literature describing such changes should be made readily available to students as they seek to satisfy their often-burgeoning interest in the development and implementation of forest resource policies.

The preparation of a book is never the product of a single mind. Many students have reviewed and challenged the manuscript, and have suggested alternative approaches to the study of forest resource policy. Their insights have been invaluable. In addition, the author appreciates the contributions of many practitioners of policy development with whom he has become acquainted over the years. There is no substitute for suggestions from active participants in policy development and implementation. Many fine suggestions have also been made by the following formal reviewers of early drafts of the book, who sharpened the focus of many chapters and enriched the presentation of concepts and principles: Thomas M. Bonnicksen, Texas A&M University; David K. Lewis, Oklahoma State University; Kerry F. Schell, University of Tennessee; and Karen Potter-Witter, Michigan State University. The University of Minnesota's Department of Forest Resources and Agricultural Experiment Station also deserves recognition. The author assumes full responsibility for flaws contained in the book.

Paul V. Ellefson

ONE

INTRODUCTION

The physical presence of forest resources on the nation's landscape is vast and richly diverse. But physical presence alone is not a virtue. Forests must be transformed by society according to important social and political values which reflect a broader interest in assuring citizens of healthy and comfortable lives and surrounding them with ample opportunity for leisure pursuits. The transformation of the physical presence of forests into something of value to society is often accomplished by a willingness to establish and carry out forest resource policies. An understanding of how such policies are developed and implemented and the topics toward which they are directed is crucial to the activities of professional managers of forests and related natural resources.

1

PRESENCE AND IMPORTANCE OF FORESTS

Forests are an important source of a rich variety of benefits sought by nearly everyone in the world. Forests are a far-reaching presence on worldwide landscapes. They make considerable contributions to the social and economic fabric of many societies. Further, many citizens (or their agents) have an intense interest in the ways forests throughout the world are used and managed. The importance of forests is commonly identified by a number of measures, including area of land covered by forest vegetation, extent of forest-based employment and income, and public recognition of important forest resource issues. Some of these measures and their magnitude are discussed below.

PHYSICAL PRESENCE

Forests are a major land use, occupying over 10 billion acres or nearly one-third of the world's land area (Table 1-1). They are especially common in South America, where they occupy over half the land area, and in the U.S.S.R., where 4 out of 10 acres of land are forested. Combined, these two areas account for 45 percent of the world's forestland. In North and Central America, forests are located primarily in Canada and the United States. Canada has nearly 870 million acres of forests. Forests in the United States occupy 731 million acres and are located in all 50 states (Table 1-2). They are especially prevalent in the eastern part of the nation, which has over half the nation's forests and where they are in close physical proximity to nearly three-quarters of the country's

TABLE 1-1
Forest and woodland area in the world, by region, 1985–1987

	Area of forest and woodland			
			Proportion of	
Region	Hectares (thousands)	Acres (thousands)	Region's total land area (percent)	World forest area (percent)
Africa	689,254	1,701,768	23	17
North and Central America	684,544	1,690,139	32	17
South America	904,628	2,233,526	52	22
Asia	539,890	1,332,988	20	13
Europe	157,164	388,038	33	4
U.S.S.R.	942,667	2,327,445	42	23
Australia and Oceania	156,280	385,855	20	4
Total	4,074,427	10,059,759	—	100

Note: Forest and woodland is land containing natural stands of woody vegetation in which trees predominate.

Source: World Resources: 1990–1991, by World Resources Institute. Reprinted by permission of Oxford University Press, New York, 1990.

population—especially to persons living in major urban centers (Forest Service 1980, 1989; World Resources Institute 1990).

Forests are richly diverse ecosystems, the exact composition of which is greatly influenced by climate, soil, and topographic conditions. In the frostfree equatorial regions of the world, for example, tropical broadleaf forests abound, while in areas adjacent to tundra in the north, coniferous forests are commonplace. Between such extremes there often exists an incredible mixture of trees and related species. The diverse nature of forest ecosystems means that they are physically able to supply an abundant assortment of goods and services, which in turn are capable of satisfying an equally diverse collection of human needs. Of special importance is the often unique ability of forests to simultaneously produce or influence the production of a broad assortment of benefits, including wood, water, recreation, wildlife, forage, and aesthetic beauty. The physical prominence of forests is also heightened by their often complementary association with and support of closely related vegetative types, such as natural grasslands, savannas, coastal marshes, tundra, and wet meadows. Too, forests are often found in close proximity to valuable mineral resources which often lie beneath them. Forests are physically widespread on the geographic landscape of the world. Nearly all the world's more than 5.2 billion people are in some ways dependent on forests.

TABLE 1-2
Forestland area in the United States, by region, 1987

Region	Forest land				Total land area	
			Proportion of			
	Hectares (thousands)	Acres (thousands)	Region's total land area (percent)	U.S. forest area (percent)	Hectares (thousands)	Acres (thousands)
Northeast	34,501	85,251	67.4	11.7	51,205	126,526
North Central	32,465	80,221	28.0	11.0	155,844	286,247
Great Plains	1,712	4,229	2.2	0.6	78,527	194,037
Southeast	35,510	87,744	59.8	12.0	59,362	146,682
South Central	46,840	115,741	29.8	15.8	156,907	387,713
Pacific Northwest	72,424	178,958	38.4	24.4	188,556	465,916
Pacific Southwest	16,645	41,129	39.6	5.6	42,042	103,884
Rocky Mountains	55,891	138,104	25.3	18.9	221,215	546,614
United States, total	295,987	731,377	—	100.0	953,658	2,257,619

Note: Forestland is land at least 10 percent stocked by forest trees of any size, including land that formerly had such tree cover and that will be naturally or artificially regenerated.

Source: Forest Statistics of the United States: 1987 by K. L. Waddell, D. D. Oswald, and D. S. Powell. Resource Bulletin PNW-RB-168. Pacific Northwest Forest and Range Experiment Station, USDA-Forest Service, Portland, Oreg., 1989.

SOCIAL AND ECONOMIC IMPORTANCE

Wood Products

Forests are also important for economic and social reasons. They are an especially abundant source of wood fiber, which in one form or another—housing, furniture, containers, writing paper, books, newspaper—affects the quality of life of nearly all human beings. In the late 1980s, roundwood production from the world's forests exceeded 3,255 billion cubic meters, a total which was divided nearly equally between fuel and charcoal uses and industrial roundwood uses (sawnwood, panels, pulp, and paper). The single largest 1987 producer of roundwood in the world economy was the United States (15 percent), followed by China (8 percent), India (8 percent), and Brazil (7 percent) (World Resources Institute 1990).

In 1986, the United States harvested 510 cubic meters (nearly 18 billion cubic feet) of roundwood products from the nation's forest-growing stock—an amount expected to climb to 765 cubic meters (27 billion cubic feet) by the year 2040. Over 1.6 million persons (8.2 percent of all manufacturing employees) are employed in the manufacturing of wood-based products; they are annually paid nearly $27 billion in wages. The value added by the nation's more than 49,000 wood-based establishments exceeds $60 billion—approximately 7 percent of the national total. Over 6 percent of the United State's gross national product (GNP) can be traced to the manufacture and use of wood products (Forest Service 1980, 1989).

Watersheds

Forests are also important as a source of most of the high-quality water in the United States (over 60 percent of all stream flow). Mostly located at high elevations, in favorable topographic positions, and in areas impacted by major moisture-producing weather systems, the forests of the United States receive nearly twice as much average yearly precipitation as does land allocated to other uses, and yield upward of four times the annual runoff produced by land that lacks a forest cover. Water flowing from forested watersheds is used for a variety of in-stream purposes (e.g., navigation, fish and wildlife habitat, hydropower generation, water-oriented recreation) and is commonly transported for use in irrigation, municipal consumption, and industrial manufacturing. By slowing rates of overland flow and enhancing infiltration rates, forests also play a significant role in enhancing surface water quality and the recharge of groundwater. Freshwater (water from a stream or an aquifer) withdrawals in 1985 totaled some 344 billion gallons per day in the United States (259 billion from surface water), of which 79 percent was used for irrigation and thermoelectric cooling. Withdrawals are expected to increase more than 50 percent over the next 50 years (Forest Service 1980 and 1989).

Fish and Wildlife

The extensive nature of forest ecosystems also makes forests an especially important source of habitat for fish and wildlife. Their variety, successional stages, and arrangement across landscapes enables them to fulfill at least part of the life-cycle requirements of nearly 90 percent of the resident common migrant vertebrate species in the United States. At least 90 percent of the total bird, amphibian, and fish species and at least 80 percent of the nation's mammal and reptile species are represented in forest ecosystems. As an example, big game (especially deer) associated with forests in the United States are in large measure why over 12 million people annually go hunting for sport. In 1985, big game accounted for 60 percent of all hunter-related expenditures (an extremely important source of revenue for many local communities). Likewise, nearly 11 million people (5 percent of the U.S. population) annually hunt small mammals (e.g., squirrels, rabbits) and resident birds (e.g., grouse, quail)—fauna which are largely dependent on forests and associated rangeland for their existence. Another significant statistic is that forests provide habitat to over 80 of the nation's more than 300 threatened or endangered species of flora and fauna (Forest Service 1980 and 1989).

Recreation

Forests also provide opportunities for virtually all interested citizens to enjoy recreational pursuits in dispersed or relatively undeveloped and solitary settings. Such pursuits can take a number of forms, ranging from camping to hiking, from skiing to snowmobiling, and from horseback riding to driving for pleasure. In the United States, lands subject to the administration of the U.S. Department of Agriculture (USDA)—the USDA-Forest Service—and the U.S. Department of the Interior (USDI)—the USDI National Park Service—were visited by nearly 342 million recreational visitors in 1986. Nearly all these visits were associated with forestland. Camping at developed federal and state areas is among the more popular outdoor recreation activities associated with forests: nearly 35 percent of the nation's population spent approximately 7 days camping in these areas in 1987.

Activities of similar prominence that likewise benefit from forest environments are sightseeing, picnicking, and walking or driving for pleasure. More than 4 out of 10 individuals participate in each of these activities one or more times annually. Forest-oriented recreational sightseeing, backpacking, and day hiking are expected to double by the year 2040.

A unique and often forest-based recreational resource is wilderness, especially the 89 million acres of federal public land in the National Wilderness Preservation System. In the USDA-Forest Service portion of the system, over 12 million recreation visitor days occurred in 1986, much of which was in or closely associated with forests. Outdoor recreation is possible in many situations other than forests, and yet a forested recreational area oftentimes possesses special attributes that are uniquely important to the recreational pleasures of millions of Americans (Forest Service 1980 and 1989).

Vegetative Support

Forests are also socially and economically important to the extent that they are associated with or supportive of other types of vegetation. An important example is range vegetation (e.g., grasses, forbs, shrubs), which is often supported as an understory within forest ecosystems or which may appear as part of the mosaic of vegetative types on larger forested landscapes. In the United States, rangeland vegetation supports over 10 million animal unit months of livestock grazing in the National Forest System and over 12 million such units on rangeland administered by the USDI-Bureau of Land Management. It is also an important source of sustenance for wild herbivores such as elk, deer, antelope, burros, and wild horses. In addition to grazing, rangeland is prominent in the production of significant quantities of high-quality water, serves as a home to a number of threatened and endangered species of plants and animals, and provides the basis for a significant variety and number of recreational experiences (Forest Service 1980 and 1989).

Minerals

Forests are also associated with a variety of minerals, including coal and oil used for energy; metallic minerals such as lead, copper, and cobalt; precious metals and gems; and common building materials such as sand, gravel, and clay (e.g., one-quarter of all the U.S. potential energy reserves lie within the National Forest System). Such resources often occur in the same locale as forests and therefore must be appropriately considered in making decisions concerning the use and management of forests. Unlike forests, most minerals are finite, nonrenewable, and difficult to inventory, which does not, however, detract from their importance to local and national economies. For example, National Forest System receipts for rents, royalties, and sales of minerals totaled nearly $150 million in 1987. In 1985, the U.S. minerals industry contributed $122.8 billion to the nation's GNP. An important portion of this contribution can be traced to locations covered with forests (Forest Service 1980 and 1989).

POLITICAL AND INSTITUTIONAL PROMINENCE

Public Acknowledgment

The enormous expanse of forests and the importance attached to them by society has not gone unnoticed. They are, for example, acknowledged by and of meaningful concern to a large portion of the general public in the United States.

Nationwide polls in 1989 found that 97 percent of the population had formed an opinion on "maintaining trees and forests": 43 percent believed the country was doing a good job in such an undertaking, while 54 percent thought it was not. In the same year, 86 percent of those polled nationally had formed an opinion on "cutting trees down and leaving dry land": 54 percent indicated that this practice was a very serious problem for the nation, whereas 32 percent thought that it was not. Similarly, 94 percent of the population recognized deforestation as a forestry issue in 1989; 78 percent judged the issue to be of sufficient concern to warrant at least prompt action by the government.

Nationwide polls in 1978 acknowledged the general public's ability to relate to specific forestry issues: 59 percent were of the opinion that National Forest reforestation efforts should be expanded, 47 percent wanted more incentives for private forest landowners, and 21 percent urged restrictions on National Forest timber harvesting. In the same year, 81 percent of those polled recognized and viewed wildlife habitat as an important use of National Forests, and 70 percent considered timber production to be an important use. Similarly, in 1988, a large proportion of the general public was cognizant of forest and related wilderness issues: 47 percent argued that wilderness should be preserved at the expense of developing new energy sources (42 percent said it should not). In 1985, 55 percent of the public viewed acid rain as a serious problem for forests (Americans for the environment 1989, American Forest Institute 1985, Cheek 1979).

Environmental Linkages

Forests are also prominent as a result of their often close association with intensely controversial and highly visible environmental issues. Such issues have frequently propelled forests and a variety of forestry practices to the forefront of media and citizen attention. The use of pesticides in forest environments, the allocation of public forests to formally designated wilderness status, the forestry consequences of atmospheric pollutants, the role of fire in forest environments, the trade of wood-based products between nations, and the role of forests in sustaining endangered species of plants and animals are examples of forestry issues which are often in the national spotlight of public affairs because of their close association with broadly defined environmental issues. To the extent that forests are party to continuing high-level public interest in environmental matters in general, they will also continue to be highly visible in regional, national, and worldwide political settings.

Legislative Recognition

The political prominence of forests is reflected by the willingness of state and national legislative systems to address important issues concerning their use and management. The U.S. Congress, for example, often brings forests to public attention. They do so by sponsoring public hearings on such topics as tropical deforestation, public-land wilderness, herbicide use in forests, timber sale contract relief, and clear-cutting practices on national timberlands. They also enact important, far-reaching laws (e.g., Wilderness Act of 1964, Land and Water Conservation Fund Act of 1965, National Forest Management Act of 1976, Cooperative Forestry Assistance Act of 1978, Federal Water Pollution Control Act Amendments of 1972) which address concerns of citizens in general and of organized interest groups in particular.

Government-Established Agencies

The prominence of forests is also shown in society's willingness to establish and promote government agencies that are responsible for their protection and management. In the United States, the USDA-Forest Service, the USDI-Bureau of Land Management, and the USDI-National Park Service are examples of large and costly bureaucracies which are responsible for conducting forestry business on behalf of society. Interest in forests is also reflected by the existence of agencies that have broader responsibilities than just forestry but whose programs directly impact the use and management of forests. Examples in the United States are the Environmental Protection Agency (EPA), the USDI-Fish and Wildlife Service, the USDA-Soil Conservation Service, the Tennessee Valley Authority (TVA), and the Internal Revenue Service (IRS) of the U.S. Department of the Treasury. Not to be overlooked are the more than 100 similar agencies and bureaus that exist in nearly all 50 state governments in the United States.

Financial Investments

The importance of forests is also reflected in the willingness of public and private organizations to make substantial sums of money available for the management and protection of forests. For example, in the United States the federal government's investment in natural resources and the environment totaled more than $15 billion in 1988—a goodly portion of which was focused on forests and closely related natural resources. The U.S. Congress appropriated over $2.5 billion in 1988 to the USDA-Forest Service (which has nearly 32,000 employees) for purposes of carrying out a variety of forest policies and programs. Federal financial support afforded to other agencies involved in forestry is likewise significant. In 1988, for instance, the USDI-Bureau of Land Management received $842 million in federal funds, the USDI-National Park Service $956 million, and the USDI-Fish and Wildlife Service $646 million. At the state level, public investment in forestry programs (exclusive of federal cost share) totaled over $775 million in 1987 (General Services Administration 1988, and Office of Management and Budget 1989).

HISTORICAL SIGNIFICANCE

Complementing the physical, economic, and political importance of forests is their historical significance. In North America, Native Americans in pre-Columbian times secured food and building materials from forests. Their conduct toward forest resources was often guided by self-imposed rules and customs, including a conscious effort to foster the establishment and growth of certain tree species. Whether or not explicitly recognized as such, these customs and rules were probably among the earliest forest policies in North America. They provided Native Americans with the guidance necessary to survive in a variety of often harsh and unkind natural environments (Bonnicksen 1982).

Contemporary policies on the use and management of forests in the United States are the products of gradually evolving attitudes toward forests and the practice of forestry in general. They reflect the nation's response to major forest resource issues of the times and represent a willingness of individuals and institutions to negotiate their interest in solutions needed to address such issues. Forestry issues requiring policy responses have not occurred in isolation; most often they have been thrust on the national scene as part of broader national events, including economic depressions, armed conflicts, political turbulence, and energy crises. Observed from a historical perspective, forest resource policy in the United States can be described as evolving through a number of broad stages. These stages have been well described, especially by Dana and Fairfax (1980), Nash (1990), and in various issues of *Forest History*.

National Treasure

Forests from the late 1700s through the early 1890s were viewed as a national treasure to be used in building a nation. Natural resources of all types—including forests—were considered inexhaustible and were for the most part available for the taking. Especially noticeable was the rush to dispose of the public domain and its timber resources. In this respect, the Free Timber Act of 1878 and the Timber and Stone Act of 1878 were important.

Exploitation Challenged

Toward the end of the nineteenth century, prevailing laissez-faire attitudes toward natural resources began to be seriously challenged, especially by persons such as John Muir, Frederick Olmsted, and Henry Thoreau. These challenges helped to nourish the establishment of the nation's first national park, the Yellowstone National Park, in 1872. The myth of inexhaustible resources was eroding throughout the late 1800s. Countrysides denuded of forests, especially in the East, made it obvious that something had to be done. Thus began the period of awakening as epitomized by creation of federal forest reserves.

The period 1891 to 1911 was one of the most colorful eras in American forest history. Legal authority to reserve forestland in the public domain was combined with the vision of leaders such as Theodore Roosevelt and Gifford Pinchot. The results were the creation of the USDA-Forest Service, the establishment of the National Forest System (composed of federal forest reserves), and the crystallization of a wise-use conservation philosophy: natural resources were to be used, not preserved.

Deliberate Management

This heady era was followed by one involving less crusading and more maturing, from 1912 to the late 1960s. Policies fostering the deliberate management of forests and an accompanying recognition of the importance of privately owned forests were gradually developed. Cooperative arrangements for forest management were established among state, federal, and private forestland owners, and a sound and comprehensive program of forestry research was set in place. The wise-use conservation philosophy was increasingly challenged by people who were more interested in the natural beauty and scenic attractiveness of forests than in their ability to provide timber. The challenge fostered establishment of the USDA-National Park Service and foreshadowed the political turbulence which was to engulf the forestry community in the 1970s and 1980s. Other forestry interests besides timber emerged, especially interest in recreational use of forests and a budding interest in establishment of formally designated wilderness areas. A multiple-use strategy toward forestland management developed as an appropriate response to these fundamental changes. An especially important organization which converted much of the adversity associated with the Great Depression of the 1930s into extremely positive contributions to the development of the nation's forests was the Civilian Conservation Corps.

Environmental Concerns

The period from the late 1960s to the present encompasses policy responses to important concerns about environmental quality in general. Such concerns have in large measure been fostered by often extraordinary citizen activism and by the involvement of important institutions which heretofore had been only modestly involved in the development of forest resource policies (legislatures, judicial systems, interest groups). More users competing for additional benefits from a limited forest resource base led to a passage of a plethora of environmental quality laws that were not specifically focused on forestry but did have an often dominant influence on the use and management of forests. Examples are the National Environmental Policy Act of 1970 and the 1972 amendments to the Federal Water Pollution Control Act.

In many cases programmatic efforts to attain environmental quality goals led to the establishment of complex and detailed federal and state processes and resulted in greater exercising of state and federal government police powers over private forestry matters. Growth in citizen activism in forestry resulted in significant growth in the number and membership of organized interest groups with an interest in forests, and very often led as well to establishment of universal requirements for citizen participation in decisions regarding the use and management of forests, especially public forests. Intense political battles over the use of public forests (especially efforts to formally designate wilderness areas) and a broadening of public and professional concern for forests in a global setting have also been hallmarks of the current era.

REFERENCES

American Forest Institute. *The Forest Products Industry in Perspective: Public Opinion and the Forest Products Industry.* (Outlook in 1985.) Washington, 1985.

Americans for the Environment. *The Rising Tide: Public Opinion, Policy and Politics.* (An Environmental Impact Statement.) Washington 1989.

Bonnicksen, T. M. The Development of Forest Policy in the United States, in *Introduction to Forest Science* by Raymond A. Young (ed.). John Wiley and Sons, New York, 1982, pp. 7–36.

Cheek, G. C. Understanding Public Opinion about Forests and Forestry, in *Proceeding of the Society of American Foresters National Convention: 1978.* Society of American Foresters, Bethesda, Md., 1979, pp. 158–162.

Dana, Samuel T., and Sally K. Fairfax. *Forest and Range Policy: Its Development in the United States.* McGraw-Hill Book Company, New York, 1980.

General Services Administration. *The United States Government Manual 1987–88.* Office of the Federal Register, National Archives and Records Service, Washington, 1988.

Nash, Roderick F. *American Environmentalism: Readings in Conservation History.* McGraw-Hill Publishing Company, New York, 1990.

Office of Management and Budget. *Budget of the United States Government: Fiscal Year 1990.* Executive Office of the President, U.S. Government Printing Office, Washington, 1989.

USDA-Forest Service. *An Assessment of the Forest and Range Land Situation in the United States.* FS-345. U.S. Department of Agriculture, Washington, 1980.

USDA-Forest Service. *Analysis of the Minerals, Timber, Water, Range Forage, Wildlife and Fish, and Recreation and Wilderness Situation in the United States.* (Multiple volumes.) U.S. Department of Agriculture, Washington, 1989.

World Resources Institute. *World Resources: 1990–91.* Oxford University Press, New York, 1990.

Worrell, Albert C. *Principles of Forest Policy.* McGraw-Hill Book Company, New York, 1970.

2

FOREST RESOURCE POLICIES

Forests occupy vast areas of land and are capable of producing a variety of valued goods and services. They also are frequent focal points for intense political discussions. Given such circumstances, how does a society composed of enormously diverse interests develop policies and programs appropriate to the forestry concerns of its citizens? How does a society organize itself to address the range of far-reaching and socially important issues concerning the use and management of forests which always seem to be present on political agendas? What social customs and formalized rules does a society apply when faced with important decisions about alternative policies and programs? How are diverse, often intensely held, convictions accommodated by these decisions? Once made, how are such decisions sustained over time? Under what conditions can they be changed or totally abandoned? Who participates in the selection of major and far-reaching policies? Who has the authority to decide which forestry conditions are most appropriate to society's best interest? Do the participants in the policy-making process owe allegiance to a formally organized group or bureaucracy, or are they individual citizens concerned with immediate access to forests and the goods and services forests are capable of producing? Once the "how" and "who" concerns of policy development have been understood, what can be said about the substance of the policies and programs that result?

Such questions provoke exploration of a variety of topics in forest resource policy. They inspire an interest in forest resource policy, as well as in the political events that affect the development and continuance of such forest policies.

Such questions also lead to consideration of the individuals and organizations that participate in the development and implementation of forest resource policies, and of specific policies and programs which are currently in force. This book elaborates on these spacious subject areas, with the intention of fostering a better understanding of how forest resource policies are developed and ultimately carried out in society's best interest.

DEFINITIONS

Policies are pervasive throughout society; when acted upon, their substance has commanding effects on the very essence of citizen's lives and aspirations. They are the mechanisms by which society tries to promote a more positive human condition, including attempts to thwart the trauma of poverty, to promote education, to curtail use of harmful pollutants, to ensure equal opportunity, and to instill a sense of national safety. Reflecting upon and questioning the generic nature of policies may be useful in this context: What are policies? How can they be described? What characteristics give them meaning and importance?

The meaning of the term "policy" is elusive because it is used in many different ways. For example, "policy" is often used as a label for a particular field of government activity—economic policy, foreign policy, environmental policy. It is also used to mean a particular collection of laws and programs (agency policy, legislative policy), or to apply to a specific government decision. It is used in reference to a specific law or judicial ruling, as well as for a general description of a state of affairs that might prevail if some suggested action were undertaken. To further complicate matters, "policy" is often used interchangeably with terms such as "goals," "programs," "standards," "rulings," "law," "plans," "orders," "regulations," "opinions," "statutes," and "legislation." Many such terms are simply the formal written expressions of the substance of a particular policy.

The term *policy* and the phrases *forest policy* and *forest resource policy* have been defined in a number of ways in forestry and related fields. For example, Worrell (1970, p. 2) suggests that a forest policy "specifies certain principles regarding the use of society's forest resources which it is felt will contribute to the achievement of some of the objectives of that society." In addition, Worrell suggests a distinction between *forestry policy,* which focuses on policy concerns involving the practice of forestry (scientific management of forests for continuous production of goods and services), and *forest policy,* which encompasses all policy considerations associated with and having an impact on forests (e.g., land-use policy, interest-rate policy). *Forest policy* has also been described as "a settled or definite course or method adapted and followed by a governmental institution, body, or individual" (Sharpe et al. 1986, p. 19), and as "a series of negotiated settlements resulting from interaction among competing interest groups, among competing regions, and among agencies competing for the support, interest, and attention of the public'' (Dana and Fairfax 1980, p. xii). The literature of public affairs also yields many interpretations of *policy,* including: a projected program of goals, values, and practices; a purposeful course of action followed by one or more people to achieve some objective; a standing decision characterized by behavioral consistence and repetitiveness both by those who make it and those who abide by it; and the sum of government activities that influence the lives of citizens.

Consistency is needed in application of the term ''policy'' in a study of development and implementation of forest resource policies. Hence, in this book, *policy* is defined as:

> A generally agreed-to and purposeful course of action that has important consequences for a large number of people and for a significant number and magnitude of resources.

In this sense, the term is reserved for actions of notable substance that have far-reaching effects. Neither trivial actions that demand little study and analysis nor actions commonly recognized as routine or tactical are to be considered policies. ''Those parameter-shaping acts which are taken most seriously, which are presumably most difficult to arrive at, and at the same time most difficult and most important to study'' are the essence of policy (Bauer and Gergen 1968, p. 2). In a more specific sense, policy involves:

Direction. It provides guidance for actions, and it eliminates potentially random and haphazard actions.

Agreement. Persons who establish policy and those toward whom policy is directed are in general agreement with its intent (even though the threat of sanctions has led to agreement).

Important consequences. Accomplishment of policy purposes has wide ramifications for many people and for numerous resources over long periods of time.

Significant analyses. Establishment of policy frequently requires large quantities of information and significant examination and study.

Multiple cases. A course of action established by policy applies to numerous situations and involves more than applying a single rule or course of action to a single case or situation.

Action and inaction. Policy consists of deliberately established courses of action as well as actions resulting from failure to act. It includes what is not being done.

More than one decision. It involves numerous decisions, both preceding and following critical moments of choice. It typically results from many, more or less closely related, decisions.

Public forest resource policies (those generated or at least processed within the framework of government) are of major concern here; however, the term ''public'' is implied when the phrases ''forest resource policy'' and ''forest policy'' are used here.

Many terms and concepts are important to an understanding of forest resource policy, according to Kruschke and Jackson (1987) and others. Refer to the Glossary for definitions.

CHARACTERISTICS

Forest resource policies are often presented in *vague and general terms* (Table 2-1). For example, that ''the public lands be managed in a manner that will protect the quality of scientific, scenic, historical, ecological, environmental, air and atmosphere, water resource and archeological values'' (Federal Land Policy and Management Act of 1976) is of significant interest generally but is of limited value for providing administrative direction. Similarly, ''the policy of the Congress that National Forests are established and shall be administered for outdoor recreation, range, timber, watershed, and wildlife and fish purposes'' (Multiple-Use Sustained Yield Act of 1960) provides general guidance to the uses of National Forests but leaves opportunity for a substantial amount of interpretation as to the prioritizing of uses in specific cases.

TABLE 2-1
Forest and related resource policy statements

Multiple-Use Sustained Yield Act of 1960

Administer the renewable surface resources of the National Forests for multiple use and sustained yield of the several products and services obtained therefrom.

Forest and Rangeland Renewable Resources Planning Act of 1974

Make and keep current a comprehensive survey and analysis of the present and prospective conditions of and requirements for renewable resources of the forest and rangelands of the United States.

Field offices . . . and regional office of the Forest Service shall be so situated as to [give] priority to . . . location of facilities in rural areas near National Forest and Forest Service program locations.

Presidential Policy Statement of 1980 as Required by the Forest and Rangeland Renewable Resources Planning Act of 1974

Forest and rangeland protection programs should be improved to more adequately protect forest and rangeland resources from fire, erosion, insects, disease, and the introduction or spread of noxious weeds, insects, and animals.

Resources Planning Act Program 1985–2030: Final Environmental Impact Statement. USDA-Forest Service

Growing domestic and export demands for wood products [will be met] by using economic opportunities for management and assistance, and by development of new technology to increase and extend timber supplies while protecting the environment and providing opportunities for wildlife, recreation, and other multiple uses of forest lands.

(*Continued*)

TABLE 2-1 (Continued)

Alabama Forestry Commission (*Policy Number 9, 1985*)

Employees of the Alabama Forestry Commission will not use political endorsements for the improvement of the conditions of employment (e.g., salaries, promotions, job assignments). Employees will also refrain from actively participating in political campaigns in a manner that might be interpreted as an act of the Alabama Forestry Commission.

California Department of Forestry and Fire Protection (*November 1989*)

Reduce or eliminate physical fire hazards through activities such as removing hazardous vegetation along roadsides, developing standards for machines that reduce the probability of causing fires, and inspecting the outside of structures in wildland areas for fire safety.

Central Forest Service, Ministry of Agricultural of Greece (*1981*)

Recognize the social value of forests and prevent speculation in forest land; protect forests and natural environments in general against fire, pests and damage by man; increase forest production, especially of industrial wood; and improve the economic and social situation in mountain areas via creation of new jobs in the forest-based sector.

Oregon Board of Forestry: Goals, Objectives and Missions (*1987–89 Biennium*)

Board of Forestry Lands will . . . generate revenue for county governments and local taxing districts . . . make raw materials available on a sustained yield basis to help meet demands for forest products . . . provide for community stability . . . encourage efficiency in harvesting and processing . . . provide for employment.

Michigan Division of Forest Management, Statewide Forest Resources Plan (*1983*)

Strengthen and diversify economy through forest resource development; increase productivity of nonindustrial private forests consistent with landowner objectives; increase public forest outputs for all forest uses while protecting forest resources; improve energy situation through energy-responsive forest management.

Tennessee Valley Authority (*1933*)

Provide for reforestation and the proper use of marginal lands in the Tennessee Valley . . . industrial development . . . [and] to further the proper use, conservation, and development of natural resources.

USDA-Bureau of Land Management, Strategic Recreation Plan (*1988*)

USDA-Bureau of Land Management will . . . provide and maintain a diversity of recreation opportunities on public land . . . issue special recreation permits in an equitable manner for specific recreational uses of the public lands . . . maintain recreation facilities that protect the resource, the public, and the public investment . . . expand and strengthen cooperative partnerships with federal, state, and local agencies and the private sector . . . assure protection of sensitive resources and the continued availability of quality outdoor recreation opportunities.

Code of Federal Regulations (*Part 22, Range Management, 1975*)

All grazing and livestock use on National Forest System lands and on other lands under Forest Service control must be authorized by a grazing permit . . . grazing permits and livestock permits convey no right, title, or interest held by the United States in any lands or resources.

USDA-Forest Service Manual: Civil Rights (*Title 1700, 1986*)

Ensure that no person is denied participation in Forest Service programs because of race, sex, color, natural origin, age, handicap, creed or marital status . . . Forest Service employ-

TABLE 2-1 (Continued)

ees shall conduct official business so that [the agency] eradicates all forms of discrimination from its programs and activities . . . , all levels of the [agency] are supportive of affirmative action . . . , no economic or social barriers limit program participation . . . , [and] programs and services are equally available to all persons.

United States Court of Appeals for the Fourth Circuit (*1975*)

It is evident that Congress intended the Forest Service designate the area from which the timber was to be sold and, additionally, placed upon the Service the obligation to mark each individual tree which was to be cut. . . . The judgement of the district court is affirmed.

Act on Forestry and Forest Protection of Norway (*1987*)

National interest is to increase the productivity of forest land and promote afforestation and forest protection. As long as forest land is managed in accordance with intentions of Act, owners shall have the right to manage their forests without interference from the authorities.

Shoshone National Forest, Land and Resource Management Plan (*1986*)

Ensure that National Forest developed [recreation] sites are appropriate for the surrounding forest setting and do not compete with the private sector or unnecessarily duplicate other public land facilities and services.

State of Washington, Forest Practice Rules and Regulations (*1976*)

Reforestation plans must be submitted with the [harvesting] application . . . , the landowner . . . shall file a report with the Department [of Natural Resources] either at the time of completion of planting or at the end of the normal planting season.

Stocking levels are acceptable if 300 well-distributed, vigorous seedlings per acre . . . have survived on the site at least one growing season.

National Park Service, Management Policies (*1988*)

In natural zones, landscape conditions caused by natural phenomena, such as landslides, earthquakes, floods, tornadoes, and natural fires, will not be modified unless required for public safety or for necessary reconstruction of dispersed-use facilities, such as trails.

Unclear and indefinite policies exist for many reasons. They may be so designed in order to accommodate a wide and highly variable range of resource and administrative conditions; for example, narrowly specified national policy may not be applicable to site-specific conditions which vary from one forested environment to another. Indefinite policies may also assure agreement among a variety of interested parties that hold dissimilar views about or approaches to the use and management of forests; for example, agreement over narrowly defined proposed policies may be impossible to attain, or explicit proposals may be divisive among divergent groups. Although vaguely stated forest policies may be a necessity from a resource, administrative, or political perspective, they can be difficult to implement (unclear as to what is to be accomplished and how such is to be attained) and hard to evaluate (unclear as to what costs and benefits are to be assigned to a policy).

Forest resource policies may also embody *inconsistencies* (within a policy and between policies) and can be *limited in comprehensiveness*. For example, a forest policy may call for an impossible combination of forest benefits, such as timber and wilderness recreation from the same forest site. One forest policy may be diametrically opposed to actions called for by another forest policy, as when a habitat-improvement policy for rare and endangered species conflicts with a policy of intensive timber management. Forest resource policies may call for a focus on one portion of a much broader combination of resources or administrative responsibilities. An agency, for example, may be asked to implement policies concerning surface resources (e.g., wildlife, water, timber, aesthetics) while another policy assigns responsibility for subsurface resources (i.e., minerals) in the same locale to another agency. The federal government may implement wildlife habitat policies in the same forested area for which a state government has the responsibility for establishing appropriate levels of wildlife harvest.

DEVELOPMENT

Democratic Principles

Forest resource policies are usually the products of highly complex political systems. The average citizen often views them as the result of *democratic principles* in action. Citizens are considered politically equal, in that they may have differing preferences about forest policies but all citizens have equal rights to place their preferences on the agenda of government. Information about forest policies is thought to be fully available to all; policies are proposed and fully aired in a marketplace of free-flowing ideas. Citizens act on their preferences for forest policies by voting for a policy or for a person to whom authority for voting or acting is delegated. Majority rule prevails in the selection process, in that a plurality is required for selection of a policy or a representative. The choices of policymakers and the preferences of citizens are highly correlated, in that the representatives' ultimate responsibility is to enact citizen forest policy preferences.

Ideally, forest resource policies are proposed by citizens to elected representatives, who in a body of representatives (such as a legislature) give legitimacy to a particular policy. Bureaucracies then carry out the will of the citizens. When major interpretation of a forest policy is required, judicial systems enter the scene. The public policy finally implemented is like a completed puzzle, pieces of which have been contributed by each major branch of government.

Group Competition

An alternative explanation of how forest resource policies are developed is *group competition.* Political processes leading to the selection of forest policies are thought not to involve individuals, but rather to involve organized groups of individuals—interest groups. The latter are viewed as organized groups of people with common interests (such as recreational use of forests, or water flowing from forested areas). Members of interest groups share public policy objectives and generally agree on the means of achieving them. Public forest policies are viewed as the result of political struggles between interest groups. Which forest resource policy ultimately prevails depends on the groups' relative power, influence, and ability to press claims for their interests (usually at the expense of competing groups' interests). The role of legislature is to ratify (enact into law) forest policies that result from the political equilibrium that is established through struggles between groups. The influence of any one interest group is controlled by the countervailing power of other groups. Groups tend to check and balance one another's abilities to prevail on a matter of policy.

Powerful Elites

Yet another view is that forest resource policies are developed by *powerful elites.* All political institutions of society are believed to be dominated by small groups of skillful individuals (agency heads, interest-group leaders, executives in the private sector) who know how to manipulate instruments of power in order to secure forest policies favorable to their interests. They are presumed to have the means for gaining control over avenues of power in government (wealth, prestige, knowledge) and the motivation for doing so (ideological commitment, heightened concern for wealth and power). From an elitist perspective, the role of individuals in forest policy development is not denied; they are simply presumed not to wish to exercise leadership skills. They are thought to want to be left alone and to delegate to others the responsibility for selecting forest policies that affect their lives.

Policy Process

The most recently suggested approach to an understanding of how policies are developed is to view them as the result of a sequence of activities which in total constitute a process—the *policy process,* a sequence of political events leading to policy outcomes. This approach is preferred in this book and is shown in Figure 2-1.

The *agenda-setting* phase of the policy process commences with a fuzzy perception by someone or some organization that an important forestry problem or issue exists. This individual or organization discovers that others share an interest in the same problem or issue. They discuss the issue or problem, and may form new organizations or seek alliances with existing organizations as a means of satisfying their concerns. Eventually they demand action. Their demands may lead to formal recognition of the issue by government—the setting of the agenda.

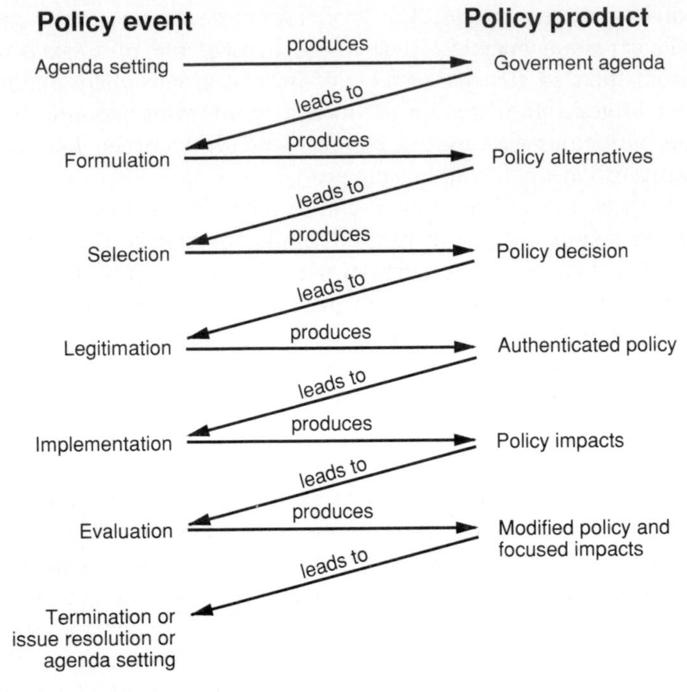

FIGURE 2-1
Policy development and implementation process, by policy event and policy product.

Recognition of an issue by government leads to *formulation* of possible policy responses considered capable of addressing the forestry issue which has achieved agenda status. Implied is identification of goals that might be sought in order to resolve the issue and the specification of alternative forest policies that might be capable of achieving such objectives. Once policy options have been defined and assessed in a technical, an economic, and a political sense, a *selection* is made. The forest policy option considered most meritorious is chosen. Policy decisions are based on criteria of numerous types and frequently are made only after intense bargaining among competing interests. *Legitimizing* a selected policy provides it with an aura of respect and authority, in that a law is enacted or a regulation is established.

The carrying out of a selected and legitimized policy option is its implementation. Legitimized policy choices are further clarified and interpreted, organizations are established and charged with management of people and materials, and the routine of delivering goods and services to those interested in the products of a particular forest policy is set in motion. Unlike previous phases of the policy process, *evaluation* is retrospective. It involves specification of which portion of a complex forest resource policy is to be evaluated, the criteria that are to be used to judge its character, the application of analytical tools, and the setting forth of recommendations about how to make the forest policy more effective. Forest policies that are found to be redundant or outmoded or to have accomplished their purpose are subject to *termination,* which in turn may lead to development of new issues for government attention. Thus, the policy process may begin a new cycle.

Process Operation It would be inappropriate to describe the workings of the policy process as a well-tuned, internally rational sequence of events, free of institutional influence. In reality, the process is often truncated: the policy option selected may be to do nothing. The dividing lines between events are often blurry: agenda setting may blend with formulation. The events may not be performed in logical order: implementation may occur without careful formulation of options. Furthermore, some of the events in the process may not occur at all.

Similarly, presuming that the policy process is the exclusive domain (or under the control) of a single individual or institution is also inappropriate. Implementation, for example, is not suggested by the policy process as being the sole domain of bureaucracies. Rather, implementation is viewed as a collection of activities that are actively engaged in by a number of institutions, including bureaucracies, legislatures, interest groups, and courts. Individuals or forestry organizations may exercise special power over certain segments of the process, as when policy is implemented by a bureaucracy, but other individuals or organizations may have similar influence over other segments of the process, as when an agenda is set by an interest group. Interest groups may foster concern over a particular forest resource policy (e.g., application of even-aged silvicultural systems), which can lead to federal district and appellate court decisions. These decisions can ultimately require the involvement of legislatures, bureaucracies, and a plethora of additional interest groups. The notion that one person is in command of the policy process is decidedly absent in nearly all policy development cases. The policy process perspective encourages interest in discovering the reasoning behind the actions of those involved and discourages common tendencies to label such participants as irrational or uninformed.

Modest Policy Changes The process by which policies are developed is not inclined to produce large or dramatic changes in forest resource policies. Incremental policy changes built on existing policies that have experienced some degree of success are most often favored. A common argument is that forest policies that are politically feasible are usually only slightly different from existing policies, because political support can be more easily assembled for small changes than for large ones. Then too, retreat from forest policies that are only slightly different from existing policies is usually less costly to society. Making small changes in policy avoids the potential for large-scale adverse consequences of a major but faulty change. For example a policy of abandoning wilderness status for all forests within the federal National Wilderness Preservation System and designating them for timber production or developed recreation uses could prove especially costly to society if the policy of abandonment were later determined to be a mistake and actions had to be undertaken to reverse the consequences of the mistake. Reconstructing such forests to wilderness standards would be impossible in a short time. Society would have to be willing to accept the long periods of time that natural ecosystems frequently require to achieve the characteristics of a wilderness. Likewise, a policy of transferring all public forestland to the private sector could be a costly social error if later opinion found it in society's best interest to return such forestland to public ownership.

Multiple Decision Criteria Use of a wide variety of criteria or standards in making policy decisions is another important trait of the policy process. Seldom is a single set of narrowly defined rules used in deciding whether an issue deserves agenda status, in judging the merits of a formulated policy, or in evaluating the effectiveness of an implemented program. Technical and ecological criteria are used to judge the physical feasibility of implementing a policy option, while economic criteria are often used to judge the efficiency and effectiveness of existing and proposed policies. Procedural criteria are used in determining the appropriateness of administrative processes, and ethical and equity criteria are used to judge the distribution of benefits among clients of forest resource policies and the incidence of the costs to various groups of implementing such policies.

Power and Influence The policy process is also characterized by the liberal use of *power,* defined here as the capacity to control or change the behavior of others. By exercising power at various stages in the process, individuals and organizations can facilitate, delay, or prevent the development of forest resource policies. Power comes into play when a public forestry agency refuses to acknowledge or address an issue concerning the management of public forests. Power over policy selection is exercised when a legislator informs a public agency that funds for capital improvements will be withheld unless recommendations for forest road construction are forthcoming. Similarly, power to terminate a policy occurs when a district court evaluates the use of pesticides and decides to issue a restraining order prohibiting their use in certain forest environments. Fame, charisma, status, and special expertise can also be sources of power. Power over the formulation of forest resource policies is exercised when a forestry professional, who is highly regarded for knowledge about forest ecosystems, testifies on a legislative proposal concerning the use of even-aged silvicultural systems in forest environments.

To effectively wield power in the policy process, professionals must be perceived as being very credible; they must be thought to actually possess unique knowledge or skills. They must also be perceived as extremely trustworthy, not as trying to deceive someone. Their knowledge and skills must also be considered relevant—useful to a policymaker. If a policy is to survive the process, the exercise of power in the right place at the right time is often paramount.

Competing Values The policy process usually includes intense contests between advocates of competing *values*. Often involved are unending struggles between individuals and organizations whose perceptions of the use and management of forests are at odds. At issue are beliefs (values) that certain conditions are inherently good while others are inherently bad. Here are some statements that convey beliefs about a particular circumstance:

"Public forestland should be designated for wildlife habitat."

"Human population levels should be reduced to ensure higher-quality forest environments."

"Public investment in forestry should have priority over investment in programs that benefit the unemployed."

These statements cannot be judged either right or wrong, or not in the way that facts can. Examples of facts are:

The National Forest System is composed of 191 million acres of land.

1.4 million people visited national parks in 1987.

The USDI-Bureau of Land Management 1988 budget was $842 million.

Values are a basic currency of the policy process. To enlist support for a specific policy (e.g., "We will double the allowable timber harvest rates for a specific public forest"), a clear link to values that many people share (e.g., "Housing is necessary for human comfort and survival") must be established. Values must be alluring enough to convince political leaders, organized interest groups, and the general populace to become involved in the long and often risky business of converting values into specific forest policies.

Political Strategies The policy process is energized by a variety of political strategies that are undertaken in order to win support and ultimately attain victory in agenda setting, selection, implementation, and the like. Such strategies include mobilization of grass-roots support, initiation of analyses used to support favored policy options, and distribution of messages that advocate a proposed forest policy and define its linkage to a significant public concern. They also involve identification of important participants in the policy process (those likely to aid or to hamper a policy's development), and the gathering of appropriate resources (such as finances and knowledge) that might be required to counter threats from opposing organizations or individuals.

Strategy development also entails assessment of alliances, or seeking help from others with similar views. In addition, participants must come to a clear understanding about when compromise with the opposition is appropriate, and about what changes in position will be acceptable, if loss appears inevitable or if the cost of establishing a favored policy soars beyond acceptable limits. Finally, political strategies implemented within the policy process often pay significant attention to leadership. People with proven track records must be identified—people who can make political strategies work in the rough-and-tumble world of politics. The leaders chosen must be able to accomplish such tasks as assuring that limited resources are not misapplied, that situational advantages are not dissipated, and that political opponents are not underestimated. Struggles between rivals in the policy process never cease; the leaders who have well-designed political strategies for dealing with such struggles are usually the most successful.

Multiple Participants The process by which forest resource policies are established cannot be divorced from the individuals and institutions which make it function. Agenda setting, for example, does not occur in the abstract but is made to happen by interest groups and others concerned about the manner in which existing policies are being carried out (or not carried out) by government. Similarly, implementation of forest policies cannot be divorced from the multitude of bureaucracies which have responsibility for delivering the goods and services called for by an established policy. Some participants have official responsibility to enter the policy process and to recommend, ratify, and implement public policies. Under the U.S. Constitution, for example, the legislative branch of government is responsible for enacting laws which the executive branch must administer and the judicial branch must interpret. Although separate in spirit, the three branches of government share power in many areas, thus making for a system which involves a number of checks and balances.

Government is not the only source of participants in the policy process; the private sector is also actively engaged in public policy development activities. Among those involved are organized interest groups, political parties, the mass media, individual private citizens, and corporations and related businesses.

The combination of public and private sector participants makes for a heterogeneous lot of actors, all of whom are seeking to influence the flow of policy products. Van Horn et al. (1989) have argued that the actors can be better understood if they are organized into various political settings, for example:

- *Boardroom politics.* Policy is made by business elites and professionals, but has important public consequences.
- *Pressure group politics.* Policy is made by interest groups.
- *Bureaucratic politics.* Policy is made and adjudicated by bureaucrats, with input from clients and professionals.
- *Cloakroom politics.* Policy is made by legislators, who are constrained by their various constituencies.

- *Chief executive politics.* Policy is made by presidents, governors, mayors, and their advisers.
- *Courtroom politics.* Judicial decisions are made in response to aggrieved individuals and organizations.
- *Living room politics.* Public opinion is galvanized, usually by the mass media.

To attempt to divorce the policy process from its multitude of participants would be a grievous error. Just as forest resource policies are only as good as the process which produces them, so too the process is only as good as the ideas of the many participants who activate the process.

Decision-Making Tenets

Development and implementation of forest resource policies involves highly pluralistic decision-making processes characterized by compromise, incrementalism, and continual adjustment. Though political theories have been proposed to explain such processes, the common realities of politics and administration often defy easy explanation. Consider the following tenets, which were suggested in a different context yet most certainly apply to the forest policy arena (Jones 1984). Keep in mind that these tenets should not be accepted without critical appraisal.

- Issues, problems and demands are constantly being defined and redefined by the policy process.
- People most often do not maintain an interest in other people's problems or issues.
- People have varying degrees of access to the policy-making process.
- Issues are interpreted in different ways by different people at different times; things that are issues to some people are not issues to others.
- Many problems may result from the same issue.
- Policymakers are seldom faced with a single, clearly defined issue.
- Policy making is most often based on limited information and inadequate communication.
- Policies and programs reflect an attainable consensus rather than a substantive conviction.
- Policies and programs are often developed and implemented without a clear definition of the issue or problem being faced.
- People prefer small—not large—changes in policies and programs.
- Private sector issues are often acted on by government as though they were public issues.

Part I has set the stage for a more in-depth assessment of the process by which forest resource policies are developed, as well as the roles of various participants. What follows in the rest of this book is a more detailed discussion of these important aspects of forest policy development, as well as highlights of selected public forest policies that have been generated by the policy process. The highlights are not intended to be a definitive assessment of the nation's myriad of forest resource policies and programs. Rather, they are presented in the spirit of encouraging appreciation of the many, highly diverse forest policies and programs that have found their way through the policy process over the years.

REFERENCES

Bauer, Raymond A., and Kenneth J. Gergen. *The Study of Policy Formation.* The Macmillan Company, New York, 1968.

Dana, Samuel T., and Sally K. Fairfax. Forest and Range Policy: Its Development in the United States. McGraw-Hill Book Company, New York, 1980.

Jones, Charles O. *An Introduction to the Study of Public Policy.* Brooks/Cole Publishing Co., Monterey, Calif., 1984.

Kruschke, Earl R., and Byron M. Jackson. *The Public Policy Dictionary.* ABC-Clio, Inc., Publishers, Santa Barbara, Calif., 1987.

Sharpe, Grant W., C. W. Hendee, and W. F. Sharpe, *Introduction to Forestry.* McGraw-Hill Book Company, New York, 1986.

Van Horn, C. E., D. C. Baumer, and W. T. Gormley. *Politics and Public Policy.* Congressional Quarterly, Inc., Washington, 1989.

Worrell, Albert C. *Principles of Forest Policy.* McGraw-Hill Book Company, New York, 1970.

TWO

POLICY PROCESS

Society establishes forest resource policies in order to capture important social and economic opportunities that are often represented by forests; it may also develop such policies in response to deep-seated concerns over the use and management of forests. The means by which society achieves its interest in forests requires policies that are products of a sequence of political events which constitute the policy process. The events are agenda setting, formulation, selection, legitimizing, implementation, evaluation, and termination. Although each involves a highly complex set of political activities in its own right, such events must be viewed as important ingredients of the broader policy process which they create.

3

POLICY AGENDA: ISSUES TO GOVERNMENT

The number of public problems seeking the attention of government at any one time is nearly incalculable. Not all forest resource issues are acknowledged by nor acted upon by government. But why are some issues recognized while others are not? Who determines which forestry issues will be addressed? What criteria are used to judge an issue's worthiness for agenda status? Do some forestry issues have special traits which give them an advantage in the struggle for government recognition? Is the system of issue recognition a manageable one —a system that can be manipulated by political strategists intent on seeing a policy change in their favor? Questions of such a nature direct attention to subjects embraced by *agenda setting,* a topic that confronts the seemingly obvious subject of deciding what to make policy about. In this chapter, therefore, we shall consider the nature of issues and agendas, the approaches commonly used to explain agenda-setting processes, the disposition of issues once they have achieved agenda status, and the role of various participants, including forestry professionals, in agenda-setting activities.

ISSUES AND AGENDAS

Issues

The focal point for agenda setting is concerns or issues that are cause for discomfort or disagreement among individuals or organizations. Specifically, *issues* are conflicts between groups of individuals or organizations over policies of procedure or substance concerning the distribution of power or resources (Cobb and Elder 1983). Key portions of the definition are groups, policies, and distribution. Even-aged silvicultural practices as applied on public forests, for example, frequently involve groups (e.g., user groups interested in timber as a forest use versus groups interested in uses such as recreation, water, and scenic beauty) that disagree over a policy (e.g., clear-cutting as a means of harvesting and regenerating forests) involving the distribution of resources (e.g., what benefits are to be produced by public forests and who is to receive them). Procedures can also be at issue. In the early 1970s, regulation of pesticide use became an issue involving two organizations, the U.S. Environmental Protection Agency (EPA) versus the U.S. Department of Agriculture, that disagreed over a matter of procedure involving the distribution of power. The disagreement was about authority to approve the use of certain types of pesticides, and the question was which agency should have authority to decide.

Agendas

An *agenda* is commonly regarded as an enumeration of issues which are of concern to someone or some organization. For purposes of understanding, a distinction is made between two types of agendas, namely systemic and institutional agendas.

Systemic Agendas A *systemic agenda* is ''all issues commonly perceived by members of a community as meriting public attention and as involving matters within the legitimate jurisdiction of government authority'' (Cobb and Edler 1983, p. 85). The agenda represents a community's collective assessment of what ought to be acted upon by government—a summing of the lists of demands for government action by all concerned parties. Appearing on systemic agendas are broadly defined issues such as pollution, crime, education, poverty, unemployment, energy, discrimination, taxes, and environmental quality. All local, state, and national political communities have a systemic agenda; each such agenda may be a subset of a larger more broadly defined systemic agenda. Issues of which systemic agendas are composed are seldom directly addressed by government; their abstract and indefinite nature makes the initiation of government action to resolve them extremely difficult (often impossible).

Institutional Agendas In order to be effectively addressed, issues appearing on systemic agendas must be refined and moved to *institutional agendas,* namely, agendas composed of explicit issues that are up for active and serious consideration by government policymakers that have authority to deal with them (Cobb and Elder 1983). To be on an institutional agenda, a forest resource issue must be defined in such a manner that specific policies to deal with the issue can be formulated, adopted, and ultimately implemented. Examples are "Add 2,000 acres to the Bob Marshall Wilderness" (not "More wilderness is needed to save humanity"); "Increase the budget of the state division of wildlife management by $500,000 for purposes of deer habitat improvement" (not "Hunters demand greater deer-hunting success"); and "Build 100 additional miles of hiking trail in a specific public forest" (not "Greater recreation opportunities are needed for backpackers"). Issues appearing on an institutional agenda are sure to be considered by government, and action may even be taken to resolve them—although there is no guarantee of such an outcome. Institutional agendas are controlled by government officials or government organizations that are empowered to address an issue, including authority to authorize a policy change if such is judged to be necessary.

Institutional agendas exist in concrete and readily identifiable forms. Very often they take the form of a calendar of topics to be considered by decision-making bodies such as legislatures, judicial systems (i.e., courts), or government executive agencies. Announcements of forthcoming public hearings as made by legislative committees commonly describe forest resource issues which have attained institutional agenda status. Similarly, the list of cases to be decided by a district or appellate court very often reflects issues that compose an institutional agenda. Some institutional agendas take the form of issues identified for action by the senior staff of a forestry agency. In a more tangible sense, long-range plans prepared by a public forestry agency may identify issues toward which the substance of plans are focused. Such issues most definitely are the substance of an institutional agenda.

AGENDA-SETTING PROCESSES

The establishment of forestry agendas can be viewed as a *process,* a logical series of actions that ultimately lead to recognition of a forest resource issue by a unit of government that is charged with the development and implementation of forest policies. Unfortunately, the process is not always logical or neat, nor is there general agreement on its content. Some argue that agenda setting is much like a process of natural selection, comparing potential issues and problems to molecules floating in primeval soup. Just as relatively few molecules actually combine to form life, relatively few issues make it to an institutional agenda. The forestry community can be envisioned as a soup of issues in which members of the judiciary, bureaucracies, interest groups, legislatures, and the mass public are continually bringing to the surface issues deemed to be in need of government attention.

Problems and issues are in a constant turmoil within the community. Each segment of the community has opinions about important forestry issues for the sake of which new programs should be formulated or the direction of existing programs should be changed. Issues are presented in various forums, such as luncheon discussions, memos, articles, hearings, testimony, inspections, and regulations. Issues often conflict with one another, and when this happens, they may be combined or recast into more acceptable forms. New issues float to the surface, only to be confronted by issues already enjoying a degree of respect. Most issues fade into oblivion or are put on hold, but a few survive and prosper. The survivors of the natural selection process gain formal recognition on an institutional agenda. However long and arduous the process may have been, the advocates of a successful issue are rewarded with the knowledge that action will probably be taken to address their concerns (Kingdon 1984).

The process by which institutional agendas are established is the subject of continuing debate and research (Cobb and Elder 1983; Kingdon 1984). It is helpful to consider the process from four perspectives: as functional activity, power and influence activity, opportunity activity, and strategic activity.

Functional Activity

Agenda setting can be viewed as a series of functional activities. The perspective is one of a sequence of actions considered necessary for a forest resource issue to be placed on an institutional agenda. The following functional activities have been suggested (Jones 1984):

Perception. An individual or organization recognizes that an event has occurred for which there is a need for relief (e.g., large openings have been observed in the forest landscape—openings that impair opportunity for enjoying the scenic beauty of a forest).

Definition. The individual or organization assesses the event so as to gain understanding and a sharper focus on the event (e.g., large openings many be caused by wildfire, road construction, or timber harvesting; large clear-cuts are judged to be the issue).

Aggregation. The individual or organization discovers that others have come to similar conclusions about the issue and are also interested in seeking relief (e.g., a number of others are found to have perceived and defined the issue in the same fashion).

Organization. The individual or organization concludes that relief from the issue can ultimately be best obtained by establishing a formal organization (e.g., depending on leadership and resources, an organization is formed to address the issue).

Representation. The individual or organization, through the newly formed organization, seeks to place the issue on the institutional agenda of government (e.g., the organization, or its representative, meets with a public forester and presents the issue in expectation of favorable action).

The functional activities identified above represent potential events in the agenda-setting process. There is no guarantee that they will occur in sequence, or at all. Some individuals who perceive a problem may go directly to the public forester with a request for action. Others may go directly to a well-established organization (e.g., an existing interest group), which in turn may successfully place the issue on the agenda of the public forester.

Power and Influence Activity

Agenda setting may also be viewed as an activity that involves the exercise of power and influence. This is in contrast to a pluralistic view of the marketplace of forest resource issues and its solutions as an open arena, in which any or all interested parties (organized or unorganized) are given an opportunity to influence the makeup of an institutional agenda. Not all groups can expect their policy preferences to prevail on all issues; all, however, are assured of continuing access to the process. As a group's political fortunes enlarge and its ability to mobilize them improves, it may come to be in a position to place a new forest resource issue on the agenda of government.

Achievement of agenda status via the exercise of power and influence implies the existence of a powerful elite that dominates the agenda-setting process. Adherents of such an approach believe that institutional agendas do not represent the competitive struggle of relatively equal groups but rather the systematic use of power and influence to decide which issues the political system will or will not consider. The argument is that nearly all agenda-setting and policy-making activity is limited to a relatively small number of active participants and that a tremendous amount of this activity is controlled by keeping it so private that it is almost completely invisible (Jones 1984). In this quiet setting, composed of agency heads, legislators, and interest-group leaders, debates occur, bargains are made, and agenda items are agreed upon. Participants who find fault with such actions must look to the outside for help; in fact, they may implement a strategy of deliberately making an issue visible to other audiences and participants in hopes of reordering issues placed on the institutional agenda. Losers in such internal battles become desperate; they must seek help from the outside.

Opportunity Activity

Agenda setting can also be viewed as a function of the times. Social and political circumstances make a forest resource issue ripe for appearance on an institutional agenda. Political moods become correct, consensus occurs among important political actors, conditions within government change significantly, windows of opportunity emerge, and entrepreneurs arise to manage issues toward institutional agenda status.

Political Moods A *political mood* exists when the general citizenry are thinking along certain common political or social lines. For example, the mood of the citizenry and its leaders may be conservative, or antigovernment, or development-oriented, or distrustful of big business, or environmentally sensitive. The mood of the times can facilitate as well as detract from successful agenda status for certain issues. The pro-environment mood of the early 1970s and late 1980s, for example, greatly facilitated the eventual establishment of important state and national environmental laws. Similarly, the rational planning mood of the forestry community in the 1970s and 1980s facilitated the government's creation of large-scale planning programs. In contrast, the inflation, wasteful government, and Proposition 13 outlook of the late 1970s was a political mood that steered many financially demanding forest resource issues into obscurity. The political mood of the citizenry is important to agenda setting. According to Kingdon 1984, p. 156), it

> makes some proposals viable that would not have been viable before, and renders other proposals simply dead in the water. Advocates from the newly viable proposals find a receptive audience, an opportunity to push their ideas. Advocates for the proposals out of favor must adapt to their unfortunate situation, present their ideas for consideration as much as is possible under the circumstances, and wait for the mood to shift once again in their direction.

Political Consensus Complementing political moods in the agenda-setting process is *political consensus* among important and especially knowledgeable political actors within a community. If understanding persons assess the political scene and find that interest groups and political parties, for example, view a forest issue as commonly important, a powerful impetus exists to move the issue to prominence as an institutional agenda item.

Changes in Government Similarly important are significant *changes in government,* especially changes in an agency's jurisdiction or changes in strategically located people. A new agency head, a change in legislative leadership, a newly appointed state forester, or the retirement of an agency's regional forester can quickly propel a forest resource issue toward institutional agenda status. From an agenda-setting perspective, new leaders and administrators are motivated by an urge to distinguish themselves from their predecessors and by an interest in furthering the forest resource issues which they personally favor.

Windows of Opportunity The political mood of a community, the strength of concensus existing among important political participants, and changing conditions within government are all factors that affect an issue's agenda status. When such conditions become properly aligned, a *window of opportunity* occurs, an opportunity which an issue's advocates are well advised to exploit. Windows of opportunity are opportunities for action ''which present themselves and stay open for only short periods. If the participants do not take advantage of these opportunities, they must bide their time until the next opportunity comes along''

(Kingdon 1984, p. 174). An environmental window of opportunity was open wide in the early 1970s: citizens' interest in the environment was high, important interest groups were in agreement on the need for additional environmental laws, and government was experiencing changes in leadership that facilitated environmental initiatives.

Issue Entrepreneurs Windows of opportunity and the confluence of important political moods are not entirely sufficient for successful agenda setting. Also needed are *issue entrepreneurs* who can manage or exploit an agenda-setting opportunity; persons or organizations that can invest the time and energy required to promote an issue. To be successful, individuals as entrepreneurs must have some claim to standing. For example, an agency head has authority. A specialist on pesticides is known to possess certain expertise. An interest-group leader is in a position to speak for others. A well-respected ex-legislator has superb political connections and shrewd negotiating skills. Another leader or activist may have a history of persistence in the agenda-setting arena.

Implementation Good issue entrepreneurs attempt to link issues of immediate concern with more broadly defined issues. They seek out a popular and highly visible issue (such as global warming) with the intent of placing their more narrowly defined issue (say, reforestation) on the same institutional agenda as the popular concern. Too many issues may, however, overload an institutional agenda. The agenda's keeper (e.g., an agency head or the chair of a legislative committee) may jettison all issues, arguing that they are too numerous and too complex, and that the possible solutions are too overwhelming to deal with. Successful entrepreneurs are especially astute about such matters and about issue management in general. They wisely wait for that all-important window of opportunity—and then make their move.

The confluence of political moods, political consensus, and changes in government with the occurrence of a window of opportunity and the availability of an issue entrepreneur is frequently very short-lived. Windows of opportunity often close because participants have successfully placed their issue on an institutional agenda, or because an issue's advocates are reluctant to invest additional time and effort after failing to secure agenda status for their issue, or because the events which initially led to the opening of the window suddenly change (e.g., a change in leadership may occur, or public concern and interest may decline. Some windows open with great predictability, because of recurring budget cycles, annual reports, or renewals of legislative authority. The natural resource assessment, which is renewed every 10 years, and the program document, which is required by the Forest and Rangeland Renewable Resources Planning Act of 1972 to be renewed every 5 years, are classic examples of predictability. In most cases, however, windows of opportunity are difficult to foresee. Their short duration adds support to the advisability of striking while the iron is hot.

Institutional agenda setting viewed as an opportunity activity highlights the complexity of the agenda-setting process. Combinations of many events occur, few of which can be consistently identified as responsible for successful placement of a forest resource issue on an institutional agenda. When asked why certain forest resource issues achieve institutional agenda status, people intimately involved with the politics of such issues often respond, "It was a combination of things," "Several things came together at the same time," or "A number of factors blended into the mix."

Strategic Activity

Agenda setting may also be viewed as a strategic activity undertaken with the intent of getting a forest resource issue on the institutional agenda of government. What is involved is a conscious decision to place an issue on an agenda by exploiting triggering events, manipulating political symbols, seeking out sympathetic gatekeepers, and defining issues in a manner that elicits sympathy among persons responsible for ultimate action on the issue (Cobb and Elder 1983). The basic proposition underlying agenda setting as a strategic activity is that the greater the size of the audience to which an issue can be enlarged, the greater the likelihood it will eventually achieve access to an institutional agenda. The events involved in the process can be portrayed as shown in Figure 3-1.

Triggering Event Of initial concern to an individual or organization interested in placing a forest resource issue on an agenda is identification of a *trig-*

FIGURE 3-1
Sequence of events involved in institutional agenda setting undertaken as a strategic activity.

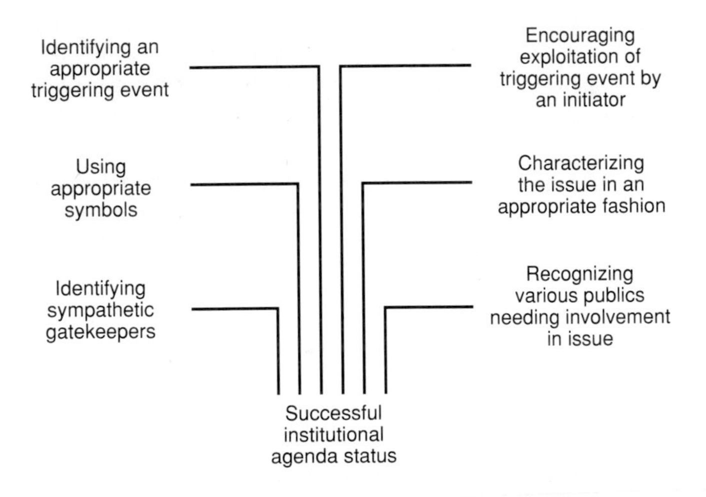

gering event. Such an event may erupt unannounced during seemingly routine administrative activities. A triggering event can sweep an issue to widespread public recognition. As general social concerns, these events take the form of military conflicts, riots, strikes, stock market crashes, and massive unemployment. In forestry a triggering event may take the form of a catastrophe. A major wildfire, for example, can trigger issues such as the adequacy of public investments in fire control and the location of residential dwellings in forested areas. Similarly, major outbreaks of destructive forest insects, such as spruce budworms or tussock moths, can trigger issues concerning pest-management practices—chemical versus biological control, for instance. Widespread floods and resulting loss of property and human life can trigger issues concerning the adequacy of watershed-management practices, the nature of timber-harvesting activities, and the location of commercial establishments in forested floodplains.

Triggering events may also be spawned by new technologies. New and more efficient means of transportation, for example, may facilitate access to forests for recreational purposes, resulting in significant recreational pressures on forest ecosystems. This happened when automobiles became popular earlier in this century, leading to widespread development of road systems. The advent in the 1920s and 1930s of advanced timer-harvesting equipment (crawler tractors) had by the 1960s spawned significant concerns about the impact of such equipment on the production of quality water flowing from forested watersheds also devoted to timber production.

Human events may also trigger issues. Tragic deaths of firefighters battling wildfire in forested areas invariably raise concern over the morality of losing human lives in order to save forests from destructive wildfire. Individual acts of vandalism, as when large iron spikes are driven into trees, and the resulting injuries in sawmills can likewise heighten the visibility of certain forest resource issues; an example is allocation of timber versus wilderness land in public forests. The isolated yet misapplied clear-cut, or the leaving behind of a large amount of wood residue after harvest in a public or private forest, may fuel public concern about the practice of even-aged silvicultural treatments for timber production purposes.

Initiators The impact of a triggering event can be heightened by the use of *initiators,* individuals or organizations that can skillfully bring an event to a wider audience. Clear-cutting in a public forest may initially be of limited concern, but the situation may change markedly when an interest group that wishes to place clear-cutting on an institutional agenda publicizes details by using vivid photos and provocative language. Similarly, use of herbicides in the management of forest plantations may be of limited concern until a helicopter applying herbicides is fired upon and the event is widely publicized by the media and various interest groups. Public timber allegedly sold below the cost of production may not be perceived as an important issue until a respectable interest group studies the matter and gives it widespread publicity. Without initiators standing readily to publicize triggering events, implementation of many agenda-setting strategies would fail in infancy. Initiators add fuel to issues that are smoldering but might well die out if they were not publicized.

Gatekeepers Placement of a forest resource issue on an institutional agenda also requires strategic decisions about *gatekeepers*. Gatekeepers are individuals or organizations (public or private) that have authority or influence to determine which issues will be acknowledged by government. They serve as vehicles for entry into government, and they also prevent public agencies from becoming overloaded with issues. In a democratic system, gatekeepers are widely dispersed; they include chairs of legislative committees, interest-group leaders, forest supervisors, wildlife area superintendents, district court judges, legislative staff, and state foresters. Their sympathy must be obtained if an issue is to be successfully placed on an institutional agenda. Gatekeepers' power may emanate from various sources. Judges acquire their power from legal mandates, National Forest supervisors from delegated authority, legislative staff from association with powerful persons, and experts in a profession from their prestige and the high esteem in which they are held.

Gatekeepers can assume many modes of operation. Some can be *promoters,* in that they strive to get forest resource issues of importance to them on the agendas they oversee. Rather than being simply arbiters of forestry issues demanding attention, they aggressively advocate issues and fight for their placement on institutional agendas. For example, a legislative committee chair may demand that wilderness designation proposals be presented by an executive agency. Gatekeepers may also act as *vetoers,* who exercise authoritative power and simply refuse to grant agenda status to forest resource issues. Others, who operate as *bargainers,* become expert at negotiating the content of an agenda (''I'll permit your issue to be considered, if you agree to support my issue''). Still other gatekeepers are relegated to the role of *neutralists*. An example would be an influential interest group that supported free trade in softwood lumber, only to find that an important segment of its membership was demanding import restrictions. With its values in conflict, the influential group would decide not to actively oppose legislative proposals for such restrictions. Gatekeepers who wish to curb government acknowledgment and involvement in a forest resource issue (e.g., a proposal to restrict uses of private forestland) often argue in favor of values such as individualism, free enterprise, and local initiative. In contrast, gatekeepers who wish to facilitate placement of a forest resource issue (e.g., public purchase of private forestland) on an institutional agenda often use arguments based on equality, justice, civil rights, and the needs of future generations.

Development of a strategy to assure institutional agenda status must recognize that not all groups are looked upon equally by gatekeepers. The agenda-setting arena is not a field on which all who seek representation play on equal terms.

Some gatekeepers may be indebted to a particular group or sympathetic to their cause. A supervisor of a public forest, for example, may have a special fondness for a particular use of forests (e.g., as wilderness, as a source timber, or as a wildlife preserve) and would look favorably upon the issues of a group which represents that interest. In other situations, certain individuals and organizations may possess an abundance of resources and an ability to organize them so as to secure a gatekeeper's favorable attention. Involved may be better leadership, greater access to information, more financial resources, and a positive track record in organizing people to pressure for recognition. Also meaningful to gatekeeper access is the social and economic importance of groups or individuals. A group may be so important within an economy that it commands attention. For example, representatives of a wood-based economy of a medium-sized community dependent on a National Forest for an assured timber supply may have ready access to the supervisor of the forest. Also, groups that are prestigious and held in great esteem may command easy access to gatekeepers.

Relevant Publics Strategic decisions necessary to place a forest resource issue on an agenda often entail a search for support of *relevant publics* or the organizations that represent them. Support is not an easy commodity to come by; in large measure it develops because a forest resource issue becomes relevant to the interests of a segment of the public. Note the phrase "a segment of the public." The public is not a single homogeneous blob; an issue's public is always specific (relevant) to the particular situation or issue. For example, not all the citizens of the nation are interested in environmental issues, nor are all that claim allegiance to environmental issues interested in every environmental issue that happens to surface. At any point in time, interests in forests will be composed of an overlapping complex of specific and general publics that may or may not be relevant to a given problem or issue. Since success in achieving agenda status often depends on the support of relevant publics, it is useful to identify the types of publics from which support might be sought—ranging from a few very involved people to the mass public. Again, the presumption is that the greater the size of the audience, the more likely it is that the issue will achieve institutional agenda status.

Identification Groups Beyond those initially involved in a dispute (possibly a forester applying pesticides and a citizen who detests such action), the first to become involved in an issue may be *identification groups*. Their involvement stems from a concern about peers (individuals or organizations) who have become embroiled in a forest resource issue; the issue per se is of secondary concern. For example, if the practice of clear-cutting as prescribed by foresters is challenged, peer groups likely to come to the rescue of foresters might include the Society of American Foresters, the Association of Consulting Foresters, the National Association of State Foresters, and the Forest Service Retiree Association (U.S. Congress 1971). Of immediate concern to identification groups is the forester; of secondary concern is the practive of clear-cutting.

Attention Groups Second only to identification-group involvement in a forestry issue is the involvement of *attention groups,* which become involved because of an issue's substance and the proposals suggested for its resolution. The focus is on the issue, not the parties involved. Attention groups are generally disinterested in most forest resource issues, although they may be reasonably well informed about and interested in specific forestry matters. For example, if clear-cutting as an issue expands beyond identification groups, attention groups such as the following could become involved: the National Association of Home Builders, the United Brotherhood of Carpenters and Joiners, The Wildlife Society, the Soil Conservation District Society of America, the U.S. Chamber of Commerce, and the Wildlife Management Institute (U.S. Congress 1971). Home-builder involvement would stem not from concern over the forester's predicament with regard to clear-cutting but from concern over the effect that clear-cutting restrictions could have on the supply of wood products required for home construction. Likewise, labor unions would become involved because of concern about possible reductions in forestry's labor force.

Mass Public Once identification and attention groups have become involved in a forest resource issue, interest turns from groups to the mass public. The first such portion of the mass public to become so involved is the *attentive public*. The attentive public is composed of broad-gauged individuals who consistently express concern over public affairs. Though far from homogeneous, the attentive public consists of the better-educated and higher-income segments of society. Opinion leaders arise from the attentive public, and so do persons who spread the word to the less active and less interested segments of society. Less than 10 percent of the general populace is estimated to be part of the attentive public (Cobb and Elder 1983). Members of the attentive public write letters to editors of newspapers and periodicals, and they also submit statements for inclusion in the hearing records of legislative committees. For example (U.S. Congress 1987, 1988 pp. 639–640, 997):

Chairman Bruce Vento
Subcommittee on National Parks and Public Lands
Committee on Interior and Insular Affairs
U.S. House of Representatives
Washington, DC

Dear Congressmen Vento:

I understand the Subcommittee recently held hearings on proposed legislation that would dictate how grazing fees for the use of public rangeland are determined. . . . The dispute, as I understand it, is whether to increase grazing fees for holders of about 31,000 grazing permits to levels that equal or at least closely approximate fair market value for the forage consumed . . . permitees have historically paid only about 20 or 25 percent of what the forage is really worth.

It seems to me that this historical practice is unfair and unwise. . . . Sale of forage on rangeland may not be a major revenue source for the federal government . . . but every bit of revenue helps. Consider the current federal budget and our national debt! As a taxpayer, I feel cheated. . . . Grazing fees should be increased immediately.

Sincerely,

P. H. E.

Honorable Harold Volkmer, Chair
Subcommittee on Forests, Family Farms and Nutrition
Committee on Agriculture
U.S. House of Representatives
Washington, DC

Dear Representative Volkmer:

I have been a resident of Alaska for 19 years [Juneau and Tenakee Springs]. . . . I have been appalled by the course of the management of the Tongass National Forest. . . . It has always been apparent that this management has been for one use only – logging, and even this would have been somewhat acceptable if the management for logging had been based on the true timber economics of the forest, was more responsibly freely competitive allowing the growth of non-subsidized timber enterprises, and more cognizant of other economic and non-economic development of the forests . . . .

I would ask and encourage your committee to not linger on history and to not take on the overwhelming task of separation of fact from fancy because in this case it doesn't lead anywhere. I ask the committee to immediately engage the foresight to end the obvious polarization of residents in Southeast Alaska and to solve the dilemma for their Congressional Delegates with a renewal and fresh outlook on the Tongass timber resource.

Respectfully,

J. J. E.

The second portion of the mass public that may become involved in a forestry issue is the *general public*. On matters of public affairs, the latter is composed of the least active, least interested, and least informed segment of the citizenry. For the general public to become involved, an issue must be highly generalized and very symbolic. For example, assassination of prominent political figures frequently results in bitterly contested issues concerning government control of handguns. When so involved, their response is highly unorganized and short-lived. It is highly unlikely that the mass public has ever mobilized to express an opinion about the use and management of forests.

Character of an Issue Another important element in the setting of an agenda that is subject to the control of an issue manager is the *character of an issue*. Involved is an ability to define an issue along dimensions such as ambiguity, social significance, complexity, precedence, and related implications (Cobb and Elder 1983).

Ambiguity If a forest resource issue can be characterized in an abstract, vague, or general fashion, there is greater likelihood that it will expand to additional segments of the public and ultimately achieve institutional agenda status. Such a strategy rests on the assumption that larger segments of the public can identify with broadly defined issues. In attempts to secure public support, for example, there is a tremendous difference between groups that state their interest as ''removing chemicals from the nation's natural environments'' and groups that list specific herbicides to be banned from use in a specific portion of a wildlife reserve. A strategy of defining a forest resource issue in vague and general terms is easily recognizable by the widespread use of vague political symbols such as ''rape of the land,'' ''environmentally degrading,'' ''excessive pollutants,'' ''impairment of aesthetic qualities,'' ''multiple use,'' and ''sustained yield.'' Waging an agenda-setting war with generalities is not without risk. If a group is successful in placing an issue on an institutional agenda, it must be prepared to list specific demands to which government can respond. If no list of specific demands is made available, a gatekeeper may sweep the forest resource issue aside with a comment such as, ''They don't know what they want, so how can we act?''

Social Significance Another way to enhance an issue's chances of achieving agenda status is to portray the issue as having broader social significance—to construe the forest resource issue as fundamental to the basic welfare of a particularly sensitive portion of the public. The implication must be that the issue and the means suggested for its resolution have consequences for such matters as housing, hunger, health, personal income, the needs of the physically handicapped, and equity (both wealth and ethnic).

The more socially significant an issue can be made to appear, the larger the portion of the public that can identify with it. Basic human needs are common to all; an issue that concerns such needs affects many. For example, if restrictions are proposed on the practice of clear-cutting or on the application of pesticides, more people are likely to become involved if the restrictions are publicly portrayed as having an adverse affect on available housing. Similarly, although advocates of clean-air legislation may be primarily interested in the beauty of an unpolluted atmosphere, their cause will not be hurt by pronouncements that polluted air is a major cause of poor health. Pointing to relaxation as important to a healthy mind and body will further the cause of expanding the funding of recreational programs managed by public agencies. Many links between the character of a forest resource issue and human welfare are factually correct; properly cast, they can become powerful allies for the agenda-setting strategist operating in a political setting.

Technical versus Nontechnical Another dimension important to an issue strategist is how a forest resource issue is defined along a continuum from highly complex and technical to simple and easily understood. The more an issue is cast as being nontechnical, the more likely a wider audience is to favor involvement in the issue. Highly technical issues such as military weapons systems and chemical pesticides are understood by relatively few people. Very often such issues are encased in technical jargon; "mean annual increment," present net worth," "PCB's," "dioxin," "successional stage," "climax species." Issues concerning the technology of even-aged forest management, for example, are frequently expanded to larger audiences by assigning the technology a term that is easily understood—"clear-cutting" instead of "even-aged silvicultural systems." The publics that do become involved in technical issues have broad policy preferences but often very little understanding of what is actually involved in applying the technology.

Novelty Portraying a forest resource proposal as being extraordinary in the sense of lacking precedence in either substance or solution can also be an effective strategy for expanding issues to larger publics. Can policy proposals for addressing a forest resource issue be made to appear as though they are new and precedent-setting, rather than routine or familiar? The more an issue can be portrayed as lacking a prior occurrence or being solved in a suggested manner, the more opportunity there exists for additional segments of the public to become involved. If there is no history of similar forest resource issues and their resolutions, another source of disagreement surfaces for the issue in question and another segment of the public may become involved. Proposals in the 1940s for federal regulation of forestry practices applied by private owners of forestland were a major event—because federal involvement in such matters had no prior history. Even in the 1970s, restrictive regulations of a similar nature at the state level (California, Oregon, and Washington) became major issues—because the scope and severity of proposed regulation had no precedent. Heightened judicial involvement in significant matters of public forestry was viewed as highly inappropriate by many. The judicial systems of the nation had not been involved to such an extent in the past. Bringing forest resource issues to the judiciary for resolution was deemed inappropriate by many.

Link to Related Issues An agenda-setting strategist may also choose to spur interest in a forest resource issue, with the ultimate intention of setting an institutional agenda, by linking it with another issue of greater importance. Such a linkage is predicated on the assumption that the issues are closely related, so that addressing one may pose implications for another. If an issue has implications (good or bad) for other issues about which the public is highly sensitive, the opportunity for involvement of additional segments of the public is facilitated. For example, in the early 1970s clear-cutting was supposedly the forest resource issue before the U.S. Congress (U.S. Congress 1971). A careful review of the record reveals that the vast majority of the witnesses used clear-cutting as applied in the National Forests as a vehicle for expressing concerns about related issues (e.g., road construction, wilderness land allocations, pesticide use, budgets). Assessing consequences for related issues can lead to effective linkages that expand public involvement in a forest resource issue, and that may help the issue to achieve institutional agenda status.

Orchestration of Characteristics As suggested, the manner in which a forest resource issue is characterized has much to do with its successful placement on an agenda. It is important to recognize, however, that such characteristics are not implemented in isolation; all may be implemented at the same time. As one grand strategy, a forest resource issue may be cast as a matter of general environmental concern (it may be ambiguous); defined as creating a significant adverse impact on human health (as being socially significant); portrayed as easy to understand, even though highly complex (as being technical in nature); set forth as lacking prior government involvement (lacking precedence); and suggested to have consequences for other issues of popular concern (to have related consequences). Operated in a well-orchestrated manner, such characteristics can greatly facilitate successful agenda status for a forest resource issue.

Symbols Agenda building as a strategy can be substantially facilitated by the language or *symbols* used to attract the attention of various publics. A symbol such as "endangered species" or "timber famine' can be an important rallying cry for advocates of an issue and can enliven groups with the enthusiasm necessary to walk the long road to successful agenda status.

Symbols are words or phrases that generate an emotional response—either positive or negative—to an issue. The response may be far removed from the original meaning of the symbol (Edelman 1974). For example, "communism" as a symbol used by democratic governments evokes concern about political oppression, notorious dictators, and civil strife—all in contradiction to the technical definition of the word. Similarly, the term "wilderness preservation" evokes notions of solitude, natural beauty, leisure activities, and everlasting endurance. Quite literally, however, preservation of natural ecosystems is impossible; they are subject to natural forces that change their makeup on a continuing basis. Other phrases commonly used in connection with forest resource issues include "fair trade," "biological diversity," "profit-oriented," "rabid environmentalist," "ancient forest," "sustainable development," "law and order," "unfair competition," and "police state."

To be effective, symbols must be appropriate to the forest resource issue in question. For example, private-sector advocates of increased allowable cuts from National Forests would not refer to anticipated increases as profits. Rather they would mention the creation of additional jobs and greater income and economic stability within a timber-dependent community. Similarly, advocates of funding for a forestry cost-share program would refer to such programs as "forestry incentives," not as "government subsidies."

Symbols must also be applicable to the group toward which they are directed and must be of an appropriate weight or potency. For example, the phrase "oppressive dictator" might be considered excessively harsh if it were used by an interest group to describe a pollution-control agency official who is responsible for issuing regulations implementing a state's clean-air act. Ill-chosen language may be so offensive that other groups may lose interest in forming alliances with the offending parties.

Symbols vary in their effectiveness as tools for moving forest resource issues toward agenda status; they must be chosen with care. Some symbols are effective because they have a long and enduring history of popularity; examples are "liberty," "individual rights," "federal aid," "federal control," "timber famine," "wilderness preservation," and "environment." Others, such as "fire-contained" and "insect-controlled," rapidly lose their effectiveness if they are not appropriate to the times or if no action is taken. Care must be taken not to saturate an audience with a symbol, which may cause it to lose its utility as a tool of issue expansion. Symbols that may have lost their effectiveness because of overuse include "police brutality," "national interest," "public interest," "multiple use," "sustained yield," and "the greatest good for the greatest number of people over the long run." Symbols that are extremely effective convey a sense of urgency about a forest resource issue. "Timber famine" and "endangered species" are examples. Symbols play a definite role is issue expansion and should be looked upon as one of the many tools in the arsenal of the agenda-setting strategist.

Summary

Agenda setting can be viewed from four perspectives. As a *functional activity*, agenda setting can be viewed as a series of actions beginning with an individual's concern over an event that is personally troublesome and culminating in formation of an aggregation of individuals or organizations that also find the event to be troublesome. Unified, the individuals or organizations are in a better position to secure government recognition of concerns that have become common to all. Agenda setting can also be viewed as a *power and influence activity* in which persons who are in positions of power or who have special influence over government conduct, can simply decide which issues a government organization will acknowledge and ultimately act on. As an *opportunity activity,* agenda setting can be viewed as the unique confluence of political moods, changes in government, issue entrepreneurs, and agreement among important political figures that an issue should be addressed by government. Agenda setting can also be viewed as a *strategic activity* in which an issue is strategically managed to the point of acknowledgment and action by government. Implied are an issue entrepreneur's wise judgments about triggering events, political symbols, gatekeepers, and issue characteristics, all of which are implemented with a focus on expanding an issue through various groups and mass publics.

DISPOSITION OF ISSUES

The appearance of a forest resource issue on an institutional agenda implies that an agency or related unit of government was willing to recognize and address the issue. Most issues are not so fortunate. Many fade from the agenda-setting process because their advocates become weary of repeated denials of their requests for agenda status. For some issue advocates, government's failure to respond positively becomes sufficient reason to cease expenditure of the political resources needed to keep a forest resource issue before government. People and organizations become fatigued. Issues and their advocates slide into oblivion. Matters of concern in this context are the diversion and displacement of issues, the outright disregard of issues by government, and actions frequently taken by opponents seeking to keep an issue off an institutional agenda.

Diversion, Deferment, and Displacement

A forest resource issue that does not achieve a favorable government response and agenda status may be *diverted* for further definition by its advocates. This usually happens because government's response to the issue as initially defined was considered incomplete. For example, advocates of a ban on the use of all chemicals for forest management purposes may secure an agency's attention; the agency's response may be to establish guidelines for the use of a single chemical. Undaunted, the advocates may return to demand that a group of five other chemicals be governed by the same guidelines.

Still other issues are *deferred,* to again be addressed by government. In such instances, all involved agree that government's response to the issue was unsatisfactory. For example, advocates of intensified timber management on public forests may be given agenda status by an administering agency. The agency's response may be to increase reforestation efforts by diverting funds from timber sale preparation, campground maintenance, wildlife habitat improvement, and fire prevention and control. All interested parties may be discontented and may again seek an audience with the responsible agency.

Some forest resource issues are *displaced,* as when the focus of controversy shifts from the original issue to an entirely new issue. For example, clear-cutting may be alleged as a major cause of soil erosion, only to be displaced as an issue when construction and maintenance of roads are found to be the real culprits.

Issues and Demands Ignored

Forest resource issues may be ignored—or may seem to be ignored—by public agencies. Typically, this is not a willful action but rather a difference in perceptions about an agency's ability to successfully address an issue. Certain forest resource issues may appear to have been ignored, yet in reality may have received the attention, possibly quite sparingly, of an agency. The disparity often occurs because of unrealistic expectations of groups pressing for change in forest policies and programs. For example (Eyestone 1978):

- *Action expected, not merely a policy decision* Issue advocates may want a visible response to an issue, not a series of written regulations or lofty platitudes speaking about intent. Realistically, however, an agency may not have the capacity to respond as expected. For example, a policy decision to respond to concern over diminishing forest wildlife habitat may not be implemented because the financial and professional resources simply are not available. Likewise, advocates of more roads in public forests cannot expect action if funding to build such roads is not legislatively appropriated.
- *Significant commitment wanted, not a token effort* Issue advocates may want an all-out commitment to programs focused on an issue, not just a limited or an isolated response. Realistically, however, an all-out commitment may be impossible because of limited financing, bureaucratic inertia, and lack of leadership. For example, state-mandated timber-harvesting standards in private forests may be enforceable in only a limited number of private forest ownerships, not all private forest ownerships, because of limited financial resources and limited availability of professional talent.
- *Predictable results looked for, not unexpected consequences* Issue advocates may expect a certain result from policies designed to address their issues. Realistically, however, results other than those expected may occur. For example, forestland may be designated as wilderness with an expectation of the solitude forest wilderness can provide. Ironically, wilderness designation may lead to use by more recreationists and an erosion of the solitude originally sought.

• *Immediate response wanted, not program delays* Issue advocates may want an immediate program response to an issue, not prolonged delays. Realistically, however, limited resources, untested technologies, and procedural requirements may frustrate attempts to respond quickly. For example, curtailing the use of chemicals for timber-management purposes may require long and lengthy hearing procedures, designating public forestlands for certain uses may require comprehensive and very thorough environmental impact statements, and funding of new public programs requires involvement in a multiyear budget cycle.

• *Response looked for everywhere, if possible anywhere* Issue advocates may want a policy to respond to an issue's occurrence everywhere, if the issue is capable of being responded to anywhere. Realistically, however, diverse forest resource conditions, differing styles of administration, and varying political beliefs make uniform application of a policy throughout a large management system nearly impossible. For example, a Pacific Northwest reforestation standard of 300 seedlings per acre may be totally inappropriate to other regions of the nation.

Forest resource issues may only appear to be ignored by a public agency, for the reasons listed above. Under some circumstances, however, issues are in fact ignored. Consider the following:

• *Trivial demands* Issue advocates may make demands that are trivial in the context of broader social and resource-management issues. Inaction may occur because more important forest resource issues are being processed by government. For example, concern over capital gains treatment of income from the sale of timber was a relatively minor issue within the context of major federal tax reform which occurred in the mid-1980s. The latter had sweeping social significance; the former did not and was given commensurately less attention.

• *Inappropriate government unit* Issue advocates may make demands of the wrong unit or level of government. The unit or level being petitioned may have limited or no authority over an issue. For example, petitioning the federal government to establish limits on the harvesting of forest wildlife is wrongfully directed; state governments typically establish such limits. Or petitioning the chief of the USDA-Forest Service to invest a larger portion of the agency's budget in recreation and wildlife programs instead of timber production programs may be misdirected; Congressional action is required to make such budgetary changes.

• *Political bias* Issue advocates may make demands of political processes that are inherently biased against some issues. A conservative government bent on reducing government involvement in the affairs of private citizens would not look with favor on the transfer of private forestland to public ownership or on public regulation of the forestry practices of private landowners. Such political phenomena are perfectly normal and legal facets of governments—and are capable of derailing an issue.

- *Displacement* Issue advocates may find their issue removed from an agenda because it is being replaced by another issue which has more public support or greater agency sympathy. National issues concerning energy, inflation, deficit spending, and tax reform can literally sweep agendas clean, replacing issues of long standing with items of more immediate concern.

Agenda Status Obstructed

Setting an institutional agenda with an important forest resource issue may well be the intent of those to whom an issue is important. As in all political situations, however, some groups may view their interests as being better served if certain forest resource issues are denied agenda status. For example, advocates of a greater timber-producing role for public forests may find their interests better served if another group's proposal for formal designation of forest wilderness fails to achieve institutional agenda status. Likewise, groups of consulting foresters may perceive their interests as better served if proposals for additional public service foresters fail to appear on the agenda of a public forestry agency. Manufacturers of chemicals used for various forest management purposes would probably not favor agenda status for proposals to severely curtail the availability of chemicals for such purposes. If circumstances of this nature exist within the forestry community, what type of actions might be expected of groups attempting to contain forest resource issues and ultimately keep them off an institutional agenda?

Opponents of issues may use a variety of strategies to keep a forest resource issue off an agenda. Perhaps most obvious is the reverse of strategies considered important to issue expansion, namely:

- If ambiguity, vagueness, and generalities are virtues for an issue's expansion, then concreteness, good definition, and high specificity may be virtues for its containment.
- If broad social significance tied to basic human needs is a virtue for an issue's expansion, then social insignificance and far removal from basic concerns over health and welfare may be virtues for its containment.
- If nontechnical, easily understood conditions are a virtue for an issue's expansion, then highly technical and very complex conditions may be virtues for its containment.
- If precedent-setting, new, and unusual are virtues for an issue's expansion, then usual, same, and continuing may be virtues for its containment.
- If linkage with another more politically important issue is a virtue for an issue's expansion, then isolation from other more significant issues may be a virtue for its containment.

Issues may also be prevented from securing agenda status by use of more focused actions. Implementation of group-oriented as well as issue-oriented strategies is implied. Opponents of a forest resource issue are employing *group-oriented strategies* if their efforts focus on the opposing group per se, not on the demand or the forest resource issue being espoused by the opposition. In contrast, *issue-oriented strategies* are being employed when the focus is on the opponent's demands. Tactically, the opposing group may be attacked directly or indirectly. In a *direct attack,* the focus is on the merits or legitimacy of the opposing group or the forest resource issue being espoused, while an *indirect attack* avoids confronting the opposition and its demands. Utilizing group orientation and mode-of-attack dimensions, four distinct strategies surface for use by groups interested in keeping forest resource issues off an institutional agenda. They are portrayed in Figure 3-2 (Cobb and Elder 1983):

	Group–oriented	Issue–oriented
Direct attack	Attack opposing group	Defuse issue
Indirect attack	Undermine opposing group	Blur issue

FIGURE 3-2
Strategies for preventing institutional agenda status for an issue, by orientation and directness of strategy.

Group-Oriented Strategies DIRECT ATTACK: Attack the opposing group directly so as to limit its appeal to potential members. This can be accomplished by:

- *Discrediting the group* Attempt to cast doubt upon the group and to dishonor it. For example, characterize the opposing group as "a bunch of timber beasts," "a communist front," "a subversive organization," or "a bunch of unrealistic environmentalists."
- *Discrediting the leaders of the group* Attempt to cast doubt upon the group leaders and to dishonor them. For example, characterize the leaders of the opposing group as "out of touch with reality," "rabble-rousers," "wilderness fanatics," or "typical timber beasts."

INDIRECT ATTACK: Undermine (erode) the opposing group's base of support and leadership. This can be accomplished by:

- *Appealing to the group's members* Demonstrate that members of the opposing group are not sympathetic to the views of the group's leaders but in fact support the views of the opposition. For example, demonstrate that the stockholders of a timberland-owning company are opposed to the company's timber-management activities involving clear-cutting.

• *Co-opting the group's leaders* Bring leaders of opposing groups into your organization in hopes of convincing them of the merits of your group's position or alleviating more serious damage to your position. For example, as an industrial group advocating reductions in softwood lumber imports, invite a well-known company representative from the exporting country to have formal membership on your committee for fair international trade in wood products.

Issue-Oriented Strategies DIRECT ATTACK: Recognize the legitimacy of the opposing group's demands or issue, but deny that the issue has any urgency or would serve any useful purposes in a contemporary setting. This can be accomplished by:

• *Providing token rewards or reassurances* Make sympathetic statements and appear to be interested in the group's issues. For example, as a public forestry official, formally accept signed petitions listing grievances about wildlife-management programs. Make sympathetic and reassuring statements about future policies and programs, such as ''Taxes need not be increased,'' ''Allocating additional land to wilderness will not be necessary,'' ''Additional roads to access additional timber supply will not be required.''

• *Taking action in a token or showcase fashion* Act in a very limited manner on a larger problem. For example, design one clear-cut in an environmentally sensitive fashion, while designing the remaining clear-cuts in a business-as-usual manner—or employ one woman or minority in a highly visible position, while failing to meet the spirit of affirmative action throughout the organization.

• *Taking anticipatory action* Create a favorable climate of opinion in which opposing groups might accept less than they would otherwise demand. Anticipate a group's demands before it can mobilize public support. For example, suggest in advance of a pending confrontation that timber-harvesting levels be reduced by half, knowing full well that not to do so could mean agreeing to two-thirds reduction as demanded by opposing groups.

INDIRECT ATTACK: Sidestep the opposing group's issue completely, making no effort to resolve the substance of the issue. This can be accomplished by:

• *Faking a constraint* Agree with the merits of the issue being advocated by a group, but inform the group that nothing can be done to address their grievances—all resources, facilities, and leadership are committed and being utilized to the fullest: ''We'd like to help, but. . . .

• *Postponing action* Take the opposing group's issue under advisement with the intention of seeking out additional information. For example, establish a commission, a committee, or a council to investigate the issue, and require it to report at a later date.

ORGANIZATIONAL AND CITIZEN INVOLVEMENT

Agenda setting involves a continually changing cast of people and organizations. At an appropriate time, the right people involved with the right organization can merge with a forest resource issue to form unique circumstances capable of fostering placement of the issue on an institutional agenda. An agency head, an interest-group executive, and a legislative leader may surface at the same time with a common interest in nonindustrial private forests. Together they may form a remarkably effective force capable of setting an institutional agenda and ultimately securing a significant change in policy directed at such landowners. Similarly, a uniquely qualified issue entrepreneur can burst upon the forestry scene with a talent for communication and a willingness to invest the necessary time and energy required to promote a forestry issue. Once the wilderness area is designated, the regulation is in place, or the endangered species protected, such entrepreneurs return to the general populace and are unlikely to be heard from again. A timely meshing of peoples' interest in an issue and the organizations they represent can be the single most significant event leading to agenda status for forest resource issues.

Governmental Participants

Participants in agenda-setting activities occur both in and outside government. In government, a chief executive officer (CEO)—e.g., a governor, the President—is often preeminent. CEOs are capable of focusing public attention on favored forestry issues because they have ready access to the media, because they can enforce loyalty to an issue among their political appointees, and because they are able to require that favored issues be addressed by civil servants. Important in their own right in agenda activities are the political appointees of chief executives. They can elevate a forest resource issue within a bureaucracy by giving it recognition and a sense of importance. Because political appointees have relatively short tenure in government positions, the agenda items they fight to establish may be displaced with the coming of a new administration.

Civil servants also play a role in agenda setting, although it has been argued that their role is primarily to develop policy and program responses to the forestry issues that have already been placed on an institutional agenda (Kingdon 1984). Bureaucrats' favored issues may be distasteful to an administration currently in power. Given such circumstances, they may keep certain issues alive but in limited circulation, waiting for a new political administration that might look upon the issues with greater favor. The strength of bureaucrats in agenda activities lies in their longevity within bureaucracies (political appointees come and go, but the bureaucracy endures), their wealth of experience in administering programs, and their enduring relationships with legislators and interest groups.

Legislators are also active participants in agenda-setting activities. They are important for a number of reasons, including the authority of legislatures to make laws addressing forest resource issues and the special ability of legislators to garner publicity by holding hearings, introducing bills, and making speeches. A legislator's drive to affect the makeup of institutional agendas can flow from a desire to satisfy constituents, to enhance political reputations, or to shape the nature of public policy by addressing an issue because of an interest in its substance. Often facilitating a legislator's agenda-setting ability are members of a legislative staff, especially committee staff and the staffs of independent legislative agencies (the Congressional Office of Technology Assessment and the Congressional Research Service of the Library of Congress). They can be prominent in the agenda-setting process because they, unlike legislators, can devote significant attention to a specific issue or public problem.

Nongovernmental Participants

Participants without formal positions in government are also active in agenda-setting processes. Included are interest groups, researchers, academics, consultants, media, political parties, and various segments of the general public. In agenda setting, the differences between those inside and those outside government often become fuzzy. Just as interest groups lobby government, so government also lobbies interest groups. In part this is true because the communication channels between the two are extraordinarily open to the free flow of ideas about current and prospective agenda items. However, the distinction is important: persons in government have been granted formal authority to propose or dispose of issues; those outside government do not enjoy such status.

Interest groups loom large in agenda setting because of their ability to mobilize support for an issue by writing letters, sending delegations, and forming coalitions. To some government officials, "the louder [interest groups] squawk, the higher [an issue] gets" (Kingdon 1984, p. 52). Although an interest group may successfully secure agenda status for a forest resource issue, it is not always in a position to control the policies formulated to address the issue or to control the selection of a favored policy. For example, although funding of a forestry program may become an agenda item, the source of the program's funding may be a reduction in another program of interest to a group. Interest groups do not always seek placement of issues on institutional agendas; their interests are sometimes better served by working to abort agenda status for issues that would curtail government benefits currently being enjoyed by their members.

The Media The media can also play a significant role in agenda setting. Although some question the media's ability to actually establish the institutional agendas of government, their action as a facilitator is important (Cook 1989). For example, the media can act as communicators within the forestry community, bringing forest resource issues and options for their resolution to the attention of government. Then too, the media can affect agendas by magnifying issues which have been set forth elsewhere—by accelerating public interest in them and enlarging their importance. As one journalist stated, "Media can help shape an issue and help structure it" (Kingdon 1984, p. 63). The media can also influence public opinion, which in turn may influence the type and nature of issues appearing on an institutional agenda. Investigative reporting may arouse public concern about wildlife management, for example—concern which is ultimately acknowledged by government.

It can be argued, however, that the media simply report about issues on government agendas; they do not independently affect placement of issues on institutional agendas. Similarly, the media's impact on agendas may well be diluted by a tendency to cover a given issue for only a relatively short period of time. Short attention spans may not be useful in a political setting in which long-term perseverance is a virtue. Likewise, impact may be diminished by a tendency to focus on the end of the agenda-setting process rather than the beginning. For example, the media may do well in covering a public hearing on a proposed forestland-management plan; ironically, however, the institutional agenda for the hearing was probably set weeks or months prior to the hearing with no media influence.

Public Opinion Public opinion also plays a role in setting institutional agendas. Exactly how and to what extent is subject for debate. The debate is heightened by the fact that the public is really composed of a vast array of publics, each concerned with different issues and each focusing on issues with varying degrees of intensity. Government officials contemplating placement of an issue on an institutional agenda surely give thought to the general mood of the citizenry—its public's tolerance for new taxes, acceptance of government regulations, or interest in environmental quality. If, however, there is general public concern, real or imagined, about an aspect of natural resources (say, the air is polluted, the forests are depleted, or the wildlife is threatened) and a belief that collective action should be undertaken to improve or correct the situation, a more focused segment of the general public (an interest group) may well be the vehicle by which such concern is transformed into specific issues that eventually appear on an institutional agenda. More often than not, government officials and activists concerned about a public problem set the agenda of the mass public rather than the other way around. One public official observed that "well-known politicians make an issue, and the 'people' come aboard" (Kingdon 1984, p. 70). The general public is a broad-based and often unwieldy group; problems with focus and a limited attention span do not facilitate its efficiency in agenda-setting activities.

FORESTRY PROFESSIONALS AND AGENDA SETTING

Agenda-setting processes involve collective behavior; any one individual's control of an agenda tends to be limited. Even officeholders (say, chairs of important legislative committees), who supposedly have ultimate control over institutional agendas, are subject to strong political pressures to favor some issues at the expense of others. Among the many participants in the agenda-building arena are forestry professionals. Owing to their significant expertise in forest management and the premium that political and social systems place on rational problem solving, such professionals can play a significant role in the process of institutional agenda setting. Their influence is often greatest when forest resource issues are defined as technical matters requiring some special understanding or knowledge. Most often, however, the professionals' influence over institutional agendas is indirect, arising from the general climate of informed opinion which they help to create.

Forestry professionals may act to counter the biases of some political groups involved in institutional agenda setting; they may also introduce biases of their own. In doing the latter, publicly employed professionals are recognizing the importance of issue definition to maintenance of their own self-interest—a less than honorable perspective. Professionals may also detract from efficient agenda setting by making debates over forest resource issues more specialized and esoteric, thus creating barriers to popular participation and rendering a forest resource issue's ultimate definition less intelligible to wider audiences. For example, they may focus discussions on the technicalities of clear-cutting when the use of public forest land is the fundamental issue of concern. The ultimate result of narrowing an issue in a technical sense can be the establishment of policies skewed in directions contrary to popular preferences (Elder and Cobb 1984).

Passive Involvement Professionals can approach agenda-setting processes in a number of ways. Strategically, they may be passive participants, facilitators and encouragers, or initiators and decision makers (Jones 1984). In practice, a combination of all such strategies is most likely. A posture of *passive involvement* implies that public forest resource issues will be developed and prioritized by current and potential forest users in a freewheeling democratic fashion, with a minimum of professional or public agency involvement. Such an approach requires the existence of free and open communication between public organizations and interested individuals and groups. It denies opportunity to systematically determine the wishes of forest users or to assist such users in their efforts to define and prioritize issues and demands. This approach is based on a highly pluralistic view of government, one that presumes that individuals and groups are able and willing to define forest resource issues, to organize, to seek access, to influence decision making, and to monitor implementation of forest policies. It also ignores the reality of unequal distribution of resources and influence among groups and individuals—a distribution which favors the resource wealthy over the resource poor.

Facilitate and Encourage A strategic option that acknowledges the bias of a totally free system toward the strong over the weak is to *facilitate and* *encourage* the weak. To correct the disparity, concerted efforts are made to assist the public in the process of defining and articulating wishes for the use and management of forests. Implied are professional and public organization roles of assisting individuals and groups in the definition of forest-management opportunities and in bringing publicly chosen opportunities to government for action. The emphasis is on equipping people to participate—not on assuming for them the tasks of defining issues and setting forth policy options. Facilitating places professionals and public organizations in the difficult position of determining which individuals and groups need assistance. It also raises the troublesome question of whether public forestry organizations should be building their own agenda of issues by choosing the groups that receive assistance.

Initiate and Decide A more aggressive role for forestry professionals in agenda setting is to *initiate and decide.* Such an approach implies a proactive role for professionals and public organizations. Forestry professionals move to systematically implement issue-defining and priority-setting mechanisms within an organization charged with the management of forests; they do not wait for the results of democratic processes. Professionals determine the land-use and management issues that are worthy of attention. A professional plays both the role of the technical professional and the role of the citizen. This strategy places enormous burdens on forestry professionals and public forestry organizations. Surveying events, judging consequences, and establishing issue priorities are demanding responsibilities. If accomplished too well, they raise an obvious concern about whether a democratic government can remain true to principles of democracy.

CONTEMPORARY FOREST RESOURCE AGENDAS

Contemporary forest resource issues range from global concerns about atmospheric pollution to political struggles over the allocation of public forestland to formally designated wilderness, and from volatile debates over ability to export wood products to certain foreign countries to conflicts over appropriate allocation of federal monies to forestry programs. The issues are many; they range widely in scope and importance. Some are on the institutional agendas of public agencies with national and international responsibilities, such as the United Nations, the USDI-Fish and Wildlife Service, and the USDA-Forest Service. Others are agendas of concern to agencies with a more narrow geographic focus, including state departments or divisions of forestry, and municipal or county forestry units. The nature of contemporary forest resource issues appearing on the institutional agendas of public agencies, or the issues private organizations wish to place on such agendas, is best understood by examples. Consider the following.

Congressional

Below is a selected list of forestry and closely related issues which the 94th through 98th U.S. Congress made conscious decisions to consider, by conducting hearings:

- Forest ecosystems and atmospheric pollution
- United States-Canadian trade in forest products
- Forest productivity and the effects of acidic deposition
- Control of wildfire on federal lands
- Opportunities for reforestation of marginal croplands

- Federal timber contract modifications-extensions
- Designation of lands to the National Wilderness Preservation System
- Cooperative state forestry research and extension
- Deforestation: environmental impacts and research needs
- Phosphate leasing in Florida national forests

- Management of the Tongass National Forest
- Fees for the use of the National Trails System
- Timber harvesting receipts from military forestlands
- Urban parks and forest recreation programs
- Illegal aliens used in public reforestation programs

- Phenoxy herbicides in forest management
- Present and potential uses of wood for energy
- Proposed federal Department of Natural Resources
- Agency responsibility for environmental impact statement preparation
- Small business enterprises in outdoor recreation and tourism

- Forest pest management and control
- Mining and mineral leasing on public lands
- Forestry loans to owners of nonindustrial private forests
- Mount St. Helens National Volcanic Area
- Reforestation of public and private forestlands.

Federal Forestry Agency

As required by the Forest and Rangeland Renewable Resources Planning Act of 1974, the USDA-Forest Service prepares a renewable forest resources program every 5 years. The program identifies issues to which it is to respond. In 1990, the major policy issues identified were (USDA-Forest Service 1990, pp. 5-1 to 5-49):

Environmental protection issues Global stewardship; biological diversity; spotted owl; riparian management; range condition; water quality; air quality; catastrophic fires; and threatened, endangered, and sensitive species

Resource use issues Clear-cutting, old-growth forests, below-cost timber programs, wilderness management, changing recreational needs, minerals development, near-term softwood timber supply, and timber supply for nonindustrial private forestland

Management issues Program financing options and program appeals and litigation processes

State Forestry Agency

In recent years, state agencies have assumed substantially greater and more aggressive roles in the management of forests. In doing so, they have specified a number of forest resource issues toward which their attention will be directed. Consider some of the issues identified by California State Board of Forestry (1985, pp. 7–16):

Rural economic stability and community development Marketing of existing and new forest and rangeland products, competition between forestland-based industries and other capital investment options, production of nonmarket goods and services from forest and rangelands, and employment opportunities in rural forested areas

Protection and maintenance of the biological base Preservation and regeneration of commercial timber base, maintenance and enhancement of environmental diversity of forest and rangelands, protection and enhanced supply of noncommodity outputs from forest and rangelands, and incorporation of biological principles within fire- and insect-management programs

Social pressures on rural forestland base Representation of rural values in an urbanizing state, fragmentation of forest and rangeland base, differing perception of forestland uses and appropriate management practices, and public involvement in resource decision making

Rights and responsibilities of public and private ownership Guarantee of private rights to manage, harvest, and graze private forest and rangeland; equitable distribution of costs and benefits of private property regulation and public services; sharing of costs and revenues from publicly owned forestlands with local communities; public rights to the use of forestland

Coordination and planning Coordination of fire and pest protection and prevention; coordination of resource goal setting among various land management agencies; resolution of conflicts and crisis prevention; coordination and funding of public research and education

Agency Policy Analysis Unit

The U.S. Department of Agriculture annually faces a number of complex and politically sensitive natural resource issues. In order to effectively address such issues, the department must have available a substantial reservoir of information on the nature of the issues being faced and the policy options available for addressing them. Major responsibility for providing such information rests with the department's Office of Budget and Program Analysis staff. As of November 1987, among the issues were destined for the staff's review (Office of Budget and Program Analysis 1987) were the Tongass National Forest resource base and the impact of the proposed Canadian Stikine-LeConte Wilderness on timber and mineral resource development.

Councils and International Agencies

Organizations with a common interest in or a responsibility for forests in a global context have achieved significant worldwide recognition in recent years. The forestry issues which are the focus of their attention are large in number and broad in scope. Consider selected issues identified by the World Commission on Environment and Development (1987), the Committee on Forest Development in the Tropics (1985) and the Council on Environmental Quality's Global 2000 Report to the President (1980):

World Commission on Environment and Development Key issues concerning the global environment and sustainable development include population and human resource levels, conservation of species and ecosystems, production and distribution of energy, human settlements in forest and agricultural environments, management of industrial pollutants, international cooperation in natural resource use and management, and decision support systems for the management of natural environments.

Committee on Forest Development in the Tropics Priority issue areas include forest as a land use (transformation of forests into other land uses), forest-based industrial development (financing, fiber supplies, marketing, available labor), fuelwood and energy (scarcity of fuelwood in arid and semiarid regions), conservation of tropical forest ecosystems (sustainable production of timber, protection of rare ecosystems), and supporting institutions (education and training, planning and policy development, extension and research).

Council on Environmental Quality Deforestation in less-developed countries resulting in siltation of streams, depletion of ground water, intensified flooding, and aggravated water shortages during dry periods; increased intensity of forest management resulting in soil nutrient loss, disturbed patterns of rainfall, increased energy demands, and adverse consequences of pesticides and fertilizers; and global-scale environmental impacts of forestry activities, including changes in climate and reductions in biological diversity.

Interest Groups

Interest groups represent a very significant segment of the population which has a keen interest in the maintenance of high-quality forested environments for use by current and future generations. Consider the issues of concern to the Sierra Club in 1987–1988, the National Wildlife Federation in 1987, the Society of American Foresters, and the American Forest Resource Alliance (American Forest Resources Alliance 1989; Society of American Foresters 1988; Sierra Club 1987; National Wildlife Federation 1987):

Sierra Club Conservation Campaign priorities for 1987–1988 Reauthorization of the Clean Air Act, Arctic National Wildlife Refuge Protection, Bureau of Land Management Wilderness Designations, California Desert National Park, USDI-Bureau of Land Management Wilderness Designations, National Forest System Planning, National Park Management and Creation of Tallgrass Prairie Preserve, Implementation of Toxic Control Laws, Public Lands Leasing and Permits (oil and coal), National Forest System Wilderness Designations, High-Level Nuclear Waste Disposal.

National Wildlife Federation Wildlife: reauthorization of the Endangered Species Act, plastics in the marine environment, farm legislation (1985) conservation measures, taking of predatory or scavenging mammals and birds. Fisheries: fisheries and wildlife agencies appropriations, turtle excluder devices, national marine policy development legislation. Pollution: federal pesticide laws, groundwater, Clean Air Act amendments. Public lands: Arctic National Wildlife Refuge, on-shore oil and gas leasing, cave resources protection, oil shale and mineral extraction legislation, Tongass National Forest, USDA-Forest Service appropriations, Montana wilderness proposals. International: United Nations environmental program, African famine recovery, multilateral development bank. Water resources: central Arizona project, Soil Conservation Service small watershed legislation, Delaware river basin programs.

Society of American Foresters Below-cost timber sales, biological diversity of forest ecosystems, budgets and appropriations, clear-cutting, community stability, fire management, herbicides and pesticides, international deforestation, international trade, land-management planning, old-growth forests, professionalism in natural resource agencies, forestry research, state and private forestry, timber taxation, Tongass National Forest, wetlands protection, wilderness allocations, wilderness management.

American Forest Resource Alliance Forested wetlands, forest wilderness, global warming, clear-cutting, northern spotted owl, red-cockaded woodpecker, deforestation, old-growth forests, biological diversity, below-cost timber sales, forest roads, taxation.

Wood-Based Industry

The wood-based industry has a significant interest in the use and management of the nation's forestland, both public and private. The National Forest Products

Association, one of the industry's leading associations, identified the following issues as being of interest in 1984 (National Forest Products Association 1984):

Administration issues Federal timber sale contract extensions, USDA-Forest Service forest resources program document, forest industry private woodlands program, timber crop insurance, forest resources research

Natural resources legislative issues Forestry extension, wilderness designation proposals, appropriations for USDA-Forest Service forest resource programs

Judicial issues Oregon nonwilderness litigation; cargo preference litigation; phenoxy herbicides litigation

Environmental issues Atmospheric deposition, groundwater quality, formaldehyde used in construction materials, forestry use of 2,4,5-T and Lindane, and wood preservatives

Housing issues Mortgage and home investments, foundation-treated lumber and plywood, smoke toxicity regulations, and roof-system research programs

Economics and taxation issues Taxation of timberland estates, capital gains tax treatment of timber income

International trade issues Overseas market development, domestic international sales corporations, tariff barriers in Japan, market reciprocity, maritime regulatory reform

AGENDA SETTING IN THE POLICY PROCESS

Agenda-setting processes are the means by which forest resource issues are presented to government. People who actively participate in such processes hope to have an issue of significant concern recognized by government and ultimately addressed with a well-designed and carefully implemented policy. Whether an issue is formally recognized for action by an agency is dependent on broader circumstances (windows of opportunity), on the managerial skills of issue managers, and on the willingness of government officials to respond. Given the number of issues demanding governments attention, there is little wonder that many are diverted, ignored, deferred, or even obstructed. To understand the agenda-setting process is to understand the substance of concerns toward which government will ultimately direct its resources.

Agenda-setting processes are in operation on a continuing basis within the forestry community. Consider two issues that have been subject to such processes in the recent past:

- *Clear-cutting in government-owned forests* The practice of harvesting virtually all trees from a designated area in order to regenerate a forest with preferred tree species arose as an issue in the early 1970s. Triggering events (misapplied clearcuts, especially in the Bitterroot National Forest and the Monongahela National Forest), facilitators (various media, university faculty), organized interest groups, segments of the general public, and gatekeepers (legislators, forest administrators) were active participants in struggles to set or avoid setting formal agendas controlled by the U.S. Congress and the USDA-Forest Service (the gatekeepers). The process was expedited by a mood of environmentalism, memories of past destructive logging practices, increased demand for timber, and growing interest in recreational and wildlife uses of forests. See Dana and Fairfax (1980), especially "The Clearcutting Controversy"; Davis (1976), especially "To Cut a Tree—The Anatomy of a Controversy"; Ellefson (1972), and Fairfax and Achterman (1977).

• *Below-cost timber sales in National Forests* The practice of offering public timber for sale when the receipts from such sales will not cover the government's cost of making them available for harvest arose in the mid-1980s after allegations (the triggering event) were made by a select few interest groups (including The Wilderness Society) and subsequently expanded upon by facilitators, including other interest groups, legislators, and legislative support organizations (such as the U.S. General Accounting Office). Recognition of the issue was expedited by various analyses that identified particular National Forests that were offering timber for sale at below cost. The focus for agenda-setting activities was the USDA-Forest Service, which recognized the issue by setting an institutional agenda and subsequently developed an accounting system to track the cost of managing forests for multiple benefits (Barlow 1979, General Accounting Office 1984, LeMaster et al. 1987, Rice 1989, Risbrudt 1986).

The study of agenda setting often brings to the surface concerns about the legitimate scope of government authority to act on an issue. As expected, not all within the forestry community agree on the proper role of government in certain issues. Most often, a variety of social and political factors determine society's willingness to have government act on an issue. These factors can include the extent to which political values are commonly shared; e.g., if many people believe there is need to increase wildlife opportunities offered by nonindustrial private forests, government action to address the issue is very likely. The nature of widespread customs and traditions can also play a part; e.g., if government has customarily owned forestland, people will generally agree to additions to such ownership, even if they are leery about specific land-acquisition proposals. The occurrence of relatively short-lived yet significant events can be influential; for example, if many people become alarmed at the potential affect of atmospheric pollutants on forests, the government will in all likelihood address the issue. Changes in the way political elites think and talk about government actions can be another factor; for example, if candidates for a governorship make government involvement in the production of quality water from forested watersheds a theme of their campaign, people may eventually accept public regulation of timber-harvesting practices undertaken on private forestland.

Given such factors, intense opinions about the appropriateness of government actions are frequently formed. Some believe that whatever government does now, it ought to do, and that the scope of forestry issues addressed by government should increase because public attitudes favor greater roles for the public sector. Others hold equally strong opinions favoring a reduction in the scope of forestry issues to be addressed by government; they often feel duty-bound to reduce government excursions into private-sector matters. Whatever the prevailing trend in such opinions, fundamental questions remain: What determines when an idea's time has come? What becomes so irresistible about an issue that government action to address it becomes warranted—even demanded?

Agenda-setting activities have considerable influence over other stages of the policy process. For example, by having control over which issues appear on an agenda, gatekeepers and issue managers are able to profoundly influence the nature of policies to be addressed at later stages in the policy process. Such individuals can exercise distinct preferences for addressing particular issues—preferences that exclude certain issues deemed important by others. In addition, narrowly defined issues placed on an agenda for action may significantly constrain the scope of policies generated during formulation, may limit the number of policies available for selection, and may jeopardize the political feasibility of legitimizing a selected policy.

REFERENCES

American Forest Resources Alliance. Issue Briefs. Washington, 1989.

Barlow, Thomas J. The Giveaway in the National Forests. *Living Wilderness* 43(147):29–32. 1979.

California State Board of Forestry. Issues and Strategies Document for the Development of the Centennial Action Plan. Centennial II Conference. Sacramento, Calif., 1985.

Cobb, Roger W., and Charles D. Elder. *Participation in American Politics: The Dynamics of Agenda-Building*. The Johns Hopkins University Press, Baltimore, Md., 1983.

Cook, Timothy E. *Making Laws and Making News: Media Strategy in the U.S. House of Representatives*. The Brookings Institutions. Washington, 1989.

Committee on Forest Development in the Tropics. *Tropical Forestry Action Plan*. Forestry Department, Food and Agriculture Organization, United Nations, Rome, Italy, 1985.

Council on Environmental Quality. *The Global 2000 Report to the President*. vols. 1, 2, and 3. Government Printing Office, Washington, 1980.

Dana, Samuel T., and Sally K. Fairfax. *Forest and Range Policy: Its Development in the United States*. McGraw-Hill Book Company, New York, 1980.

Davis, Kenneth P. *Land Use*. McGraw-Hill Book Company, New York, 1976.

Edelman, *The Symbolic Use of Politics*. University of Illinois Press, Chicago, 1974.

Elder, Charles D., and Roger W. Cobb. Agenda-Building and the Politics of Aging. *Policy Studies Journal* 13(1). 1984.

Ellefson, Paul V. The Attack on Clearcutting. In *Proceedings of the 1972 Convention of the Society of American Foresters,* pp. 51–59. Society of American Foresters, Washington, 1972.

Eyestone, Robert. *From Social Issues to Public Policy.* John Wiley & Sons, New York, 1978.

Fairfax, Sally K., and Gail L. Achterman. The Monongahela Controversy and the Political Process. *Journal of Forestry* 75(8):485–487. 1977.

General Accounting Office. *Should the Forest Service Make Timber Sales Below Cost? A Policy Question for Congress.* GAO/RCED-84-96. U.S. Congress, Washington, 1984.

Jones, Charles O. *An Introduction to the Study of Public Policy,* 2d ed. Duxbury Press, North Scituate, Mass, 1977.

———. *An Introduction to the Study of Public Policy,* 3d ed. Brooks/Cole Publishing Company, Monterey, Calif., 1984.

Kingdon, John W. *Agendas, Alternatives, and Public Policies.* Little, Brown and Company, Boston, Mass., 1984.

Koenig, Louis W. *An Introduction to Public Policy.* Prentice-Hall, Inc. Englewood Cliffs, N.J., 1986.

LeMaster D. C., B. R. Flamm, and J. C. Hendee. *Below Cost Timber Sales: Conference Proceedings.* The Wilderness Society, Washington, 1987.

National Forest Products Association. NFPA Washington Report. 1984.

National Wildlife Federation. Legislative Update. Washington, August 1987.

Office of Budget and Program Analysis. Agenda of USDA Studies (November 1987). U.S. Department of Agriculture, Washington, 1987.

Rice, Richard E. The Uncounted Costs of Logging, vol. 5 in *National Forests: Policies for the Future.* The Wilderness Society, Washington, 1989.

Risbrudt, Christopher. The Real Issue in Below-Cost Sales: Multiple-Use Management of Public Lands. *Western Wildlands* 12(1):2–5. 1986.

Sierra Club. *National Conservation Campaigns: 1987–88.* San Francisco, Calif., 1987.

Society of American Foresters. *Briefings on Federal Forest Policy.* Bethesda, Md., 1988.

U.S. Congress. *Clearcutting Practices on National Timberlands.* Parts 1, 2, and 3. Hearings before Subcommittee on Public Lands. Committee on Interior and Insular Affairs, U.S. Senate, Washington, 1971.

———. *Grazing Fees and Public Rangeland Management.* Hearings before the Subcommittee on National Parks and Public Lands. Committee on Interior and Insular Affairs, U.S. House of Representatives, Washington, 1987.

———. *Management of the Tongass National Forest.* Hearings before the Subcommittee on Forests, Family Farms, and Energy. Committee on Agriculture, U.S. House of Representatives, Washington, 1988.

USDA-Forest Service. *The Forest Service Program for Forest and Rangeland Resources: A Long-Term Strategic Plan.* U.S. Department of Agriculture, Washington, 1990.

World Commission on Environment and Development. *Our Common Future.* Oxford University Press, New York, 1987.

4

POLICY FORMULATION: ACTION IN GOVERNMENT

Successful placement of a forest resource issue on an institutional agenda concludes the long and often perilous agenda-setting process. An issue has been recognized by government; efforts must now be directed to *policy formulation,* that is, creative design of appropriate policy responses to the issue. The term "policy formulation" implies the need to more clearly define an issue, to identify options for addressing the issue, and to perform analyses of the merits of these options.

Successful formulation requires that government officials and interested parties outside government discuss the technical, economic, and political consequences of the identified policy options. If, for example, an issue is defined as aerial spraying of herbicides for release of conifer plantations on a public forest, the policy options suggested could range from continuation of existing spraying programs to abandonment of spraying activities, or from adoption of mechanical release methods to reduction of the frequency and adjustment of the intensity of spraying. To ensure their technical and political feasibility, such options would be presented to a variety of program administrators and concerned interest groups for deliberation. The product of the deliberations should be a workable set of policies that could be presented to policymakers as choices. If policymakers were presented with poorly crafted policy options directed toward an ill-defined issue, they would face an unnerving prospect which could lead to a disaster: a policy recipe destined for failure.

Policy formulation is seldom a highly rational process; instead, chaos and uncertainty are some of its more prominent characteristics. Even though an issue has achieved agenda status, it may, for example, be subject to continuing debate among interested parties. It may even be resolved without need for further government intervention. For example, nonindustrial private forest owners may agree to form a cooperative to seek private-sector financing, thus negating the need for the government fiscal incentives programs that were originally considered the most appropriate response to rising timber prices. Similarly, formulation may be aborted because the policy options developed to address an issue, as defined by organized interest groups, are so broad that they are beyond an agency's ability to act in a responsible manner, because the agency may face severe financial limitations. How well policy formulation is accomplished, who participates in it, and who gains or loses in the process vary substantially from one issue to another. Some major elements of policy and program formulation—namely, common types and levels of formulation, typical sources of ideas for policy options, and frequent participants in policy formulation activities—are considered below.

FORMS AND POLITICAL CONTEXT

Policy formulation can take many forms. It can involve very little research on alternatives, passage of a meager amount of time between agenda setting and a policy decision, and very little public and professional conflict (i.e., it may be free of ideologies and involve little partisanship). Frequently, the policy alternatives formulated in such circumstances satisfy minimal criteria of tension reduction or issue abatement. For example, formulation of policies to address Dutch elm disease in urban forests implies the need for prompt action, the presence of a limited number of alternatives already well researched (infected tree removal), and the existence of minimal professional and public conflict over policy options. Under other circumstances, policy formulation may involve passage of many years between agenda setting and a policy decision, extensive research on alternatives previously undocumented, and strongly held opposing views on the policy options to be recommended. An example of such formulation is public (federal) land wilderness designations, which often appear on government agendas for decades. They may have been subject to formulation activities for similar periods of time (even before the Wilderness Act of 1964) and may have required extensive research of alternative land-use allocations. There may also have been frequent and intense conflict between opposing interests.

Policy formulation can also be viewed in the context of the policy decisions that are required. In such a context, three types of formulation have been suggested (Jones 1984, Peters 1982):

- *Routine formulation* A repetitive and changeless process of reformulation of policy options to address an issue that has long been well established on an institutional agenda (e.g., formulating budgets for the sale of timber from public forests).

- *Analogous formulation* Reliance on policies formulated for similar issues in the past (e.g., formulating state forestry cost-share programs based on previous federal experience with such programs).

- *Creative formulation* Use of unprecedented alternatives that make significant breaks with the past (e.g., formulating regulatory policies to curb the effect of atmospheric pollutants on forests).

Policy formulation is often only the beginning of a series of required formulations. An agreed-to policy selected from among a number of formulated policies typically leads to subsequent formulation of programs considered capable of accomplishing the intent of the policy. In turn, programs often lead to formulation of *forestry projects*—specific activities that are undertaken to accomplish specific objectives, have definite starting and ending points, and are considered capable of achieving a program's stated objectives or targets.

Formulation may also move through various administrative levels within an agency. For example, forest resource policies formulated in a legislative setting are likely to result in formulation of programs by an agency's headquarters office, which in turn may require formulation of projects in the agency's regional or district offices. Depending on the circumstances, formulation activities can be hierarchical within an organization as well as hierarchical in terms of policy-program-project orientation. Policies become program objectives, which in turn give guidance to targets, which guide projects (Ackoff 1981).

Policy formulation is facilitated by clear definition and careful description of an issue's substance—seldom an easy task. For example, local and national media, legislators of state and national stature, and interest groups of all types and persuasions may attack with a vengeance the application of even-aged silvicultural practices on a particular public forest. The responsible agency may formally acknowledge the issue (i.e., allegedly clear-cutting) by granting it institutional agenda status. The process of formulating policy options must then begin in earnest—but in response to what? What exactly is the issue? Is the issue a concern with clear-cuts per se or with the forest roads needed to access areas destined for timber harvesting? Are interested parties concerned about timber harvesting on public land or only on private land—or possibly on all ownerships? Is the design and layout of clear-cuts the problem, or are forest residues left after harvest the cause for concern? Is regeneration of harvested areas occurring at an unusually slow pace or is there a fundamental concern over the magnitude of public investments in timber management (possibly at the expense of recreation and wildlife management opportunities)? Is the issue impatience with a cumbersome, time-consuming land-use planning process? Whatever the issue, it must be clearly defined before an effective array of policy options can be formulated. The issue must be clarified, interpreted, and possibly quantified in terms that enable a common understanding of the concerns that are in need of formulated policy proposals.

Issue definition for formulation purposes is more, however, than simply defining the substance of an issue; invariably, there is a broader context. To effectively formulate policy options, the following queries should be addressed (Hogwood and Gunn 1984, pp. 91–99):

- *Agreement on issue's substance?* Parties may be entering the formulation process at odds over the substance, or definition, of the issue. If so, formulation of policy options may be premature.
- *Rigid policy positions adopted?* Parties may be entering the formulation process with narrowly defined, rigid policy options. If so, imaginative formulation of policy options may be an exercise of limited value.
- *Politically sensitive issue?* Parties involved in the formulation process may consider the issue to be especially sensitive politically—may feel that open, frank and objective discussion would divide the agency, the agency's clients, or the forestry professionals employed by the agency. If so, formulation of options may open dormant political cleavages and encourage unnecessary conflict.
- *Limited time available for formulation?* The issue may require immediate action. If so, systematic formulation activities involving extensive analysis and related documentation may be of limited value.
- *Issue central to agency concerns?* The issue may be central and very critical to an agency's mission. If so, well-organized and very thoughtful policy formulation may be imperative.
- *Formulated options limited by context?* The circumstances within which formulation is to occur (legal limitations, financial constraints) may severely limit the scope and type of policy options available for addressing the issue. If so, formulation of numerous broadly scoped policy options may be of limited value.
- *Issue excessively aggregated?* The issue may be aggregated to point that clear definition of substance is difficult. (An example of an aggregated issue is environmental pollutants.) If so, disaggregation may be required before effective formulation can be accomplished. (In this example, the disaggregated issue would consist of air pollutants, noise pollutants, and water pollutants.)
- *Large number of people affected?* The issue may involve a significant number of individuals and groups. If so, meaningful investment in formulation activities may well be justified.
- *Political significant groups affected?* The issue may involve politically important or especially sensitive agency clients. If so, investments in comprehensive formulation activities may well be justified.
- *Administrative flexibility restricted?* The issue may lead to policy options that foreclose agency action on other opportunities, by leaving less funding for other programs. If so, especially creative formulation must occur.

Predetermining costs, benefits, and risks of the alternative policies being suggested to address a forest resource issue can be an imposing task for formulation. For those alternatives for which there is experience, the task is relatively easy: cause-and-effect relationships are well-defined, and the relative efficiency of alternatives can be determined. In many cases, however, data are lacking, system models are poorly developed, analytical tools are inappropriate, and persons experienced with the consequences of a suggested policy option are nonexistent or in short supply. When data are limited, a tentative cause-and-effect model may be created to conform with data realities. In other cases, models of closely related policies may be injected with inferior data; analysts may advertise the limits of their results and confess them to be the best available under the circumstances.

The formulation stage of the policy process is considerably different from the evaluation stage, in which experience with a program has accumulated and postprogram evaluations can be undertaken. During formulation, the creative genius of individuals and institutions yields an abundance of new, untested options, all of which may be advocated as being the most appropriate for the task at hand. However difficult analysis is during formulation, however, analysts can perform especially valuable roles. Most important, the formulation stage can forge a meaningful linkage between inputs required of a suggested forest policy, outputs thought to flow from that policy, and the accomplishment of objectives which incorporate political values deemed to be appropriate for addressing the issue (Schultze 1968).

IDEAS FOR POLICY OPTIONS

Creativity and Imagination

Formulation of forest policies requires specification of goals and the policies to achieve them, in the context of a well-defined issue. Beyond their organizational homes, where do ideas for forest policy options come from? In a generic sense, it has been argued that well-formulated policy options (Polsby 1984, p. 166):

> arise from the intersection of three forces: (1) the interests of groups in society, (2) the intellectual convictions of experts and policy makers, and (3) comparative knowledge, usually carried in the heads of experts or subject-matter specialists, of the ways in which problems have been previously handled elsewhere.

Fundamental, however, to the source of ideas are notions of creativity and innovation: the ability to bring new forest policy options into existence and, ultimately, to bring such options into use. From a forest policy formulation perspective, creativity may be defined as "the capability to view a problem and propose alternatives in perspectives and methods that are unique and unlike any prior attempts to solve similar problems—the emphasis is upon whatever is fresh, novel, unusual, ingenious, divergent, clever and apt'' (Brewer and deLeon 1983, p. 62).

Creativity in a policy formulation context is fostered by a number of factors, including intelligence, exposure to diverse experiences, and the capacity to make remote associations (for example, to relate policy experiences in medicine or aeronautics to forestry, as appropriate). Also of obvious importance are receptivity to change, professional education, interdisciplinary activities, and organizational flexibility. Unfortunately, creativity is often absent in policy debates because the parties involved opt for safe and quick solutions rather than investing the resources needed to generate truly creative alternatives. It has been argued that creativity in policy formulation is most likely to be found in organizations losing their reason for existence: ''the arguments presented by the U.S. Army horse cavalry until about 1950 to secure its survival are excellent examples of creativity, albeit creativity misspent'' (Brewer and deLeon 1983, p. 63).

Policy alternatives resulting from formulation activities can range from doing nothing to doing something radical, with countless options in between. The process of generating alternatives is an iterative one, in which original ideas are enriched and elaborated upon (or discarded) and new ideas are found. To be avoided are forest policy options that are outside the relevant time frame, are unattainable with available resources, or are beyond the reach of a specified level of technology. It does little good to propose new investments in reforestation of public lands if the budget cycle has gone by. Nor it is wise to suggest a policy of designing all timber-harvesting operations with the aid of a landscape architect when available finances prohibit engagement of new employees. Equally unwise are proposals to achieve timber growth rates that are clearly beyond the ability of current silvicultural practices to support. More pointedly, formulators are wise to avoid policy options which involve (House and Shull 1988):

- *Unrealistic goals* The goals a suggested policy is supposed to achieve may be overly ambitious and unlikely to be met regardless of how many resources are expended. For example, excessively stringent national goals may have been set for achieving quality conditions in all the nation's waters.
- *Political and administrative resistance* The target group and the administering agency may be in opposition to a suggested policy. For example, a state forester and a state's nonindustrial private forest owners may be morally opposed to public regulation of privately prescribed forest practices.
- *Economic and financial impracticalities* The costs of implementing a suggested policy may be far beyond available or expected financial resources. An example would be a goal of doubling or tripling (in one year) the recreational trails system in a state forest.
- *Overly simplistic requirements* The simple and bold statements of a suggested policy may be administratively and technically nearly impossible to

implement. For example, a nationwide permitting system for timber harvesting would meet with untold technical and administrative complications.

- *Inequitable application* The suggested policy might be stringent on some persons while lax on others, the latter negating the achievements of the former. An example would be rigorous water-pollution standards for forestry operations and minimal standards for agricultural activities.

Knowledgeable Authorities

Forest resource policies may also result from the formulator's consultation with knowledgeable authorities, both personally and through their writings. Such individuals or organizations may have prior experience with options for addressing a forest resource issue or may be so prestigious that their suggestions are accepted verbatim.

Deductive Reasoning

Processes of deduction and association may also be the source of policy options. If a policy has been useful in addressing a particular issue, the same policy may also be effective with a closely related issue. For example, if intensive timber management on nonindustrial private forests is of interest and financial constraints have been effectively alleviated by government cost-share programs, property tax reductions may well achieve the same result. If forestry extension programs focused on owners of nonindustrial private forests have resulted in increased sensitivity to the establishment of pleasant forest landscapes, the same programs focused on such forests may be useful in achieving more productive wildlife habitat. All such forms of deductive reasoning enable the policy formulator to draw conclusions about the relationship between potential policies and the objectives sought to address an issue when statistical, inductive, or observational data are lacking (Nagel 1984).

Institutional Sources

The source of policy options can also be viewed from an institutional perspective. Legislative systems, for example, thrive on ideas as raw material. Without policy options—without ideas—legislative machinery would halt abruptly, for such machinery is designed to formulate, evaluate, select, and implement policies. What are the sources of policy options for the legislative grist mill? Many are borrowed from other states or other governments; an option which works well in one jurisdiction may be useful to a neighbor. Policy options may arise from model laws suggested by organizations such as the Council of State Governments, the American Law Institute, and the National Conference of Commissioners on Uniform State Laws. Policy options may also be sought from organized special-interest groups and from elected or politically appointed public officials (a state governor, for example, or an appointed state forester, or an appointed USDA-Assistant Secretary for Natural Resources and Environment). Agency staff (e.g., the Environmental Protection Agency's model state forest practice law of the early 1970s), legislators, and knowledgeable citizens may also be participants in the policy formulation implemented by legislators. An especially fertile source of ideas for legislation is legislative and executive study commissions. Examples are the Presidential Commission on State and Private Forests (which was established by federal law in 1990; members of this important commission have not yet been appointed), the Public Land Law Review Commission (1970), the President's Advisory Panel on Timber and the Environment (1973), the Governor's Task Force on Northern Forest Lands (1990), the President's Commission on Americans Outdoors (1986), and the Committee on Forest Development in the Tropics (1985). Here are some examples of suggestions from three of these study groups.

Governor's Task Force on Northern Forest Lands (1990)

- Landowners who agree not to subdivide or develop their land for 10 years should be given favorable capital gains tax treatment on the income from the sale of timber.
- State land gain tax should be instituted to slow land speculation. Tax should be applied to large profits when land is bought and quickly resold.
- Important resource land should be protected by acquiring conservation easements, including development rights and public recreation access rights.
- Land should be acquired in full fee to protect important resources such as threatened or endangered plants and animals. Consideration should be given to creating or expanding new national forests or creating new national parks and wildlife refuges.

President's Commission on Americans Outdoors (1986)

- Recreation agencies should charge visitors fees with the objective of covering reasonable portions of operation and maintenance costs.
- Self-sustaining, endowed trust funds should be established for purposes of meeting federal and state recreation and open-space needs.
- A national "recreation accounts network" should be established for the purpose of facilitating the collection, analysis, and sharing of statistical information about recreation.
- Older recreation facilities should be redesigned and adapted to allow access by persons with physical disabilities.
- Coalitions of recreation users and private landowners should develop and adopt a code of ethics describing acceptable behavior on all private lands.

• Greater protection should be afforded government and private recreation providers from liabilities often imposed by recreational users.

• Recreation planning activities should consider opportunities for private recreation investments in public places.

• Federal multiple-use agencies should assure that recreation is an equal partner with other uses in terms of staffing, budgets, and planning.

• A network of scenic byways, composed of scenic roadways and thoroughfares, should be established throughout the nation.

• Wildlife habitat protection on private lands should be increased by mechanisms such as reduced property taxes, low-interest loans, cost sharing, and in-kind donation of expertise from state and federal agencies.

• Mechanisms for using volunteers for recreational purposes in national parks and national forests should be enhanced.

• Coalitions for action that will lead to investment in recreation opportunities for the future should be organized.

• Communities should assess desirable support services in areas adjacent to public recreation resources.

Committee on Forest Development in the Tropics (1985)

• Develop economic incentive approaches for stimulating farmer investment in the production of fuelwood as a cash crop.

• Develop a comprehensive system of national parks and reserves wherein representative samples of flora, fauna, and natural landscapes can be protected.

• Promote research on tropical forest management with particular emphasis on compatibility between sustainable production and the conservation of genetic resources.

• Reinvest timber-harvesting revenues collected by governments (royalties, concession fees) into reforestation and subsequent silvicultural treatment of newly established tropical forests.

• Establish permanent structures within public forestry administrations to foster forestry-oriented rural associations and cooperatives.

• Administratively integrate forestry and agricultural programs to ensure attention to adequate flows of high-quality water.

• Establish planning and information-gathering systems which will facilitate and improve decision making concerning tropical forests.

• Improve the motivation and effectiveness of staff in tropical forestry agencies via improvements in personnel management, career opportunities, and in-service training.

COMMUNITIES OF FORMULATORS

Formulation of forest resource policy options is not unique to any one organization or individual. The process involves many actors, and the combining and recombining of ideas to address issues. Ideas that have been in existence for years are often reused. The sum of these actors has been labeled *communities of formulators* (Kingdon 1984). They are composed of specialists in a particular program or policy area (such as forest taxation, international forestry, or non-industrial private forest) who willingly interact and share ideas on how to address forest resource issues. They are located throughout public and private organizations. Some are specialists for legislative committees; others are assigned to planning and policy evaluation offices. Still others may be found in academia, among interest groups, and in respected nonprofit research foundations such as the Brookings Institution, the American Enterprise Institute, the RAND Corporation, and Resources for the Future. The policy community of specialists interested in endangered species, for example, may well include staff from:

- Office of Endangered Species, USDI-Fish and Wildlife Service
- Division of Wildlife Management, USDI-Bureau of Land Management
- Endangered Species Program, USDA-Forest Service
- Wildlife Research and Development Program, National Wildlife Federation
- Natural Heritage Program, The Nature Conservancy
- Committee on Interior and Insular Affairs, U.S. Congress
- International Wildlife and Conservation Office, U.S. Department of State

Within a single agency, communities of formulators may abound. Consider the Washington, D.C., office of the USDA-Forest Service (Forest Service 1987):

- *Programs and Legislation* Environmental Coordination staff, Legislative Affairs staff, Policy Analysis staff, Programs Development and Budget staff, Resources Program and Assessment staff
- *National Forest System* Engineering staff, Lands staff, Recreation Management staff, Land Management Planning staff, Minerals and Geology Management staff, Range Management staff, Timber Management staff, Watershed and Air Management staff, Wildlife and Fisheries staff
- *State and Private Forestry* Cooperative Forestry staff, Fire and Aviation Management staff, Forest Pest Management staff
- *Research* Forest Environment Research staff, Forest and Atmospheric Sciences Research staff, Forest Insect and Disease Research staff, Forest Products and Harvesting Research staff, Forest Inventory and Economics Research staff, International Forestry staff, Timber Management Research staff
- *Administration* Computer Sciences and Telecommunications staff, Fiscal and Public Safety staff, Information Systems staff, Procurement and Property staff, Human Resource Programs staff, Personnel and Civil Rights staff

Communities of policy formulators are dynamic with respect to the intensity of their work habits and their working relationships. After maintaining a subtle below-the-scenes profile for months (or even years), they spring into action when issues arise that spark their interest and need their commonly held expertise. Formulators who are interested in timber on public land, for example, communicate ideas, proposals, and research. They also generate common outlooks on public timber problems, thereby lending substantial stability to the type and nature of policies formulated to address relevant issues. More to the point, communities of formulators may look for ways of integrating policy responses into existing patterns of doing business. Change may occur only within the context of policy options unique to a particular set of formulators (Jones 1984). Stability is further ensured by the policy formulators' substantial independence of political events, such as changes in political appointees or pressure from legislators' constituencies.

Communities of forest policy formulators are held together by a common interest in forests. Beyond that common element, opportunity for fragmentation abounds. There are communities of formulators interested in specific forest outputs (water, timber, wildlife, recreation, wilderness) or in specific resource management functions (fire control, pest management, transportation). Frequently the task of senior administrators is to manage these separate fiefdoms and the competition that occurs between them. From a policy formulation perspective, it can be argued that the existence of fragmented communities of forest policy specialists can lead to fragmented and inconsistent forest policies and greater susceptibility to crisis, because agendas keep shifting. A more closely knit community of formulators, who have in common their educational background, technical language, and interest in a particular forest use, supposedly generates common outlooks, orientations, and ways of thinking (Kingdon 1984).

FORMULATION IN THE POLICY PROCESS

Formulation responds to agendas by searching for means of alleviating, mitigating, or resolving problems or issues. Emphasis is placed on defining forest resource issues in more precise terms and on determining their importance as topics for action by a government organization. Of special interest is the creative casting about for policy and program options—however ill-defined or inappropriate they may appear at the moment—as well as the initial determination of risks, costs, and benefits associated with the options identified. The process is not as tidy and well-organized as might be expected. Through time, however, and with considerable creative influence and patience, a plausible range of solutions may be identified and ultimately passed on to persons charged with selection. Some important generalizations about formulation are (Jones 1984):

- More than one set of agency actors may be involved in policy formulation, frequently producing competing proposals. For example, an agency's timber management staff and wildlife management staff may independently generate options for the management of forest habitat required by big game.

- Insufficiently defined issues may be the focal point for policy formulation, frequently producing policy options which lack effectiveness and defy evaluation. For example, tax policies, cost-sharing subsidies, and technical assistance programs may be focused on the often ill-defined forestry problems of nonindustrial private owners of forests.
- Executive agencies of government may be involved in the policy formulation process; they cannot, however, claim sole jurisdiction over formulation activities. For example, legislative systems and organized special-interest groups actively participate in formulation activities leading to the establishment of policies legitimized by laws.
- Formulation and reformulation of policy options may occur over an extended period of time without conclusion, frequently being fraught by an inability to secure political support for a set of policy options that can be offered to policymakers for a decision. An example would be issues involving the allocation of public forestland to various uses.

Formulation of policy and program options for addressing forest resource issues occurs on a continuing basis throughout the forestry community. Such activities may involve a one-time effort of limited duration or may be institutionalized for continuing attention. Consider two examples of formulation:

- *Tropical forestry action plan* In response to worldwide concern over degradation and deforestation of tropical forests, the Committee on Forest Development in the Tropics was established to formulate policies and programs capable of addressing such concerns. The committee formulated options to address forestland use, forest-based industrial development, fuelwood and energy, conservation of tropical forest ecosystems, and forestry institutions. The formulated actions were presented by the United Nation's Food and Agriculture Organization for consideration by countries throughout the world (Committee on Forest Development in the Tropics 1985).
- *Resources Planning Act goals and programs* In response to the Forest and Rangeland Renewable Resources Planning Act of 974, the USDA-Forest Service developed a national plan for forest and related resources. Major forest resource issues were identified, alternative goals describing desired future conditions were formulated, and potential program directions were conceived and suggested. Estimates were made of the possible effects of formulated goals and programs. The process is repeated every 5 years (for example Forest Service 1981 and 1986).

Formulation is linked to and has an impact on nearly all stages of the policy process. For example, formulation's efforts to define an issue may reveal other forest resource issues of public concern which in turn are susceptible to institutional agenda-setting activities. Selection, however, is probably the stage upon which formulation has the most noticeable impact. Formulation sharpens the options available for selection and enables policymakers to direct attention to a more realistic range of policy solutions. However, formulation that is incomplete because more studies may be needed can provoke policymakers to postpone selection, with the result that they may appear indecisive or hesitant to take action. In the worst of circumstances, analysts may be inclined to follow their own judgments and policy preferences, rather than waiting for the decisions of those in authority. Formulation then becomes selection.

Initial estimates of an option's resource requirements during formulation provide some appreciation of the type and magnitude of resources that will be required during implementation. Similarly, many legal and procedural matters may be addressed during the formulation process, yielding information useful to the implementor who is charged with developing rules, regulations, and guidelines. Formulation may also provide guidance as to who should be responsible for implementing a selected forest policy or program. The primary link between formulation and evaluation is the establishment of first approximations of expected performance (objectives, targets, time spans). Formulation may also describe conditions under which a forest policy or program should be terminated, as when objectives have been attained, estimated costs have been exceeded, or targeted clients cannot be reached.

REFERENCES

Ackoff, R. L. *Creating the Corporate Future*. John Wiley and Sons, New York, 1981.

Brewer, Gary D., and Peter deLeon. *The Foundations of Policy Analysis*. The Dorsey Press, Homewood, Ill., 1983.

Committee on Forest Development in the Tropics. Tropical Forestry Action Plan. Forestry Department. Food and Agriculture Organization, United Nations, Rome, Italy, 1985.

Governor's Task Force on Northern Forest Lands. The Northern Forest Lands: A Strategy for Their Future. Rutland, Vt., 1990.

Hogwood, Brian W., and Lewis A. Gunn. *Policy Analysis for the Real World*. Oxford University Press, London, England, 1986.

House, P. W., and R. D. Shull. *Rush to Policy: Using Analytic Techniques in Public Sector Decision Making*. Transaction Books, New Brunswick, N.J., 1988.

Jones, Charles O. *An Introduction to the Study of Public Policy*. Brooks/Cole Publishing Company, Monterey, Calif., 1984.

Kingdon, John W. *Agendas, Alternatives, and Public Policy*. Little, Brown and Company, Boston, Mass., 1984.

Nagel, Stuart S. *Public Policy: Goals, Means, and Methods*. St. Martin's Press, New York, 1984.

Peters, B. Guy. *American Public Policy: Process and Performance*. Franklin Watts Press, New York, 1982.

Polsby, Nelson W. *Political Innovation in America: The Politics of Policy Initiation*. Yale University Press, New Haven, Conn., 1984.

President's Advisory Panel on Timber and the Environment. *Report of the President's Advisory Panel on Timber and the Environment.* U.S. Government Printing Office, Washington, 1973.

President's Commission on Americans Outdoors. *Report and Recommendations to the President of the United States.* U.S. Government Printing Office, Washington, 1986.

Public Land Law Review Commission. *One Third of the Nation's Land.* U.S. Government Printing Office, Washington, 1970.

Schultze, Charles L. *The Politics and Economics of Public Spending.* The Brookings Institution, Washington, 1968.

USDA-Forest Service. Alternative Goals. 1985 Resources Planning Act Program. Program Aid No. 1307. U.S. Department of Agriculture, Washington, 1981.

———. Resources Planning Act Program 1985–2030. Final Environmental Impact Statement. FS-403. U.S. Department of Agriculture, Washington, 1986.

———. Organizational Directory. FS-65. U.S. Department of Agriculture, Washington, 1987.

5

POLICY SELECTION: ACTION IN GOVERNMENT

An effectively organized policy formulation effort produces a variety of well-conceived policy options. Faced with a menu of forest policy proposals, policymakers must confront the stark reality of *policy selection*—choosing from among the many, or the few, policy alternatives that have been suggested. Will public forestland be allocated to wilderness or to timber? Will additional finances be allocated to large-scale road construction or to a major wildlife habitat improvement program? Will a new tropical reforestation policy be initiated, or should policy commitments to forestry research be continued? Policy selection is the most overtly political stage in the policy-making process. Solutions formulated for a forest resource issue must somehow be narrowed to a single alternative, or a select few. Many individuals and groups will not be happy with the outcome; some will obtain substantially modified versions of their policy preferences, whereas others will receive nothing at all.

The process of selection often involves more irrationality than rationality—more art and craft than technique or science. Policymakers strive to strike a balance between the options recommended by policy formulators and the multiple, changing, and conflicting interests of persons and organizations that have a stake in the outcome of the selection process. Partial insight into the process is described by this emotional accounting, where the path to selection is (Lasswell 1960, pp. 184–185):

> the transition between one unchallenged consensus and the next. It begins in conflict and eventuates in a solution. But the solution is not the ''rationally best'' solution,

> but the emotionally satisfactory one. The rational and dialectical phases of [selection] are subsidiary to the process of refining an emotional consensus.

The portrait of the policymaker shows a person driven to secure agreement on policies for addressing important forest resource issues, knowing full well that in the wings lie other issues begging for attention. The transition from one issue to another may well be bumpy—a challenge to be leveled by the artful policymaker. In this context, we shall consider the nature of selection and the conditions that influence policy decisions, the differing models and concepts that are often used to explain selection processes, the art and craft of bargaining and negotiating as used during selection, and the various rules and criteria that are often used to guide selection from among a group of formulated policies.

SELECTION IN CONTEXT

Programmed versus Unconventional Selection

Selection of policies and programs ranges from the commonplace to the unusual. Seldom do policymakers face the selection process with certainty—with complete knowledge of the probable outcome of formulated policies. Uncertainty is equally rare; rarely does a policymaker have no knowledge of the probable outcome of formulated policies. Most common are situations in which some risk is involved, though a probabilistic estimate of the outcomes of formulated policies does exist. The following types of formulated policies often placed on the doorstep of decision makers have been identified in studies of decision-making processes (Reitz 1987):

- *Good alternative* High probability of positively valued outcomes and low probability of negatively valued outcomes.
- *Bland alternative* Low probability of both positively and negatively valued outcomes.
- *Mixed alternative* High probability of both positively and negatively valued outcomes.
- *Poor alternative* Low probability of positively valued outcomes and high probability of negatively valued outcomes.
- *Uncertain alternative* The relative probabilities of positive and negative outcomes cannot be, or have not been, determined.

Some of these alternatives involve relatively easy choices. For example, if one forest policy alternative is good and the others are bland, mixed, poor, or uncertain, the selection is easy. If, however, there are only two policy options, of which one is bland and the other poor, selection is a problem. Similarly, it is difficult to choose between a mixed and a poor alternative, or between a bland and a mixed alternative.

Programmed Selection Forest resource issues that occur with some frequency and have a well-known cause can usually be addressed by routine selection procedures. Here the selection process is *programmed;* it relies on previously defined policies, criteria, and procedures. For example, a decision about a proposed modest increase in the budget for the management of designated wilderness areas on a public forest can probably be based on customary policies, rules, and procedures for agency budgeting in general. Likewise, a decision regarding strategies for managing a forest insect, such as the spruce budworm, that annually damages forests may also be made within the confines of existing policies, rules, and procedures. In both cases, the nature of the issue, the frequency of its occurrence, and the degree of uncertainty about the policy alternatives are such that the decision becomes programmed, or routine. Programmed decisions are typically the domain of an organization's middle managers (Gibson et al. 1988).

Unconventional Selection Some policies formulated for decision makers' attention are unique; they may be unstructured, and substantial uncertainty regarding their outcome may be involved. These are *unconventional* decisions that require substantial intuition and tolerance for ambiguity on the part of policymakers. For example, selection of a single alternative from among many approaches to management of an endangered species of fauna in a forested area may be fraught with difficulty, since little may be known about the species' habitat requirements. Similarly, a decision to implement a new, large-scale tax policy designed to encourage recreational investments by owners of nonindustrial forests is certainly unconventional, in the sense that such a program has probably never been tried before. Unfortunately, the unconventional policy decisions are the ones about which the least is known and the ones that occur most frequently during the development of policies of substantial significance to forestry. Such decisions are—and should be—the concern of top-level managers in forest resource organizations (Gibson et al. 1988).

Selection as a Nonevent

Selection of a policy to address a forest resource issue many not be a clear and identifiable event. It has been suggested, for example, that some decisions may simply be the policymaker's way of formally acknowledging a forest resource policy that has been in existence for some time. If, over 3 to 4 years, a superintendent of a wildlife refuge informally accumulates budgets and personnel for the management of a particular species of forest wildlife, the time may come when the superintendent's management policy for the species must be formally recognized by a decision and announced to the community that has an interest in the species. Such action is nothing more than the legitimizing of policy selections that have already been made.

A decision by a policymaker may be a nonevent in cases in which the selection has already been made during the policy formulation process. By the time formulated policy options reach policymakers, issue managers and policy formulators may have resolved their differences and agreed upon a specific course of action. Formal selection of a preferred alternative may not be required. For example, institutional, political, and budget constraints may so reduce the range of alternative program directions formulated by the program and assessment staff of a public forestry agency, such as the USDA-Forest Service, that opportunity for choice no longer exists by the time program alternatives are presented to high-level policymakers, such as the President's Office of Management and Budget. Also not to be overlooked are times when policy selection is embedded in the formulation process, which relieves policymakers of the responsibility for making difficult and politically unpopular choices.

Avoiding Selection Sometimes policy selections are deliberately avoided. The prevailing political atmosphere may not permit them, or the political cost of selection may be excessive. For example, if a governor favors establishment of a state forest practice law to curb undesirable timber harvesting practices on private lands, a formal decision to support such a program may be postponed until after gubernatorial elections, because announcing support prior to elections might result in the loss of a substantial number of votes. Policymakers may also avoid selection in order to buy time to gather additional information or in hopes that an issue will solve itself or simply go away. Then again, policymakers may make a selection but by various means may attempt to distance themselves from the choice or to deny association with the policy selected. For whatever reason, formal decisions are sometimes deliberately avoided, or at least made to appear as though the policymaker was not involved (Brewer and deLeon 1983).

Priorities, Scope, and Timing

Relative Importance A policymaker's assessment of the importance of a forest resource issue may weigh heavily in a judgment about which formulated alternative to select. An issue's importance governs in a major way how much time, talent, and political resources will be expended on selection. Failure to carefully evaluate the importance of a forest resource issue may result in valuable resources being spent on a problem that is trivial in comparison to more pressing matters. There may be no need, for example, to be concerned about capital gains treatment of timber income as a public incentive to industrial timberland investments when the entire capital gains tax program as related to timber is legislatively in jeopardy. Nor is there need to be concerned about options for curtailing the export of timber harvested from public forests when the nation's balance of payments and the value of the U.S. dollar in foreign currency markets are such that the incentive to export timber to foreign markets is virtually nonexistent.

Political Considerations Political considerations have special significance for judgments about the importance of a forest resource issue. Who will benefit and who will be hurt if the issue is solved in a particular way? What centers of support exist for certain formulated policies? Can interested parties be leveraged to support options not currently preferred? Can the issue be addressed by symbolic treatment, or are substantive changes in policy required? Although answers to such questions are helpful in judging which forest resource issues deserve a policymaker's attention, a rigorously formalized, concrete set of rules for judging when action is warranted may simply not be possible. "Often the decision maker makes these judgments and estimates based on informal, impressionistic sense of a situation that can best be described as visceral" (Brewer and deLeon 1983, p. 200). Willingness to decide which policy option best addresses an issue may rest less on reams of well-documented information than on past situations personally experienced by the decision maker.

Expectations of Other Parties Even when particular forest resource issues and the policies formulated to address them are judged unimportant by a policymaker, the expectations of others must be considered. For example, even a policymaker who has little sympathy for pesticide issues and the policies suggested to address them may have to be willing to face such issues and to make the necessary policy choices, in order to reduce the risk of serious morale problems among staff. Likewise, ignoring a forest resource issue or judging it unworthy of decision-making attention may be sure way of inviting criticism from various interested parties and of needlessly mortgaging political capital important for future contests. A shrewd policymaker will realize that there is no need to forsake the support of or to alienate important constituencies whose support will be needed in addressing future issues. Further, if an issue is judged as unimportant, it may nevertheless have value as a bargaining chip. Sacrificing some resources, such as time and money, in exchange for future favors may be worthwhile. For example, "We will address policy A, which is important to you but unimportant to us, if, at a later date, you agree to support policy B, which addresses a related issue of substantially greater importance to us."

Realistic Options Although forest resource issues and the policy options to address them may be judged as important and in need of decision, they must be realistically bounded by a policymaker. The issues must be manageable, and the policy options for addressing them must be realistic. If, for example, the equitable distribution of National Forest System timber among consumers of finished timber products, such as home owners, is at issue, presumptions that USDA-Forest Service decision makers can influence such an outcome with policies at their command would be highly unrealistic. From the agency's perspective, the issue is defined too broadly. What would be more realistic would be agency action to ensure an equitable distribution of timber among purchasers of stumpage from the National Forest System.

Discretion Concepts of bounded discretion also play a role in decisions. Decision makers are especially cognizant of constraints which limit their discretion to select from among formulated policies. Laws, agency rules, and moral and ethical norms must be adhered to. The USDI-Bureau of Land Management policymaker does not make policy for the USDA-Forest Service, nor do state foresters exercise jurisdiction over forestry programs in their neighboring states. Culturally, it is highly improper to even consider an illegal forest policy option, let alone select one.

Timing Timing is also critically important to policy selection. A prudent decision maker knows when to make and announce a selection and when not to take action. Passage of an appropriate amount of time can reduce the intensity of the battle, change the nature of the forest resource issue, and lead to selection of a suitable policy. Policymakers may find significant merit in postponing selection, in order to give opponents an opportunity to readjust their positions and reduce the chance of precipitous action. There is no need, for example, to hastily select a forest policy that angers legislators and leads to harsh reprisals, when delay in selection may give legislators and legislative staff the time to consider and subsequently agree with the agency's judgment. Patiently testing political winds and acting only when the time is politically correct is a highly virtuous approach to policy selection.

Leverage, Information, and Personality

Leverage A policymaker's ability to actually influence the outcome of a forest resource issue depends on access to policy options that are technically sound and conducive to implementation via the exercise of power or the use of incentives. Selection of a particular forest resource policy indicates the policymaker's faith in the ability of the chosen policy to actually accomplish agreed-to objectives—to leverage action toward their attainment (Kaufman 1981). Wildfire in forests, for example, can be alleviated by a series of major rainstorms. Knowledge of this sort, however, offers little comfort to the policymaker; nature is quite immune to bureaucratic leveraging. Before a specific policy is chosen, formulated policies must be judged implementable and actually capable of influencing the issue at hand. For example, if increased timber production on nonindustrial private forests is of interest, the forest policymaker must have access to an available range of options (points of leverage) (e.g., cost-share payments, public regulation of forest practices) that stand a good chance of getting the job done.

The technical ability of policy options to achieve objectives is not the only focus of leverage. Selection of a forest resource policy requires judging whether the option can be implemented administratively. If a particular policy option is obviously disliked by forest resource administrators, and such administrators are unlikely to respond to political incentives (such as an offer to finance a favored program) or to political power (say, a threat to transfer them to another region or office), the policy option should probably be discarded and a search for another begun. The policymaker's time and effort would be better focused on searching for formulated policies that can be successfully implemented in an administrative sense.

Political leverage to secure establishment and implementation of a program can also vary according to time and situation. A governor recently elected (e.g., within the past month) by a large majority may be in an excellent position to wield the political leverage required to secure enactment of a new forestry program by the state legislature. Such a governor most assuredly has more leverage—more influence over legislators—than a governor nearing the end of a lame-duck term.

Information Readily available information is also crucial to effective policy making. Answers must be sought to at least two fundamental questions: What technical and political information is available about formulated forest policies? Can the information be trusted? Answers to the first question highlight the policymaker's concern about having enough information, having the right kind of information, and having the information structured in a useful manner. Say, for example, that the policy option being contemplated involves a large-scale national increase in forestry extension programs focused on nonindustrial private forests. The policymaker deserves more information than might be represented by the intuitive hunches of two state extension foresters. Also needed is more relevant information than that contained in documents about the effectiveness of extension programs in an agricultural setting. Further, the information must be appropriately structured. A lengthy discourse on the value of government involvement in the private forestry sector would not constitute an effective presentation of the objectives, benefits-costs, efficiency and effectiveness measures, and clients to be served. And even if the information is sufficient, reliable, and appropriately cast, there is always opportunity for misinterpretation by the policymaker.

Trust in information about formulated policies is always important to policymakers. They recognize that virtually no information is neutral—that all information about forest policy options has been processed, interpreted, and evaluated according to someone's biased interest in the issue. Forest policy analysts may highlight only positive information about an alternative, while deemphasizing or discarding discouraging information. Political advocates of a particular forest policy may argue that the legislative coast is clear, only to discover that the proposed forest resource policy encounters enormous opposition on its way to becoming law. No wonder policymakers become extremely skeptical about the information that is provided to them. To overcome such obstacles, they often consult multiple sources of information so as to calibrate the reliability of the information received.

Personality The personal characteristics of policymakers (and others involved), though often difficult to assess, are important considerations in policy selection. Policymakers are not immune to emotional factors. According to Brewer and deLeon (1983, p. 183),

> Cool rationality, a favorite assumption of scholars of decision, is frequently tempered by—or gives way to—warmer human traits, such as blind passion, panic, hope, wariness and fear. Emotion could be considered as an integrated mind set that summarizes one's experiences and expectations and serves as an instantly accessible basis for action.

Consider four personality types that affect policy making. First, persons whose self-confidence and self-esteem are high often do not search for more information before making a decision. Rather, they frequently make hasty decisions based on inadequate information. Second, the reverse is true of policymakers with limited self-confidence and low self-esteem (Reitz 1987). Third, highly dogmatic policymakers tend to retain certain beliefs or doctrines even in the face of contrary evidence. Indications are that such policymakers select policies on the basis of endorsement by experts; a sense of technical legitimacy for a policy appears to be what they hold important. Fourth, a propensity for risk taking is also a factor in policy selection. Policymakers who are averse to risk tend to choose policies with guaranteed outcomes or with low risk of adverse outcome.

Group versus Individual Decisions

In the past, the forestry community often benefited from bold and creative decisions made by individual forestry professionals. Recently, however, *group decision making* has become more common within organizations. Rather than simply advising, groups actually make decisions. For example, a state director of the USDI-Bureau of Land Management may face a sensitive issue concerning the management of forest wildlife. For various reasons, the director may favor a group decision involving district directors within the state. Or the USDA-Forest Service deputy chief for research, in the face of congressional demands for a new research initiative focused on tropical forestry, may bring together the directors of the agency's regional research stations and give them responsibility for deciding on the composition of the new initiative. The commissioner of a state natural resources department, recognizing that natural resources can play a positive role in furthering state economic development, may call together the directors of the department's various divisions to decide whether such a role should be captured by the department and a program recommended to the governor.

Groups formed to make decisions such as the ones mentioned above are assigned various names, including "task forces," "ad hoc committees," "review boards," "commissions," and "investigating panels."

For forestry organizations, group decision making has a number of advantages over individual decision making. First, as forestry organizations become more complex, as decisions become more involved, and as uncertainties about issues and formulated alternatives become greater, it becomes unrealistic for a single individual to make far-reaching judgments about forest resource policies. Take, for example, a national recommendation for many formal wilderness designations for USDI-Bureau of Land Management lands. The expertise and shared experience of several individuals is viewed as more effective in deciding such complex, broad-scoped issues. Second, groups may be more creative than individuals. Complex decisions requiring innovative solutions may better be obtained from a group of individuals with varied backgrounds—as when a forester, an ecologist, an economist, a hydrologist, and a wildlife specialist decide on U.S. Environmental Protection Agency policy regarding atmospheric pollutants deposited on forests. Third, involvement of individuals in group decision making may be an effective way of promoting individual members' commitment to the forest resource policy ultimately selected. State foresters, for example, who suggest and agree to a regional plan of action to be implemented by a federal organization are more likely to participate in the plan's implementation. Fourth, a group decision may be less biased (may involve less favoritism) than a decision made by only one person. For instance, a strong-willed individual decision maker may be insensitive to the needs or wishes of an organization or a group.

Group Activities The events leading to a group decision about the use and management of forests vary considerably according to the nature of the forest resource issue and the extent to which policy and program options have previously been formulated. Consider a group of public-land forestry professionals charged with deciding whether a particular forest area should be designated as a recreational wilderness. As a group, they might engage in the following activities.

- *Share information* Technical facts about recreationists' expectations from forest wilderness may need to be aired.
- *Share experiences* Experiences with forests previously designated as recreational wilderness also should be known by all the group members.

• *Generate ideas* Whether the area designated will be large, or small, whether it will be devoted to hardwood or a conifer vegetative cover, and whether it will be a state park or a state forest will have to be determined by the group.

• *Make critical evaluations* The pros and cons of the policy options generated may be debated.

• *Express relevant beliefs and values* Preferences for and against wilderness designation, and for and against limitations on visitor use, as well as such matters as whether primitive or modern facilities should be provided in the wilderness, may be discussed by the group.

• *Develop a feeling of participation* Sharing information, experiences, ideas, and values often leads to a feeling of having participated in a group decision, even if the alternative selected is not consistent with the views of individual members.

• *Diffuse responsibility* Members assume mutual responsibility for the decision to designate the area as forest wilderness, and therefore they are willing to work together to implement the decision.

Advantages Substantial evidence supports claims that group decision processes are often superior to decisions made by a single individual (Reitz 1987). In large measure, such claims rest on a group's ability to access more diverse and greater amounts of information, and their built-in means for evaluating (often generating) formulated policies. Groups are particularly effective when addressing multifaceted issues, such as allocating forestland to timber, water, recreation, and range—especially when their members have complementary skills—say, when a group is made up of a timber manager, a hydrologist, a recreation manager, and a range manager.

Groups have also been found to be more willing to select risky alternatives than would be selected by individual policymakers. This finding is partly explained by the propensity of individuals within a group to spread responsibility to other members of the group.

Although the evidence on creativity is mixed, groups in some situations are more creative than individuals. Free-flowing and spontaneous group discussions may lead to development of creative alternatives and critical evaluation of their merits. Such creativity can be facilitated by Delphi and nominal group techniques, which are discussed by Gibson et al. (1988) and Delbecq et al. (1975).

Disadvantages Group decision making is not without problems. A tendency to become involved in prolonged debates and discussions often interferes with a group's ability to quickly identify and effectively address subsets of a forest resource issue, such as program goals, budgets, implementation, and evaluation. In addition, confusion can occur when the correctness of a decision is not obvious or easily demonstrated—that is, consensus may become difficult to achieve, or confidence in a consensus may fail to materialize. Additional problems can arise from the manner in which the group is organized and operated. (Reitz 1987):

- A dogmatic group leader may attempt to exert undue influence over group members.
- Some members may be reluctant to participate because of status considerations.
- Some members may be overly willing to accept the suggestions of high-status members.
- Some members may exert undue pressure on others to conform.
- Destructive competition may arise among members, especially if some members expect to achieve an unequal share of the benefits of a decision.

Inappropriate group size can also detract from the efficiency of group decision making. Communication problems can arise in large groups—those with seven or more members—and factions may develop in a group with an odd number of members.

SELECTION MODELS AND CONCEPTS

Decisions about the use and management of forests regularly flow from very complex political and administrative settings. Such complexity is more easily understood if reality is transformed into models or concepts of decision-making processes. Unfortunately, theorists have yet to provide a single widely accepted model of how policy decisions are arrived at. Several models, therefore, are discussed below: rational comprehensive, rational incremental, mixed scanning, and organized anarchy.

Rational Comprehensive

The rational comprehensive approach to policy making involves a series of logical choices designed to achieve or optimize a goal or an outcome. Using such an approach, policymakers begin by defining and ranking values or outcomes viewed as personally or organizationally important—for example, "Wood-based products should be available to consumers at fair prices." Once values have been specified, attention is focused on identifying all objectives (or goals) that are compatible with achievement of such values. Examples are increasing production of timber from nonindustrial private forests, increasing harvest of old-growth timber from public lands, and increasing the economic supply of timber via development and application of new timber processing technologies. Thereafter, policymakers proceed to formulate all relevant policy options, or means for attaining each specified objective. If nonindustrial private forests are the focus, options for increasing timber produced by such owners might include expanding the available supply of information about appropriate silvicultural technologies (via extension or service forester activities), expanding the availability of cost-share funds for reforestation activities, or mandating by force of law that timber at a specified economic rotation age be harvested and made available for sale.

Once options to achieve an objective have been specified, policymakers then identify all consequences (pro and con) associated with each option (social and private costs and benefits of cost-share programs). Employing some decision rule, policy options are then compared and that option which maximizes the values and objectives previously specified is chosen. Invariably, the decision rule involves some measure of economic effectiveness or efficiency, such as benefit-cost ratio, present net worth, or internal rate of return. Portrayed in visual terms, an example of the rational comprehensive approach appears as follows:

Values:

Fair wood-product prices for consumers

Objectives:

Increase nonindustrial private forest supply	Increase public land timber supply	Increase research and development

Alternatives:

A. Expand information	A. National Forests	A. New product
B. Expand cost-share	B. State forests	B. New process
C. Legal harvest requirement	C. County forests	C. New market system

Consequences:

Positive	Positive	Positive
Negative	Negative	Negative

Criteria:

Benefit-cost ratio	Benefit-cost ratio	Benefit-cost ratio
Net present worth	Net present worth	Net present worth
Internal rate of return	Internal rate of return	Internal rate of return

Choice:

Appropriate choice maximizes objectives and values.

The rational comprehensive approach to forest policy making is viewed as "rational" because objectives and alternatives are logically selected and their relative importance is weighed. It is considered "comprehensive" because all alternatives and values are supposedly taken into account by the policymaker. In the rational comprehensive approach, the policymaker is presumed to have a full view of relevant forest policy arenas, perfect information about values and policies, and an ability to carefully weigh and judge all policy options. Because the approach strives for comprehensive analysis and evaluation, it is often referred to as the root method (Lindblom 1959). Rational comprehensive policy selection is most appropriate to situations in which debate about values is virtually nonexistent, objectives to be sought are clearly specified, certainty over the outcomes of forest policy options is substantial, and rigorous application of economic decision rules is possible.

The rational comprehensive approach to policy selection is not without problems. One valid objection to it is that policymakers simply lack the intellect and the resources (time and finances) needed to achieve complete rationality. They are seldom, if ever, in a position to comprehensively evaluate or intellectually understand the range of policy options that could be used to address a forest resource issue. Furthermore, clearly defined values upon which all interested parties can agree are seldom available. The relative importance of the values which permeate objectives is unknown, or there is continual disagreement about values. Even if the values are known, there is no purely rational way of resolving conflicting perspectives on them. In addition, the ultimate criterion for choice among competing forest policies is not necessarily rooted in maximization of values (even if they can be identified and ranked); rather, a policy is chosen because all concerned parties judge it to be agreeable.

Rational Incremental

Policy decisions arrived at by rational incrementalism are the product of advocacy and bargaining. They result from mutually agreeable adjustments in the values, objectives, and policies that are advocated by the various parties who are interested in a forest resource issue. Progress toward a policy decision involves trial and error, successive approximations, and continuous incremental modification of objectives and policy options. A good policy is one which gains consensus rather than one which meets criteria of economic efficiency or effectiveness. Rational incrementalism (which is also known as ''muddling through,'' or as the ''branch method'') is suggested as a description of how policies are actually selected (Lindblom 1968). Rational incrementalism comprises a number of important characteristics which, in total, form an internally logical, real-life picture of policy making (Figure 5-1).

Minimal Debate about Values and Goals Of paramount importance to the process is *minimal debate about values and goals.* Specifying objectives clearly may not only be intellectually difficult for policymakers but may also be pragmatically objectionable. Values, goals, and objectives are so intertwined and so in conflict among interested persons that specifying them clearly may only pro-

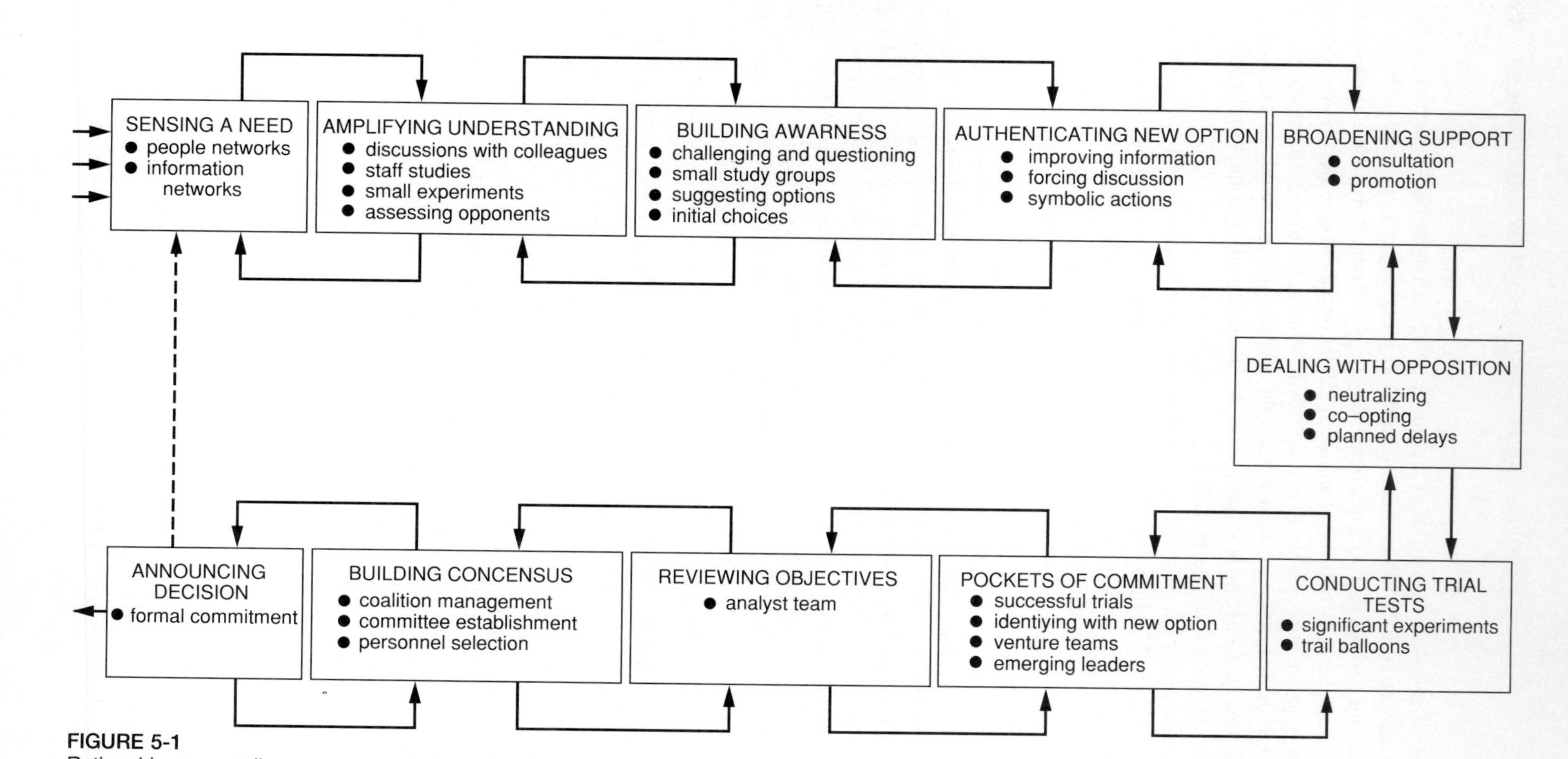

FIGURE 5-1
Rational incrementalism as a sequence of events and activities.

long ideological disputes and delay identification of a forest policy option agreeable to most parties.

For example, a proposed new state forest could bring forth from interested parties a variety of conflicting values and ideologies, such as public versus private ownership of forests, public investment in rural versus urban areas, and state versus local public ownership. Debate could quickly drift from the immediate pros and cons of the proposed forest to debate over deeply held views on values. If properly cast, however, the proposal could be appealing to a variety of interests, each with markedly different values and ideological perspectives. The timber interests could view the proposed forest as a source of valuable wood, the recreationists as a haven for leisure-time activities, the local community as a source of income and employment, the research community as a place for long-term biological studies, and state government as a means of stopping conversion of forests to agricultural uses.

In government forestry cost-share programs, there may be minimal debate over values and goals. Such programs are often supported by various interests, even though goals vary widely among interested parties. The goals represented may include employing local labor, transfering income to the rural poor, assuring timber supplies, stimulating purchase of machinery and equipment, reducing erosion of forest soils, and facilitating the production of wildlife. Topics such as government versus private responsibility for private forestry initiatives, government program competition with existing private programs, and government responsibility for assuring the availability of low-cost timber for wood-based industries are divisive and need not be raised for debate and precise specification.

Likewise, the addition of a specific area of forestland to the National Wilderness Preservation System often becomes feasible not because opposing parties suddenly agree to fundamental values that favor wilderness but because other concerns of the parties involved have been sufficiently met. For example, the wood-based industry may agree with the designation in order to secure access to additional areas of a public forest for timber-harvesting purposes.

In sum, rational incrementalism is a decision process that discourages precise and careful identification of program objectives. The process enables groups with diametrically opposed values to agree on the same forest resource policy—if their ultimate values and goals are not forced into open debate.

Continual Comparison of Policy Options Another characteristic of the incremental approach to policy selection is *successive and continual comparison of policy options*. Forest resource policies are never made one and for all, but are endlessly made and remade by comparing relatively narrow policy choices. Policymakers accept this perspective; they return again and again to issues, looking for new ways to correct policy mistakes previously implemented. For example, policies directing proper allocation of federal timber sale costs to various forest outputs (below-cost timber sale issue) were initially addressed by one set of policy proposals by the USDA-Forest Service. Then they were countered by another set of policy proposals made by the General Accounting Office of the U.S. Congress, modified to secure agreement by both parties, and implemented in a field situation by the National Forest System. Subsequently recognized as having problems when implemented, they were reformulated with a somewhat different focus, then reimplemented in a field situation, and so on (Risbrudt 1986). In rational incrementalism, decisive conclusion is a foreign concept.

Incremental Policy Changes Rational incrementalism also rests on the concept of *incremental policy changes.* When policymakers discover that existing forest resource policies are deficient, remedial policies tend to be incremental. Relatively modest adjustments are made to existing policies, and vast sweeping changes are avoided. By moving cautiously and experimentally, policymakers acknowledge the virtues of operating from a known basis of information, avoiding a giant step into an often unknown future. Predicting the entire range of technical and social consequences of a major shift in forest policy is virtually impossible; incremental change is an insurance measure against disastrous errors in policy and the high cost of correcting such errors. An example is passage of small-scale, ever-tightening industrial air-pollution regulations, rather than wholesale adoption of a massive regulatory policy and program, the effectiveness and efficiency of which would be largely unknown.

Incremental changes in forest policies are also a reality for political reasons. Political support, both within and outside government, for a massive increase in public-land ownership (such as doubling the size of the National Wildlife Refuge System) may be unrealistic, but modest additions to or sale of public forestland are common, as when inholdings are purchased or isolated parcels outside the forest boundaries are sold. Trying to muster interest-group and bureaucratic support for massive increases may be totally unrealistic, whereas modest additions and sales are within the realm of political possibility.

Strong Advocates of Policy Options Policy making in a rational incremental fashion implies the presence of *strong advocates of relatively few policy options.* Policymakers cannot possibly consider the full range of forest policy options available at any one moment, nor can they fully understand the enormous range of consequences implied by any one option. Staff, finances, intellect, and time are simply too limited. For example, the range of national policy options capable of addressing atmospheric pollutants affecting forested areas is enormous; so too are the social, economic, and biological consequences of each option.

Even narrow-scope choices about forest tax policy involve options that are not always fully understood. Federal income tax policy allowing timber income to be treated as a capital gain for tax purposes involves substantial uncertainty. It may be impossible to predict whether removal of such a policy will lead to a decline, an increase, or no change in industrial timberland investments.

Acknowledging that there are many potential policies with many implications, policymakers tend to structure decision processes so that every significantly affected interest has a voice in the process. Self-interest ensures that policy options and their consequences will be vigorously presented by vociferous advocates. Persons in and outside government join in forest policy debates, uncovering relationships between formulated policies and their own interests (values, goals, and technical consequences)—relationships the forest policymaker may not have known about (Schultze 1968).

Selection of Satisfactory Policies Policymakers operating in a rational incremental environment are often averse to a search for a single forest policy option that will maximize an abstract notion of net social benefit. Rather, a decision is made to *select satisfactory policies*. Policymakers fully expect a chosen forest policy to achieve only part of what is ultimately hoped for: maximization of net social benefits. Unexpected, adverse consequences make it impossible to attain the ideal. The policymaker therefore, for pragmatic reasons, assumes a posture of satisfaction with the chosen forest resource policy (Stillman 1983). For example, policy options reportedly able to eliminate stream sediments in forested areas might well maximize political and social interests. They would, however, be avoided as being technically and administratively unrealistic; that is, the likelihood is that the necessary financing, technology, and professional talent would not be available. Similarly, doubling timber growth on nonindustrial private forest by tripling investments in government cost-share programs may be wishful thinking if funding uncertainties for the program are great. Being satisfied with policy outcomes is quite different from searching for the one elusive forest policy thought capable of maximizing an abstract, comprehensive social interest.

Involvement of Many Interests The rational incremental approach to policy development also entails *involvement of many individuals, agencies, and interests*. Forest policies are seldom selected by a single individual or a single organization; interaction of many actors is most common. Contending interests within and outside an organization compete for influence, continually forcing policymakers to bend in many directions until a mutually acceptable policy is settled on. A USDI-Bureau of Land Management policy on the extraction of minerals in forested areas may well reflect a variety of political struggles among government units (say, the Bureau of Mines, the Office of Surface Mining, the Geological Survey, and the Bureau of Reclamation), within interests outside government (such as the National Coal Association, the Sierra Club, the American Petroleum Institute, and the Environmental Defense Fund), and between government and outside interests. Similarly, a state policy on service forestry programs may well reflect the views of interested parties in various programs and sectors, including extension foresters, industrial foresters, research foresters, forestry associations, wood-based companies, consulting foresters, district foresters, and environmental associations. Rational incrementalism scorns a single-minded approach to policy selection. It favors the interaction of many individuals and agencies, all operating in a complex network of power and influence.

Bargaining and Negotiation Rational incrementalism produces forest resource policies which have been shaped and reshaped by *bargaining and negotiation.* While policymakers and the organizations they are responsible for often have self-interests, they are not blindly partisan. Political realities force them to adjust their positions in relation to one another. Realistically, none expects the final policy choices to be the ones they most favor.

Agreement as a Decision Criterion In a rational comprehensive system, *agreement is the measure of policy goodness.* The merits and ultimate worthiness of a proposed forest policy are not judged on the basis of technical, economic, or social criteria. Rather, a good policy is one upon which most interests can agree. In pluralistic democracies, value is placed on consensus seeking. What emerges may not be the most technically or economically correct forest policy; rather, it is likely to be the most agreeable forest policy.

Application Problems Like rational comprehensive approaches to selecting forest resource policies, rational incrementalism also has difficulties in application. Small-scale, remedial changes in unsatisfactory forest policies are viewed as pointless by some interested parties. Incremental processes are seen as encouraging small-scale tinkering with existing forest policies, even when a wholesale change of a deficient policy is in order. Moreover, government may be responding to new issues so rapidly or so fundamentally that experience with past forest policies is a totally inadequate guide to future policy directions.

Under such circumstances, incrementalism becomes an inappropriate policy development mechanism. Correspondingly, policy opportunities available for addressing forest resource issues may be rapidly expanding because of new technologies, additional finances, and changes in political support. New and significantly different policy approaches are likely to be neglected on the grounds that they represent too significant a change from past direction. Policymakers whose arsenal has been filled with technologies and policies appropriate to yesterday's forest resource problems are ill-equipped to address tomorrow's issues.

Concern about rational comprehensive approaches to policy selection can also be focused on agreement as a decision criterion. In times of stability in policies, informed agreement based on past experiences may be effective. In times of rapid change, however, lessons from the past may be irrelevant; a fabricated consensus may be the result. Limited enthusiasm also exists for the role of participatory democracy in complex technical issues.

In addition to the difficulties discussed above, incrementalism is often viewed as an ineffective approach to forest policy decisions involving long-term cumulative effects. For example, errors in the design and application of silvicultural systems for timber management may not become evident until the systems have been in operation for many years. Such errors may be made because factors such as poor soils, the inappropriateness of a site, or the tendency of a site to be insect- and disease-prone are initially overlooked. Learning, a fundamental tenet of incrementalism, cannot occur until it is too late to intervene effectively. The challenge of designing an error-free timber management policy cannot always be met by incrementalism (Bass 1983, Brewer and deLeon 1983, Hogwood and Gunn 1986).

Summary In sum, rational incrementalism is especially suited to the development of forest policy in an environment in which substantial uncertainty exists about the outcome of policy alternatives, in which divisive and destructive debate over values to be achieved is a real possibility, in which the cost of an erroneous forest policy is likely to be substantial, and in which general reform in policy is neither politically feasible nor politically desirable. Rational incrementalism is a conservative approach to selecting forest resource policies; its proponents and practitioners view significant change with much concern.

Mixed Scanning

Selection of forest resource policies can also result from a process of mixed scanning. A wide range of available policy options are sought out and cursorily reviewed, and then options considered worthy of additional attention are examined in detail. The multipurpose intent of this approach is to avoid a truncated view of available policy alternatives while also avoiding an all-encompassing, unmanageable review of such options. For example, determination of an appropriate spatial distribution for timber-management investments in state-owned forests may well begin with a broad regional assessment of opportunities—a cursory view of options, deliberately omitting detailed assessment. Statewide and regionwide information will lead to forthright rejection of some areas from further consideration—areas in which, for example, forest is allocated to nontimber uses, or only low or marginal productivity is possible, or nonmarketable timber species prevail. Others will be judged to be potential opportunities requiring further investigation because of such factors as close proximity to important markets or access to major transportation systems. Policymakers are spared investment of further time and money in rejected forests. The process of scanning continues until an appropriate timber-management investment policy is settled on.

Mixed scanning distinguishes between fundamental and incremental policy decisions. Exploration of a broad range of possible policy options leads to the making of fundamental decisions. Detailed assessment of policy options is avoided. Only after broadly defined fundamental policy options have been identified and judged worthy of further attention does incremental decision making occur. For example, a logical first step in identifying forested areas that have potential for inclusion in the National Wilderness Preservation System would be a nationwide scanning of federal public forests. Once the nationwide scanning was accomplished, specific potential forest wilderness areas located in a specific region of a particular administrative unit would be assessed. Mixed scanning is dynamic in that the mix of broad and detailed scanning of policy options can be varied depending on changing circumstances. When political conditions are stable and forest policy changes are few, an emphasis on incremental decision processes becomes appropriate. When conditions are changing rapidly, more fundamental scanning for new forest policy options is in order.

Mixed scanning attempts to avoid the pitfall of rational comprehensive policy-making activities, namely, comprehensive assessment of an excessively wide range of forest policy options. By so doing it prevents the policymaker from being overwhelmed by vast amounts of enormously costly information about policies which are technically or politically not feasible. Scanning also seeks to avoid focusing excessive attention on past forest policies, or on newer policies that differ only slightly, by forcing a more comprehensive review of available policy options. Thus policymakers can avoid becoming stuck with a potentially erroneous forest policy fostered by rational incrementalism (Bass 1983).

Organized Anarchy

Forest resource policies can also be selected through processes that have little if any structure. In some cases, forest resource issues and the organizations estabished to address them are so complex that explanation of policy selection by rational comprehensive or rational incremental approaches becomes an exercise in futility. In such situations, forest resource polices can be viewed as emerging from a climate of organized anarchy, which has also been popularly identified as the ''garbage can model of organization choice'' (Cohen et al. 1972).

In the organized anarchy process, a hodgepodge of issues, solutions, and policymakers moves from one decision opportunity to another. The forest policies ultimately selected are dependent upon the mix of forest resource issues confronting an organization, the pool of solutions available at a given time, and the nature of the participating policymakers. A policy selection is made when the following events coincide: (1) the forest resource issue requires a solution;

(2) a solution is available; (3) the solution is compatible with the organization's resources; and (4) the forest resource issue, solution, and available resources are all known to a policymaker who has the time, energy, and interest needed to address the issue. A neat and highly rational process it is not. Even so, organized anarchy is a condition from which important forest policies often flow.

Characteristics Organized anarchy has a number of fundamental characteristics.

First, *preferences for goals, objectives, and policies are not clearly defined.* There exists within an organization a loose collection of ideas rather than a coherent ordering of preferred policy options. Policymakers, administrators, and analysts disagree about what the organization should accomplish—whether it should emphasize timber, water, wildlife, or recreation programs, for example. Further, they are often obliged to become actively involved with a policy before their individual preferences become known. Even if their preferences are known, the means of achievement may be elusive.

Second, *policymakers have an uncomfortable understanding of the organization within which they operate.* Immediate responsibility for programs may be known, and the operation of the organization as a whole may be understood, but there exists only fragmented and rudimentary understanding of why certain policies exist and how such policies relate one to another. Policymakers ''don't necessarily understand the organization of which they are a part; the left hand doesn't know what the right hand is doing'' (Kingdon 1984, p. 90). Policymakers rely on trial and error; they learn from everyday experiences and intervention during crisis.

Third, *policymakers drift in and out of policy-making roles.* Participation in policy making is definitely very fluid. Who is invited to and who shows up for critical meetings can make a tremendous difference to policy selection. Even in hierarchical organizations, some individuals take on an importance that may not be consistent with their formal roles. A researcher with special expertise on pesticides, for instance, may be assigned to advise top-level policymakers. Formal organizational boundaries are further clouded when persons from outside government (such as advisory committees, consultants, and lobbyists) become involved in the policy selection activities of a government agency. Involvement of other branches of government, such as legislatures and other bureaus and departments, may further complicate the selection process, and the normal turnover of personnel (as when the state director of the USDI-Bureau of Land Management is transferred to the agency's Washington office) can also be a factor. Despite these seemingly chaotic conditions, agencies do function, do select forest resource policies, and do implement program preferences.

Operation Organized anarchy as a decision structure is composed of certain fundamental and readily identifiable phenomena that flow on a continuing basis through an organization: issues (public problems), policy options (solutions), participants (policymakers, analysts, administrators), and opportunities for choice. These phenomena are largely unrelated to one another, and each has a life all its own. For example, participants may generate forest policies because of their own self-interest—say, they want to expand a program—not because policy options are needed to address an anticipated forest resource issue. Participants may drift in and out of decision situations because they have a vested interest in an issue or because they want to promote a pet solution, not because superiors demanded their participation. The situation has been aptly described as a (Cohen et al. 1972, p. 2):

> collection of choices looking for problems, issues and feelings looking for decision situations in which they might be aired, solutions looking for issues to which they might be the answer, and decision makers looking for work.

As an example, consider the selection of a new state forester—an event of substantial importance to a state's forestry programs. An opportunity for choice (the need for a new state forester) occurs within the organization (the state division of forestry). Various participants (division staff, departmental staff, legislators, and division clients) become involved, and all are promoting their own solutions (their own candidates for the position). Various issues (program expansion, affirmative action, division visibility, and sensitivity to all division clients) are introduced, and various solutions are considered (candidates from within and outside the division; candidates with forestry experience or with general management skills). The outcome (the individual selected) depends on the mix of options available and the participants involved in the decision process. If the opportunity for choice is coupled with the stream of solutions and participants at a particular point in time, a certain individual will be selected, whereas at another time, a different individual would be chosen. Who participates in the process (who is invited to attend the candidate-evaluation meetings) and the solutions suggested (the candidates themselves) are critical. A candidate who appears to be the ideal solution at one point may fade from consideration at a later date because an important participant fails to attend a subsequent decision meeting.

Organized anarchy is a policy selection process that reflects the often chaotic conditions from which forest policies can emerge. Structurally important to the process is the independent flow of events through an organization (issues, policy options, participants, and opportunities for choice), the merging of which results in a unique policy choice. The process is unlike rational comprehensive approaches to forest policy development and selection in that policymakers do not willingly set out to solve an issue in a logical and carefully prescribed fashion. Nor is it like rational incrementalism, for streams of solutions and participants may be coupled in a previously untried combination. The product may well be an abrupt change in forest resource policy.

Summary

Policy selection is a complex process that can be explained in a number of ways. A common view is that selection is a *rational comprehensive procedure* involving clearly specified goals (apart from values) and well-documented policy options that are amenable to selection with the aid of efficiency or effectiveness criteria. Another model is *rational incrementalism,* wherein values and objectives remain murky, the number of policy options considered are few, strong advocates of each option are ever-present, agreement is the measure of a policy's worth, and new policies are not much different from old policies. A third explanation is *mixed scanning,* which, in reality, is a hybrid of the rational comprehensive approach and rational incrementalism. The fourth suggested explanation of policy selection is *organized anarchy.* An appropriate mix of issues, policy options, and policymakers is thought to surface at an appropriate time and to culminate in selection of a policy or program to address a passing but pressing issue.

BARGAINING AND NEGOTIATION

One of the most notable characteristics of political and administrative systems charged with policy selection is an explicit reliance on bargaining and negotiation. In large measure, bargaining and negotiation are a result of a deliberate historical effort to diffuse authority among competing government institutions—legislative versus judicial versus executive, or federal government versus state governments. Bargaining and negotiation also encourage cooperation and consensual decision making among persons within organizations that perform government services. Policymakers who rise to high-level and influential positions in government usually do so because of their ability to persuade and bargain—not because they freely exercised authority and force. They become party to the process of policy selection wherein the measure of success is a keen ability to bargain, make deals, and seek consensus. In such a context, we shall consider bargaining as a process of cooperation and mutual adjustment among policymakers. Various strategies and tactics which are often suggested as means for facilitating the bargaining process are also discussed below.

Cooperation and Mutual Adjustment

Policymakers strive to be keenly aware of a policy option's technical feasibility, economic efficiency, and social equity. However, the policymaker's drive to identify the formulated policy on which all can agree—the forest policy which is agreeable to subordinates and superiors, to lawmakers and their staffs, to organized interest groups, and to judges who rule on the legalities of forest resource policy—is pervasive. To achieve agreement, parties interested in a for est resource issue must be willing to adjust their views on preferred policies. For example, the chair of a legislative budget committee, the director of a government wildlife division, and the vice-president of an interest group may be unable to agree on a funding level for a new forest wildlife program that will meet the self-interest of each party. If they adjust their positions, however, funding at a level agreeable to all may well be possible. Persuasion and influence must be applied in measured amounts until *mutual adjustment* culminates in agreement (Lindblom 1968). Agreement will be facilitated by various forms of bargaining and the strategies and tactics which accompany bargaining (persuasion, obligations, reciprocity, mediation, logrolls, and side payments).

Bargaining is an activity in which two or more persons or organizations with conflicting interests seek a joint agreement on how they will behave to one another. The agreement, or bargain, may specify an acceptable distribution of rewards (for example, more public land will be allocated to wilderness if it is agreed that more public land will also be allocated to timber management) or a division of contributions (say, greater public as well as private investment in erosion control). Not to mutually adjust positions via bargaining can lead to deadlock: no action taken, the status quo preserved.

The process of bargaining can be illustrated by a two-party situation involving public versus private investments in timber-management practices to improve the quality of water flowing from forested areas (Figure 5-2). A state pollution-control agency, for example, may initially have an interest in securing from owners of private forestland at least a minimum level of investment in sediment-reducing forestry practices. This level would be called ''level A'' and would require installation of water bars at regular intervals on forest roads. Similarly, the private landowner's association may expect a minimum level of agency investment in sediment-reducing practices. The agency's participation would be called ''level B'' and would entail 20 percent agency cost sharing of practices undertaken by private landowners. Through the exercise of power and persuasion, the agency may convince the landowner association to participate at level A. The equally powerful and persuasive landowner association may in turn challenge the agency, demanding agency participation at level C, defined as 40 percent agency cost sharing. The agency may agree but quickly respond with a demand that landowners expand pollution abatement action to include more intense activities: level D, which would prohibit timber harvesting within a specified distance of streams. Again the association would respond with a counterdemand, and the process would continue along similar lines until mutual adjustments in positions resulted in a mutually agreeable solution: level X, involving investment levels E and F.

The hypothetical example shown in Figure 5-2 could easily have involved other actors and other issues—for example, bargaining over public land-use allocations by conservation groups, wood-based industrial groups, legislative committees, and a public agency. Interagency bargaining over program fund-

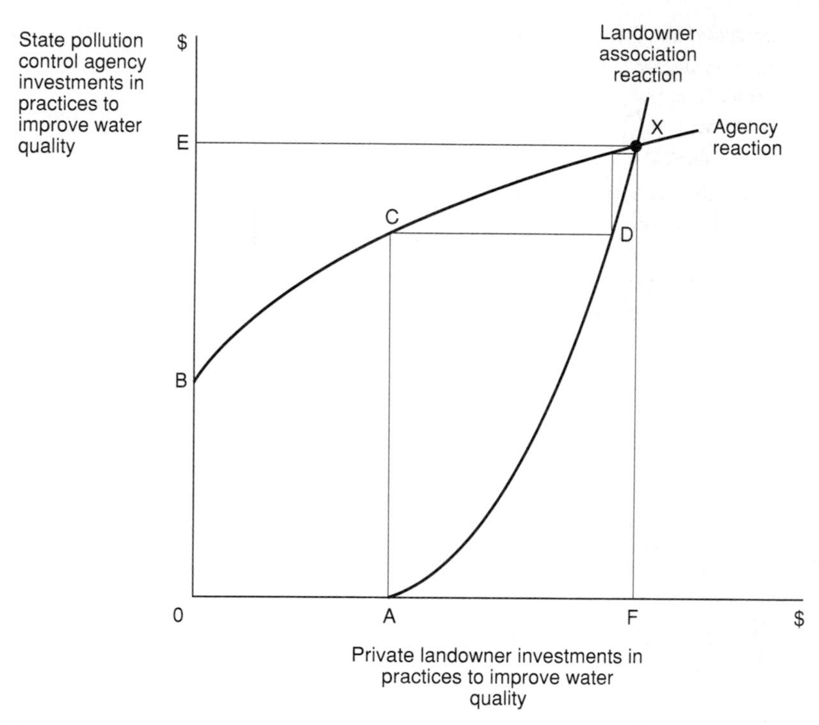

FIGURE 5-2
Bargaining relationship example involving government and private owners of forestland.

ing—public-land management versus service forestry versus research—could have been involved. Bargaining within a coalition of conservation interest groups over an appropriate coalition position on proposed amendments to an important air quality law might have been a factor. Another possibility could have been bargaining among geographically disbursed regional foresters about which region would secure what portion of a legislatively appropriated financial pie.

Ability to secure agreement depends on the attitude of opposing interests toward bargaining. If certain beliefs are not present or are not strongly held, forest policymakers may find themselves embroiled in bitter political battles instead of constructive bargaining efforts leading to agreement. If, for example, a policymaker has great mistrust for a wood-based industrial trade association, there is little chance that he or she will freely share with the association information about formulated policies in which the association has an interest. The association may react to this mistrust with aggressive behavior, seeking to do battle rather than addressing the issue at hand. What attitudes and beliefs are important to bargaining and the cooperation and trust implied therein? Consider the following (Filley 1975, pp. 60–72):

- *Existence of a mutually acceptable policy* Policymakers and interested parties believe that a mutually acceptable forest resource policy, which will achieve the goals of all interested parties, is possible.
- *Desirability of agreeing to a mutually acceptable policy* Policymakers and interested parties find virtue in the idea of searching for a mutually acceptable policy. Agreement on a mutually acceptable policy is preferred to conditions or policies which currently exist.
- *Cooperation rather than competition* Policymakers and interested parties believe that cooperation, not competition, can lead to more complete utilization of the strengths and skills of all involved. Cooperation with opponents is preferred; competition may leave the defeated interest in a fragmented and dissatisfied condition, perhaps seeking opportunity for retaliation.
- *Equal perception of power and status* Policymakers and interested parties believe power and status among parties to be equal. Sharp distinctions between power and status tend to separate groups in ways (''we'' versus ''they'') which detract from the search for agreement.
- *Statements represent beliefs* Policymakers and interested parties believe that statements made by those involved are expressions of views actually held.
- *Differences of opinion helpful* Policymakers and interested parties recognize disagreement as an important means of fostering creativity. New policies unique to all interests may result from constructive disagreement, thus hastening agreement.
- *Trust* Policymakers and interested parties believe that they can trust one another. Trust discourages concealment or distortion of relevant information, and encourages open statements about facts, ideas, conclusions, and feelings, even though such statements may make a party vulnerable to the actions of others. Groups of policymakers with a high level of trust will be prone to exchange ideas and preferences, clarify goals and objectives, search for agreeable alternatives, and implement agreed-to-policies. They will not be prone to leaving the group, because they view themselves as part of a team.

Bargaining is also facilitated by the initiation of cooperative actions by individual forest policymakers. For example, the likelihood of agreement on a forest policy is enhanced if one party identifies characteristics and interests which are held in common by both parties. Policymakers searching for agreement on an issue concerning forestland which lacks a forest cover may well have a common interest in reforesting such land, even though substantial disagreement exists about the fate of the trees planted, once they are mature.

Similarly, policymakers may be successful in securing agreement if differences are deemphasized. If divisiveness exists between timber management and wildlife-management interests on topics such as stocking levels, herbicides, and rotation ages, it may be worthwhile for policymakers to focus on experiences which are positive to both groups, such as manipulation of vegetation via timber harvesting for purposes of improving the area as a big-game habitat.

Agreement is also facilitated when a policymaker becomes identified with something that has positive value to other policymakers. The conservation group leader who enhances the industrial association leader's reputation by informing the association's membership of positive advances in water-pollution abatement significantly improves opportunities for agreement on forest policies to address future pollution-control issues.

Development of a cooperative attitude is also encouraged when one policymaker acknowledges receiving help from another policymaker. The industrial forestry association that acknowledges the benefit of a conservation group's legislative support for additional funding of forest survey programs may well find that subsequent agreement on potentially divisive forest resource issues of the future is more easily attained. Agreement may also be fostered when a policymaker disassociates from situations viewed as objectionable by other policymakers. A state forester who withdraws from active involvement in proposals for a new and highly restrictive state forest practice law may be rewarded by industrial agreement on a new forest tax policy.

Personal styles of seeking agreement vary considerably among forest policymakers. Tough battlers seek their own goals at all costs, without concern for the needs or the acceptance of others. They stand by their convictions, failing to realize that today's crushed opponent may be in a more powerful role tomorrow. Other policymakers play the role of friendly helpers, continually setting aside achievement of personal policy goals in order to facilitate broad-scale agreement on important forest resource issues. Still others withdraw from bargaining processes, viewing conflict of any sort as a hopeless, useless, and punishing experience; if agreement is reached, their commitment will be almost nonexistent.

Some policymakers assume an integrative, problem-solving style. Such individuals make no distinction between their personal goals and the goals of others; they view conflict and debate as a natural means of generating creative forest policy options to which all can agree. There are also policymakers whose style is compromise. They enjoy the maneuvering required to attain agreement on a policy reflective of the middle ground (Filley 1975).

Strategies for Securing Agreement

A policymaker's interest in securing agreement may be intense. To achieve the desired end, it may be necessary to use various strategies and tactics, ranging from the time honored art of compromise to the legislative logroll and from discharging obligations to bluffing and threatening. Whatever the approach, strategies and tactics designed to facilitate agreement on a formulated forest policy are a vital and necessary part of the selection process. In this section we will consider compromise, logrolling, and side payments.

Compromise A fundamental approach to securing agreement is *compromise*. Uncompromising pursuit of narrow and specialized interests at the expense of agreement can jeopardize the process of policy selection. Compromise typically takes the form "You want X, I want Z, let's settle on Y," which implies that "Half a loaf is better than no bread" and that "splitting the difference" is an acceptable conclusion. In most cases, none of the parties involved is entirely satisfied with a compromise position, but all of them see no reason to continue to pursue their self-interests, given the risk of losing all. For example, an interest group that is promoting regulations for the management of air pollutants from prescribed burning in forested areas may not be satisfied with a pending legislative agreement. At some point, however, the group may realize that further pursuit of their interests could cause additional and more politically powerful opponents to enter the fray. They may also sense that loss of the regulatory programs already agreed upon is another danger. In such a situation, compromise is the only reasonable tactic.

Groups employing compromise strategies attempt to present a united front to opponents. Communication within the group is directed to the chief negotiator, who may be an agency head, a member of a legislative staff, or an interest-group leader. Communication between groups occurs between the leaders of the opposing sides.

The process of achieving compromise is often complex and difficult. Attempts to distort the images of opposing groups may be made; opponents may be stereotyped; and accusatory and judgmental statements may be hurled back and forth. A disturbing consequence of a compromise strategy may be the hoarding or distortion of information in an effort to secure an advantage in the bargaining process. When information is released, the intention may be to embarrass an opponent or force a change in position (Brewer and deLeon 1983).

Development of an enduring compromise requires flexibility. Inflexible ideologies will obstruct the road to a compromise solution, whereas initial presentation of a wide range of formulated forest policies can be helpful. If the initial range is narrow, having been suggested by relatively few interests, a compromise policy may result which will find little favor when exposed to the ever-broadening range of interests that may become involved as the struggle evolves. Generous communication can facilitate development of a mutually beneficial compromise solution.

For example, suppose that policymakers disagree over the appropriate use of a large area of public forest; some advocate timber, and others emphasize water production. If communication is sparse, the compromise agreed upon may involve dividing the forest area in half and devoting one portion solely to timber production, the other solely to water production. Accurate communication of interests, however, could lead to a combination of timber and water production on the entire tract of forestland—a compromise solution which would yield greater benefits for both parties.

Compromise often has an unfortunate negative connotation. It is often suggested that a compromise policy is not a right, proper, correct, and just alternative, or that to compromise on a forest policy that has been formulated by a rational process is somehow irrational. Altering a supposedly efficient policy option may be thought to imply agreement on a wasteful alternative. Accommodation of the views, interests, and positions of others is sometimes considered a departure from professional ethics and integrity.

However, in a political system composed of plural values, differing perceptions, and multiple goals, there is never one truly right or completely just policy option. Compromise in many respects is the key to progress in the political setting in which forest resources are immersed. Consider the following Sierra Club perspective on compromise (McCloskey 1977, p 20):

> Compromise is the key because it is the means by which legitimate interests in a democracy come to understand that they are being given fair consideration. Accommodations among contending interests build confidence that are our institutions are listening. If we want these considerations for ourselves, we must accord them to others.

Logrolling Another time-honored strategy for securing agreement is *logrolling,* which takes the form "You give me what I want, and I will give you what you want." Logrolling can increase the well-being of all parties involved. Policymakers can secure forest policies and programs for which they have strong preferences by simply making concessions on other programs and policies which they value less. More important, logrolling can facilitate agreement among interests with extreme positions. An environmentalist and a developer may initially have extreme positions on road construction in a forested area, but a logrolling process may help them to find ways to deal with their differences; each may have to give up less valued programs and plans in order to obtain the most desired programs and plans. Each party must know what support is expected by the other, and both parties must be willing to support one another after agreement has been reached. The norm is reciprocity. In order to secure its own programs and policies, each party must be willing to not interfere with the policies and programs most desired by the other party.

Logrolling is widely used in a variety of circumstances. For example:

> "You help us defeat proposals for mining in public parks, and we will help you defeat gun control legislation."

> "You support our efforts to secure wilderness designation on a public forest, and we will support your efforts to secure funding for intensified timber production on a public forest."

> "We will agree to legislative creation of a forest planning program if you agree to the legislative establishment of a transportation policy for federal public forests."

> "We will assist you in securing a state forest practices law if you call off your efforts to designate additional wilderness in a public forest."

> "You support my efforts to secure wildlife management funds from the regional forester, and I will support your efforts to obtain more road construction monies from the regional forester."

> "You agree to my legislative language discouraging the closure of a public forestry agency's field offices, and I will agree to support your legislative language establishing a national forestry planning program."

A simple logroll involves immediate rewards to both parties—for example, "We'll resolve our differences by agreeing that timber harvesting can be accelerated, but under certain strict environmental controls." However, logrolling can have an important time dimension. A time logroll may involve a promise of support at some future date. For example, "You support my forest pesticide policy which is before the legislature now, and when your proposal for a state board of forestry is introduced next year, I will support it." The promised future support may be for a definite policy or program, or the promise may be left indefinite: "Your support on the pesticide legislation will be rewarded by my promise of support on a future forestry proposal of your choice." Time logrolls must involve stable and reputable colleagues, who will still be around when the time comes to collect the debt. A legislator who sets up a time logroll with a colleague who is unlikely to be reelected could be making a serious mistake. An interest group that arranges a time logroll with a regional forester who is soon to retire or be transferred will be disappointed when attempts are made to seek support based on previous promises.

Creating and discharging obligations is often a product of logrolling. Policymakers may create among themselves a significant network of mutual obligations that can be called upon at will. For example, an industrial association may benefit from a conservation group's efforts to secure for the industry a favorable timber tax law. As a consequence, the association becomes obligated to the conservation group. At a future time when political support for a forest wildlife management program is needed, the conservation group may well collect on the obligation previously imposed on the association. The ability of some forest policymakers to exert an unusual amount of influence upon the policy selection process can often be traced to their uncanny ability to create and discharge a stock of obligations.

Logrolling has disadvantages, especially from a broader social perspective. For each party to get what it wants, rather than being forced to compromise, can undermine the collective nature of policy making. Each participant enters into a logroll for selfish reasons, such as to acquire more designated forest wilderness or more designated timberland, without considering the needs of the community as a whole. Furthermore, logrolling can dramatically increase the stock of selective government benefits: ''You give me my new bridge and I will give you the new campground.'' Little pockets of benefits proliferate, and little attention is given to the totality of their impact on budgets and related resources.

Side Payments An often highly publicized strategy for securing agreement is making *side payments,* which may take one of two forms—either ''You give me what I want, and I will reward you'' or ''If you do not give me what I want, I will punish you.'' Side payments can consist of a wide variety of favors, including personnel positions (say, membership on a legislative staff), agency assignments (as general counsel, for example, or agency director), and resources (taking the form of funding for new programs, perhaps, or campaign contributions). Securing agreement on highly controversial formulated policies frequently involves side payments. In such situations, if something must be taken from one group and given to another, pressure to replace what was taken away may become intense. A side payment is ideally suited for such replacement. Employment of a side-payment strategy implies that the bargainer has resources either in hand or readily accessible. For example, since conservation groups do not generally own public forestland, they cannot reward or punish their opponents by granting or withholding public forestland. Public land-owning agencies, however, can use forestland as a side payment in order to secure agreement on a formulated policy.

Here are some examples of side-payment strategies:

- A wood-based industrial association might agree to establish a program for forested natural areas involving company lands. It could designate ecologically important timberland areas (the side payment) as part of the program, in return for a conservation group's assurances that efforts to designate additional forest wilderness in public forests would be severely curtailed.
- The chair of a conservation organization's political action committee might inform a legislative candidate that failure to continue strong support for environmental forestry programs would result in drastic curtailment of contributions (the side payment) to the candidate's reelection committee.
- Legislators involved in a bitterly contested proposal for designating additional forest wilderness in a public forest might agree to the additional wilderness if funding for intensified timber management (the side payment) on adjacent state and county forestlands would then be forthcoming.
- A federal agency's proposal for funding a cost-share timber management program focused heavily on the South might not be agreed to by state foresters in the North unless significant investments in research on timber utilization of northern species (the side payment) were promised.

Combining Strategies Strategies such as compromise, logrolling, and side payments often occur in sequence. If, for example, an agency director's threat goes unheeded ("We demand a $10 million increase in the budget for the reforestation cost-share program"), compromise typically follows ("Let's all give in and agree to a $5 million budget increase"). Failing at compromise, a logrolling strategy may next be exercised ("We will agree to a $5 million budget increase in exchange for reducing our opposition to your proposed forest tax policy"). If logrolling fails, a side payment may be offered ("We will agree to a $5 million budget increase in exchange for our reallocation of $2 million in agency funds for your new forest tax program").

Personalities also play a role in determining which strategy is employed. Individuals and organizations strongly tied to an ideology may find compromise difficult but logrolling much to their liking. Highly moral persons may have great difficulty with side payments. Persons who operate with high regard for the state and its institutions may find logrolling difficult (though each proposal must be examined on its own merits) but may exercise compromise with great ease (Froman 1967).

Tactics for Securing Agreement

Tactics for securing agreement can involve harsh manipulations. Actions that result in an opponent's acceptance of a formulated policy without accommodation of the opponent's interests are often involved. *Inducement* is one example. A policymaker may attempt to secure agreement by improving an opponent's circumstances ("If you agree to my policy preference, we will provide substantial new funding for your program"). In *coercion,* a policymaker threatens to aggravate an opponent's circumstances in order to force the opponent to agree to the policymaker's preferred forest resource policy: "If you fail to agree to my policy preference, I will make sure that your preference is never realized." *Persuasion* is a strategy in which a policymaker presents arguments, often using flowery rhetoric, demonstrating that an opponent actually would be better off agreeing to the policymaker's preference: "If you agree to my policy preference [for specified reasons], you will be better off than if your policy preference were adopted." *Bluffing* and *threats* may also be used: "If you fail to support my policy preference by noon tomorrow, we will not provide the program funding you requested" (MacMillan 1978).

Mediation

Securing agreement on a formulated policy may entail confrontation with interests that have differing and intensely held views. A means of resolving such disputes is needed, and it must be a means that is short of full-scale engagement in a judicial or a comparable adversary process. One possibility is *mediation,* which has been defined as (Meeks 1985, p. 15):

> a structured process in which the mediator assists the disputants to reach a negotiated settlement of their differences. Mediation is usually a voluntary process that results in a signed agreement which defines the future behavior of the parties. The mediator uses a variety of skills and techniques to help the parties reach a settlement but is not empowered to render a decision.

A number of forest resource issues have been subject to mediation in recent years. The issues mediated have ranged from disagreements over recreational trail locations in forested areas to differing opinions on forest roads as proposed in federal forestland management plans, and from concern about aerial application of pesticides for timber-management purposes to intense controversy over forested areas proposed for inclusion in a federal wild and scenic rivers system. Agreement on actions to resolve such issues has been achieved in large measure through the intervention of distinterested third parties, namely mediators, who have the ability to structure and direct mediation processes. Mediators shoulder significant responsibility for encouraging parties to articulate their concerns, for assisting in the formulation of policy options for addressing such concerns, and for fostering access to important technical information. Mediators also play an important role in drafting final agreements and designing provisions for their enforcement (Bacow and Wheeler 1984; Bingham 1986; Mernitz 1980).

Mediation offers an opportunity to involve multiple interests in the process of policy selection. Rather than a state division of forestry and a single interest group (say, a group that is interested only in fish habitat) attempting to seek agreement on a pesticide issue, mediation facilitates involvement of a broader range of potentially affected parties (perhaps a wildlife interest group, a timber interest group, a preservation group, a pollution-control agency, and an agricultural agency). When involvement in a forestry issue is broadened, a sense of sharing power and responsibility can be fostered, a wider range of formulated policies can be presented and subsequently debated, an atmosphere of trust among interests can be nurtured, a pool of information previously not available to the parties involved can be brought forward, and a broader base of political support conducive to long-term agreement can be achieved.

Circumstances appropriate for mediation vary according to the forest resource issue at hand and the parties involved. The following have been suggested as criteria which must be met if mediation is to be successful (Cormick 1982):

- All parties having a stake in the outcome of the mediation process must be represented. No party that could prevent an agreement from being carried out can be excluded.
- All parties must have reached agreement on the scope of the issues to be addressed.

• Negotiators for each party must be able to speak for their constituencies. There must be reason to believe an agreement will be honored by the groups that the negotiators represent.

• Immediate parties and eventual decision makers must have committed themselves to a good-faith effort to reach a consensual agreement.

• Realistic deadlines must have been set for the mediation process.

• Reasonable assurance must be offered that affected governmental agencies will cooperate in carrying out an agreement.

• The mediator must operate from a base that is independent of both the immediate parties and the decision makers who have jurisdiction over the dispute.

• All parties involved must trust the mediator to carry messages when appropriate and to honor confidential remarks.

CRITERIA GUIDING CHOICE

Selection of a forest policy from among a range of formulated policies implies that there is something especially good, virtuous, or meritorious about the selected policy. But by what standards should such judgments be made? Some would argue that only when benefits exceed costs should a forest resource policy be selected. Others assert that only forest policies which maintain diversity within forest ecosystems should be selected. Still others contend that only forest policies that serve many people rather than few should be chosen. Selection of an appropriate forest policy quickly brings to center stage the subject of *criteria,* or the standards against which a range of formulated policies are compared. Without criteria, the policy selection process becomes prone to haphazard, random judgments, which works to the detriment of well-designed, effective forest policies.

Criteria that are clear, consistent, and generally applicable are useful in the selection process. They must be *clear* so that they can be applied with minimal opportunity for confusion. Of little value, for example, are criteria involving an abstract test for general welfare, the public interest, or the greatest good, all of which have a multiplicity of meanings. Criteria must be *consistent* so that they can reconcile various technical and political values. Seldom, for example, can policies produce the greatest total good and at the same time distribute that good over the greatest number of people. Criteria must also withstand the test of *general applicability* so that they can facilitate comparative judgment of a wide range of policy alternatives. Economic efficiency rules may well be capable of narrowing a range of competing timber-management policies—a field in which outputs are uniform and output measures are comparable—but may encounter obstacles when applied to policies producing a broader range of forest outputs, such as timber, water, wildlife, wilderness, recreation, and forage.

Agreement is an obvious criterion for judging the worthiness of a forest policy: if a policy is not agreed upon by the parties concerned, failure will likely occur, no matter what other merits the policy possesses. In reality, however it seldom happens that a single criterion serves as the standard for policy selection. Policies must also be meritorious in terms of technical feasibility, ecological integrity, economic efficiency, and distributional equity. Identifying such criteria, however, can be difficult. Advance presentation of standards used to judge the merits of competing forest policies is not always a policymaker's practice. Moreover, selecting a forest resource policy and then searching for criteria that support the selection occurs in the world of policy development.

In this section, we shall consider five broad categories of criteria: technical and ecological, efficiency and effectiveness, equity and ethical, values and ideologies, and procedural criteria.

Technical and Ecological Criteria

An extremely important group of criteria addresses technical and ecological relationships. Of interest is the physical ability of a forest policy to actually accomplish some desired goal or objective. Obviously, there are technical and ecological limits to what society can do with forests. Forests themselves determine the limits of policies. For example, a policy of reforesting Lake States forests with various species of southern pine is not possible: fundamental biological and ecological laws prohibit such action. Nor is it physically possible to grow sawtimber-sized trees in 1 year: to adopt such a public policy would be folly for biological reasons. A policy of introducing to Alaskan forests an exotic wildlife species native to the tropics would surely meet with disaster. Technical and ecological criteria enable policymakers to filter out formulated policies which have no biological or physical likelihood of succeeding.

Consider the following as examples of biological and ecological criteria (Worrell 1970, pp. 53–56):

- Forest policies that maintain the stability of forest ecosystems are preferred. Serious instability in forest ecosystems—overabundance or disappearance of animals, large periodic insect outbreaks, sudden mortality of a large number of trees—is not considered desirable.
- Forest policies that maintain a diversity of plant and animal life in forest ecosystems are preferred. Simplified ecosystems are not desirable because they are prone to instability.
- Forest policies that prevent nonreversible changes in forest ecosystems are preferred. Permanent destruction of flora or fauna is not socially desirable.

Technical and ecological criteria abound. They are truly the domain of the technical professional. Although such criteria provide substantial guidance to whether formulated policies are physically capable of success, they provide little guidance to the social goodness or badness of a proposed forest policy

Efficiency and Effectiveness Criteria

Forest policies must frequently meet tests of economic efficiency and effectiveness. Resources are presumed to be scarce and therefore must be allocated in an efficient manner. In addition to being efficient, formulated policies must also be effective; that is, they must accomplish the goals for which they are intended. Examples of efficiency and effectiveness criteria are (Worrell 1970, pp. 41–49):

- Forest policies that result in total net benefits exceeding costs are preferred.
- Forest policies that result in the greatest difference between total benefits and total costs are preferred.
- Forest policies that result in the greatest possible output with a given input, or that achieve a given output with the least input, are preferred.
- Forest policies that result in accomplishment of desired goals are preferred.

Efficiency criteria become operational when they are put into the form of benefit-cost ratios, internal rates of returns, present net worths, and marginal analyses. Expressing criteria in such forms is not easy, because adequate consideration of externalities and specification of all benefits and costs in comparable units of measure must be included. Despite such complications, merely arraying known benefits and costs in whatever units are feasible is of great value to policy selection.

Equity and Ethical Criteria

Forest resource policies invariably have distributional implications: some individuals or organization receive more (or fewer) benefits or pay more (or fewer) costs than others. Distributional consequences can be contemporary in nature, in that one segment of society may realize immediate gains while another becomes an instant loser. Examples are public-land allocations that benefit one group at the expense of others and public timber sales that benefit small enterprises at the expense of larger enterprises. They can also be temporal in nature, in that a disproportionate share of policy benefits is received by one or more generations. For example, postponement of old-growth timber harvesting may benefit the timber interests of future generations. They can also be geographical in nature, in that a person's proximity to the implementation of a forest policy dramatically skews benefits received or costs paid. For example, persons in close proximity to a forest recreation area in a public forest are more apt to benefit from the area than are those living at greater distances from the designated recreation area. Furthermore, they can involve an uneven distribution of the cost of implementing forest policies, as when nonresident hunting fees are higher than resident hunting fees.

The task of ethical and equity criteria is to provide a sense of reason to judgments involving the incidence of benefits and costs associated with formulated forest policies. Invariably such criteria rest on powerful ethical standards concerning race, creed, sex, and national origin. A preference for equality in terms of justice, fairness, and distributive rights is often implied or stated. These terms can have multiple meanings which are strongly held by various segments of society.

From a forest resource perspective, consider the following examples of equity criteria (Clawson 1975, pp. 112–122; Worrell 1970, pp. 49–53):

- Forest policies that provide more people an opportunity to achieve their goals are preferred to policies that give few people an opportunity to do so.
- Forest policies that result in costs being borne in the same proportion as benefits received are preferred.
- Forest policies that maximize societal net benefits (social benefits less social costs), while providing a minimal level of benefits to all, are preferred.
- Forest policies that result in everyone being better off than—or at least as well off as—they formerly were are preferred.
- Forest policies that result in improvements to some that are sufficiently large to compensate for expected losses to others are preferred.

Suggestions have been made for imposing more rigorous distributional criteria on proposed policies. Such criteria take the following form (Frohock 1979, pp. 240–250; Wenz 1988):

- *Strict equality criterion* Select a combination of forest resource policies that results in equal net social benefits for each segment of society.
- *Productive criterion* Select the combination of forest resource policies that results in the greatest total net social benefit to society.
- *Worst-off criterion* Select a combination of forest resource policies that includes that policy having the highest net benefits for the worst-off segment of society.
- *Average criterion* Select the combination of forest resource policies that has the highest average net social benefits for society.

The implications of such criteria can be demonstrated by application to selection from among three policy options, as shown in Table 5-1. Each is capable of producing varying amounts of benefits to three interested segments of society: timber interests, recreation interests, and water interests. The number in each cell represents units of value; a higher number is better than a lower number.

If a strict equality criterion were employed, policy option II would be selected. Each interested user would receive an equal share of the value produced. The total net social benefits, however, would be small (30 units) in comparison to that obtainable if policy option III were selected, according to productive criteria (215 units); but timber interests would be at a decided disadvantage, with only 5 units. All would be better off with policy option I, selected by the worst-off criterion. There would be greater total amounts of value

TABLE 5-1
Hypothetical net social benefits produced by forest resource policies, by policy option and society beneficiary

	Segment of society (Units of net social benefit)		
Policy options*	**Timber interests**	**Recreation interests**	**Water interests**
Policy option I	100	60	40
Policy option II	10	10	10
Policy option III	5	10	200

* Option I involves moderate investments in timber management, recreation management, and watershed management. Option II involves low-level investments in timber management, recreation management, and watershed management. Option III involves very low investments in timber management, low investments in recreation management, and very high investments in watershed management.

to distribute (200 units), and the worst-off segment (water interests) would be better off (with 40 units) than if a strict equality criterion were used to judge the options (which would give it only 10 units). Application of the average criterion would guide the policymaker back to policy option III—again an unsatisfactory option to timber and recreation interests, both of which have potentially better forest policy outcomes.

Within the framework of equity and ethical criteria, the concept of public interest is frequently suggested. When employed in the selection process, a policy is judged to be in the public interest if the ends of the whole public are served rather than the ends of some segment of the public. For example, a government policy requiring reforestation of all private lands might well be consistent with public-interest criteria since the policy would be beneficial to all in society even though it would not be in the interest of individual landowners, who would incur significant reforestation costs. Concepts of public interest are predicated on some definition of the ends or general goals of society. The question is, are such ends simply the summation of individual interests of all members of society, or are there some ends and interests that are unique to society as a whole—interests quite separate from interests of individuals? As might be expected, public-interest criteria are often the subject of lively policy debates (Worrell 1970).

Values and Ideologies

Formulated forest policies survive in the policy selection process in large measure because they are compatible with the values and ideologies of policymakers and the clients such policies are designed to serve. In this respect, values and ideologies become criteria for judging the worth of formulated policies. Obviously, not all interests or policymakers concerned with the use and management of forests have the same values. In many respects, however, they do share common thoughts, feelings, attitudes, and beliefs; they have feelings about public ownership of forests, government intervention in private forest management, state control of forest wildlife, and higher social values of wildlife over other forest benefits. Values and ideologies in the policy selection process cannot be ignored, nor should they be discounted. They provide a stable base against which goals and objectives of public policy can be assessed, and they set forth a historical framework within which future events can be judged.

Ideologies important to forestry professionals abound in number. Some have been systematically identified as doctrines or tenets of faith. Arguably, they include (Duerr and Duerr 1971):

- *Doctrine of timber primacy* Timber is the chief product of the forest, and all other products are secondary. Wood is necessary; it has no substitutes. Consumption of wood is assured; consumers of wood will always exist. There will be a timber shortage; central to forest management are biological and engineering problems of growing more wood. *Possible result:* a bias in favor of timber as a use of forests.
- *Doctrine of sustained yield* Current generations have an obligation to pass on to future generations forests managed at a high level and in an undiminished state. Forests should be managed so as to produce large physical quantities of forest benefits (e.g., timber), period after period into the indefinite future. *Possible result:* investment in forest practices regardless of the economic efficiency of doing so.
- *Doctrine of the long run* Forests change very slowly; they require long periods of time to fulfill human expectations. Society must adapt to this fact and curb selfish, short-sighted interests, especially those of private enterprise. The long-run future character of forests will not be much different from their past character. In fact, the forest of the future should look like the forest of the past. *Possible result:* public forest ownership favored over private ownership.
- *Doctrine of absolute standards* Forests are living enterprises with their own natural laws that must operate over long periods of time. The forest manager must be guided by such laws. People should not be trusted with natural resources; they are at a loss to understand natural laws. *Potential result:* clearcut timber harvesting practices which mimic nature.
- *Doctrine of science application* Forest-management problems can be solved by searching for and applying scientific facts. New technologies offer the only hope for resolving problems. Required is the application of tools and techniques of learned technicians. *Potential result:* application of technical solutions to potential problems.

Procedural Criteria

Criteria may be used to judge the process by which a policy is formulated, rather than the merits of a forest policy per se. If the process does not meet a specified criterion, the policy will be rejected. Examples of procedural criteria are:

- *Equal influence* Forest policies resulting from political processes in which citizen preferences are given equal weight are preferred to policies resulting from processes in which citizen preferences are differentially weighted.
- *Appropriate involvement* Forest policies resulting from processes in which all persons to be affected have been given an opportunity to express their views are preferred to policies resulting from processes in which some have not been given such an opportunity.
- *Due process* Forest policies developed by due process are preferred to policies resulting from random and arbitrary processes.

The types and number of criteria used by policymakers to judge the worthiness of formulated policies and programs can be overwhelming. Criteria exist, either explicitly or implicitly, in every organization that is responsible for the selection of forest resource policies. In large measure, an appreciation of criteria used to judge competing policies can be obtained only by review of selected examples.

From a broad policy and programmatic perspective, the criteria listed below have been suggested for judging alternative forest policies (Marty 1975). Preferred forest policies should be:

- *Congruent with established forest resource policy* Legal authority must exist, and action must be within legally set limits.
- *Neutral or beneficial in their environmental effects* Physical productivity of forest must not be impaired.
- *Economically efficient* Benefits must outweigh costs, and the scale of the policy must be appropriate.
- *Feasible within existing budget and personnel allotments.*
- *Beneficial to many clients rather than few* Those who benefit must bear an appropriate share of costs.
- *Minimal in external impacts unless compensated for by other benefits* Adverse impact on safety, health, and the environment must be minimized.
- *Low in risk of an unacceptable result* There must be a low risk of failure or significant adverse consequence.

Policies guiding the use and management of public lands should adhere to the following criteria, according to the Public Land Law Review Commission (1970):

- Protect environmental values as major permanent elements of public-land management.
- Provide for priority-setting mechanisms which will reduce conflict over the use and management of public lands.
- Ensure increased efficiency in the allocation of public-land resources and investments therein.
- Assure that no party benefits unreasonably and to the detriment of others, unless there exists an overriding national interest.
- Protect the rights of individual citizens, and assure that each is dealt with fairly and equitably.

Policies and plans selected to guide the management of public land administered by the USDI-Bureau of Land Management (1983, p. 6) should be:

- Consistent with principles of multiple use and sustained yield
- Products of processes which give the public early and frequent opportunity to participate in policy development
- Fully cognizant of physical, biological, economic, and social consequences of public land management
- Considerate of federal land impacts on adjacent nonfederal lands
- Based on reliable inventories of public-land resources, to the extent that such inventories are available
- Congruous with state and federal laws addressing air, water, and noise pollution

Nationally, USDA-Forest Service program directions for the years 1981 through 2030 should, from a social perspective, be consistent with the following criteria (USDA-Forest Service 1980):

- Maintain and enhance community viability, as indicated by changes in community income, employment, and population.
- Maintain opportunities to help the disadvantaged through programs which develop and utilize human resources.
- Dampen increases in housing costs through programs that extend and increase supplies of wood.
- Improve health and safety through programs that reduce hazards on forest and rangeland.
- Maintain outdoor recreation opportunities through programs that provide for access, activities, and resource protection.

Policies and programs selected for the management of a specific forest (for example, Beaverhead National Forest) within the USDA-Forest Service's National Forest System must be consistent with criteria such as the following (USDA-Forest Service 1986a):

- Maintain quality elk habitat capable of providing high-quality hunting opportunities and supporting herd populations without conflicting with the interests of private landowners.
- Ensure sufficient, high-quality water necessary to promote fish habitat, water-based recreation, and sustainable water supplies for municipalities.
- Assure quality primitive or semiprimitive recreational opportunities via forestland maintained in a roadless status.
- Enable opportunity for use of forage by domestic livestock.
- Facilitate development of a healthier distribution of timber age and size classes on forestlands suitable for timber management.
- Ensure road systems to be at the minimum density needed to carry out scheduled management activities.
- Protect cultural resources that have significant scientific and historical value.
- Assure opportunity for mineral exploration, consistent with applicable laws and regulations.
- Coordinate land-planning and resource-management activities with other federal, state, and local governments and with private landowners.

SELECTION IN THE POLICY PROCESS

Policy selection encompasses an incredibly complex set of activities through which policy alternatives are narrowed to the one or the select few that can be readied for implementation. In general, the process is multifaceted and thus calls for a variety of disciplines to aid understanding. It is rife with suggested decision rules and thus precludes reliance on any single set of choice criteria. It is cooperative in nature and thus relies on notions of agreement and the strategies and tactics that agreement implies. It is heavily value-oriented and thus impinges on virtually everyone's sense of well-being. The entire process is heavily involved with politics; few participants obtain the forest policy of their first choice.

The selection of forest resource policies from an array of formulated policy options occurs within government on a continuing basis. Consider a select few aspects of the selection process involving:

- strategic planning for the USDA-Forest Service Resources Planning Act Program.

As required by the Forest and Rangeland Renewable Resources Planning Act of 1974 (RPA), the USDA-Forest Service formulates a number of program alternatives or long-term strategies from which a single program (strategy) must be selected. The alternative selected is the product of significant technical analysis, extensive public review and comment, careful specification of decision criteria, and considerable negotiation within the agency and between the agency and its various clientele. Examples of the criteria guiding choice are: technical and physical attainability, efficient investment opportunity, ability to sustain community employment, and recognition of historical budget trends. The program ultimately selected is very often only slightly different from previously selected programs (Fedkiw 1982; USDA-Forest Service 1980, 1981, 1986b, 1990).

Policy selection has considerable influence over all segments of the policy process. For example, persons responsible for selection may demonstrate a clear preference for certain policy options and thus may have a significant influence on the nature of the policies formulated. In some circumstances, the reverse may be true: formulators may suggest their own forest policy preferences; formulation, in fact, may become selection.

Selection can have an impact upon implementation to the degree that policymakers wish it to do so. Implementation may be overlooked, however, because the details are too many, the issue being addressed is unfamiliar, or the fear of criticism for faulty implementation is large. This is unfortunate, since failure to adequately consider the ability to implement a proposed forest policy can result in a significant divergence between what was desired of a policy and what actually happens.

Evaluation is influenced by selection to the degree that the policy option chosen includes clear statements of intent, well-defined clients, and suggested criteria for evaluation. When such information is lacking, the task of the evaluator becomes that much more difficult.

Although the need is real, the link between selection and termination is seldom made. Termination activities are greatly facilitated if a policy option's ability to be modified or discontinued is explicitly considered during selection. Unfortunately, the thought of discussing termination, while at the same time seeking agreement on the selection of a particular forest policy, brings to the surface notions of political strife, which may stifle discussion of termination plans.

REFERENCES

Bacow, L. S., and M. Wheeler. *Environmental Dispute Resolution.* Plenum Press, New York, 1984.

Bass, Bernard M. *Organizational Decision Making.* Richard D. Irwin, Inc., Homewood, Ill., 1983.

Bingham, Gail. *Resolving Environmental Disputes: A Decade of Experience.* The Conservation Foundation, Washington, D.C., 1986.

Brewer, Garry D., and Peter deLeon. *The Foundations of Policy Analysis.* The Dorsey Press, Homewood, Ill., 1983.

Clawson, Marion. *Forests for Whom and for What?* The Johns Hopkins Press, Baltimore, Md., 1975.

Cohen, Michael, James March, and Johan Olsen. A Garbage Can Model of Organizational Choice. *Administrative Science Quarterly.* 17(March):1–25. 1972.

Cormick, Gerald W. The Myth, the Reality, and the Future of Environmental Mediation. *Environment* 24(7):14–20, 36–39. 1982.

Delbecq, A. L., A. H. Van de Ven, and D. H. Gustafson. *Group Techniques for Program Planning: A Guide to Nominal Group and Delphi Processes.* Scott, Foresman and Company, Glenview, Ill., 1975.

Duerr, William A., and Jean B. Duerr. The Role of Faith in Forest Resources Management, in *Man and the Ecosystem.* College of Agriculture and Home Economics, University of Vermont, Burlington, 1971.

Fedkiw, John. The USDA Decision Process for the 1980 RPA, in *Forests in Demand: Conflicts and Solutions,* by C. E. Hewitt and T. E. Hamilton (eds.). Auburn House Publishing Company, Boston, Mass., 1982, pp. 163–168.

Filley, Alan C. *Interpersonal Conflict Resolution.* Scott, Foresman and Company, Glenview, Ill., 1975.

Frohock, Fred M. *Public Policy: Scope and Logic.* Prentice-Hall, Inc., Englewood Cliffs, N.J., 1979.

Froman, Lewis A. *The Congressional Process: Strategies, Rules and Procedures.* Little, Brown and Company, Boston, Mass., 1967.

Gibson, James L., J. M. Ivancevich, and J. H. Donnelly. *Organizations: Behavior, Structure and Process.* Business Publications, Inc., Plano, Tex., 1988.

Hogwood, Brian W., and Lewis A. Gunn. *Policy Analysis for the Real World.* Oxford University Press, London, England, 1986.

Kaufman, Herbert. *The Administrative Behavior of Federal Bureau Chiefs.* Brookings Institution, Washington, D.C., 1981.

Kingdon, John W. *Agendas, Alternatives, and Public Choices.* Little, Brown and Company, Boston, Mass., 1984.

Lasswell, Harold D. *Psychopathology and Politics.* Viking Press, New York, 1960.

Lindblom, Charles E. The Science of Muddling Through. *Public Administration Review* 19(2):79–88. 1959.

———. *The Policy-Making Process.* Prentice-Hall, Inc., Englewood Cliffs, N.J., 1968.

Marty, Robert. Comprehensive Analysis of Public Forestry Project and Program Alternatives. *Journal of Forestry* 73(10):701–704. 1975.

MacMillan, Ian C. *Strategy Formulation: Political Concepts.* West Publishing Company, St. Paul, Minn., 1978.

McCloskey, Michael. Are Compromises Bad? *Sierra Club Bulletin* 62(2):20. 1977.

Meeks, Gordon. *Managing Environmental and Public Policy Conflicts.* National Conference of State Legislatures, Denver, Colo., 1985.

Mernitz, Scott. *Mediation of Environmental Disputes: A Sourcebook.* Praeger Publishers, New York, 1980.

Public Land Law Review Commission. *One Third of the Nation's Land.* Washington, 1970.

Reitz, H. Joseph. *Behavior in Organizations.* Richard D. Irwin, Inc., Homewood, Ill., 1987.

Risbrudt, Christopher D. The Real Issue in Below-Cost Timber Sales: Multiple-Use Management of Public Lands. *Western Wildlands.* 12(1):2–5. 1986.

Schultze, Charles L. *The Politics and Economics of Public Spending.* Brookings Institution, Washington, 1968.

Stillman, Richard J. *Public Administration: Concepts and Cases.* Houghton Mifflin Company, New York, 1983.

USDA-Forest Service. *A Recommended Renewable Resources Program—1980 Update.* FS-346. U.S. Department of Agriculture, Washington, 1980

———. *Alternative Goals: 1985 Resources Planning Act Program.* Program Aid Number 1307. U.S. Department of Agriculture, Washington, 1981

———. *Forest Plan: Beaverhead National Forest.* U.S. Department of Agriculture, Dillon, Mont., 1986a.

———. *1985–2030 Resources Planning Act Program.* Final Environmental Impact Statement. FS-403. U.S. Department of Agriculture, Washington, 1986b.

———. *The Forest Service Program for Forest and Rangeland Resources: A Long-Term Strategic Plan.* U.S. Department of Agriculture, Washington, 1990.

USDI-Bureau of Land Management. *BLM Planning: A Guide to Resource Management Planning on the Public Lands.* U.S. Department of the Interior, Washington, 1983.

Wenz, Peter S. *Environmental Justice.* State University of New York Press, Albany, N.Y., 1988.

Worrell, Albert C. *Principles of Forest Policy.* McGraw-Hill Book Company, New York, 1970.

6

LEGITIMIZING POLICY CHOICES: ACTION IN GOVERNMENT

Forest resource policies may be viewed as desirable, may be agreed to, and may be selected. But selection is no guarantee of implementation and ultimate attainment of laudable goals. If an agreed-to policy is to be placed on the track to success, it must be given some official status—the policy must be *legitimized.* It must be formally approved by a person—a forest supervisor, an agency head, a legislator—or an organization in a position of power. It must also be made official—say, as a statute, a regulation, or a judicial ruling. Once legitimized, the selected policy becomes genuine or authentic, and the organization responsible for the forest resource issues to which the policy is addressed can proceed with implementation. For many, legitimation is a major payoff in the often long and arduous process of policy development. It is the point at which reference can be made to a specific course of action to which an agency will direct attention in earnest—maybe. This chapter describes the different styles and forms for legitimizing policies, as well as various approaches to assuring the political feasibility of legitimizing a forest resource policy.

STYLES AND FORMS

Forest resource policies can be legitimized in various ways; the manner employed determines the actors involved. If legislative legitimization is required, authority rests with the unified action of legislators. Although legislators may be subject to persuasive actions by bureaucrats, lobbyists, and legislative staff, ultimate responsibility for legitimizing a policy selection in law rests with the lawmakers themselves. As an example, the following portion of the Wild and Scenic Rivers Act of 1968 legitimized in statute a policy on the preservation and management of free-flowing, wild rivers and their forested environs (Shannon 1983, p. 231):

> Selected rivers of the Nation which, with their immediate environments, possess outstandingly remarkable scenic, recreation, geologic, historic, cultural, fish and wildlife, or other similar values, shall be preserved in free-flowing condition, and they and their immediate environments shall be protected for the benefit and enjoyment of present and future generations.

Forest resource policies may also be legitimized by action of judicial systems, notably local, state, and federal courts. Because of increased litigation involving social and natural resource issues, courts have set forth an ever-increasing number of decrees that legitimize forest policies. Such decrees may further legitimize actions of other policymakers in government, or may declare selected policies to be unconstitutional or in violation of existing law, thus relegitimizing previously established policies. A far-reaching judicial finding of error in a policy and a reaffirmation (relegitimizing) of originally intended timber management policy for the National Forest System as administered by the USDA-Forest Service resulted from litigation involving the West Virginia Division of the Izaak Walton League of America and the U.S. Secretary of Agriculture. A portion of the ruling which legitimized a revised timber-management policy states (Shannon 1983, p. 437):

> The court declare[s] that the practice, regulations, and contracts of the Forest Service which (1) permit the cutting of trees which are not dead, mature or large growth, (2) permit the cutting of trees which have not been individually marked and (3) allow timber which has not been cut to remain on the site violate[s] the provisions of the Organic Act. The court enjoins the Forest Service from contracting or otherwise allowing the cutting of timber in the Monongahela National Forest in violation of the Organic Act. The order further requires the Forest Service to revise its regulations. . . .

Forest resource policies can also be legitimized through regulations developed by administering agencies. As the result of a threefold process, federal regulations, for example, are published in the Code of Federal Regulations. An agency first publishes in the Federal Register a notice of intent to issue a regulation. A 3-month period is allowed during which affected parties may suggest content for the regulation. The agency then issues a draft regulation; again affected parties have 3 months to comment. After considering the comments received, the agency issues final regulations which have the force of law, a force derived from statutes enacted by the U.S. Congress. Examples of policies legitimized via regulations are those prepared by the Council on Environmental Quality for purposes of implementing the National Environmental Policy Act (NEPA) of 1970. A portion of the regulations require agencies to (Shannon 1983, pp. 252–283):

- Integrate the NEPA process into early planning.
- Emphasize interagency cooperation before an environmental impact statement is prepared.
- Issue swift and fair resolutions of agency disputes.
- Employ scoping processes for early identification of issues requiring environmental impact statements.
- Integrate NEPA requirements with other environmental review and consultation requirements.
- Define categories of actions which do not individually or cumulatively have a significant effect on the human environment and thus are exempt from environmental impact statement preparation.

Policies may also be legitimized via executive order at both state and federal levels. In 1965, for example, President Lyndon B. Johnson issued an executive order establishing reasonable user fees for certain recreational areas and facilities managed by federal agencies. Recreational user fees are to be imposed if the following conditions exist (Shannon 1983, pp. 211–214):

- The area is administered primarily for scenic, scientific, historical, cultural, or recreational purposes.
- The area has recreational facilities or services provided at federal expense.
- The area is such that fee collection is administratively and economically practical.
- The area is administered by any of the agencies specified in the Executive Order, including the USDI-National Park Service, the USDI-Bureau of Land Management, the USDI-Fish and Wildlife Service, the USDA-Forest Service, and the Tennessee Valley Authority.

An especially abundant source of legitimized forest resource policies exists in the USDA-Forest Service Manual and handbooks. The Manual presents the authorizations, objectives, policies, and responsibilities that govern USDA-Forest Service functions. The handbooks contain administrative and technical procedures for performing such functions. Examples are the Land Use Planning Handbook, the Wildfire Prevention Handbook, and the Timber Sale Accounting Handbook. The Manual contains a wide variety of policy statements—for example, policies on how grants are to be made to states for Youth Conservation Corps purposes, policies governing the sale and disposal of timber produced by the National Forest System, policies on the use of Woodsy Owl and Smokey Bear symbols, and policies guiding the management of areas designated as part of the National Wilderness Preservation System. Here is an example of a legitimized policy—the National Wilderness Preservation System policy concerning the reintroduction of wildlife species (USDA-Forest Service 1988, FSM 2323.33a):

> *Reintroductions.* Reintroduce wildlife species only if the species was once indigenous to an area and was extirpated by human induced events. Favor federally listed threatened or endangered species in reintroduction efforts. Reintroductions shall be made in a manner compatible with the wilderness environment. Motorized or mechanical transport may be permitted if it is impossible to do the approved reintroduction by nonmotorized methods.

Another example of a forest resource policy legitimized in the Manual concerns use of prescribed fire on National Forest System lands (USDA-Forest Service 1988, FSM 5140.3):

> *Prescribed Fire.* The following statements apply to all users of prescribed fire . . . , regardless of the type of ignition: 1. Prepare a prescribed fire plan for all prescribed burning projects in advance of ignition; 2. Conduct each prescribed fire in compliance with approved plan; 3. Use only trained and qualified personnel to execute each prescribed burn; 4. Base size of organization needed to safely achieve prescribed fire objectives on the size and complexity of each project; and 5. Address [in plan] appropriate actions to take if on site conditions change. A prescribed fire that . . . cannot be returned to prescription with project funds is a wildfire, take appropriate suppression action.

The appropriateness and character of material—including statements of policy—to be included in the USDA-Forest Service Manual and handbooks are guided by an elaborate agency directive system. The directive system is the primary basis for the management and control of all internal programs and the primary source of administrative direction (including legal authorities, management objectives, policies, standards, and procedures) for USDA-Forest Service employees. It incorporates policies originating outside the USDA-Forest Service as well as internally generated statements of policy. Material to be included in the Manual must clearly set forth desired results (while limiting procedural details); must rely on the judgment of field professionals; must limit explanatory material, philosophy, and history; must encourage a management-by-objectives approach; must be the minimum necessary to provide administrative control and accountability; and must not constrain the formulation of alternatives as called for by USDA-Forest Service planning programs. Inclusion of a forest policy statement in the Manual implies that the statement is needed, that it conforms to current management policy and direction, and that it does not impose an unreasonable administrative burden (USDA-Forest Service 1988, FSM 1100). Supplemental Manual and handbook directives are prepared for various administrative levels within the agency, including the Washington office as well as regional, forest, and district offices.

Forest policies are also legitimized in forest resource plans. From a statewide perspective, consider the following examples of policies legitimized within the forest resource plan for New Hampshire (Division of Forests and Lands 1982).

- Inventories of the state's productive forestland should be carried out, and an accurate assessment of the state's forest base should be made.
- Forest landowners should be informed about the benefits of woodland management, and the general public should be informed about the many values provided by forest resources.
- Wood-based industry should be strengthened by increasing secondary manufacturing within the state, by expanding markets for low-grade wood, and by stimulating a short-term export market for quality material.
- State government should assume a leadership role in improving interagency coordination and forest resource management and planning.
- Forestry research efforts should be channeled to areas of greatest need, and the results of research should be applied as appropriate.

Land-management plans as required by the National Forest Management Act of 1976 contain innumerable statements of forest policy. For example, among the legitimized policies set forth in the forest plan for the Deerlodge National Forest are (USDA-Forest Service 1987, p. II-1):

- Restore damaged riparian zones.
- Facilitate development of mineral resources.
- Provide for the geographic distribution of a variety of hunting recreation opportunities.
- Follow management practices that will contribute to the longevity of threatened and endangered species.
- Maintain adequate areas for quality motorized and nonmotorized recreation.
- Emphasize cost efficiency in timber management and engineering activities.
- Protect resource values through integrated pest management.

The advent of modern state forest practice laws often entails the promulgation of rules and regulations that legitimize agreed-to policies concerning timber-harvesting practices on private forestland. Consider the following legitimated policy concerning satisfactory reforestation of clear-cuts in Washington (Washington Forest Practice Board 1976, p. 38).

> Satisfactory reforestation of a clearcut harvest occurs if within three years of completion of harvest, or a period of one to five years . . . in the case of a natural regeneration plan, the site is restocked with . . . 300 well-distributed, vigorous seedlings per acre of commercial tree species.

Similarly, the California Board of Forestry legitimizes forest policy concerning the conversion of timberland to nontimber uses by stating (California Department of Forestry 1983, p. 80.40):

> Any person, firm, corporation, company, partnership or government agency owning timberland for which the timberland owner proposes conversion [to nontimber uses shall apply to the Director, Department of Forestry] for issuance of a Timberland Conversion Permit. . . . No timber operations or other conversion activities shall be conducted on timberland which is proposed to be converted to a use other than the growing of timber unless a conversion permit has been issued by the Director [Department of Forestry] or the Board.

As can be seen from these examples, forest policies can be legitimized in various forms by a wide variety of actors. Typically, however, legitimation is performed by persons in positions of power, such as legislators and executives. Via constitutional mechanisms, citizens grant such persons authority to legitimize policy and to carry out the programs which flow therefrom. A number of states, however, have mechanisms by which citizens can take direct action to legitimize policy selections. For example, states employ referenda for some policy decisions (e.g., bonding authority for investments in forest recreation). In such an instance, a legislature or other authoritative body asks for citizen vote on an issue. This happens when an issue is viewed as so important, or so politically charged, that voters must make the decision to legitimize. A more extreme action than a referendum is an initiative, which enables voters not only to judge the merits of an issue put to them by government but also to place issues on the ballot (Forestry Initiatives Study Group 1990).

POLITICAL FEASIBILITY

Legitimizing a forest resource policy can be a routine, nearly mundane phenomenon—or it can a highly charged political event requiring careful management. Buried within a large agency, a natural resource professional assigned the task of writing conventional and highly predictable regulations may see little need to be concerned with lobbyists, political appointees, and legislative aides. Such people are unlikely to upset the legitimizing of routine regulations. However, routine is not always the case. The legislator, the bureau chief, the forest supervisor, the state forester, or the district ranger may find that the legitimizing of forest policies is hindered by political difficulties. To the political strategist interested in legitimizing a policy, concerns about the efficiency and effectiveness of a forest resource policy may be less important than knowledge of who will support the proposed policy and who will oppose it.

The concept of *political feasibility* evokes many definitions (May 1986). Political feasibility can be viewed as the probability that a selected forest resource policy will receive sufficient political support to be legitimized and subsequent implemented. Like boating enthusiasts, policymakers wish to know the chances for clear sailing. Probability assessments can be made in the context of policy formulation, which often results in reformulation to improve chances of approval. They can be made in the context of timing; that is, might the chances of successful legitimation be greater if a forest resource policy were pushed at a different time? An example of probability calculations entering into the legitimation process might be a state forester who queries the forestry voting records of legislators prior to proposing a new and quite significant policy focused on nonindustrial private forests. The administrator of the U.S. Environmental Protection Agency might assess the political feasibility of congressional approval of policies for the management of atmospheric pollutants by preparing a map depicting which policies might be favored by each state's congressional delegation. A forest supervisor charged with preparing a forest land-management plan might informally suggest to a community of interest groups a number of strategic management directions, with the intention of assessing the intensity of support for and opposition to the informally proposed alternatives.

Political feasibility can also be viewed in terms of a political price or opportunity cost—actions that diminish a policymaker's store of political capital, such as prestige, influence, commitments, and friends. Actors in the legitimation process may impose an excessive number of commitments or obligations on a policymaker, or the cost to a policymaker of not giving attention to more profitable issues may become too high, or the risk of irritating dependable supporters and generating new political enemies may become excessive. A political price may also be paid when a policymaker agrees to support related, less desirable policies in order to secure legitimation of a favored forest resource policy. For example, a policymaker might agree to maintain a specific forestry field station in return for approval of a plan to manage water pollutants originating from forestland. A selected policy may be feasible in terms of its probability of being legitimized, but the political costs of legitimation may be so large that the policymaker's support must be withdrawn; the path to legitimation may become too conditional.

Assessing the political feasibility of legitimizing a selected forest policy is not easy. Unfortunately, the task is not facilitated by the fear of appearing to be excessively political—a fear that is prevalent among policymakers. Because a rational management image is the ideal, policymakers attempt to separate political knowledge from economic and program analysis. In many natural resource agencies, policy analysis staff are separated from political appointees, legislative liaison staff, and legislative strategy groups. Though such separation may encourage thorough analysis of a policy's technical and economic merits, it does little for careful assessment of political feasibility—a crucial ingredient of the policy development process.

There are several criteria for judging political feasibility from an agenda-setting perspective, including: (1) whether relevant officials pay attention to a forest resource policy that has been selected, (2) whether political circumstances are right (i.e., whether a window of opportunity exists), and (3) whether the source of the policy is deemed to be of high quality. If a policy proposal originates solely from within a public agency, failure to include outside interests may doom the legitimizing process. In the early 1980s, for example, federally initiated proposals for selling certain federal lands were vigorously opposed by various interest groups (Lewis 1983). A slightly different scenario might have occurred if such proposals had originated in response to group initiatives.

Feasibility of legitimation can also be judged by assessing the likelihood that interest groups will become involved in the process. Such an assessment requires knowledge of the position each relevant group is likely to assume toward the selected policy, the ability or power of a group to influence the process of legitimation, and the degree to which the group cares about or has an interest in the policy. Not all interest groups will have a stake in the issue; therein lies the hazard of generalizations about interest groups' motives and intentions. (Use of terms such as "the environmentalists," "the developers," and "the timber interests" often indicates excessive generalization). Likewise, an interest group's ability to influence legitimation may vary from one policy subject to another. A wood-based interest group capable of derailing a proposed federal land-management policy proposed by the USDA-Forest Service may well be stymied when efforts are directed at federal environmental policy proposed by the U.S. Environmental Protection Agency (EPA).

Organizing Information

Ability to wisely judge success or failure in legitimizing a forest policy hinges on access to readily available political information. Such information can take many forms and will vary in content and importance from one circumstance to another (Meltsner 1972).

Actors and Organizations Of special importance is identification of *actors and organizations* likely to be involved. If a policy to discourage the use of chemicals for forest insect management is proposed for legitimation, who will support the proposal and who will oppose it? What organized interests are likely to become involved? Who within government might have a stake in the issue? What about the elected official, the newspaper editor, the head of a chamber of commerce, the educational administrator, or the leader of a professional association? "At first there are friends, enemies, and fence sitters. Soon the dynamics of politics pushes the actors to take sides. Those who are for and those who are against become identified" (Meltsner 1972, p. 860). Identifying actors is not always easy. Clues to the reputations of actors may often be obtained from informal contacts, records of past decisions, and presentations at public hearings.

Motives Knowledge of the *motives of actors* is another type of information needed in assessing the possibility of legitimizing a policy. All actors, organizations, and individuals have a range of goals, objectives, needs, and desires on which preferences for or against forest resource policies are based. Motives are highly individual and vary with time and place. Organizationally, interest groups may seek greater influence over the development of policies, while bureaucracies may covet new programs and the resulting public visibility. Individuals may have well-defined career expectations within an organization, as well as a desire for greater access to power and a larger share of an agency's budget. Individual motives are also driven by ideologies and preferences for certain policy directions (e.g., reduction in governmment's presence in private forestry). If the motives of organizations and individuals can be identified, they may well be subject to manipulation in manners favorable to legitimation. Political strategies of various sorts (such as compromise, logrolling, and side payments) can be used to secure an opponent's support for legitimizing a selected forest resource policy.

Available Resources Once actors are identified and motives known, the *resources available* must be appraised. Which actors have access to significant material and human resources—say, interest groups capable of mobilizing a large number of members or political action committees with substantial financial reserves? Do the actors include people with important symbolic powers, such as a professional held in high esteem or a greatly admired administrator? Are some actors—for example, the chair of a legislative committee, the head of an agency, or the supervisor of a forest—able to use their positions to advantage? Do some actors, such as expert lobbyists, control access to special information? Are some actors especially skillful in gathering support and assembling coalitions?

Site Another important factor in organizing information about political feasibility is *site*—where the legitimizing activity is to take place. Is the legitimizing process a function of a legislative committee, an agency head, or an area forestry director? Or will responsibility rest with many individuals, all of whom are geographically disbursed? An example would be the legitimizing of a multitude of individual forest plans by an agency with national responsibilities. The site provides a focus for the key actors, for their motives, and for their resources. If the selected policy is to be legitimized by a state legislature, a particular set of actors and their associated traits will be involved. If the policy is to be legitimized by five different district foresters within a state, geographically local actors and traits will become important.

The gathering of information about political feasibility can be viewed as an exercise in political mapping. Attempts are made to portray factors and relationships which are likely to affect the process of legitimizing a selected forest policy. Because such information usually depicts alignments at a point in time, it often fails to portray the dynamic nature of political feasibility. For example, what happens to political feasibility when the substance of a selected forest policy is changed? If the substance remains the same, what strategies are needed to change the landscape of political feasibility? Access to models that predict the political feasibility of policy proposals are yet a dream. Efforts to develop such models are, however, under way (Kelman 1981, May 1986).

Strategies and Tactics

The legitimizing of forest resource policies requires deployment of various *strategies,* defined as large-scale, future-oriented plans or actions, and *tactics,* defined as specific narrowly focused political actions.

Consider a plan involving a perfect forest policy and an imperfect strategy for legitimation. A policymaker may select a forest resource policy that meets all technical and economic standards but may fail to select a complementary strategy for legitimizing the policy. For example, the policy might be, "We will manage forest insects via aerial application of insecticides," and an appropriate strategy would be to build coalitions of individuals and organizations necessary to secure approval of the policy. If such a strategy is not included in the plan, perhaps the policy will be legitimized on its own merits; more likely, however, legitimacy will not be granted.

Equally disastrous is a plan that combines an imperfect forest resource policy with a perfect strategy for legitimation. The policymaker may implement well-designed political strategies that result in policy acceptance with little or no opposition. Unfortunately, the technically flawed policy may be totally inadequate to deal with the issue at hand. When this happens, a catastrophe may result.

The combination of a perfect policy with a perfect legitimizing strategy is much preferred. However, in the world of politics, this perfect combination is seldom encountered (Jones 1984).

Political tactics for legitimizing can take various forms. If, for example, legislative action is required and opposition by key actors within and outside the legislative area is expected, political tactics involving compromise, logrolling, and side payments may be in order. A wise legislative tactician may be able to redefine a proposed law so as to avoid hearings by unfavorable legislative committees, and may be able to include language to make the proposed forest resource policy agreeable to interest groups that threaten to prevent enactment. The astute tactician may also be able to combine groups into coalitions that can achieve policy legitimization. Designing effective coalitions requires patience and a unique ability to anticipate the precise makeup of coalitions that can achieve legitimization at a decisive moment. For instance, a legislative committee's vote and a bureau chief's approval may be needed at just the right time. Recognizing and addressing political feasibility early in the process of policy formulation is also wise. Implied in the selection of a forest policy may be agreement among various actors that the policy should (and will) be legitimized. Such agreement may result from intense bargaining, whereby policy formulators, legitimizers, and other relevant groups and individuals hammer out differences and ultimately agree upon a salable product.

Political strategies may entail the construction of an organizational framework that facilitates legitimization of a forest resource policy. For example, an interest group wishing to legislatively legitimize a series of economic development policies based on a state's timber-based sector may propose to a key state legislator a number of arrangements, including (Buckler 1980):

- *A special joint legislative committee on timber development, assisted by a citizens advisory group.* The committee would be composed of the chairs of policy and appropriation committees of the state house and senate. The function of the committee would be to draft omnibus timber development legislation.
- *A citizens task force on timber development, convened by the legislature* The committee would be composed of broad-gauged representatives of the forest resources community. The function of the task force would be to design comprehensive timber-development legislation for recommendation to legislative policy committees.
- *A forest resources consultant, assisted by a citizen's advisory group* The function of the consultant would be to develop suggested timber development legislation for recommendation to legislative policy committees.
- *A division of Forestry, Department of Natural Resources, normal budget and policy development process* The function of the division would be to design timber-development legislation and related appropriations for recommendation to the legislature.
- *A legislative committee hearing process* Appropriate legislative committees and staff would gather public testimony and design timber-development legislation for processing by the legislature.

As implied by such arrangements, the process of legislatively legitimizing a forest resource policy may be long and cumbersome. A policymaker's job would be substantially simplified if a very powerful legislator agreed to deliver the votes needed for legislative enactment. Unfortunately, the world of policy development is not so simple.

LEGITIMIZING IN THE POLICY PROCESS

Legitimizing forest policies and programs provides the proper or lawful base for their existence. When policy statements are embodied in rules, statutes, or legal judgments, they are given an aura of authenticity which facilitates their travel through later stages of the policy process. The legitimizing process is not for the politically naive. In fact, it makes the politics of formulation and selection seem tame by comparison. Administrators who are able to wisely judge the political feasibility of legitimizing a forest resource policy or program have a significant edge in assuring that later stages of the policy process will be effectively carried out.

Forest resource policies and programs are legitimized in a variety of manners by various participants and organizations. consider the following examples of legitimizing processes and products:

- *Legitimizing the proposed Small Tracts Act of 1983* In response to the need for administrative guidance concerning the sale or exchange of certain parcels of national Forest System land, proposed policies were set forth and ultimately agreed to and legitimized in law. The process of legitimizing involved consideration by various congressional committees and required testimony by USDA-Forest Service and U.S. Department of Agriculture officials. The legitimized law was amended numerous times, involved numerous committee mark-up sessions, and was further subject to legitimizing action via agency rules that defined public interest, procedures for determining value, and location consideration for lands to be exchanged (USDA-Forest Service 1986).
- *Legitimizing the National Forest Management Act of 1976* In response to concerns over the use and management of National Forests, various laws were proposed to address such concerns. Actions by various congressional committees reflect the variety of proposals and counterproposals which were suggested and ultimately disposed of during the process of legitimizing. Support necessary for legitimizing the latter had to be secured from the USDA-Forest Service, the forest products industries, various organized interest groups, and parties within Congress that often had opposing views (American Enterprise Institute for Public Policy Research 1976 and Committee on Agriculture 1976).
- *Legitimizing rules and regulations to implement the National Forest Management Act of 1976* In response to the National Forest Management Act of 1976, the USDA-Forest Service undertook to legitimize rules and regulations required to implement the act. Working with the formally designated Committee of Scientists and providing significant opportunity for public comment, the agency developed the agreements and political support necessary to formally issue (legitimize) rules via the Code of Federal Regulations (USDA-Forest Service 1989).

The legitimizing of policies and programs has implications for other parts of the policy process. Most obvious is the reality that a poorly designed approach to legitimizing—one which fails to recognize political feasibility—can force return of a proposed forest resource policy to the selection, the formulation, and even the agenda-setting stage. If a policy cannot be legitimized politically, additional alternatives must be formulated or persons responsible for selection must reconsider policy options previously discarded. As for implementation, legitimizing actions can be facilitators, since people who are responsible for implementation are likely to respond positively to a "law," an "executive order," or a "judicial ruling.," Legitimizing can also foster evaluation of programs by giving license and administrative force to the work of analysts.

REFERENCES

American Enterprise Institute for Public Policy Research. *National Forest Management Proposals: Legislative Analyses.* Washington, 1976.

Buckler, Robert C. *Strategic Options for Developing Timber Policy Legislation by the Minnesota State Legislature.* Minnesota Forest Industries Association, Saint Paul, 1980.

California Department of Forestry. *Forest Practice Rules and Regulations.* California Administrative Code, Title 14. Natural Resources, Sacramento, 1983.

Committee on Agriculture. *Business Meetings on National Forest Management Act of 1976.* U.S. House of Representatives, 94th Congress, Washington, 1976.

Division of Forests and Lands. *Forests and Forestry in New Hampshire: Action Program for the Eighties (Forest Resources Plan).* Department of Resources and Economic Development, State of New Hampshire, Concord, 1982.

Forestry Initiatives Study Group. New Directions for Private Forests in California. Choices Offered by 1990 Ballot Propositions. Report No. 1. Department of Forestry and Resources Management. University of California, Berkeley, 1990.

Jones, Charles J. *Introduction to the Study of Public Policy.* Brooks/Cole Publishing Company, Monterey, Calif., 1984.

Kelman, Steven. *What Price Incentives? Economists and the Environment.* Auburn House Publishing, Boston, Mass., 1981.

Lewis, Bernard J. *The Reagan Administration's Federal Land Sales Program: Economic, Legal and Jurisdictional Issues.* Staff Paper Series Number 37. Department of Forest Resources, University of Minnesota, Saint Paul, 1983.

May, Peter J. Politics and Policy Analysis. *Political Science Quarterly.* 101(1):109–125. 1986.

Meltsner, Arnold J. Political Feasibility and Policy Analysis. *Public Administration Review.* 32 (November/December). 1972.

Shannon, Richard E. *Selected Federal Public Wildlands Management Law,* volumes I and II. Montana Forest and Conservation Experiment Station, University of Montana, Missoula, 1983.

USDA-Forest Service. *A Case Study of Forest Service Legislation in the Legislative Process: The Small Tracts Act.* Legislative Affairs. U.S. Department of Agriculture, Washington, 1986.

———. *Forest Plan: Deerlodge National Forest.* U.S. Department of Agriculture, Butte, Mont., 1987.

———. *Forest Service Manual.* U.S. Department of Agriculture, Washington, 1988.

———. *National Forest System Land and Resource Management Planning.* Chapter II, Title 36, Code of Federal Regulations. Washington, 1989.

Washington Forest Practice Board. *Washington Forest Practice Rules and Regulations.* Olympia, Wash., 1976.

7

IMPLEMENTING POLICIES AND PROGRAMS: GOVERNMENT TO ISSUES

The formulating and legitimizing of forest resource policies are necessary for the accomplishment of agreed-to ends. They are not, however, processes to be viewed as sufficient in and of themselves. If forest policies and programs are to create lasting or fundamental changes in the use or management of forests, they must be successfully *implemented.* For example, an agreed-to policy of increasing the export of timber products to foreign countries is effective only to the extent that someone or some organization effectively carries out the policy. Similarly, the interest group that successfully secures enactment of a state forest practice law that calls for reforestation of private land is successful only to the extent that a sympathetic agency accomplishes the intent of the law. An administrator who secures agreement on policies contained in a statewide plan for forest wildlife is successful only to the extent that execution of the plan is satisfactory to interested clients.

In this chapter, we shall consider the implementation process, especially interpretation, organization, and application; the conditions that foster successful policy and program implementation; and the nature of individuals and organizations involved in implementation.

INTERPRETATION, ORGANIZATION, AND APPLICATION

Implementation involves the translation of broad statements of policy into specific actions and activities. A variety of activities and a number of program elements (e.g., design of new and overhaul of existing administrative structures, reinforcement of political support, attainment of clearances and permits, and hiring or reassignment of personnel) must be gathered and meshed. The process can be viewed from a number of vantage points. It can be considered an evolutionary process in which problems and new circumstances are addressed and readdressed until the intent of a forest resource policy has been achieved. Conditions not foreseen during policy formulation and selection are accommodated as they occur, and the implementation evolves as specific implementing actions are undertaken.

Implementation can also be viewed as a rigorous planning and control activity that involves administrative superiors and their subordinates. Goals are perceived as clear, the means of achieving them as highly rational, and implementors as incapable of misinterpreting instructions. Commands are given, actions are taken, and accountability is rigorously enforced.

From a third vantage point, implementation can be viewed as a highly charged political process involving bargaining, accommodation, and extensive give-and-take. The forest resource policy to be implemented is simply a point of departure for bargaining among many implementors. The importance of goals and plans is minimized, and accountability is assured by public outcries of displeasure (Pressman and Wildavsky 1984).

Regardless of the vantage point chosen, implementation can usually be separated into three major activities: interpretation, organization and application (Jones 1984).

Interpretation

Translation of often vague and ambiguous policy language into acceptable and workable directives involves *interpretation,* the process by which administrators seek to decipher the specific intent of laws, rules, and regulations. Because the intent is often expressed in lofty terms and ambiguous language, successful interpretation relies on careful assessment of records compiled during the formulation and legitimizing processes, such as public hearing records and legislative conference reports. Contacts with the actors involved in such processes must also be assessed, and in fact, such contacts are often made for the purpose of seeking advice and determining intent. The task of interpreting significantly amended or newly established forest resource policies may be considerable, because the interpreters have no experience with the policy. Compounding the task may be the recurrence of conflicts that were resolved during the formulation, selection, and legitimizing phases—conflicts that may grow in intensity as the implementing language becomes progressively clearer. A major task of the forestry professional as implementor may be to curb deterioration of the political consensus that produced the initial forest policy—not an easy task, since building consensus is often easier than maintaining it.

Organization

Designating administrative units to put the interpreted language into effect is the *organization* phase of implementation. The organizational assignment continuum may include several phases. The first is reliance on an old-line agency that continues to administer increments of forest policy as interpreted over time, as when state departments of conservation implement modifications in public timber sale policies. Second, a new unit may be created within an existing old-line agency—for example, a new bureau of program planning within an existing state division of fisheries and wildlife. Third, the status of an agency may be upgraded, in line with added or refocused responsibilities; departmental-level status may be assigned to a pollution-control and environmental protection bureau, for instance. Fourth, reorganization of existing agencies is a possibility—say, when the forest-based economic development responsibilities of several agencies are combined into a single agency.

Assignment of an interpreted forest policy within an agency or among a group of agencies can make a great deal of difference in how the policy is implemented. Promoters of a particular policy are continually on the lookout for sympathetic agencies and sympathetic actors within agencies, because victories gained in the selection and legitimizing processes can be short-lived if the interpreted policy is assigned to an unsympathetic organization.

Application

Once a forest resource policy has been interpreted and an administrative unit assigned to its implementation, the process of *application* begins. This process involves the actual provision of services, payments, management activities, or other agreed-upon actions called for by the specifics of the implementing language. Application is a dynamic process in which the forestry professional is guided by agreed-to program directives and by circumstances encountered in real-life situations. For example, the rules and regulations that interpret policies concerning prescribed wildfire in a forested setting of a public forest may have to be revised after the occurrence of large, uncontrolled wildfires. Similarly, the forest landowner's failure to respond to an offer of cost-share payments for reforestation may be of sufficient concern to warrant an upward revision in the government-offered payments.

Guidelines

Interpretation, organization, and application invariably call for the development and legitimization of *guidelines,* or rules and regulations. Guidelines translate general statements of forest policy into specific prescriptions for administrative action. They define procedures and standards of behavior to be followed by those responsible for application of a forest policy. Legitimized guidelines take many forms, including formal rules and regulations (the Code of Federal Regulations is an example), handbooks and manuals (for example, the USDA-Forest Service Manual and handbooks), agency memoranda, and opinions of an agency's legal counsel. Preparation of guidelines often requires the time and effort of forestry professionals throughout a natural resource agency, as well as people not specifically charged with rule-making responsibilities, such as interest groups and individual citizens.

Implementation presented as a series of logically correct, sequential actions can convey a false impression of the process. Implementation is best regarded as a process of interaction and negotiation between parties who are seeking to put a forest resource policy into effect and those who are likely to depend on the outcome of the policy. For example, charged with implementing a newly enacted timber sale law, the central office of a state forestry agency may well be faced with a lengthy process of proposing, negotiation, and revising rules and directives, working with a multitude of administrators at various levels within the agency. A regional forester may react negatively and may negotiate a satisfactory change with the state forester, only to discover that the district forester who reports to the regional forester expresses great concern over the new rules. A district forester who is comfortable with the new timber sale rules may be unpleasantly surprised to find out that the timber sale officers in the district are unable to live with the new rules. Even this may not be the end of the episode: the loggers who must ultimately abide by the implemented policy may well find cause for taking action to modify or redirect the policy or the rules designed to achieve its intent.

Given such circumstances, implementation is best viewed in the context of interactions among administrators of forest resource programs–administrators who may have different values, perspectives, and priorities from one another and from those who advocate the forest policy in question.

SUCCESSFUL IMPLEMENTATION

Successful program implementation is dependent on a number of conditions: vague and general goals must be specified in operational terms, action programs must be designed and responsibility for their execution assigned, and sufficient resources must be allocated to support the actual accomplishment of policy intent. Such conditions are easy to list but difficult to accomplish. Although the manipulation of physically and ecologically complex forests is an obvious challenge to implementation, challenges of an institutional nature are equally awesome. For example, public agencies and the forestry professionals employed therein may resist implementation because they view the substance of a forest resource policy as unimportant, the ends to be achieved as inappropriate, and the means to accomplish them as inefficient. To some the implementation process is a creative act, while to others it is an act of destruction. Approval of a permit for strip mining in a forested area is creative in the eyes of geologists, and mining engineers, but wildlife and timber managers consider it destructive. Failure to achieve effective implementation can also be attributed to attempts to do too much or to reach too many goals, to assignment of implementing activities to ill-suited or unsympathetic agencies, to resources woefully inadequate for effective implementation of a forest resource policy, to imposition of crippling constraints such as fiscal austerity, to adverse judicial ruling, or to an array of detailed operational level problems that simply overwhelm the forestry professional's ability to carry out important tasks in the field.

Implementation is no easy task. Considered below are some major conditions that can influence its successful accomplishment (Brewer and deLeon 1983, Hogwood and Gunn 1986, Bardach 1982).

Clarity of Intent

Successful implementation implies *clarity of intent*. The more precisely a forest resource policy's intention is stated—whether in statute, judicial rulings, or administrative directives—the more likely it is that the policy will be implemented as originally desired. From a legislative perspective, the difference can be glaringly obvious. For example, the Multiple Use-Sustained Yield Act of 1960 directs the USDA-Forest Service to "administer the renewable surface resources of the national forests for multiple use and sustained yield of several products and services therefrom." (Obviously a very ambiguous directive.) In contrast, the National Forest Management Act of 1976 requires of the USDA-Forest Service that, "prior to harvest, stands of trees through out the National Forest System shall generally have reached the culmination of mean annual increment of growth." (Obviously a more direct statement of intent.)

That such differences occur should be no surprise. Long and often heated political debates in a legislative setting are prime producers of ambiguously stated policy; indeed, clarity of intent is often sacrificed to attain agreement. The lack of clarity in policy intent can have many consequences. Forestry professionals may be delegated considerable discretion to design programs, rules, and regulations. Substantial negotiation over intended goals may occur, possibly resulting in a significant narrowing or enlargement of the original intent. A multiple-use forest plan, for example, may become a timber-management plan. Onlookers may see lack of clarity as an opportunity to attach their goals and objectives to the rules and regulations, as when a wildlife manager appends wildlife-management practices to a timber-management plan for public forests.

Clarity of intent can be a two-edged sword for high-level policymakers. They may express frustration at the ambiguous nature of the policy statements they are asked to implement, but they may be inclined to prescribe excessively detailed rules and regulation for use by those farther down the chain of command. The problem is that forest and administrative conditions vary widely; uniform rules are not always appropriate to the multitude of site-specific conditions that are likely to be encountered. In addition to fostering specification of erroneous rules and guidelines, excessive detail can destroy local initiative and creativity. ''Rather than being concerned with the minutiae of policy implementation strategies, the policy maker needs to make certain that the policy objectives are clear through careful specification of intent and well-considered plans for feedback and evaluation'' (Brewer and deLeon 1983, p. 268).

Cause-and-Effect Linkage

Successful implementation implies a valid *linkage between cause and effect.* The policy must be based on a valid theory of what will happen if certain actions are taken. If the theory is fundamentally flawed, the forest resource policy will fail, no matter how well it is implemented. For example, if nonindustrial private landowners are reluctant to undertake timber-management investments because they lack the financial wherewithal, a forest policy that provides them with only technical forestry advice is unlikely to succeed—no matter how well the policy is implemented. The constraint on landowners is available cash, not information. Similarly, a forest resource policy calling for large-scale investments in public timber-management activities as a means of encouraging near-term economic development may be useless if the link between such investments and the production of jobs, income, and economic stability is dubious. No matter how effective the implementation process, ultimate policy success will not occur if underlying cause-and-effect relationships are in error or largely unknown.

The problem of poorly specified cause-and-effect linkages is best addressed at the policy formulation stage. Rigorous analyses that improve the definition of the problem or issue and enhance understanding of various approaches that might be used to attack it are needed. For example, careful assessment of constraints on nonindustrial private forest landowner interest in timber management and the merits of various policy options (technical education, cost-share payments, regulatory actions) for addressing such constraints are necessary prerequisites to achieving public interest in private timber-management activities. Some modest experimenting with alternative hypotheses may also be in order. What can be learned from various state and foreign experiences with alternative policies and programs focused on nonindustrial private forests? The problem of cause and effect can be addressed by continuing review and feedback during implementation. If the results desired of a particular forest resource policy are not forthcoming, adjustments can be made and implementing activities changed accordingly.

Support of Intent

Successful implementation requires *understanding, agreement, and support of intent.* Sympathy and support from implementors and those affected by a forest resource policy are crucial to implementation. For example, if a legally mandated forest policy requires that private forest landowners reforest their land after harvesting and that publicly employed foresters enforce such requirements, implementation of the policy may flounder if neither party is sympathetic to regulatory approaches. Before implementation, potential clients of a forest resource policy should be identified, the intensity of their opposition determined, their willingness to commit resources in opposition defined, and the possible duration of their opposition clarified. Such factors of support must be estimated and plans made to address them. Similarly, the understanding and support of a policy's intent by implementors should be assessed. Forestry professionals may outright oppose a forest resource policy, may become confused by objectives that lack clarity, or may become perplexed by objectives that are not compatible. Such circumstances can lead them to establish their own unofficial goals and objectives, much to the dismay of upper-level policymakers. It is important to remember, however, that forestry professionals are human and suffer from the general failings of the species. Even the most carefully worded directive is open to misinterpretation.

Sufficient Resources

Successful implementation requires *sufficient resources*—time, finances, machinery, organizations, professional talent, and the like—properly focused on the task at hand. Not to have such resources available in the amounts needed and at the appropriate times can spell disaster for implementation processes. If a forest resource policy requires preparation of numerous forestwide plans but adequate resources are not available, the policy cannot be successful. Similarly, a policy of offering a specified level of public timber for sale will not be implemented if the professional talent needed to design public timber sales is in critically short supply. Research programs focused on a policy of assessing the effects of atmospheric pollutants on forests will not be implemented if the necessary research equipment is not available at the right place and the right time.

Resource scarcities affecting implementation can be brought on by a number of circumstances. High-level policymakers are not averse to specifying policy intent while at the same time overlooking the resources needed to achieve that intent. Legislators, for example, may prescribe policies calling for a reduction to acceptable levels of water pollutants originating from timber-harvesting activities on private land. If only modest amounts of financing are provided, implementing activities will not be undertaken on a large enough scale and the policies will fail.

Implementation processes may also fail because too much is expected by the originators of a policy. The federal Cooperative Forestry Assistance Act of 1978 (Section 4, Forestry Incentives), for example, sets out on a national scale to "encourage the development, management and protection of nonindustrial private forest lands." There are more than 7 million such owners in the United States; annual cost-share funding of $12 to $18 million enables the accomplishment of timber-management activities on only a few ownerships.

Implementation may also fail if results are expected too soon. Trees take time to grow, wildlife habitats must mature over time, and environmental assessments must often be prepared before significant actions can be undertaken. Agencies forced to meet unrealistic deadlines are often destined for inefficient program implementation. Equally important, an appropriate combination of resources must be available at the correct time if implementation is to be effective. If only one needed resource is delayed, the result can be a lengthy setback. Such complications are significant, given the variety of resources—money, land, equipment, professional talent—that are often required to implement a forest resource policy.

Administrative Simplicity

Successful implementation requires *administrative simplicity*. Implementation of nearly all forest resource policies involves a number of administrative units (both horizontally and vertically), each of which has its own set of expectations, interests, and perceptions. The more units that are involved in the implementation of a forest policy, the more difficult the implementing task becomes. For example, a state forester's directive to increase the sale of public timber may become blurred and distorted as it is subjected to administrative actions down the chain of command. The implementing activities of the lowest administrative units may well include actions that have little or no relationship to the state forester's original intent. Policymakers cannot become involved in the day-to-day details of implementation. They can, however, take actions early in the implementation process that will facilitate administrative simplicity.

Policymakers can facilitate administrative simplicity by designing programs that have few administrative links, so as to minimize the number of clearance and veto points. Forest resource policies which depend on a long sequence of cause-and-effect relationships for their implementation are at significant risk because there are many opportunities for administrative breakdowns. If, for example, the U.S. Environmental Protection Agency (EPA) sought to reduce air pollutants emitted by wood-based firms, its chances of successfully doing so would be substantially greater if the agency policy were delivered to the offending firms via few rather than many layers of administrative structure. Delivery of the policy through a chain consisting of the national office, a regional office, and the individual firm would provide fewer opportunities for breakdown than would delivery from the national office to a regional office to a subregional office to a district office to a local office to the individual firm. In general, the farther removed a policymaker is from the target of a policy's intent, the greater the opportunity for distortion of the policy's original intent.

Administrative simplicity will also be enhanced by keeping dependency relationships minimal in number and importance. The number of agencies with which authority or activities are shared should be minimized. A single agency responsible for implementation has a greater chance of success than has an agency that is dependent upon a number of agencies or private organizations. If, in the above example, the EPA relied on a multitude of outside organizations (state governments, regional compacts of state governments, industrial cooperative units, federal coordinating committees) to implement clean-air policies, managerial and coordination activities could overwhelm the implementation process and detract from its success. An agency that chooses to internally implement forest resource policies through a large number of intervening links is at a disadvantage. The agency's problems will be compounded by involvement of other agencies and organizations in the implementing process.

Administrative Compliance

Successful implementation requires *administrative compliance*. Policy implementation presumes that administering agencies and the personnel employed therein will carry out the required tasks. The exercise of authority and the application of power in ways that will secure the compliance of others—their willing consent and cooperation—are required. However, reality dictates that there will always be a high probability of suspicion, recalcitrance, and outright resistance to new policies, especially if insufficient time has been allowed for explanation and consultation or if previous experience with a similar forest resource policy has been unfortunate.

The process of securing administrative compliance is more an art than a science. Some guidelines, however, are helpful. Administrative compliance is more likely to be obtained if the forest resource policies to be implemented are congruent with the interests of those responsible for implementing them. If an agency is required to implement a forest policy it considers to be outside its primary role—for example, if a wildlife agency must oversee the extraction of minerals located within wildlife refuges, or a forestry research organization must assure dissemination and adoption of newly developed technologies—implementation will be less than effective. Similarly, implementation may prove difficult if an agency is required to carry out professionally distasteful or bureaucratically unpopular activities. Requiring state foresters to exercise police power in order to secure private landowner compliance with legally mandated reforestation standards is an example.

A program designed to suit an agency's best interest can be an incentive to administration compliance. For example, allowing a forestry agency to retain a portion of its timber sale receipts for purposes of reforestation may spur the implementation of timber sale policies and also foster heightened interest in reforestation activities. Another way to facilitate administrative compliance is to make a significant investment in communication activities. Though perfect communication of intent and the flow of information regarding its attainment are impossible, efforts should be made to assure that data, advice, and instructions are understood by all involved in the implementation process.

SIGNIFICANT PARTICIPANTS

Numerous individuals and organizations, many of them active participants in previous stages of the policy process, are involved in the implementation of forest resource policies. These individuals and organizations often spawn a web of relationships among legislators, administrators, and interest-group leaders. Such relationships monitor the implementing process and serve as the communication channels through which interests act either to facilitate or to inhibit implementation of a particular forest resource policy.

General Public

The breadth of individual or organizational involvement in implementation is often determined by law or judicial ruling. The National Forest Management Act of 1976, for example, specifically calls for the public's involvement in the development of the USDA-Forest Service's National Forest System plans: ''The Secretary shall provide for public participation in the development, review, and revision of land management plans.'' It also requires development of a means by which other governmental units can participate in the plan development: ''The Secretary, by regulation, shall establish procedures . . . to give the Federal, State and local governments . . . an opportunity to comment upon the formulation of standards, criteria, and guidelines.''

Government Agencies

The most common participants in implementation are government agencies that have exclusive or, in some cases, partial responsibility for implementation. Very often, a law will call for a policy to be implemented by a single agency. A state division of forestry, for example, may be designated by law as the lead agency charged with the development of an elaborate process for implementing a statewide forest practice law. The implementing rules will be developed and applied by that division; other administrative divisions will not be directly involved in the process. Implementation by a single organization may also occur at the federal level. For example, the USDA-Soil Conservation Service implemented the Soil and Water Resources Conservation Act of 1977, the EPA implemented the Clean Water Act of 1977, and the USDI-Bureau of Land Management implemented the Federal Land Policy and Management Act of 1976.

Far more common are laws that require a number of agencies to undertake implementing activities that will lead to achievement of the laws' intent. For example, the National Environmental Policy Act of 1970 is implemented to some degree by virtually all federal agencies, even though the Council on Environmental Quality promulgates the rules for implementing the act. Similar implementation arrangements exist for the Endangered Species Act of 1973, the Land and Water Conservation Fund Act of 1965, and the Wilderness Act of 1964. At least four federal agencies—the USDA-Forest Service, the USDI-National Park Service, the USDI-Bureau of Land Management, and the USDI-Fish and Wildlife Service—are responsible for implementing actions pursuant to achievement of the intent of the Wilderness Act.

The most obvious participants in the implementation process are the agencies or bureaus charged with interpreting, organizing, and delivering the goods and services called for by a forest resource policy. They are the cutting edge of rule making, adjudication, law-enforcement, and program operations. In a forestry agency, the sale of timber, the collection of recreational user fees, the enforcement of air-pollution standards, the management of insects and diseases, the purchase of capital equipment, and the hiring of forestry professionals and related staff are carried out at this level. In large measure, the effectiveness of a forest policy is a function of how smoothly an agency assumes responsibility for the various implementing tasks.

As important as specifically designated agencies are in implementing forest resource policies, they are not the only actors in the process. For example, often of equal influence in the implementation process are other executive agencies, including those that oversee the implementing agency (at the federal level, this means, for example, the departmental-level offices of the Office of Management and Budget) (Sample 1989) and those that have an interest, legal or otherwise, in the implementation process. For example, a pollution-control agency would be responsible for laws concerning use of pesticides by a reforestation program being implemented by a state forestry agency. Special counselors, assistants, and liaisons from such agencies influence the implementation process on behalf of the chief executive and on behalf of the policies they are legally required to implement.

Legislatures

Legislative systems also play a significant role in implementation, especially in terms of formally prescribing directives in law, a willingness to participate in interpretation activities, and an inclination to eagerly review (to assume oversight responsibility for) agency actions involving implementation. As individuals, however, legislators may express only a modest interest in implementation. They may view the process as less visible than policy making; as more protracted and plodding than legislative goal setting; as beset with politically unrewarding problems; and as apt to require that political rewards associated with successful implementation be shared, especially with bureaucrats.

Although legislators may be inclined to remove themselves from implementation, their staff members are often in daily contact with agencies that are responsible for implementing forest resource policies and are often assisted by various specialized arms of the legislature. Examples at the federal level include the Congressional Budget Office, the Library of Congress, and the Office of Technology Assessment. In some cases, legislatures will require such interaction by law or by other actions having the force of law. For example, in a conference report of FY (final year) 1985 Forest Service Appropriations Law, the U.S. Congress required that the following interaction occur as part of the USDA-Forest Service's efforts to develop a timber cost accounting system for the national forest system: ''The Forest Service should develop proposals for a reasonable but complete system and should work with the accounting systems division of the GAO [General Accounting Office] in developing the system. . . . The [Congressional] Committees expect to be kept fully informed of the progress on this effort.'' (USDA-Forest Service 1987a). Such language makes it obvious that Congress is interested in having the USDA-Forest Service work with the General Accounting Office to develop and implement the system.

Judicial Systems

Judicial systems also become involved in implementation via the interpretation of statutes, regulations, and administrative decisions involving matters of implementation. In ruling on cases brought to their attention, judges have frequent opportunities to prod or redirect implementing agencies toward judicially perceived legislative and constitutional intent. Some of the most far-reaching judicial actions have involved interpretation of the National Environmental Policy Act of 1970. The act requires, for example, the preparation of a detailed environmental impact statement for ''major Federal actions significantly affecting the quality of the human environment.'' In addressing pertinent cases brought before them, the federal courts have played a major role in interpreting what constitutes a major federal action and when such actions are likely to have a significant environmental effect. Judicial interpretation of these matters set the stage for ascertaining the circumstances under which a federal agency must prepare an environmental impact statement. Implementation of the Act was facilitated considerably (Schoenbaum 1982).

Special-Interest Groups

Also active in the implementation process are privately organized special-interest groups. Once a law, for example, has been enacted, interest-group struggles may shift from the legislative to the administrative arena. Here the groups' concerns are forcefully made known to forestry professionals responsible for implementation within an agency. Indeed, some groups that are opposed to a forest policy may remain quiet during the legislative legitimizing process, with the intention of achieving more decisive and less publicized victories during struggles over implementation. Such groups may even have the will and power to redefine the policy intentions of the originators of a forest resource policy. Administrative agencies may actually welcome the role of interest groups in implementation, and may appoint such groups to advisory groups or advisory boards. Such an appointment is a virtual guarantee that the nongovernmental group will participate in implementation.

Policy Beneficiaries

Also to be recognized in the implementation process are the ultimate beneficiaries or recipients of a forest policy's outcomes, such as a wilderness recreation enthusiast who consumes the products of a national wilderness policy and a homeowner who benefits from the shelter provided by wood fiber grown as a result of public timber-management policies.

Implementation does not always command the attention of citizens to the extent that legitimizing processes do. The excitement of enacting a statute is apparently more alluring than the supposedly mundane process of carrying out the intent of a statute. Citizens who ignore implementation, however, may find themselves deceived by legitimized policies that are merely symbolic. To avoid such an outcome, interested citizens should become acquainted with agency administrative processes and should regularly check on progress toward implementation. Most often, however, the ultimate beneficiaries of a policy's outcome rely on government agencies or organized special interest groups to represent them in implementation; they are reluctant to take part in implementation directly.

IMPLEMENTATION IN THE POLICY PROCESS

Implementation is rarely a swift or rational process. Only in rare situations, in which the effectiveness of a newly established forest resource policy is obvious and in which commitment to its virtues is widely shared, will implementation proceed according to a rational plan. More common are situations in which the forestry goals are uncertain and in which the various means of attaining them are little known. Faced with multiple and conflicting renditions of what is to be accomplished and how such vague and conflicting intentions are to be achieved, forestry professionals responsible for policy implementation confront one of the more challenging phases of the policy process.

Implementation is a fundamental, ongoing activity in the forestry community. Some of the more significant implementing actions in recent years include the following:

- *Log export restrictions on federal public timber harvesting* In response to congressional riders attached to annual Department of the Interior and Related Agencies Appropriations Act that have banned (since 1974) the export of logs harvested from federal lands, federal agencies developed detailed implementing procedures, including specification of responsibility at nearly all administrative levels. Required was definition of terms such as "export," "private land," "unprocessed timber," and "historic sale levels." Responsibility was assigned, reporting systems were developed, and surveillance mechanisms were established (for example, Darr 1980; USDA-Forest Service 1983, 1989a, 1989b). The riders were formally legitimized in law by the Customs and Trade Act of 1990.
- *Timber Sale Information Reporting System* In response to congressional directives, the USDA-Forest Service developed and implemented the Timber Sale Program Information Reporting System (TSPIRS) for the National Forest System. The reporting system was designed to provide three types of information: financial, economic, and income and employment opportunities. The system was developed and tested in selected National Forests and later applied to all National Forests (Forest Service 1987a).
- *National Forest land-management plans* In response to the National Forest Management Act of 1976, each National Forest prepares a land-management plan, a segment of which must deal with implementation. This segment addresses selection and geographic identification of the management practices to be applied, designation of administrative responsibility for the work, accomplishment of the management practices, and monitoring and evaluating of the results (for example, Forest Service 1987b, 1988).

Implementation is inseparably intertwined with other segments of the policy process. For example, in the early stages of implementation, difficulties (e.g., inappropriate cause-and-effect relations) may be encountered. A return to formulation may be warranted, or implementation may produce new or unexpected problems of such scope and importance that a return to the agenda-setting process is necessary. Similarly, implementation may be linked backward to selection for answers to sensitive political questions such as: How serious are administrators about implementing a program? What exactly is the intent of a program? Who is and who is not to be serviced by the program? Implementation may also bring to the surface a number of concerns relevant to evaluation, such as: When is an implemented program to be considered a success? What criteria should be used to judge program performance? When are program outputs to occur and at what cost are they to be produced? Finally, implementation may often imply termination, in that new or modified programs are substituted for existing programs.

REFERENCES

Bardach, E. *The Implementation Game: What Happens after a Bill Becomes Law.* The Massachusetts Institute of Technology Press, Boston, 1982.

Brewer, G. D., and P. deLeon. *The Foundations of Policy Analysis.* The Dorsey Press, Homewood, Ill., 1983.

Darr, D. R. Softwood Log Export Policy: The Key Questions. *Journal of Forestry* 78(3):138–140, 151. 1980.

Hogwood, B. W., and L. A. Gunn. *Policy Analysis for the Real World.* Oxford University Press, London, England, 1986.

Jones, Charles O. *An Introduction to the Study of Public Policy.* Brooks/Cole Publishing Company, Monterey, Calif., 1984.

Pressman, J. L., and A. Wildavsky. *Implementation: How Great Expectations in Washington are Dashed in Oakland.* University of California Press, Berkeley, Calif., 1984.

Sample, V. A. National Forest Policy-Making and Program Planning: The Role of the President's Office of Management and Budget. *Journal of Forestry* 87(1):17–25. 1989.

Schoenbaum, T. J. *Environmental Policy Law: Cases, Readings and Text.* The Foundation Press, Inc., Mineola, N.Y., 1982.

USDA-Forest Service. *The Principal Laws Relating to Forest Service Activities.* Agricultural Handbook No. 453. U.S. Department of Agriculture, Washington, 1983.

———. *Timber Sale Program Information Reporting System: Final Report to Congress.* U.S. Department of Agriculture, Washington, 1987a.

———. *Land and Resource Management Plan: 1986–2000.* Natahala and Pisgah National Forests. U.S. Department of Agriculture, Ashville, N.C., 1987b.

———. *Forest Plan Implementation: Our Approach.* Northern Region. U.S. Department of Agriculture, Missoula, Mont., 1988.

———. *Forest Service Handbook.* Timber Sale Administration. U.S. Department of Agriculture, Washington, 1989a.

———. *Forest Service Manual* (and Regional Supplements). U.S. Department of Agriculture, Washington, 1989b.

8

EVALUATING POLICIES AND PROGRAMS: PROGRAMS TO GOVERNMENT

Implementation of forest policies and programs soon generates tangible and discernable results. Forestland is allocated to specific uses, habitat for an endangered species is intensively managed, cost-share payments are made to private landowners, public timber is sold to loggers and processors associated with small enterprises, and regulations to reduce air pollutants are imposed on manufacturers of wood-based products. Policymakers and important program clients watch the implementation process to see how a policy and its implemented programs are progressing. Are the selected policies that were diligently legitimized and carefully implemented actually accomplishing what was intended?

Evaluation is the examination of policy and program effects on the targets or goals they were designed to achieve. During evaluation, several things happen:

- Issues, suggested policies, and experience with policies converge.
- Knowledge about a program's application is collected and calibrated.
- Judgments about merit are made.
- Changes in direction and approach are considered.

Evaluation is a vital element in the search for forest resource policies that will have favorable outcomes for all involved. It has been argued that selection of a policy should be predicted on its susceptibility to evaluation.

Evaluating policies and programs requires an understanding of evaluation processes, including purpose and likely outcomes; possible approaches to evaluation, especially potential analytical techniques and standards of performance; role and expectations of policymakers and policy analysts; and major obstacles to effective evaluation of policies and programs. In this chapter, we shall consider these topics and related areas.

CONCEPT AND PROCESS

Concept

Though evaluation is often portrayed as a continuing activity carried out by expert analysts in a systematic and reasoned fashion, this is typically not true. Most often, evaluation is a disjointed process involving one-time assessment of a single isolated program. It involves actors and organizations beyond the trained analysts situated in administering agencies. Several procedures, including (but not limited to) rigorous analytical tools, are applied. Evaluation has an applied orientation; that is, it does not attempt to unearth new theories. Most often it involves the application of expertise associated with a variety of disciplines.

Evaluators are usually sensitive to the political environment within which they operate, because they understand and sympathize with the political nature of the policy process. They are usually oriented to the service of a particular client, knowing that the results they obtain will be used by an individual or an organization to judge a program and improve its performance.

Process

Evaluation can be viewed as a process involving selection, measurement, analysis, and judgment.

Selection *Selection* entails choice of a policy or program to be evaluated—setting the evaluation agenda. Selection may be influenced by various circumstances, including concern expressed by program administrators (who may feel, for example, that recreation management costs are rising rapidly) and apprehension voiced by the beneficiaries of a forest resource policy (who are unhappy, for instance because sufficient public timber has not been offered for sale). In some cases there is no choice: evaluation is required by the force of law. The decision to evaluate may be guided by explicitly stated objectives, such as a requirement that the efficiency of service forestry programs in a particular geographic location be determined. It may be fostered by interest in exploring whether very ambiguous and imprecise purposes have been attained, as when there is a need to determine whether the lost confidence of legislators important to forestry has been restored.

Measurement *Measurement* activities involve collection of data, solicitation of opinions, and documentation of impressions. Such activities may range from the statistically correct gathering of data about day-use campers in a state park system to a casual visit from legislators interested in how a nonindustrial private forest landowner views the services provided by a forestry extension agent.

Analysis *Analysis* can involve significant differences in method and style. Data may be analyzed with powerful quantitative techniques that enable rigorous comparison of program benefits and costs. On the other hand, analysis may simply involve roundtable discussions of opinions and impressions about a program's effectiveness.

Judgment Evaluation entails *judgment* about a program's virtues or detractions. Recommendations for change may be made to a policymaker, a program administrator, or a client of the forest resource policy or program in question (Jones 1984).

Scope

Evaluation is programmatically quite specific. The question is usually whether the policy or program in question is working, not with whether there are alternative programs that should be surfaced. If, for example, a forestry extension program is destined for evaluation, attention will be focused on extension as a program area, not on taxation or cost-share policies as alternative extension.

The alternatives typically addressed during evaluation are whether the program should be expanded, continued, curtailed, modified, or terminated. Ironically, not all programs and policies receive evaluations in such a context. Policies and programs that are new or that come under intense political attack are likely to be probed by evaluators. For example, National Forest System timber sale accounting policies implemented by the USDA-Forest Service quickly became the subject of evaluation after the Natural Resources Defense Council and The Wilderness Society expressed concern (Shands al et. 1988).

PURPOSE AND INTENTION

Evaluation is common in forestry and related communities. Legislators actively engage in the practice, large staffs in forestry agencies exist for evaluation purposes, and foundations and universities expend considerable energy on evaluation.

Accountability

Why is evaluation of so much interest? It is undertaken for numerous reasons. Perhaps the most obvious reason is that evaluation is the means by which those entrusted with the implementation of forest policies are held *accountable* for their actions. Without competent evaluations, implementation may drone on in privacy, may be carried out in highly inefficient ways, and may involve flagrant expenditure of resources. Without evaluation, policymakers, legislatures, and citizens would remain ignorant of the performance of those responsible for implementation of forest resource policies. Thus opportunities to judge and reward (or punish) implementors of forest policies would be lost.

Control

Evaluation can be viewed as a means of assuring *control* over policy implementation and direction, through a process in which the motives, talents, and inclinations of forestry professionals responsible for implementation are continually being steered by orders and instructions. The result is policy outcomes that are favored by policymakers and high-level administrators.

Evaluation is one of many ways that administrators can exercise control. Employing persons with "built-in controls" is another way. When a policymaker appoints subordinates who are loyal to the policymaker's beliefs and willing to be obedient implementors of a policy or program, the need for control through evaluation is lessened. For example, the commissioner of a state department of natural resources may appoint a director of a wildlife division who is in total agreement with the commissioner's views on game versus nongame wildlife management in forested areas. As a practical matter, however, it is not often that the number of political or personal loyalists in large natural resource agencies is high enough to ensure complete and faithful compliance during implementation.

Fostering Change

Evaluations are also valuable in *fostering change and accountability.* Forestry agencies may be bound by tradition and vested interests, and for that reason, evaluations performed by persons or organizations outside an agency may be especially valuable. Persuasive outside evaluators, working with an aroused citizenry, can do much to make implementation of existing agency policies more effective. They can also generate interest in formulation of other policies and programs that may be able to accomplish agreed-to goals in a more forthright fashion.

Clarification

Not to be overlooked is the important role that evaluation can play in further *clarifying* the intent of a forest resource policy. The objectives of a policy may be muddled by the compromises needed to secure legislative enactment. Carefully crafted evaluations can push, probe, and suggest alternative directions and their consequences. As a result, hazy objectives and unclear beneficiaries of policy may become clearer targets for the implementing actions of forestry professionals.

Ritual and Symbolic

Evaluation can also be a ritualistic means of allaying citizens' doubts about a program's effectiveness. A political leader such as an agency head or a legislator may find such a ritual helpful. Furthermore, even when there is no fundamental problem that calls for evaluation of a program, an evaluation may be undertaken, largely as a symbolic exercise (Koenig 1986).

Legal Requirement

Agencies and forestry professionals may have no choice but to undertake evaluations. Laws, rules, and regulations may simply state that they will be carried out. For example:

> To provide information that will aid Congress in its oversight responsibilities and to provide accountability in implementing this Act, the Secretary [of Agriculture] shall prepare an annual report [that] shall set forth accomplishments of the Renewable Resources Extension Program, its strengths and weaknesses, recommendations for improvement, and costs of program administration (Renewable Resources Extension Act of 1978).

> Every twenty-four calendar months . . . , the Secretaries of the Interior and Agriculture will submit to Congress a joint report on the administration of this Act, including a summary of enforcements . . . [and] costs (Wild Horses and Burros Protection Act of 1971 as amended).

> For purposes of providing information that will aid Congress in its oversight responsibilities and improve the accountability of agency expenditures and activities, the Secretary of Agriculture shall prepare an annual report [that] . . . shall include a description of the status, accomplishments, needs, and work backlogs for the programs and activities conducted under the Cooperative Forestry Assistance Act of 1978 (Forest and Rangeland Renewable Resources Planning Act of 1974 as amended).

> The Secretary of the Interior and Secretary of Agriculture shall annually prepare a joint report detailing the activities carried out under this act and providing recommendations. Each report . . . shall be submitted . . . not later than April 1 (Youth Conservation Corps Act of 1970).

> The Secretary of Agriculture shall . . . provide a report to Congress which sets forth the scope of the national forest reforestation needs, and a planned program for reforesting such lands, including a description of the extent to which funds authorized by this Act are applied to the program (Supplemental National Forest Reforestation Fund Act of 1972; reporting provision repealed in 1980).

Obstacles

There are undeniably strong reasons for evaluating forest policies and programs. However, there are equally well-grounded reasons for not undertaking evaluations, and some of them are described below.

Expense Evaluations are frequently very expensive. A legislative committee that conducts investigative hearings concerning an agency's policy on recreational user fees will command the time and talents of legislators and their staffs, as well as the efforts of a multitude of agency and public witnesses. Detailed analyses may be called for, travel will be required, data will be gathered, and salaries will be paid—all at great expense. Similarly, a large-scale and very costly evaluation of a specific program that consumes very little of a forestry agency's finances or professional talent is probably not warranted. Detailed and thorough probing of all forestry programs simply does not stand the test of efficient resource use.

Destructiveness Evaluation activities often disrupt ongoing administrative activities, disclose institutional weaknesses, and threaten the views and positions of individual managers of policies and programs. The problem has been well-stated by Szanton (1981, p. 38):

> Since nothing works as well as it might, and since evaluators demonstrate their acuteness most readily by finding fault, program evaluations are almost always critical. Even when they propose correctives, evaluations focus mainly on fault: questionable policies, probable inefficiencies, inadequate foresight, perhaps a taint of fraud.

Evaluation can be viewed as a destructive process that stimulates hostility among those whose programs are being appraised. A claim of program inefficiency, for example, may prompt demands for an evaluation. The very thought of an evaluation in such a context may bolster an administrator's fear of failing and of being charged with fault. The administrator's resistance to evaluation is understandably intense. Whether such a disruption has merit in the larger framework of administering forestry programs is debatable. Its cost may be greater than the supposed gains to be achieved from evaluation.

Error and Bias A further deterrent to evaluation is the risk that a completed evaluation may be erroneous or inaccurate. An evaluation that reports a program's internal rate of return to be below acceptable levels may do profound damage to a forest wildlife-management program that is actually quite efficient. Errors in specifying the value of forest wildlife benefits may be so large that the conclusions are questionable. Obviously such conclusions should not be used as a basis for decisions about the program's future.

Even in the relatively specialized world of analytical evaluations, serious differences exist about how evaluations are to be accomplished. Qualified experts may disagree on techniques, different policymakers may value a program's outputs quite differently, and unscrupulous outsiders may present only partial evaluations of forest resource policies that they dislike and are intent on destroying. The rarity of technically correct and bias-free evaluations can be a strong reason for minimizing evaluation activities (Brewer and deLeon 1983).

OUTCOME AND EFFECT

Policy and program evaluations can lead to a variety of outcomes, depending on why an evaluation is carried out, who undertakes it, and how it is done. If, for example, evaluators are intent on justifying a forestry program and find evidence to support such an end, the outcome of the evaluation will be supportive of the forest policy and the means by which it was implemented. If evaluators determine that minor adjustments in targets and implementing procedures are needed, corrective action may include minor reorganization, personnel changes (such as reassignment, removal, or reprimand), or clarification of policy intent. Evaluators of a forest resource policy may, however, find that a policy is simply not accomplishing its objectives, implying a need for extensive change. Forest administrators may or may not have the authority to carry out such changes. If they do not, new authorization must be sought and the coalitions necessary for legitimizing such authorization may have to be reconstructed.

Redefining an Issue

Extensive revision of a policy and its implementing procedures may be precipitated by discovery that the policy or program is based on a policy formulator's erroneous definition of a forest resource issue or on an incorrectly specified link between a forest policy and its ability to cause change toward an objective. Policy formulators, for example, may assume that timber-management clear-cuts are the cause of sediment in waterways. In response, administrators may implement forestwide policies that limit the size of clear-cuts. Evaluation activities, however, may subsequently reveal that waterways continue to carry unusually heavy sediment loads and that the source is poorly designed and inadequately maintained roads. Extensive redirection is called for, including a review of formulation, legitimizing, and implementing activities.

Additional Policies Needed

Evaluation may also lead to the discovery of new problems requiring the attention of another forest policy, or a group of policies. Evaluation of government policies involving cost-share payments to private nonindustrial forest owners may find that, in addition to lacking the financial wherewithal to prepare sites and plant trees, the landowners also lack technical understanding of which tree species to plant under what circumstances and for which of the landowner's many forestry objectives. An additional forest policy is needed—one that calls either for direct delivery of technical forestry information to such landowners (e.g., through a forestry extension or a service forestry program) or for support of a forestry consulting community that can provide such information for a fee. The evaluation of cost-share payments may well have spun off another topic that must find its way through another cycle of the policy process.

APPROACHES TO EVALUATION

Forest policies and programs can be evaluated in a number of ways using various standards of performance and tools of analysis. The process can involve the unstructured workings of a legislative committee or the procedurally bound application of the scientific method by a university analyst. Similarly, the performance of a program can be judged in light of various criteria, including efficiency, equity, and due process. The methods or tools of evaluation can be as simple as an experienced administrator's impression or premonition that something is wrong with a program or as complex as highly sophisticated, data-demanding mathematical models.

Though evaluations range from highly focused and formal to unstructured and informal, informal evaluations are by far the most common form of performance assessment (Brewer 1983). Considered below are formal versus informal evaluation, policy and program performance standards, and methods and techniques used for evaluation.

Informal Evaluation

Unstructured *informal evaluations* can involve a legislative committee conducting oversight hearings on forest tax policy, a daily newspaper's investigation of timber-management policies as implemented by a specific state forest, or the often value-laden judgments of organized interest groups that decry an agency's implementation of big-game wildlife-management policies. Informal evaluation most certainly occurs when citizens express dissatisfaction with the implementation of important forest policies by voting to remove elected public officials. Informal evaluations are seldom scientifically rigorous. They do, however, have an unusual capacity for affecting decision making and institutional behavior. They can be especially useful in signaling a need for more formal and detailed evaluations, especially in situations in which programs are highly complex and simple corrective actions do not yield improved performance.

Formal Evaluation

As compared to informal evaluations, *formal evaluations* are more rigorous in experimental approach and more demanding in terms of the methods and techniques applied. Their focus can be on process, response, or impact (Brewer 1983).

Process Evaluation *Process evaluation* concentrates on the internal performance of a forest resource program or the agency responsible for a program. The often rigorous standards of accounting, internal auditing, and fiscal analysis are applied. Enhancement of administrative practice and control is emphasized by focusing on questions such as:

- Are patterns of authority and responsibility clear?
- Is the behavior of administrators predictable?
- Is communication within the program adequate?
- Have employment opportunities been offered to members of minority communities?
- Are enough resources being devoted to the program?
- Are marginal costs increasing?
- Could higher levels of program success be achieved by other means at the same or less cost?

In short, process evaluation of a forest policy's implementation is concerned with notions of effective management, especially with regard to structure, budgeting, personnel, and decision making.

Response Evaluation *Response evaluation* is concerned with a program's ability to react to an externally imposed problem or issue. The question is whether a policy or program demonstrates an ability to increase, accommodate, or otherwise address a forest resource problem or opportunity that suddenly (or slowly) appears on the horizon. Here are some examples:

- If state policy required control of wildfire in forested areas, response evaluation would focus on an agency's performance in responding to a wildfire that nearly engulfed a small community.
- If an agency was responsible for implementing a forest policy that called for the protection of forests from the ravages of insects and diseases, response evaluation would determine by what means, at what costs, and how efficiently the policy was implemented when it was applied to a major outbreak of destructive insects.
- From a more subtle perspective, response evaluation would determine whether an agency charged with economic development in the forestry sector responded in an aggressive fashion to firms that periodically expressed an interest in expanding operations in a forested region.

The key ingredient in response evaluation is whether a forest policy was implemented in timely fashion so that major problems were diverted or significant opportunities accommodated.

Impact Evaluation A program's ability to achieve previously agreed-to goals, objectives, or targets is of concern in *impact evaluation.* Once implemented, for example:

- Did a timber sale program that specified the sale of a specific timber volume actually result in the specified volume being offered to potential purchasers?
- Did a research program develop the much-wanted scientific information that would enable ultimate placement of genetically superior trees in the hands of timber managers?
- Did a habitat-improvement program provide the additional days of hunter satisfaction that were sought by wildlife managers and their clients?

Impact evaluation is typically concerned with absolute measures of performance, such as whether targets were achieved, and if so, whether they were achieved in a timely fashion and at the lowest possible per-unit cost.

Performance Standards

Evaluation of an implemented policy or program implies an eventual judgment about its success, failure or need for fine-tuning. Judgment implies comparison with a standard of performance—a criterion. Perplexing questions that can arise during evaluation include the following:

- What criteria should be applied?
- Who should apply them?
- When should they be applied?
- At what parts of a complex program should they be directed?

Perspective Perspective is an important consideration in seeking the answers to questions such as those listed above. From a managerial perspective, criteria may emphasize concerns about efficiency, effectiveness, and economy: Did implementation of a forest resource policy accomplish an agreed-to target in an efficient manner with minimal mistakes and waste? From a political perspective, evaluation criteria may stress participation, public desires, and accountability: Did implementation of the policy allow for participation by important segments of society, did it actually deliver the types of goods and services wanted by clients, and did it provide opportunities for holding administrators and elected officials accountable for their actions? From a legal perspective, standards of performance may be entwined with matters of equal protection and procedural due process: Was the implemented forest policy applied equally to different ethnic and racial groups? Did persons excluded from the products of the policy have access to procedure for redress (Rosenbloom 1986)?

What quickly becomes apparent from a recognition of these differing perspectives is the breadth of interests that are concerned with judgments about performance. Indeed, the type of criteria used and the manner in which they are applied during evaluation depends on who one is, where one sits, and what one intends.

Source Criteria for judging performance arise from a number of sources, such as:

- *The opening paragraphs of a law* For example, "A cost-share program shall focus on nonindustrial private forests."
- *The hearings and conference reports preceding enactment of a law* "The productivity of forestland should be a consideration in cost-share disbursements."
- *The rules and regulations governing implementation of a law* "Cost-share payments are limited to $10,000 per person per year."
- *Subsequent legal rulings concerning a law* "Landowners of an ethnic minority are to be given full consideration in cost-share disbursement."
- *Formal and informal statements made by those granted power to administer the implementation of a law or program* "The cost-share program will be operated efficiently."

Performance criteria are not always explicitly prescribed by law or rule. Social custom and historical tradition may imply that a forest resource program should be judged in terms of its efficiency and effectiveness, and in terms of the distributional consequences for various segments of society. Similarly, criteria may not surface until after a forestry program has been in effect for an extensive period of time. In such circumstances, criteria may be defensive rationalizations created by the administrators or clients of program.

Effectiveness Evaluation of forest resource policies and programs requires the use of a variety of criteria, ranging from efficiency to due process and from equity to abstract values and ideologies. However, evaluation is fundamentally historical in perspective, concentrating on what has happened or is happening in specific settings. Therefore, effectiveness criteria—not efficiency criteria—often become paramount. For example:

- Did implementation of the forest policy lead to accomplishment of previously agreed-to goals?
- Were air- and water-pollution standards met?
- Did targeted recreational use occur?
- Were timber-harvesting goals met?
- Were affirmative action goals realized?

Although effectiveness may be emphasized, judgments about program performance must be based on the blending of various criteria. As Brewer and deLeon stated so well (1983, p. 339):

> Guard against emphasis on any one [criterion]: none will be best-in actual circumstances, and all may be helpful in gaining an appreciation of a program's many outcomes and effects. . . . One should resist a temptation to show consistent preference for any single criterion over others. The matter of weighting or valuing is a highly subjective issue not made any easier by misguided efforts to stress precision in measurement.

Methods and Techniques

The methods and techniques available to evaluators of forest resource policies and programs abound in number. Many of the primary analytical tools encountered during policy formulation are also used to conduct evaluations. Where differences do occur, they surface because the evaluator is dealing with after-the-fact conditions—that is, a program is in place and measurable consequences are occurring—whereas the policy formulator must deal with artificial, before-experience situations. While a diversity of specific methods and techniques may be available to an evaluator, two fundamental types of designs stand out: with-and-without design and experimental design (Brewer and deLeon 1983; Dye 1981; Jones 1984; Hogwood and Gunn 1986).

With-and-Without Design Evaluations involving comparison of conditions before and after a forest resource policy is implemented involve *with-and-without design*. Before the program is implemented, variables upon which it is designed to have an impact are measured, so as to establish baseline conditions. After the program is implemented, equivalent measurements are made to determine the program's effects. Whatever changes have occurred are then attributed to the implemented program. For example, if a public cost-share program designed to augment future supplies of timber from nonindustrial private forests were to be evaluated using a with-and-without design, the evaluator would attempt to determine the type, volume, and quantity of timber that results from such ownerships before and after implementation of the cost-share program. Once the amount of additional (if any) timber attributable to the program is known, values are ascribed to it and then compared to program costs. Measures of efficiency are then calculated (e.g., internal rate of return and benefit-cost ratio). The critical aspect of with-and-without designs for evaluation is the presumption that differences in outcome are attributable to the policy or program being implemented.

There are a number of varieties of with-and-without designs, several of which are shown in Figure 8-1. The most straight-forward is the before versus after approach (design 1). If conditions prior to a program's application are known to be at level A_1 and conditions after its application are found to be at level A_2, the difference (A_2 minus A_1) is judged to be the program's effect (example, Ellefson and Risbrudt 1987).

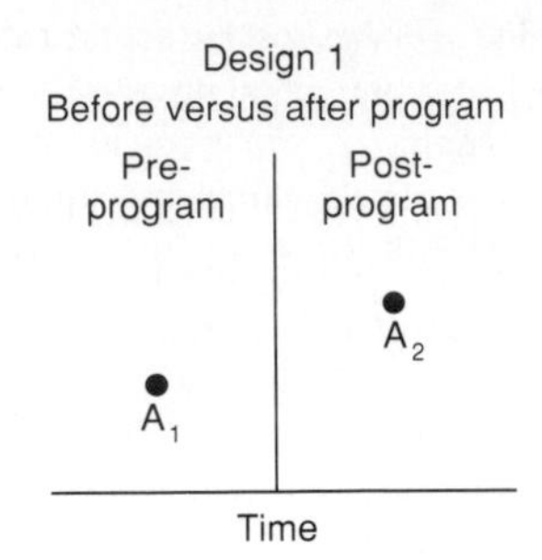

- $A_2 - A_1$ = estimated program effect.

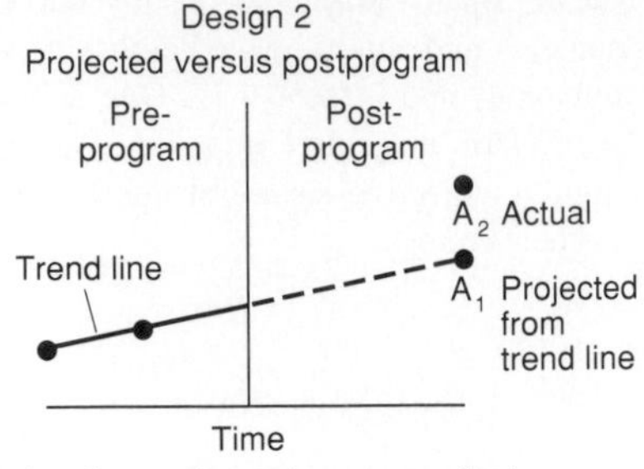

- $A_2 - A_1$ = estimated program effect.

Design 3
With versus without program

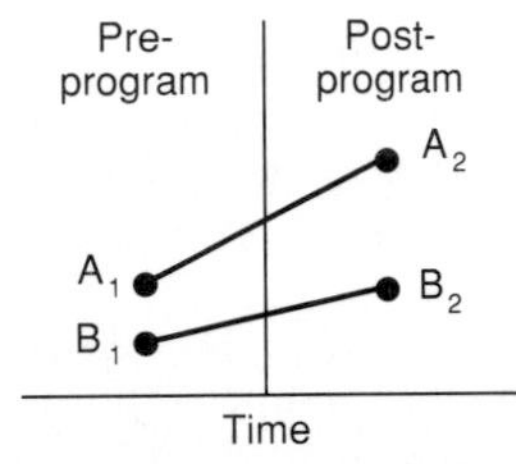

- A has program; B does not.
- $(A_2 - A_1) - (B_2 - B_1)$ = estimated program effect.

Design 4
Experimental design

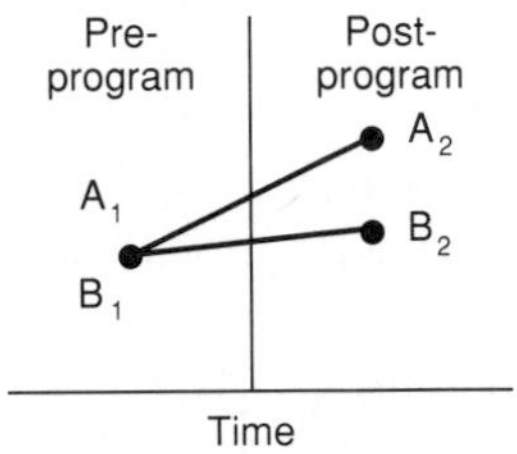

- A has program; B does not.
- $A_2 - B_2$ = estimated program effect.

FIGURE 8-1
Research designs for policy and program evaluation. (*Source: Thomas R. Dye,* Understanding Public Policy, *6th ed., © 1987, p. 375. Reprinted by permission of Prentice-Hall, Englewood Cliffs, New Jersey.*)

Comparison of the projected trend line with the post-program trend line is another variation of with-and-without evaluation (design 2). Preprogram trends are projected into a postprogram period and subsequently compared with what actually happened after the program was implemented. If, for example, the trend in day-use camping in a public forest was projected to be at level A_1 without a new public information program and was determined to be at level A_2 with the program, the difference would be attributed to the new program.

A third approach to with-and-without design is comparison between jurisdictions with and without programs (design 3). The intent is to compare individuals or administrative units that have participated in a program with those who have not. For example, if only one district (district A) of a public forest implements an intensified watershed management program, the program could be evaluated by comparing preprogram and postprogram water-flow conditions in district A with conditions in an adjacent area, district B. The difference between the net change in water flow from the two districts would be attributed to the intensified program $[(A_2 - A_1) - (B_2 - B_1)]$ (example, Cubbage et al. 1985). To attribute the difference to the intensified program entails a significant assumption if conditions have not been rigorously controlled—say, if the districts differ in vegetative cover or receive different amounts of rainfall because of their locations.

Experimental Design The most rigorous approach to evaluation is *experimental design,* in which experimental (treatment) and control groups are carefully selected. See Figure 8-1, design 4. The program or policy in question is applied to the treatment group but not to the control group. Any measurable differences that exist between the experimental and the control group are attributed to the program. Key characteristics of experimental design include:

- *Random, statistically valid selection* Exceptional care is given to selection of experimental and control groups.
- *Unambiguous definition of variables, capable of measurement with a high degree of precision* Output variables are specified and measured ahead of time and with much care.
- *Exclusion of all foreign influences and external factors* Conditions external to the application of the program in question are rigorously controlled.

The rigorous demands of experimental design make its application to evaluation of policies and programs extremely difficult, especially for programs that are large in scale and very complex, and for programs that contain multiple variables and are evolutionary in nature.

Selection of an appropriate means of evaluating forest resource policies and programs is no easy task. Matching an evaluation design with the conditions faced by a policymaker or evaluator is of paramount concern. If, for example, a policymaker is planning next year's program funding and would like a quick assessment of the short-term impacts of a forestry program that is in its early stages of implementation, there is little reason for a large-scale, costly evaluation that involves a rigorous experimental design. Similarly, a policymaker who must implement broad and diffuse policy objectives (often evolving as the program is implemented) will not appreciate an evaluator who demands clear and constant objectives against which before and after program estimates can be made.

Likewise, policymakers must be cautioned against generalizing based on evaluation of a single forest resource program. Evaluations do not generalize very well in practice. Conditions that make for a successful forestry program in one situation may not exist in another situation. For example, a wilderness-management program that is administered by highly motivated professionals who are given ample resources and rewards may be judged highly successful. The same program could fail, however, if it were applied elsewhere by apathetic administrators who were given very limited professional and financial resources. Evaluators generally seek to answer specific questions about a program under unique conditions; they are often of dubious value as a basis for broad generalizations.

Evaluation Tools Evaluators of policies and programs rely on a variety of tools and techniques. Oftentimes staggering in number and diversity, they include the gamut of techniques commonly found under the umbrella of operations research, systems analysis, and policy sciences. Some techniques, such as mathematical and linear programming, focus on optimality as a solution, while others, such as benefit-cost analysis, are designed to explore notions of efficiency. Approaches such as input-output analysis focus on the structural nature of economies and institutions, whereas simulation of anticipated conditions via the manipulation of models composed of mathematical and econometric equations is the strength of others. Nearly all such tools rely on efficiency as a measure of worth. Because of their frequent statistical and mathematical nature, they demand much data and information. Especially useful in situations involving substantial uncertainty are a variety of group decision-making procedures (including the Delphi technique and nominal group technique) and related future-forecasting procedures.

CLIENT-EVALUATOR RELATIONSHIPS

Evaluation of forest resource policies and programs does not occur in a vacuum. It is the product of a healthy relationship between a policymaker who is interested in furthering the performance of a program and an evaluator (analyst) who has certain technical and intellectual capacities which can complement the policymaker's interests. The relationship may best be compared to that between a consumer (the policymaker) who is seeking a product (advice) from a producer (the evaluator or policy analyst). The policy-maker consumer of advice about the performance of a policy or program can hold any of a number of positions, such as legislator, agency head, or program director. The advice being sought can take many forms; examples are political, efficiency, and effectiveness advice. It can be presented in various ways including confidential conversations, formal speeches, polished written reports, and articles in academic journals. Similarly, the role of evaluator, or producer of advice, can be performed by a person in any of several positions, including member of program operating staff or specialized evaluation staff, university researcher, investigative reporter, or staff member or organized interest group. The number and variety of possible client-evaluator relationships are staggering.

Client and Administrators

Clients of policy evaluators are as diverse as they are numerous. They range from the head of a forest resource agency to the administrator of a forested park system and from the leader of an important legislative committee to the manager of a public wildlife refuge. They are the people who demand information that enables them to make informed judgments about forest policies and programs. Clients and policy evaluators differ widely in their desire for information about the performance of forestry and related programs. Some want information to reduce uncertainty surrounding a pending policy decision. Others want explanations or justifications of program decisions that have already been made. Still others want sophisticated evaluations simply because they are fashionable.

From a political perspective, it has been found that policymakers look favorably upon evaluations that have the following characteristics (Asher 1978):

- *Impartiality* The evaluators who are responsible for the evaluation should be known to have no vested interest in the outcomes of actions based on the evaluation.
- *Simplicity* The evaluation should be presented in plain language with few qualifications.
- *Diversity* The evaluation should present a number of options for change, not a single or best recommendation.
- *Consensuality* The evaluators who are responsible for the evaluation should uniformly agree with the conclusions of the evaluation and should approve of the degree of analytical rigor upon which they are based.

In contrast, an evaluation's influence upon a policymaker may be slight (even though well conceived and rigorously developed) for various reasons, including (Ellefson 1986):

- *Hard-to-understand results* The evaluation is not focused, is very complex and poorly documented, and has limited relevance to the program in question.
- *Incomplete results* An important dimension or factors are excluded (e.g., evaluation of market outputs but not nonmarket outputs).
- *Politically unacceptable results* Objective estimates of a program's virtues do not fit the interest of politically important groups.
- *Political consequences of technical error* Uncertainties in estimating benefits and costs are large; the evaluation could be wrong, with politically unacceptable consequences.
- *Reduced opportunity for qualitative negotiation* Rational and explicit evaluations leading to a single option or a small group of options reduce the softness of a policymaker's position and limit ability to negotiate workable compromises.
- *Inappropriate timing* The evaluation fails to arrive in sufficient time to mesh with other administrative processes (e.g., budget cycles, planning cycles).
- *Inappropriate administrative level* The evaluation is not relevant to the management level to which it is presented (e.g., national evaluation applied to district level).

- *Lack of logic and analytical rigor* The evaluation is insufficiently robust to withstand intense intellectual criticism.

Policymakers expect many things from evaluations per se; they also expect much from the persons (evaluators or analysts) who prepare such evaluations. They expect evaluators to keep them out of trouble by providing information that will prevent them from taking ridiculous and indefensible positions; policymakers wish to appear knowledgeable and imaginative. Evaluators are also held responsible for anticipating conditions that would warrant a full-scale evaluation as well as situations in which the results of an evaluation could cause substantial turmoil and discord in a political setting, and thus jeopardize the existence of efficiently operating, related programs.

Policymakers also expect evaluators to provide them with a sense of direction—the level of direction depends on a policymaker's style of management. At the least, the evaluator should provide a menu of imaginative options that will enhance the performance of the program in question. Policymakers also often expect evaluators to act as coordinators of the flow of information deemed necessary for well-informed decisions about a forestry program's future. An evaluator may be required to enter into contracts with other parties, to carry out data-gathering missions, and to conduct portions of major evaluations.

Evaluators and Analysts

Like the clients of evaluators, persons who conduct evaluations are diverse in education, position, and interests. The titles assigned to evaluators by their agencies reflect this diversity. For example, analysts are frequently assigned titles such as "policy analyst," "program examiner," "systems planner," "budget analyst," "operations researcher," and "management analyst."

Depending on their political and analytical skills, evaluators can be grouped into three major types as shown in Table 8-1. Objective-technician analysts are pleased when peers laud their work. Client-advocate analysts are content with a satisfied client. Issue-advocate evaluators are not complacent until resources have been allocated and a program successfully implemented (Ellefson 1986; Meltsner 1976; Weimer and Vining 1989).

The specific activities engaged in by analysts depend on a variety of circumstances. Some have suggested, however, that the role of policy and program analysts has four fundamental aspects, as follows (Schultze 1968, pp. 55–76):

- Systematically clarifying policy objectives into meaningful targets for accomplishments
- Encouraging (forcing) consideration of a wider range of implementing options for accomplishing agreed-to policy objectives
- Relating, in a meaningful manner, program inputs to program outputs, as well as to policy goals or intents

TABLE 8-1
Types of policy analysts, by political and technical behavior

Character	Objective-technician	Client-advocate	Issue-advocate
Motive	Opportunity to do evaluations per se ("intellectual in residence"); clients are necessary evil.	Opportunity to advance cause of clients; very loyal to clients.	Opportunity to achieve own policy preferences; clients are selected opportunistically.
Standard of success	Quality and analyses that satisfy intellectual peers.	Ambiguous and carefully crafted analyses that support interest of clients.	Ambiguous and carefully selected analyses that result in acceptance of own policy preferences.
Resources	Command of detailed facts; and knowledge of evaluation techniques.	Command of communication skills.	Command of communication skills.
Timing of impact	Long-term impacts (prefer having analytically correct analyses to being on time).	Short-term impacts and immediate acceptance of analyses.	Short-term impacts and immediate acceptance of analyses.
Attitude toward evaluation	Analyses should be apolitical, separate facts from values, focus on consequences of alternatives, and often be an end in themselves.	Analyses should be used as a means of legitimizing the position of clients.	Analyses should be used as a means of securing acceptance of one's own policy preferences.

Sources: Adapted from Decisions about Public Forestry Programs: The Role of Analysts and Analytical Tools by Paul V. Ellefson, *Journal of Resource Management and Optimization* 4(1):65–77, 1986; *Policy Analysts in the Bureaucracy* by A. R. Meltsner, University of California Press, Berkeley, Calif., and *Policy Analysis: Concepts and Practice* by David L. Weimer and Aidan R. Vining, Prentice-Hall, Inc., Englewood Cliffs, N.J., 1989.

• Conducting postprogram evaluations to facilitate decisions on whether implemented programs should be expanded, contracted, or terminated

The environment within which evaluators operate can be demanding in many respects. At times they must deal with a very open political setting plagued with conflict, as when a legislative analyst is absorbed in public forestland-use issues. Other circumstances may force them to address very technical forestry issues involving technical persons. Some evaluations must be carried out in an environment of crisis. For example, a very powerful interest group may suddenly charge inefficiency in the implementation of a forest policy. The environment of evaluators may be chronic in nature. Some policies and programs in need of evaluation never seem to go away, as when evaluators must do annual evaluations of agency program budgets.

Probably the most perplexing of all difficulties faced by evaluators is the necessity of coping with a limited information base from which to carry out evaluations. The legislator who asks a legislative staff member to make a recommendation about a forestry program's effectiveness in 2 days or less is not being realistic about the pitifully small information base from which such recommendations may have to flow.

Evaluators and their clients may have differing views on which forest resource policies and programs should be evaluated, but the importance of maintaining a healthy working relationship cannot be overemphasized. A good relationship is especially important during the problem-definition stage of evaluation, when the question "what's the problem?" is answered and reanswered many times. The evaluator and client must engage in an iterative open-ended discussion that leads to continuing revisions in judgments about the need for evaluation and the approach that it might take. Initial judgments about the need for an efficiency-focused evaluation may prove incorrect; an equity and effectiveness orientation may be more appropriate. This process has been described by Brewer and de Leon (1983, p. 154) as follows:

> The analyst probes to determine the origins of the [program], who the various participants are within the context of the [program]—including what seems to be their interests and objectives; whether there are existing programs and policies that bear on the situation as presented by the client; who has responsibility for decision alternatives that could in time flow from the [evaluation]; what solutions or answer might look like and . . . whether they could be feasible, appealing, or even relevant.

Evaluators have significant (but not total) influence over the programs they choose to evaluate. What prompts an evaluator to focus on a particular forestry program? Why, for example, would an analyst choose to evaluate a recreational user-fee program at the expense of a forest road program? Why address a watershed-management program about which agreement as to intent is obvious while avoiding with significant rigor a timber sale program the implementation of which has been beset with controversy? One obvious reason for evaluating a particular program is that a policymaker or client demands that it be done. Evaluators may attempt to screen out low-payoff and high-risk policies and program with criteria of the following nature (Ellefson 1986; Meltsner 1976);

TECHNICAL CRITERIA Programs with the characteristics listed below are often selected for evaluation.

- *Large-budget programs* These offer greater opportunity for meaningful impact.
- *Programs about which the evaluator has prior knowledge and experience.*
- *Programs with many internal options for change* Programs with single-option solutions are often avoided.
- *Programs which can be evaluated with available agency resources* Funds, personnel, and equipment must be available.

ORGANIZATIONAL CRITERIA Programs with the characteristics listed below are often selected for evaluation.

- *Programs over which an agency has direct control* Interagency jurisdictional battles over program responsibility are avoided.
- *Programs which involve supportive clients and sympathetic policymakers.*
- *Programs for which cooperation among important actors is assured for carrying out an evaluation* Statisticians, economists, a modeler, and a budget expert are needed.
- *Programs for which there is sufficient time and information to carry out an evaluation.*

POLITICAL CRITERIA Programs with the characteristics listed below are often selected for evaluation.

- *Programs for which there is consensus on the intent (objectives) of the policy being sought* Lengthy debates over goals or program direction are avoided.
- *Programs which are being addressed by relatively few evaluators.*
- *Programs for which there is likely to be widespread support for change that might be recommended by an evaluation.*

Although evaluators are commonly perceived as apolitical, they may well be advocates of a particular forestry program and may use a variety of means to support their interests. They may undertake only partial evaluations of an implemented forest policy. For example, in a project that involves building additional forest roads, they may fail to undertake analyses of the adverse impact of road sediments on fish. Evaluators may choose a planning period that excludes (or includes) important benefits or costs; short periods, for instance, ignore both cumulative effects of pesticide use and the fire hazards of not using prescribed burning to reduce fuel levels. Analysts may employ analytical techniques that provide high (or low) proxy prices or interest rates for nonmarket forest outputs, or they may choose a discount (interest) rate that favors their preferred alternatives (Davis 1980).

Evaluators also express their value preferences through their career and administration decisions. For example, an analyst may choose to work in a renewable resources agency instead of in a minerals agency. Another analyst may select a timber policymaker as a client rather than a recreation policymaker. Still a third may choose or suggest a program or problem for analysis by anticipating policy payoffs, such as potential budget increases (Meltsner 1976).

An evaluator may even discourage the undertaking of an evaluation, or may depending on the circumstances, at least decline to become involved. This can happen for a number of reasons, including the following:

- Plans and commitments to modify the implemented forest policy in a particular way have already been made. The evaluation is being sought to simply support such preferences.
- Concern is being expressed about only a small portion of a forest resource program, when a more comprehensive evaluation would be in order. For example, perhaps forest pests are being controlled only by chemicals when biological control options exist.
- The likelihood that evaluation results will cause a change in a forest resource program is virtually nil, because the program is thoroughly entrenched in an agency and is vigorously supported by external constituencies.
- The resources needed to undertake an evaluation are woefully inadequate.
- The analyst is offended—morally, ethically, or otherwise—by the client or the program in question.

OBSTACLES TO EVALUATION

Implemented forest resource policies are not evaluated in a dream world void of obstacles, barriers, and difficulties. The sad reality is that many evaluations, once completed, suffer from benign neglect; they fail to exert any noticeable effect on the implementation of policy (Brewer and deLeon 1983; Koenig 1986; Hogwood and Gunn 1986; Wholey 1979). Of the numerous reasons for this state of affairs, a few will be considered in this section, along with their relevance to the implementation of forest policies and programs.

Conflicting Expectations

A significant portion of the dilemma involves the conflicting expectations of the policymaker and the assumptions of the evaluator. Evaluators are usually interested in providing a single, in-charge policymaker with the products of sophisticated analytical techniques applied to a program that has well-defined objectives. In reality, most forest resource programs are not of that character. Several policymakers may be in charge, and the objectives sought may be unclear and far from fully developed. The administrators in command may be less interested in an evaluation's validity and integrity than in relevant information that will enable them to make judgments about whether to expand, maintain, curtail, or terminate a program. The choice dilemma is between interest in rigor and demand for relevance. Stressing rigor and the quantification that often goes with it may well force evaluation of narrow segments of a forest resource program while neglecting broader, more comprehensive concerns.

The result: Forest policymakers become frustrated because evaluations—even though scientifically they are sound—fail to address their needs. Evaluators express discontent because the complexities of an implemented forest resource program cannot be sufficiently explored with crude and unsophisticated tools.

Official Resistance

Evaluation should, ideally, be a cooperative effort between policymakers and evaluators, but in actuality, official resistance to evaluation may be a problem. Since evaluation calls for the measurement of a forestry program's impacts and ultimately a judgment about its merits, program officials may be less than enamored with proposals or requirements to evaluate—and with good reason. If the results do not please an authorized administrator, the program, as well as the reputations and careers of the policymakers and implementors, may be in jeopardy. Assuming the worst, program officials may discourage evaluations, refuse access to data, keep incomplete records, and alert program constituencies to the possible harmful effects of a pending evaluation. Such circumstances have obvious implications for evaluators' ability to carry out evaluations with ease. They can seriously limit evaluators' ability to recommend program changes for the good of all involved (Hogwood and Gunn 1986). Again, the disparity between the world of the policymaker and that of the evaluator becomes apparent.

Mistrust of Results

Evaluation can also suffer because the products of evaluation are oversold as a basis for judgment. Evaluators can easily slide into a position in which recommendations based on evaluations appear valid and trustworthy because they are somehow based on supposedly scientific and objective approaches. In fact, however, errors can be made in the course of evaluations, and quantitative techniques—however powerful they may be in theory—can fall far short in practice. Mistrust of the products of evaluation among administrators is often the consequence. The reaction of a distributed program administrator suggests that evaluators should be more mindful of the intrinsic limitations of their craft (Kent 1971, p. 62):

> I am not so sure that analysis as a credible ingredient in decision-making will necessarily have a brilliant future [because] decision-makers are becoming increasingly annoyed that different analysts get quite different answers to the same problem . . . there must be something wrong when quantification of some particular problem produces radically different results.

Administrative Disparities

Program evaluation activities are also deterred from being effective by administrative disparities. For example, program-implementing activities may be fragmented among several forestry agencies. A comprehensive legislative program for the management and use of forest residues may be established and implemented by three, four, or more land-management agencies. For an evaluator, the task of tracking down measurable impacts of the forestry program as administered by multiple organizations may be next to impossible. The situation is further confounded when no single policymaker is in charge of the forestry program as a whole.

Similarly, the effectiveness of evaluation is hampered when it is treated as a single discrete event rather than an integrated part of the implementing process. A forested watershed-management program that is intensively evaluated and then ignored by evaluators for many years may breed ineffectiveness worse than a policymaker's wildest nightmares.

Furthermore, evaluation may go for nought if its recommendations are not implemented. Few evaluations make suggestions for implementing their recommendations. Most are presented to a client, a conference, or a journal with little attention to necessary follow-up actions. If an evaluation suggests a need to target timber management intensification to high-site growing conditions, investments in the evaluation will have been squandered if the program's implementation continues as in the past. What is even more disturbing is that fundamental inefficiencies in the program will continue.

Operational Problems

Although efforts to evaluate forest resource policies are deterred by large-scale issues, evaluation must also confront operational problems. For example, unclear, diffuse, or diverse program goals or objectives can wreak havoc on the process of evaluation. An evaluator may well lament, ''How will we know when we have arrived if we don't know where we are going?'' Although program goals are frequently made unclear for the purpose of securing agreement during the process of legitimacy, evaluation procedures and analytical tools require some indication of targets to be accomplished. An evaluator can help clarify objectives by submitting the question to implementing personnel, by drafting goals and targets for review by implementors, or by asking the clients of a forest resource program what they consider the objectives to be. An important question, however is: How far should evaluators go in drafting program targets? More important, how are weights to be assigned to conflicting objectives?

The lack of readily available information in usable form can also be an operational deterrent to successful evaluations. Information may be available at the local level, but not at the regional level. Information may vary from one jurisdiction to another, making comparisons impossible. Information may be available about the people who actually receive the products or services of a program but not about those specified to be targets of a program. Furthermore, official flows of information, such as technical reports, conference presentations, scheduled inspections, and public hearings, may be far less reliable than unofficial information sources. Unofficial sources may include informal discussions with implementors, the candid reactions of program clients, impressions of new employees, and idle chatter or gossip—which is especially likely to take place when concern about a program's future is high. Information problems become especially acute when evaluators rely heavily on sophisticated quantitative techniques (for example, linear programming and input-output analysis). Although such approaches force specific consideration of important variables, they often require significant investments in data-gathering activities.

Operationally, evaluation efforts are also hampered by difficulties in separating consequences attributable to a forestry program from consequences that would have normally occurred. A policymaker can be seriously misled if the observed effects are not direct products of the forestry program being implemented. For example, are increases in observed wildlife on a state forest the product of wildlife-management programs focused on the forest, or are they the result of increased development and urbanization in adjacent forests—development that results in wildlife migration to the forest in question? Similarly, what portion of an implemented program's side effects should be attributed to a forest policy's intent? How should the side effects be accounted for? A timber-management program often has positive side effects (such as creation of a wildlife habitat, or recreational access) which, if not accounted for, can place the program in serious jeopardy (Shands et al. 1988).

Definitions of standards or program performance (criteria) also pose difficulties for evaluation. Often, an evaluator is driven by an interest in efficiency while implementors are concerned with less harsh measures of program success. The danger is that relatively hard, or measurable, criteria may be used when qualitative indicators of program performance would be more valid and appropriate indicators of a forest resource program's accomplishments.

In sum, evaluation of policies and programs is often confronted by a number of obstacles. Among the more significant impediments are:

- Differing expectations evaluators and policymakers
- Policymakers and program clients' resist evaluation

• Unclear and diffuse policy and program goals
• Administrative disparities (especially in instances involving multiagency program responsibility)
• Lack of quality data and related information
• Errors in completed evaluations
• Problems in defining acceptable standards of policy and program performance

COMMUNITIES OF EVALUATORS

Evaluation of forest resource policies and programs is carried out by a variety of organizations and institutions. It is not the sole responsibility of an obscure analyst in a natural resource agency, nor is it the exclusive custody of high-level government commissions. The institutional landscape of evaluation is incredibly diverse. As might be expected, the products of evaluation in such a landscape are often a set of discontinuous judgments about the character of public and private forest policies. Some of the more significant participants in the process of evaluation are considered below.

Legislatures, Commissions, and Citizens

Legislatures The legislative branch of government is one of the more visible evaluators of policies and programs. Charged by the Legislative Reorganization Act of 1946, Congress, for example, is to ''exercise continuous watchfulness of the execution of laws.'' It does so via a variety of means, including through oversight gained at public hearings conducted by standing and select committees. Oversight is more likely to be obtained if the subject matter is nontechnical, the activity is concentrated in a single agency, and a committee chair accords the subject high priority and is aided by a capable staff (Ogul 1976). Among the forestry and natural resource topics subjected to oversight hearings during the 99th and 100th Congresses (1985–1988) were:

• Small Business Timber Sale Set-Aside Program
• Entrance Fees and Resource Protection for Units of the National Park System
• Conservation Reserve Program Review
• Management of Old-Growth Forests
• Western Forest Fires in 1987
• USDA-Forest Service Silvicultural Program in Region Five
• Management of the Tongass National Forest
• Forest Service Plans for Managing the National Forest: A Review
• Land Acquisition Policy and Programs of the National Park Service
• Forest Service Practices in New Mexico: A Review

Evaluation activities may also be carried out by subsidiary agencies of legislatures. For example, the U.S. Congress has at its disposal the General Accounting Office, the Office of Technology Assessment, and the Congressional Research Service of the Library of Congress. Examples of recent topics subject to evaluation by such organizations are:

U.S. GENERAL ACCOUNTING OFFICE

- *Tongass National Forest* Timber Provisions of the Alaska Lands Act Needs Clarification.
- *Federal Land Management* Consideration of Proposed Alaska Land Exchanges Should be Discontinued.
- *Wetlands* The Corps of Engineers' Administration of the Section 404 Program.
- *Parks and Recreation* Park Service Managers Report Shortfalls in Maintenance Funding.
- *National Forests* Timber Utilization Policy Needs to be Reexamined.
- *Resource Protection* Using Semipostal Stamps to Fund the Nongame Act.
- *Endangered Species* Management Improvements Could Enhance Recovery Program.

U.S. CONGRESS OFFICE OF TECHNOLOGY ASSESSMENT

- Wood Use: U.S. Competitiveness and Technology.
- Technologies to Sustain Tropical Forest Resources.
- Energy from Biological Resources.
- Technologies to Maintain Biological Diversity
- Renewable Resource Planning Technologies

CONGRESSIONAL RESEARCH SERVICE, LIBRARY OF CONGRESS

- Federal Grazing Fees on Lands Administered by the Bureau of Land Management and Forest Service.
- Wild Horses and Burros: Federal Management Issues.
- Merging of the USDA-Forest Service and the USDI-Bureau of Land Management.
- Impact of Presidential Tax Reform Proposal on Timberland Owners and Forest Products Industry.
- Dominant Use Management in the National Forest System.
- National Forest System Receipts: Sources and Dispositions.
- Tropical Deforestation and the International Tropical Timber Organization.

Commissions and Review Boards Evaluations are also within the realm of specialized commissions and review boards appointed by legislatures or high-level policymakers in government. Often composed primarily of persons from the private sector, commissions are given a variety of roles (including fact finding, advisory, and public relations roles) but are invariably drawn into the evaluation of existing policies and programs. Despite being ad hoc in nature and seldom able to enforce the results of their evaluations, commissions can serve various purposes, such as focusing public attention on ineffective forest resource programs, providing evaluations that are independent of executive agencies and legislative systems, and bringing differing viewpoints to the evaluation process. Examples of commissions or panels that have served the forestry and natural resources communities in recent years include:

- Outdoor Recreation Resources Review Commission, 1962
- Public Land Law Review Commission, 1970
- President's Advisory Panel on Timber and the Environment, 1973
- National Commission on Materials Policy, 1973
- National Water Commission, 1973
- Committee on Forest Development in the Tropics (Food and Agricultural Organization of the United Nations), 1985
- Commission on Environment and Development (United Nations), 1977
- President's Commission on American's Outdoors, 1988
- Governors' Task Forest on Northern Forest Lands, 1990
- Committee on Forestry Research (National Academy of Sciences), 1990
- Presidential Commission on State and Private Forests, 1991

The Media Some of the most dramatic evaluations of policies and programs come from outside government. Evaluation of forestry and natural resource programs is almost a daily occurrence in some newspapers and occurs with substantial frequency in popular periodicals such as *American Forests, Audubon, Forest Industries, National Parks, National Wildlife Magazine, Sierra,* and *Wilderness.* Especially compelling is television's ability to undertake investigative reports in which policy and program deficiencies are alleged and subsequently transmitted into the homes of thousands of citizens. Timber-harvesting practices (for example, clear-cutting) and the management of catastrophic wildfire and especially amenable to evaluation of this type.

Scholars and Free-Lance Writers Scholars and free-lance writers can also assume the role of evaluators of implemented forest resource policies. The 1970 evaluation of forest-management practices on the Bitterroot National Forest in Montana in an especially pointed example of evaluation activities undertaken by a university faculty. The evaluation's statement of findings alleged, for example, the following (U.S. Senate 1970, p. 13):

- Multiple use management, in fact, does not exist as the governing principle on the Bitterroot National Forest.
- Quality timber management and harvest practices are missing. Consideration of recreation, watershed, wildlife and grazing appear[s] as afterthoughts.

• The management sequence of clearcutting–terracing–planting cannot be justified as an investment for producing timber on the Bitterroot National Forest. We doubt that the Bitterroot National Forest can continue to produce timber at the present harvest level.

• Manpower and budget limitations of public resource agencies do not at present allow for essential staffing and for integrated multiple-use planning.

• [The] research basis for management of the Bitterroot National Forest is too weak to support the management practices used on the forest.

• The USDA-Forest Service as an effective and efficient bureaucracy needs to be reconstructed so [that] substantial, responsible, local public participation in the process of policy formation and decision making can naturally take place.

The charges made by the evaluation were forthright, to say the least. In combination with other evaluations, they led to significant changes in reforestation and timber-harvesting activities throughout the National Forest System.

Consulting Organizations Agencies frequently request that outside consulting or research organizations evaluate the implementation of a policy or program. A forest resource agency may take this route for any of a number of reasons, including obtaining access to skills and techniques not available within the agency, acquiring a fresh and bureaucratically unencumbered view of a forestry program's implementation, or building greater public confidence in the results of an evaluation that is focused on a particularly sensitive program. Among the organizations called upon by agencies to evaluate policies and programs are private management consultants; university centers of expertise; and nonprofit foundations and institutes such as the Conservation Foundation, Resources for the Future, and the Brookings Institute.

Interest Groups Interest groups are active participants in the evaluation of forest resource programs. Often armed with skilled analysts, they have scrutinized a number of public forestry programs. For example, the National Forest Products Association assumed an especially important role in evaluating and informing policymakers and the public about the productivity of forests for timber production (Forest Industries Council 1979). Likewise, the Society of American Foresters has evaluated the implications of acidic deposition in forested environments and the scheduling of old-growth timber harvests (Society of American Foresters (1984a, 1984b). With much intensity, The Wilderness Society has undertaken evaluations of old-growth forests in the Pacific Northwest (1988a), and has also evaluated USDA-Forest Service policies (e.g., biological diversity, water quality and timber management, reforestation, and timber suitability) for the National Forest System (1988b, 1988c, 1988d). The Natural Resources Defense Council has been similarly active on subjects concerning cost accounting systems applied to the sale of public timber from public lands (Barlow 1980).

Citizens Citizens can evaluate forestry and natural resource programs by simply casting ballots. A city council member who fails to support a popular and publicly desired urban forestry program may not be returned to office by a disgruntled citizenry. Similarly, the power of citizens' votes can force a change in personnel in positions outside the competitive civil service. When a new national administration takes over, for example, there will be new occupants of positions such as USDA-Assistant Secretary for Natural Resources and Environment, USDI-Assistant Secretary for Fish, Wildlife and Parks, and USDI-Director of Bureau of Land Management (U.S. House of Representatives 1980). The Senior Executive Service, created in 1978, accomplishes much the same end without the prod of the voting public. In a variant on the theme of rotation, a disgruntled bureaucrat may elect to go public with specific details of a forestry program's inefficiency, incompetence, or corruption. Although such actions may lack the clarity of formal evaluations, the consequences are much the same.

Operating Staff and Specialized Staff

Self-Evaluation Evaluation of implemented programs can be carried out by forestry professionals employed by an implementing agency. In some situations, those responsible for implementing a forest resource program may also assume the task of assessing its performance. For example, the director of recreation and lands for a public forestry agency may launch an evaluation of a program concerning trails and dispersed recreation. Budgets may be reviewed, interviews conducted, achievement of targets appraised, and changes recommended. With detailed knowledge and understanding of the program, the director is able to focus quickly on parts of the program that are in need of special scrutiny. Because of close working relationships, the director is in a position to create a nonthreatening (or at least less threatening than if an outsider evaluated the program) environment among program implementors. Program implementors might even be more receptive to program changes suggested as a result of the evaluation.

However, personnel (such as an economist, a planner, and a soil scientist) with the specialized skills needed for an effective evaluation may not exist on the recreation and trails staff. If the program area has repercussions for other program areas (such as timber management, watershed management, and forest planning), a comprehensive, truly programwide evaluation will not be achieved by reliance on a single operating unit (i.e., recreation and trails).

A matter of more fundamental concern, however, is whether it is appropriate for a unit within an organization to evaluate its own implemented programs. Understandably, a unit (or for that matter, an entire agency) cannot rely on outsiders to evaluate all implemented programs, and yet "no matter how good its internal analysis, or how persuasively an organization justifies its programmes to itself, there is something unsatisfactory about allowing it to judge its own case'' (Wildavsky 1972, p. 518). There are advantages in periodically subjecting programs to independent evaluation, either by other units within a large organization or by evaluators from outside an organization.

Specialized Evaluation Staff Evaluation activities can also be undertaken by a separate, specialized evaluation staff within an agency, which has the skills needed to undertake specialized evaluations. Such a staff may not have direct links to specific programs and thus may not have a vested interest in continuation of a program in a particular manner. In state natural resource agencies, specialized staff may be found in offices of planning and research; resource policy units; environmental review offices; and the like. Examples of federal agency offices or units in which evaluators practice their trade are:

- *USDI-Bureau of Land Management* Located in Washington: Division of Program Evaluation (Management Services); Division of Planning and Environmental Coordination (Support Services); Division of Legislation and Regulatory Management (External Affairs); Division of Policy Analysis and Program Coordination (Minerals Resources).
- *USDI-National Park Service* Located in Washington: Office of Planning and Development; Office of Policy, Budget and Administration.
- *USDI-Fish and Wildlife Service* Located in Washington: Office of Policy, Budget and Administration; Program Development Staff (Wildlife Resources Program); Office of Program Development (Fishery Resource Program); Office of Program Development (Habitat Resources Program); Division of Program Analysis, Division of Program Planning (Planning and Budgeting).
- *USDA-Forest Service* Located in Washington: Programs and Legislation Staff (Policy Analysis Staff, Program Development and Budget Staff, Environmental Coordination Staff, Resources Program and Assessment Staff); National Forest System Staff (Land Management Planning Staff, especially analysis and special studies). Located in regional offices: Regional Office Staff (Planning, Programing and Budgeting Staff).

For example, projects undertaken or reported on by the USDA-Forest Service Policy Analysis Staff during fiscal year 1986 include (Forest Service 1985):

- *Policy Reviews* Community Dependency on Federal Timber Sales; Comparing Resources Planning Act Goals and Land Management Plan Targets; Timber Contract Modification Act Costs; Developing Directions for Forest Economics Research; Timber Sale Program Accounting; Extensive Margin for Timber Production on National Forest System Lands; Long-Term Leasing of National Forest System Lands for Timber Production; Non-project Costs of Reforestation
- *Program Evaluations* Econometric Forecasting of Recreation Maintenance Backlogs and Annual Funding Needs; Evaluation of Graduate Study in

Systematic Analysis and Future Training Needs; Categorizing Research into Enhancing and Sustaining Activities; Evaluation of 1985 Farm Bill Conservation Reserve Forestry program; Evaluating the Management of Knutson-Vandenberg Program Funds; Assessing Impacts of Reduced Road Programs on National Forests

During fiscal year 1990, the Policy Analysis Staff focused on evaluation of the following programs (examples): critique of forest planning, below-cost timber sale guidelines, Greater Yellowstone Area Management, community dynamics, rural development, timber supply disruption, and pesticide policy and programs (Forest Service 1989).

Appointed Task Forces Evaluation activities may also be undertaken by task forces or investigative teams composed primarily of agency employees appointed by high-level policymakers. Examples are:

- The 1988 Fire Management Policy Review Team, which was appointed by the Secretary of Agriculture and the Secretary of the Interior to review national fire-management policies and their application in National Parks and wildernesses
- The 1978 Interagency Task Force on Tropical Forests which was co-chaired by the U.S. Department of State and the U.S. Department of Agriculture and had as its purpose providing a focal point for policy review and coordination

The nature of the evaluation efforts undertaken by high-level task forces and investigative teams can be appreciated by naming some of the recommendations of the National Fire Management Policy Review Team (1988):

- Line officers certify daily that adequate resources are available to ensure that prescribed fires remain within prescription.
- Develop regional and national contingency plans to constrain prescribed fires under extreme conditions.
- Consider opportunities to use planned ignitions to complement prescribed natural fire programs and to reduce hazard fuels.
- Utilize the National Environmental Policy Act requirements in fire-management planning to increase opportunity for public involvement and coordination with state and local government.
- Review funding methods for prescribed fire programs and fire suppression to improve interagency program effectiveness.
- Investigate and as appropriate act upon allegations of misuse of fire-management policy.

Professionalism Professionalism within a forestry organization is also a means by which demands for evaluation may be lessened. It is sometimes argued that if well-understood and generally accepted norms for professional behavior (as transmitted by academic and on-the-job training) are enforced on the job, patterns of performance become predictable and control via evaluation activities is of less concern. In this view, high-level administrators are presumed to be in the unique position of having a cadre of subordinates whose policy-implementing activities are guided by the ideology of professionalism—an ideology attuned to the views of such administrators. Landscape architects, for example, would facilitate a policymaker's interest in fostering aesthetically pleasing forest landscapes much more than would civil engineers, who might have an entirely different concept of how to construct a pleasing landscape.

Critics of the professionalism perspective, however, argue that members of a profession are far more diverse in their views on policy implementation than is commonly acknowledged, and that control of implementation via professionalism is not always a viable option (Brewer and deLeon 1983).

EVALUATION IN THE POLICY PROCESS

Properly conceived and carefully dispatched, evaluation can make significant contributions toward improving the performance of forest resource policies, programs, and projects. Evaluation takes place in many forms, including the rigorous analytical exercises carried out by agency policy analysts, as well as legislators' informal queries—such as ''How are things going?''—to program clients. The practice of evaluation is often frustrated by ambiguous notions of program success or failure: what one person views as success may be seen as failure by others. Too, evaluation activities are often thwarted by the size and complexity of forest resource programs. At times, all that may be said with confidence is that such programs have a positive, or negative, impact on intended goals—which may be sufficient, depending on the information needs of administering officials.

Forest resource policies and programs subjected to evaluation are numerous in number and very diverse in character. Consider the following as examples of such diversity:

- *Evaluation of Forestry Incentives Program* In response to concerns over the efficiency and effectiveness of federal cost-share payments to nonindustrial private owners of forests, evaluations of the Forestry Incentives Program were undertaken. The evaluations were designed to estimate the increased timber yield likely to result from such payments, and to determine the financial return associated with such increases. Among the recommendation for improving the program's future performance was more careful targeting of cost-share assistance, in terms of requirement for minimum are and a minimum financial return (for example, Ellefson and Risbrudt 1987; Mills and Cain 1978).
- *Evaluation of private forest-management assistance* In response to concern over the efficiency and effectiveness of federal and state forest-management

assistance to nonindustrial private owners of forests, evaluations of state-administered programs were carried out. The evaluations were designed to determine the economic and environmental consequences of forest-management advice provided to landowners by service foresters. Using a with-and-without model, landowners receiving forestry advice were compared with those who had not received such advice. The former generally received higher prices for their timber and usually were left with more productive stands after timber harvest (for example, Budelsky et al. 1989; Cubbage et al. 1985; Henly et al. 1988).

• *Evaluation of cooperative fire protection programs* In response to concern over criteria used to judge investments in the protection of forests from wildfire (e.g., reduction of acreage burned, acceptable area burned), an evaluation was made of the economic efficiency of investments in fire protection focused on nonfederal forest lands. The analysis identified opportunities for improving protection activities and defined the general distribution of benefits that result from fire protection investment (Bellinger et al. 1983; Bellinger 1983).

Evaluation activities are linked to various segments of the policy process. For example, evaluation can bring to the surface heretofore unknown problems and can thus stimulate interest in the formulation of additional, more appropriate policy options. The evaluation process can even unearth concerns of sufficient magnitude to foster interest in a return to agenda-setting activities. By making administrators better informed and more realistic in their expectations, evaluation can improve the implementation of policies and programs. Evaluation also has obvious implications for termination activities. Once evaluation has demonstrated that a program is inefficient and dysfunctional, the process of termination may begin. Indeed, the numerous feedback mechanisms between evaluation and implementation may be the strongest link in the policy process.

REFERENCES

Asher, W. *Forecasting: An Appraisal for Policy-Makers and Planners.* Johns Hopkins University Press, Baltimore, Md., 1978.

Barlow, T. J. *Giving Away the National Forests: An Analysis of U.S. Forest Service Timber Sales Below Cost.* Natural Resources Defense Council, Washington, 1980.

Bellinger, M. D. The Effectiveness of Cooperative Fire Protection Programs, pp. 272–277, in *Nonindustrial Private Forests: A Review of Economic and Policy Studies,* by J. P. Royer and C. D. Risbrudt (eds.). School of Forestry and Environmental Studies, Duke University, Durham, N.C., 1983.

———, A. A. Dyer, R. A. Chase, and H. F. Kaiser. *Economics Efficiency Procedures Used to Evaluate the National Fire Management Program on Non-Federal Lands.* Technical Bulletin 149. Experiment Station. Colorado State University, Fort Collins, Colo., 1983.

Brewer, G. D., and P. deLeon. *The Foundations of Policy Analysis.* The Dorsey Press, Homewood, Ill., 1983.

Budelsky, C. A., J. H. Burde, F. H. Kung, and others. *An Evaluation of State District Forester Timber Marking Assistance on Nonindustrial Private Forest Lands in Illinois.* Department of Forestry, Southern Illinois University, Carbondale, Ill., 1989.

Cubbage, F. W., T. M. Skinner, and C. D. Risbrudt. *An Economic Evaluation of the Georgia Rural Forestry Assistance Program.* Research Bulletin 322. University of Georgia Agricultural Experiment Station, Athens, Ga., 1985.

Davis, L. S. Analyst or Advocate? The Role of Natural Resource Professionals in Policy Formulation. *Edge: Natural Resources and People* 3(2):50–56, 1980.

Dye, T. R. *Understanding Public Policy.* Prentice-Hall, Inc., Englewood Cliffs, N.J., 1981.

Ellefson, Paul V. Decisions about Public Forestry Programs: The Role of Policy Analysts and Analytical Tools. *Journal of Resource Management and Optimization* 4(1):65–77, 1986.

——— and C. D. Risbrudt. Economics Evaluation of a Federal Natural Resources Program: The Case of the Forestry Incentives Program. *Evaluation Review* 11(5):660–669, 1987.

Fire Management Review Team. *Report on Fire Management Policy.* U.S. Department of Agriculture and U.S. Department of the Interior, Washington, 1988.

Forest Industries Council. *Forest Productivity Project Reports* (various states). National Forest Products Association, Washington, 1979.

Henly, R. H., P. V. Ellefson, and M. J. Baughman. *Minnesota's Private Forestry Assistance Program: An Economic Evaluation.* Msc. Pub. 58-1988. Agricultural Experiment Station. University of Minnesota, St. Paul, Minn., 1988.

Hogwood, B. W., and L. A. Gunn. *Policy Analysis for the Real World.* Oxford University Press, London, England, 1986.

Interagency Task Force on Tropical Forests. *The World's Tropical Forests: A Policy, Strategy and Program for the United States.* U.S. Department of State and U.S. Department of Agriculture, Washington, 1980.

Jones, C. O. *An Introduction to the Study of Public Policy.* Brooks-Cole Publishing Company, Monterey, Calif., 1984.

Kent, G. A. Decision-Making. *Air University Review.* May/June 1971.

Koenig, L. W. *An Introduction to Public Policy.* Prentice-Hall, Inc., Englewood Cliffs, N.J., 1986.

MacRae, D., and J. A. Wilde. *Policy Analysis for Public Decisions.* University Press of America, Lanham, Md., 1985.

Meltsner, A. R. *Policy Analysts in the Bureaucracy.* University of California Press, Berkeley, Calif., 1976.

Mills, Thomas J., and Daria Cain. *Timber Yield and Financial Return Performance of the 1974 Forestry Incentives Program.* Research Paper RM-204. Rocky Mountain Forest and Range Experiment Station. USDA-Forest Service. Fort Collins, Colo., 1978.

National Fire Management Policy Review Team. *Report on Fire Management Policy.* U.S. Department of Agriculture and U.S. Department of the Interior, Washington, 1988.

Natural Resources Defense Council. *Taxing the Tree Farm: Sensible Policies for Sensible Private Forestry.* Project on Agricultural Conservation and Tax Policy, Washington, 1988.

Ogul, M. *Congress Oversees the Bureaucracy*. University of Pittsburgh Press, Pittsburgh, 1976.

Rosenbloom, D. H. *Public Administration: Understanding Management, Politics, and the Law in the Public Sector*. Random House, Inc., New York, 1986.

Schultze, C. L. *The Politics and Economics of Public Spending*. The Brookings Institute, Washington, 1968.

Shands, W. E., T. E. Waddell, and G. Reyes, *Below-Cost Timber Sales in the Broad Context of National Forest Management*. The Conservation Foundation, Washington, 1988.

Society of American Foresters. *Acidic Deposition and Forests*. Washington, 1984a.

———. *Scheduling the Harvest of Old Growth*. Washington, 1984b.

Szanton, P. *Not Well Advised*. Russell Sage and Ford Foundation Press, New York, 1981.

The Wilderness Society. Old Growth in the Pacific Northwest: A Status Report. Washington, 1988a.

———. Water Quality and Timber Management, vol. 1 of *National Forests: Policies for the Future*. Washington, 1988b.

———. Protecting Biological Diversity, vol. 2 of *National Forests: Policies for the Future*. Washington, 1988c.

———. Reforestation Programs and Timberland Suitability, vol. 3 of *National Forests: Policies for the Future*. Washington, 1988d.

U.S. House of Representatives. *United States Government: Policy and Supporting Positions*. Committee on Post Office and Civil Service. Washington, 1980.

U.S. Senate. *A University View of the Forest Service* (*Bolle Report*). Senate Document 115. 91st Congress, 2d Session. Washington, 1970.

USDA-Forest Service. *Plan of Work. Policy Analysis Staff*. U.S. Department of Agriculture, Washington, 1985.

———. *Policy Analysis Recommended Program of Work:* FY 1990. Policy Analysis Staff. U.S. Department of Agriculture. Washington, 1989.

Weimer, David L., and Aidan R. Vining. *Policy Analysis: Concepts and Practice*. Prentice-Hall, Inc. Englewood Cliffs, N.J., 1989.

Wildavsky, A. The Self-Evaluating Organization. *Public Administration Review* 32:509–520, 1972.

Wholey, J. S. *Evaluation: Promise and Performance*. The Urban Institute, Washington, 1979.

9

TERMINATING POLICIES AND PROGRAMS

Forest resource policies and programs are designed and implemented with the intent of achieving laudable goals and objectives which have been prepared in response to problems or issues. For example, a sharp increase in the price of timber may result in the design of policies and programs to accomplish sustained high-level flows of timber at acceptable prices. A significant demise in a fishery may cause development of forest land-management policies to foster the production of water quality viewed as conducive to the propagation of fish and related aquatic species. Public outcry over timber harvesting's adverse impact on scenic beauty may influence the development of harvesting regulations to promote forest landscapes considered pleasing to the eye of the forest visitor. A decline in hunting success rates may lead to design of habitat-improvement programs that will please hunters of big-game wildlife species. Even though such statements of problems, goals, and policy intent are laudable, questions can be raised about them. Here are some examples of valid questions:

- Are the goals and objectives toward which such policies and programs are directed ever accomplished?
- Can policies and programs outlive their usefulness as means of addressing an issue or a problem? Need forest resource programs live on forever?
- How should policymakers deal with programs that evaluation finds to be redundant, outmoded, and dysfunctional?

• By what means should an agency, a bureau, or a work station be dismantled when the forestry program it implements is no longer demanded or supported by an important clientele?

Answers to questions of this nature are the substance of policy *termination*. In this chapter, we shall consider the nature and intent of termination, the differing kinds and levels of termination, and the strategic and tactical means by which termination activities can be carried out.

CONCEPT AND PURPOSE

Issues

Termination is sometimes viewed as the deliberate conclusion or cesssation of a policy, a program, or an organization. Conscious decisions are made to cease the production of certain goods and services, to stop public investments, to discharge employees, and to abandon or transfer ownership of facilities and equipment. The issues raised by termination are many:

• How can a program or policy be rationally and humanely adjusted or terminated?

• Who will suffer from such adjustment or termination, and in what ways?

• What obligations do authorities have to such individuals—compensation, moral obligation, access to appeal process?

• Who should be responsible for terminating actions? A targeted agency or a detached agency?

• Why is it so difficult to terminate certain policies and programs?

• Can forest resource programs be assembled so they can be easily terminated, if they are later found to be defective?

• What incentives might be created to encourage termination of outmoded forest policies, programs, and agencies?

• What political dangers might be posed by termination? Would support for other programs be lost, for example?

• How might the expectation of permanence found in many public forestry programs be shifted to favor flexibility and greater adaptability?

Degrees

The word ''termination'' often is taken to convey a sense of finality. People think of termination as the final step in the policy process. A policy, a program, or an organization is viewed as being abandoned because it was ineffective or because its intended purposes were not accomplished. In reality, however, termination is seldom so final. For instance, suppose a forest policy was initiated in response to a problem; if evaluation suggests that it is doing more harm than good, termination is begun. At the other extreme, a forest policy may be so effective that it erases the problem; in this instance also, termination is begun. Between these two extremes, there may occur any number of fine-tunings or partial terminations—adjustments and refinements that make forest policies more responsive.

The process of termination is more successional than final. Complete terminations of forest policies and programs are rare. Administrators are seldom presented with a clean slate upon which to formulate and implement a new set of forest policies or organizations. Most often they are presented with a program or organization that has been steadily fine-tuned and adjusted (in other words, that has been subjected to a succession of partial terminations) in response to both experience and new conditions. In this context, it is obvious that final large-scale terminations and clean breaks with the past are rare. Subtle partial-termination activities, however, are almost routine in forest resource administration.

Purpose

Termination of a forest resource policy or program may be suggested for any of a number of reasons. For example, sometimes advocates of terminating a policy or an organization (i.e., its opponents) truly believe that the substance of the policy being implemented is in error. Others (i.e., economizers) seek to terminate a policy because they view it as inefficient, ineffective, or expensive. Yet others (i.e., reformers) urge termination because they consider it a prerequisite to installation of a better, replacement program. However, the process of termination is usually initiated for more fundamental reasons.

Termination may be initiated because of an adverse evaluation of a policy or a program's performance. For example, the public subsidies designed to enhance economic development in the forestry sector may be found inefficient and ineffective. The policies, programs, and administrative units involved will be dismantled and the resources directed elsewhere. In another example, a policy of banning exports of logs from public lands may be adversely affecting the nation's balance of payments. The policy is stopped, and other options are considered. Termination may also be undertaken in response to a political climate of budget retrenchment. A national or state recession may strain public finances to the point that agency budgets must be reduced, severely cut, or eliminated. Field stations may have to be closed, research programs abandoned, implementation of forest road plans postponed, and wildlife habitat-improvement programs dismantled.

Termination also becomes a reality when political support for a policy or program's continued existence begins to wane. Legislators' ardent enthusiasm for the imposition of fees for use of public campgrounds may rapidly diminish with the advent of strong public opposition to such fees. With legislators' support gone, the fees and all activities attendant to their collection are abandoned.

Another instance in which termination takes place is when the private sector either gradually or suddenly assumes the production of goods and services that were previously produced as a result of a public policy or program. For example, the private sector may begin to offer overnight recreational camping facilities. The need for the public program disappears.

TYPES AND LEVELS

Termination is not a uniform exercise. Rather, it is a process that varies according to the activity or administrative level in question, whether it be an organization, a policy, or a program (Brewer and deLeon 1983).

Agency or Organization

Termination of an *organization* (say, a forestry agency, or its parts) is often very difficult. Organizations, which are usually designed for long-term operations, have significant strength and ability to resist termination. An example is the failure of efforts to terminate the USDA-Forest Service and transfer its activities to a proposed federal department of natural resources (Bor 1979). More common than outright termination is the merging or splitting of an organization, as when a state forestry agency is combined with a pollution-control agency to form a department of the environment. Administrators prefer to forsake some of an organization's programs and policies rather than to allow the organization itself to be terminated.

Policies

Termination may also be directed at a *policy* (here defined as an aggregate of programs) which sets forth guiding themes and objectives. For example, termination may be focused on policies assuring the economic stability of communities located in close proximity to a state forest, or on policies that assure forest recreational opportunities for disadvantaged persons. Policies are easier to terminate than sponsoring organizations, for a number of reasons. They are likely to have fewer political allies than an organization (some legislators or interest groups, for example, may favor an organization but oppose the implementation of certain policies). If an organization seems to be threatened with termination, it may willingly sacrifice (or terminate) selected forest policies rather than face extinction. In power politics, policies have less clout than organizations. In addition, because specific policies are easier to evaluate than is a multifaceted, multiobjective organization, solid reasons for termination are easier to come by.

Programs

Termination may also focus on a specific *program,* such as a law-enforcement program, a legislative affairs program, a day-use recreation program, or a watershed-management program. Programs often represent the smallest investment on the part of a sponsoring agency and may have the fewest political resources with which to protect themselves. They are closest to the forest resource problem being addressed; their impact can be directly measured, and if it is found lacking, blame can be easily fixed (Brewer and deLeon 1983).

Partial Termination

Termination need not be viewed as wholesale dismissal of a forestry organization, policy, or program. More commonly, termination is partial or incremental. Changes in context (such as political or economic conditions) or enabling technologies (such as better communication technologies) supplant existing programs or processes. In such a setting, various types of termination can be identified, including the ones listed below (Brewer and deLeon 1983, p. 396).

- *Replacement* Something old is replaced with something new to satisfy the same objectives or goals. Replacement is the consequence of innovation in which old methods or procedures are displaced by new more efficient and effective ones. For example, when programs involving detection of wildfire via sighting from lookout towers are replaced by programs involving detection from aircraft, lookout tower programs are terminated.
- *Consolidation* Existing administrative arrangements are brought together, and redundant or less efficient parcels are eliminated. For example, a forestry organization's operations are centralized in order to achieve economies of scale. Less efficient units are terminated.
- *Splitting.* An existing program or administrative unit is divided into many separate administrative units which separately cannot logically or financially support certain activities. For example, a state forest is split, and the resulting units cannot support a program-planning team or a watershed-management team. Planning and watershed teams are terminated.
- *Discontinuation* An existing program or administrative unit is replaced with something new, and at the same time program or unit goals and objectives are altered. For example, state seed-tree laws focused on legally requiring private landowners to regenerate forestland for timber production purposes are succeeded by modern comprehensive state forest practice laws focused on the promotion of various aspects of environmental quality.
- *Deincrementation* Long-term marginal deemphasis of a policy or program is the response to long-term changes in political or economic conditions. For example, long-term economic stagnation leads to gradual termination of a spe-

cific public forest resource program; or intentional (and extended) deemphasis of the federal role in facilitating state forestry programs leads to gradual termination of federal programs focused on state forestry activities.

STRATEGIC AND TACTICAL DESIGNS

Termination of programs, policies, and organizations is seldom a pleasing event. "Organizations resist and fight back. Confusion, indignation and pathos reign; a sense of betrayal pervades" (Brewer and deLeon 1983, p. 428). People and organizations do not expect termination to happen to them; when it does happen, they seldom accept it with grace. The fear of loss is a more powerful stimulus than the prospect of gain.

The question is, what can be done strategically and tactically to ease the process of termination? Some suggestions are presented in this section (DeLeon 1978; Brewer and deLeon 1983).

Termination and Policy Development

Termination should be made an integral part of the policy-development process and should be deliberately addressed during the process of policy formulation. This has been well stated by Hogwood and Gunn (1984 p. 250):

> It would be advisable for those who advocate solutions to the problems faced by society to begin to design their policies with a greater sensitivity to the problems which future generations of policy-makers will have when they will almost inevitably attempt to alter a policy. . . . One generation's monuments may be the next generation's mausoleums.

Forest resource programs may accomplish their intended purposes or they may fail to do so; either situation represents an opportunity to employ termination as a means of allocating resources for a more productive use. To facilitate such allocation, termination should be explicitly considered during formulation activities. It is useful to consider which of the policy and program options could most easily be terminated and how such termination could be accomplished. Once policymakers have the answers to such a question, they can base selection of a forest resource policy in part on the ease with which termination can be invoked, when the need arises. Termination contingency plans may have to be developed, specifying what constitutes failure, what constitutes accomplishment of policy intent, and what procedures are to be followed when termination becomes desirable. Knowing what constitutes program success or failure enables analysts and policymakers to focus on the events that will occur after selection and implementation, and to make well-informed judgments about the need to terminate a forestry program.

Positive Attitude

A receptive attitude toward termination should be developed among organizations, employees, and program clients. Termination is not facilitated by a termination culture that is replete with phrases like ''hatchetman,'' ''hired gun,'' and ''program assassin.'' Efforts should be made to instill recognition that termination is not the end of the world but rather is either the natural consequence of successful accomplishment of the program objectives or an opportunity to improve an inefficient or ineffective situation. The view that should be nurtured is that change (as reflected by termination) is a natural component of the policy process. Over time, the wisdom of addressing a particular issue through a particular forest resource policy may become questionable. In a different situation, the demand for the goods and services produced by a particular forestry program may decline or become nonexistent. In still another situation, the resources available for implementing a forest policy and subsequently confronting a forest resource problem may simply not be available—because of a nationwide economic slowdown, for instance.

Since organizations, employees, and clients like to solidify their positions and create a sense of permanence about the benefits that flow from many forestry programs—the result of an entitlement ethic—it can be difficult to obtain acceptance of such views. Administrators too have difficulty with termination, largely because they view it as a reflection of leadership failure. A more receptive administrative attitude toward termination would go far toward overcoming resistance to the ending of policies and programs, and would also further the accommodation of positive change which termination often represents.

Gradual Steps

Termination should be carried out gradually and incrementally, rather than immediately or totally, although the choice may be out of an administrator's control. Progressive termination is likely to be more successful than immediate termination because it allows program clients to gradually transfer their demands for services elsewhere and because agency and employee resistance to termination usually fades gradually, as the inevitable slowly becomes acceptable. An incremental strategy also has the advantage of enabling those affected by termination—both the employees and the targets of the program—to acquire alternative positions and resources. An immediate termination strategy is abrupt, is liable to meet greater resistance, and will require greater planning should it be selected.

Budgetary Growth Periods

Consideration should be given to terminating programs during periods of economic and budgetary growth. The relative abundance that occurs during times of economic growth can foster alternative employment and related options for those whose forestry programs are to be eliminated. Relocations, transfers, and related actions can ease the administrative and employee burdens that often accompany plans to end a policy or program. Ironically, the demand for termination of forestry programs is usually the highest during periods of financial security, when employment and related options are relatively sparse. However, if termination is recognized as a continuing process—one that is carried out during good as well as bad times—the severity of its emotional impact during periods of austerity can be lessened.

Leadership Changes

A forest policy or program should ideally be terminated during key periods in its life cycle. For example, when a change occurs in the leadership of a forestry agency, a program, or a support group (such as an interest group or a legislative committee), a distinct break with past commitments and experiences may facilitate total or partial termination of forestry programs. If a legislative leadership change is anticipated, program termination may encounter less resistance since the agency's legislative allies are weaker at such a time. Similarly, termination can be facilitated by being alert to changes in ideological and political climates. Cycles of conservatism, liberalism, or government activism can indicate conditions favoring termination. Ideological shifts toward less government involvement in the private forestry sector often bodes well for partial termination of a variety of public programs (such as cost-share programs, extension programs, and regulatory programs) that are directed at the activities of private forestland owners. Likewise, termination of policies and programs can be facilitated by taking action when an unexpected forest resource issue or problem occurs, such as an energy crisis, an economic recession, or a physical disaster (perhaps a flood or a wildfire). Such external events can provide the excuse and the motivation for policy or program termination, and events or situations in the organization can serve the same purpose. Widespread misapplication of timber-management practices for example, can become a rationale for redesigning rules and regulations, reassessing professional capabilities, and adjusting levels of financial commitment.

Administrative Design

Administrative flexibility and incentives for termination of forest policies and programs should be viewed with high regard. Organizations can be designed with a core of administrative and technical experts that can be combined in varying proportions to address problems or issues as they arise. Concerned with problems of nonpoint sources of water pollution, for example, an agency could bring together a team of administrators, hydrologists, engineers, and fisheries managers to address the problem. Once the problem was resolved, the experts could move on to other forestry issues. With such a design, an agency retains the flexibility to establish (or terminate) project teams while assuring team members of continuing employment (Biller 1976).

Financial Incentives

Allowing an agency to retain at least a portion of any financial savings realized from actions taken to eliminate a questionable forestry program can be a positive incentive for canceling or replacing questionable programs. The savings could be used by administrators to facilitate more suitable programs and would enable a smoother transition of staff, clients, and facilities from one program to another.

Zero-Based Budgeting and Sunset Laws

Termination of policies and programs can also be strategically pursued via various budgeting actions. Common suggestions include the use of zero-based budgeting or the implementation of sunset laws. Both are designed to force periodic review of an organization or an organization's programs by requiring all involved to make a renewed case for the reauthorization of a policy, a program, or a financial commitment. Both of these strategies are formidable (and often threatening) prods toward the evaluation of policies and programs. Such approaches to termination, however, have been found ineffective for various reasons, including (Carlson 1982):

- Arduous review requirements placed on agencies and legislatures
- Absence of systematic, independent, and rigorous evaluation of programs
- Intense political pressure from program clients to forgo sunset assessment of particular programs
- Notable diversion of agency resources to the defense of existing programs

Sunset laws and related strategies have an intuitive appeal. More often than not, however, they are found to be oversimplified means of addressing incredibly complex problems (Behan 1977).

Specific Tactics

In sum, a number of strategic approaches to terminating organizations, policies, and programs are possible. The possibilities include the specification of conditions for termination during formulation, the development of receptive attitudes toward termination, the initiation of termination during periods of economic and budgetary growth and during key periods in a program's life, the development of incentives and catalysts that encourage termination, and the institution of gradual termination actions during annual reviews of a program's budgetary needs.

Although strategic actions of such a nature are important to termination, they must be buttressed by specific tactics. These tactics can take many forms, some of which are highlighted below (Behan 1982; Brewer and de Leon 1983).

- *Implement a well-designed termination plan* Develop and implement a carefully scripted, step-by-step plan for termination which focuses attention on routine termination procedures (e.g., disposal of equipment, transfer of personnel) and discourages perpetuation of debate over the termination decision per se.
- *Maintain tight security about plans for termination* Although an administrator can avoid unnecessary embarrassment and antagonism by floating trial balloons for new and untested programs, trial balloons concerning termination bring forth immediate and loud resistance from program supporters. Few individuals or groups come forth to support a suggestion for termination.
- *Amass supporting information* Pay attention to details of a program's operation, since opponents of termination focus on positive specifics of a program and attempt to generalize such specifics to the entire program.
- *Act swiftly and decisively* When substantial political resistance is expected, recognize that delay in implementing a termination plan leads to an increase in such resistance and to eventual perception of administrative impotence.
- *Assault politically weak segments of a program, not politically popular segments* Refrain from indiscriminate termination of projects within a program (or programs within a policy) without consideration of political support levels for individual projects (or programs).
- *Identify conspicuous failures* Identify and widely publicize the most sensational goofs or adverse consequences of a program so as to focus attention, gain support, and embarrass a policy's supporters.
- *Terminate as an example* Single out especially vulnerable programs, terminate them, and publicize such actions as examples of what will occur elsewhere if the efficiency of other programs is not improved.
- *Bargain to remove resistance* Seek out important opponents of termination, and enter into agreements (e.g., reinforcements of other programs) that will turn resistance into agreement.
- *Employ an outsider as terminator* Select a person or an organization (such as an administrator, a commission, or the well-known leader) from outside the targeted program to make the statements and issue the directives that are necessary to ensure termination.
- *Expect and accept short-term cost increases* Recognize and be prepared to accommodate short-term cost increases due to termination. Such increases may be necessary in order to achieve long-term savings.

• *Apportion termination evenly* Indiscriminately apply a uniform percentage reduction in all program budgets and activities.

OBSTACLES AND DETERRENTS

Termination of programs, policies, and organizations is necessary. Resources need to be efficiently allocated, and the clients of forest resource programs need effective service. Unfortunately, however, termination is seldom planned for in advance, is rarely implemented in a smooth and harmonious fashion, and hardly ever results in a high degree of satisfaction for all involved. Some forestry programs and organizations put significant obstacles in the way of the administrator turned terminator.

In this section, we shall consider some major circumstances that can deter the process of termination.

Moral and Intellectual Resistance

Termination of forestry programs often meets with substantial moral and intellectual resistance. Administrators and program clients, for example, nearly always have strong negative feelings about the closing of a ranger station, the suspension of a research program, or the stopping of financial support for owners of private forests. Emotions can run high—can even be comparable to those surrounding bankruptcy, divorce, and death—and reliance on dispassionate analyses can quickly wane or become nonexistent. The reaction cycle often includes an initial shock, denial that anything so terrible could be happening, anger with survivors, and finally acceptance of the inevitable. Administrators may agonize over how to treat employees (questioning whether preference should be given to long-term, loyal employees, for instance) and affected communities (questioning what response to communitywide unemployment will be appropriate and helpful). Other problems that bother administrators are the extent to which they are responsible for employees with relatively few marketable skills, and how to minimize declining morale among a program or an agency's surviving employees. The process of termination may become so distasteful that administrators will withdraw from responsibility. They may assign the task to a faceless "they," to a personnel committee, or to a budgeting department.

Similarly, intellectual resistance to the notion of termination is often significant. Formulators and analysts design forest resource policies and programs for purposes of solving a specific problem, paying little if any attention to the eventual need to terminate the proposed policy. Especially neglected during formulation are equity and related humanitarian considerations. During termination, however, such considerations become glaringly evident; unemployed persons, loss of income, and lowering of status cannot be ignored. No matter what the reason for termination or how good the logic, most administrators and program clients find termination an unpleasant experience. Termination is not easily undertaken nor widely accepted.

Ideology of Permanence

The adaptability and seeming permanence of forestry organizations, policies, and programs are often deterrents to termination. Forestry agencies, for example, are typically designed to last. Few individuals seriously propose wholesale termination of a state department of natural resources, a federal forestry agency, or a major environmental protection office. Such organizations are established to recognize, act upon, and solve forest and natural resource problems on a continuing basis. The same is true of some forest policies and programs (e.g., insect and disease management programs). In addition, there are few political incentives to engage in wholesale termination of programs or agencies. Rewards go to those who promote innovative policies, programs, and administrative structures, not to those who terminate them.

Faced with the prospect of termination, agencies are inclined to act to ensure their own survival. They are wonderfully adept at altering their objectives, policies, and clients in order to make termination more difficult. The forestry agency whose initial role was protection of forested watersheds may, over time, become a major provider of recreational activities—if watershed management diminishes in importance while recreation assumes a dominant role. Simply attaining a policy objective may not be reason enough to disband an organization or a program. New goals can and will be defined to allow continuation of existing forestry programs and organizations.

Political Opposition

Termination can be deterred by political actions of influential groups that are opposed to the demise of a forestry program or organization. Internally, well-entrenched members of a forestry agency may take aggressive actions to demonstrate their value to society and to publicize the dire consequences of a program's demise. They may work assiduously to develop a new rationale for continuing a program. They may procrastinate and compromise in hopes of gaining new supporters, and of temporarily placating and eventually outlasting the termination initiative.

Employees of long standing in an organization or a program can be unusually adroit at mobilizing supporters, such as key legislators and powerful interest groups, to help them make an especially strong case against termination. Such employees have had ample opportunity to cultivate supporters over the years, and can usually count on enthusiastic and executive assistance. Some outside supporters, in addition, have their own bases of political support which can effectively multiply the political assets of antiterminators. An example is an interest group that calls on closely related interest groups to do battle against termination.

The ultimate weapon of employees facing termination of policies and programs for which they are responsible is procrastination. Extensive delaying actions can abort efforts to eliminate a program, a policy, or an organization.

Legal and Procedural Deterrents

Legal and procedural conditions can also be important deterrents to termination of policies and programs. Public agencies, for example, operate under civil service rules, regulations, and due process. Employees and others may argue that they would be wronged or abused by termination, in light of civil service standards. Such allegations must be given thorough review and consideration before termination activities can proceed.

Similarly, laws may prohibit termination—or at least encourage litigation. For example, Bardach (1977, p. 129) points out that the federal ''Administrative Procedures Act forbids the government to be arbitrary and capricious.'' Such wording invites legal interpretation and legal action that can forestall termination activities. Likewise, a number of concepts pertinent to contract law may inhibit efficient termination of policies and programs. Among the relevant concepts are notions of damages, restitution, relief from coerced or unfair bargains, and changes in circumstances that affect the duty of carrying out a promise. In sum, access to legal redress adds yet another layer of obstacles to the need or desire to get rid of a program (Brewer and deLeon 1983).

TERMINATION IN THE POLICY PROCESS

Termination is an important but much-neglected part of the policy process, a part that signals the beginning of the process as much as it does the end. The task of dismantling a forest resource program may be so distasteful that administrators will prefer to ignore or shun the necessary activities. The reality, however, is that most policies, programs, and agencies will eventually have to be adjusted, curtailed, replaced, or even eliminated.

Attempts to terminate policies, programs, or agencies involving the management of forest resources have occurred periodically since the turn of the century. Some terminating strategies and tactics have been carried out with little discomfort for anyone involved. For example, the Office of Management and Budget's recommended closing of the USDA-Forest Service's Central States Forest Experiment Station in the 1960s was accomplished with minimal discomfort.

In other situations, however, the orchestration of terminating activities has been poor. Intense public outcry and the wielding of political power by influential legislators has even led to the demise of certain termination efforts. An example is the 1973–1974 attempt to close, with the intention of subsequent reestablishment at other locations, the USDA-Forest Service's Intermountain Forest and Range Experiment Station in Ogden, Utah, as well as its Northern Regional Office (R-1) in Missoula, Montana. The proposed action so irritated congressional delegates that they required the agency by law to have "field supervisory offices, and regional offices . . . so situated as to provide the optimum level of convenient, useful services to the public, giving priority to the maintenance and location of facilities in rural areas and towns near national forest and Forest Service program locations" [Forest and Rangeland Renewable Forest Resources Planning Act of 1974 as amended (Section 11b)].

Here are some other examples of terminations gone awry:

- *Proposed Federal Land Sale Program of 1982* An effort to reverse (or terminate) federal policies involving land ownership and acquisition by shifting responsibility for public land management from federal agencies to state and private sectors. The program failed for many reasons, including the lack of a clearly defined constituency that was strongly supportive of a significant reduction in federal land ownership (Lewis 1983).
- *Proposed merger and reorganization of federal natural resources agencies of 1979* An effort to merge (or terminate) the USDA-Forest Service and the USDI-Bureau of Land Management (and related agencies) to form a national forest and land administration within a proposed federal department of natural resources. Lack of clearly defined constituency, a dearth of information about the proposal, and an excessively long period of time consumed by internal reassessment of the proposal were all factors that led to demise of the termination effort (Convery et al. 1979).
- *Federal Jurisdictional Land Transfer Program* An effort to exchange (or terminate) responsibility for the management of 24 million acres of federal land and 204 million acres of mineral jurisdiction between the USDA-Forest Service and the USDI-Bureau of Land Management. Proposing to eliminate duplicate staff, offices, and related facilities, the program was never accomplished for various reasons, including the length of time required to prepare the program, resistance within the agency, and significant opposition from powerful political opponents (Association of O & C Counties 1985; Bureau of Land Management and Forest Service 1986; General Accounting Office 1984; Gorte 1985).

Termination activities are linked to many other segments of the policy process. Formulation activities, for example, can guide termination by specifying under what conditions a policy or program might be concluded. For example, will it be terminated when its objectives have been accomplished? After a fixed number of years has elapsed? If questions about the ease of termination of suggested forest resource policies are stated during formulation, preference orderings and ultimate selection may go smoothly. Decision makers have a responsibility to create among recipients of a program's benefits less expectation that a program will be permanent or continuous and more acceptance of limits or ceilings, so that the eventual termination procedures can be carried out in a less hostile environment.

Implementation presents opportunities to develop rules, regulations, and guidelines with the possibility of termination in mind. In this respect, events and deadlines that force action gain prominence. Evaluation has a special link to termination, because it implies that termination could be a consequence. Since termination is often a highly charged political event, its success requires considerable emphasis on accurate and comprehensive evaluation activities. Policy goals and objects must be clearly defined and analytical procedures rigorously adhered to.

REFERENCES

Association of O & C Counties. *BLM/Forest Service Land Exchange: Response of the Association of O & C Counties.* Roseburg, Oreg., 1985.

Bardach, E. C. *The Implementation Game: What Happens after a Bill Becomes Law.* The Massachusetts Institute of Technology Press, Boston, 1977.

Behan, R. D. The False Dawn of Sunset Laws. *The Public Interest* 49:103–118, 1977.

———. How to Terminate a Public Policy: A Dozen Hints for the Would-Be Terminator, in *Cases in Public Policy-Making,* by J. E. Anderson. Holt, Rinehart and Winston, New York, 1982.

Biller, R. P. On Tolerating Policy and Organizational Termination: Some Design Considerations. *Policy Sciences* 7:133–149, 1976.

Bor, R. M. Reorganization: A Divisive Issue, in *Reorganization: Issues, Implications and Opportunities for U.S. Natural Resources Policy,* by F. J. Convery, J. P. Royer, and G. R. Stairs (eds.). School of Forestry and Environmental Studies, Duke University, Durham, N.C., 1979.

Brewer, G. D., and P. deLeon. *The Foundations of Policy Analysis.* The Dorsey Press, Homewood, Ill., 1983.

Carlson, E. Success of Sunset Laws Varies and Fights Turn to Big Targets. *Wall Street Journal,* May 4, 1982.

Convery, F. J., J. P. Royer, and G. R. Stairs (eds.). *Reorganization: Issues, Implications and Opportunities for U.S. Natural Resources Policy.* School of Forestry and Environmental Studies, Duke University, Durham, N.C., 1979.

DeLeon, P. A Theory of Policy Termination, in *The Policy Cycle,* by J. V. May and A. B. Wildavsky (eds.). Sage Publications, Beverly Hills, Calif., 1978.

General Accounting Office. *Program to Transfer Land between the Bureau of Land Management and the Forest Service Has Stalled.* GAO/RCED-85-21. Resources, Community, and Economic Development Division, Washington, 1984.

Gorte, Ross W. *Analysis of Merging the U.S. Forest Service and the Bureau of Land Management.* 85-868 ENR. Congressional Research Service, The Library of Congress, Washington, 1985.

Hogwood, B. W., and L. W. Gunn. *Policy Analysis for the Real World.* Oxford University Press, London, England, 1986.

Lewis, Bernard J. *The Reagan Administration's Federal Land Sales Program: Economic,*

Legal and Jurisdictional Issues. Staff Paper Series Number 37. Department of Forest Resources, University of Minnesota, St. Paul, Minn., 1983.

USDI-Bureau of Land Management and USDA-Forest Service. *Legislative Environmental Impact Statement: Bureau of Land Management-Forest Service Interchange*. U.S. Department of the Interior and U.S. Department of Agriculture, Washington, 1986.

THREE

PARTICIPANTS

An understanding of the process by which forest resource policies are developed and carried out is of limited value if the citizens, organizations, and institutions that actively participate in the process are ignored. The participants include legislative systems, judicial systems, bureaucratic systems, organized interest groups, political parties, the media, and the general public. Each brings to the various events of the policy process a set of unique values, interests, and abilities. All together, they give the policy process life and meaning.

10

LEGISLATIVE SYSTEMS AND PROCESSES

Legislatures are the primary means by which modern democratic societies establish and maintain legal order, crystalize and settle conflicts, grant legitimacy to policies and programs, and adapt the existing rules of society to new conditions. Their staying power—they have existed for over 200 years in the United States—attests to their ability to aggregate diverse interests and to maintain a degree of consensus within society. To be sure, legislatures are only a part of the apparatus for making authoritative social decisions. In a variety of ways, they share power and responsibility with chief executives, bureaucracies, courts, political parties, interest groups, and the like. Time, place, and leaders shape these relationships. If they so choose, legislatures may follow the lead of separate power centers, join with them, ignore them, pit one center against another, or struggle against them. Legislative systems are fascinating political entities. They have much to do with policies and programs that are focused on the use and management of the nation's forests and related resources.

In this chapter, we shall consider the functions and responsibilities of legislatures, the manner in which they are organized, the nature of the persons and administrative units that activate legislative processes, and the character of the interactions between bureaucratic and legislative systems.

FUNCTIONS AND RESPONSIBILITIES OF LEGISLATURES

Legislatures are very active participants in nearly every segment of the policy process. They formulate forest resource policies via the creative actions of legislative staff; they legitimize forest policies by enacting them into laws; they evaluate forest policies by sponsoring investigatory hearings; they terminate forest policies and programs by deciding to reduce or eliminate funding; and they establish forestry agendas by selecting issues to be granted attention by lawmaking processes. Functionally, legislatures have chosen to express their concentration on the policy process by representing various constituents, informing and educating the public, legitimizing policies and programs in law, and providing for legislative oversight.

Constituent Representation

Representation of constituent interests is often viewed as a fundamental legislative function since legislatures are assemblies that represent individuals, organized interests, and geographic localities. The concept of representation can be portrayed in a number of fashions. Some contend that legislatures should be representative in a *descriptive* sense; that is, they say that legislators should reflect, or mirror, the politically relevant characteristics (race, sex, religion, and economic status) of the citizenry. Others suggest that representation should be viewed from the perspective of *substance*; that is, legislators should reflect, or mirror, the attitudes or interests of the citizenry in their public policy decisions. Still others view representation as a *formal* relation which is constitutionally prescribed; that is, legislators are legally authorized to represent a citizenry, with accountability in the relationship being enforced by periodic elections (Hinckley 1988).

Information and Education

Legislatures also have the function of informing and educating the public. Legislatures can be focal points for debate about weighty matters of public interest. Such debate, if accurately transmitted, can be instructional to the general public. For example, legislative contests over the merits of alternative forestland uses can alert interested publics to the possible consequences of pending land-use decisions. Legislative focus on the use of pesticides as part of an intensive timber-management program can unmask health risks that need to be recognized by the general public.

Although informing and educating the public are activities that are appropriate to legislatures, they are not always successfully accomplished, for several reasons. The number and complexity of proposed laws makes meaningful public awareness of their consequences nearly impossible. Legislators' restless search to achieve several goals—finding and maintaining political security, serving constituents, and seeking negotiated settlements—simultaneously often detracts from the time they have available for public education efforts. Debating crucial policy decisions in committee sessions (often reported to limited audiences) discourages widespread understanding of important forestry issues and the means available for addressing them. A public that is typically passive and absorbed in daily living is not a receptive audience; it is not interested in most suggested public forest policies nor in the details of lawmaking.

Enactment of Law

The lawmaking function of legislatures is widely acknowledged. In its broadest sense, lawmaking consists of finding major compromises for policy actions designed to address important issues. What is involved is sifting through innumerable alternatives to find the one that can be agreed to, and eventually legitimizing the agreement into law. The overriding strategy from introduction to final vote on a proposed law is to fashion a proposal that can attract and consolidate support with a minimum number of concessions to opponents. The process of gaining supporters may lead to development of a curious assortment of provisions, many of which are included in order to gain supporters. As Keefe and Ogul aptly stated, "the end product may be a bill that under the circumstances is the best possible, a bill that no one is particularly happy about . . . rare is the major proposal that ends up in law in the same form that it was introduced" (1989, p. 15).

Ideas and concepts are the raw material of legislatures. If forestry is of importance during a legislative session, what is sought will be new thoughts on means of encouraging more intense management of nonindustrial private forests, new ways of dealing with the use of pesticides and their adverse side effects, alternative approaches to the management of urban forests, and innovative ways of directing the use of public forests. Although legislatures are empowered to make laws, they depend heavily on familiar outsiders for the infusion of ideas—outsiders, such as the chief executive, the bureaucracy, interest groups, and party spokespersons (Davies 1986). Of special importance are the chief executive's proposals for new laws, or administration bills—proposals which usually become major items on legislative agendas. Whether or not a legislature recognizes the claims of the chief executive or those of any other groups demanding legislative action depends on the extent to which influential groups can mobilize to seek legislative solutions, the degree to which the unorganized public focuses on an issue, the extent to which powerful legislators take up a legislative cause, and the degree to which opposition to a new proposal fails to materialize.

The legislative processes by which ideas are transformed into laws seems incredibly complex to the uninitiated. This is especially true of the rules and parliamentary maneuvering employed by legislatures (Oleszek 1984; Willet 1986). Although specific procedures vary from one legislature to another, nine common stages through which an idea must pass in order to achieve the status of law are described below (Rosenthal 1981). A more formal representation, depicting the U.S. Congress, is presented in Figure 10-1. Examples of forestry's involvement in these stages can be found elsewhere (USDA-Forest Service 1986, 1990).

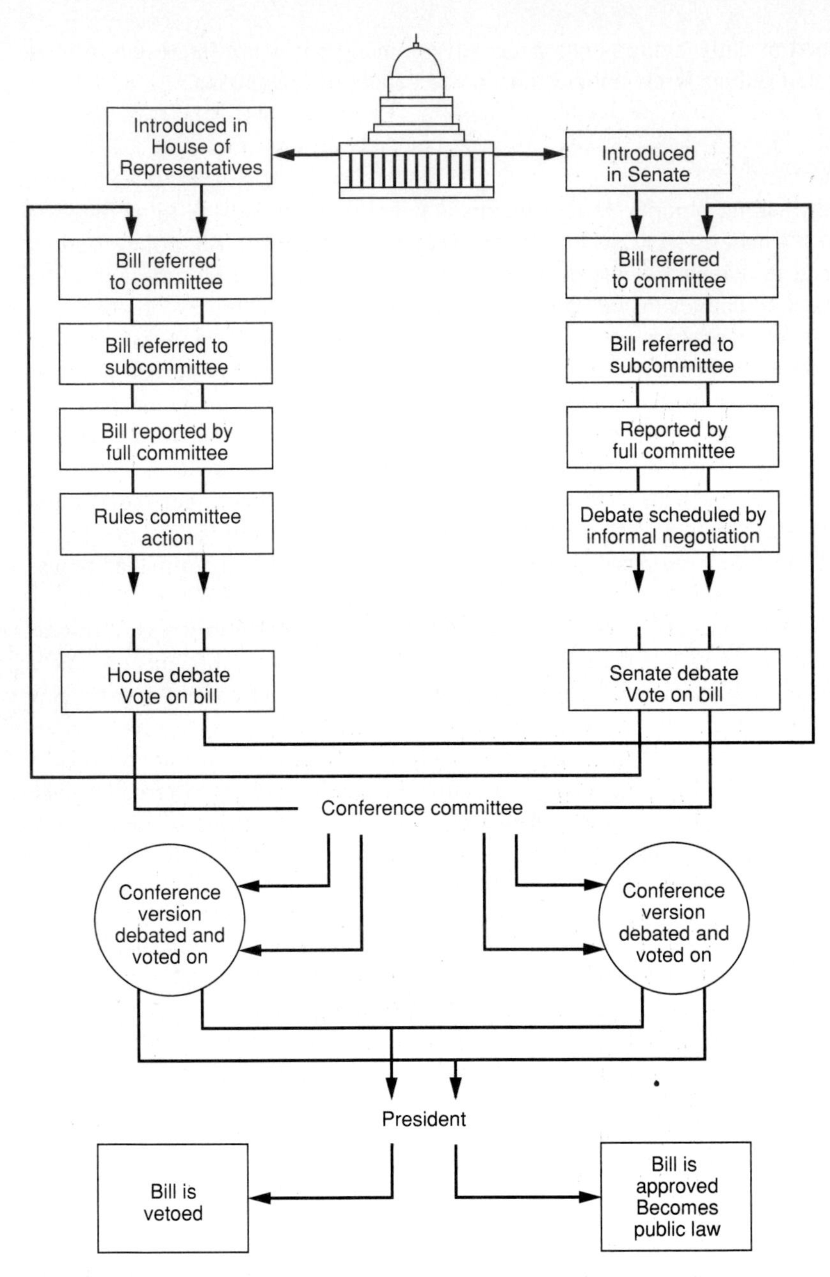

FIGURE 10-1
Sequence of events involved in enactment of a law. (*Source: From* Congress and Public Policy, *by D. C. Kozak and J. D, Macartney. © The Dorsey Press, 1982 and 1987. Reprinted by permission of Brooks/Cole Publishing Company, Pacific Grove, CA 93950.*)

Drafting A legislator who decides that a forestry idea is worthy of enactment into law must first proceed to have specific language describing the idea *drafted* into bill form. Although the draft may be initially written by any number of individuals or organizations, such as interest groups, executive agencies, and legislators, it is typically checked for form (not for substance) by a legislative agency charged with drafting. From a congressional perspective, the proposal may be designated a bill (the most common designation), a joint resolution (which requires approval by both chambers as well as the signature of the president, and which is used for specific subjects such as emergency appropriations), or a concurrent resolution (which requires approval by both chambers but not an executive signature, and which expresses an opinion on some issue) (Congressional Quarterly 1983). A draft bill will contain a title, an enacting clause, definitions, general provisions, penalties, an effective date, and a list of existing statues to be repealed or amended.

Introduction Once drafted, a forestry bill (or resolution) must be *introduced* in either the senate or the house, an action which is accomplished by filing it with the clerk of the house (or the secretary of the senate). The bill is assigned a number, after which copies are printed for distribution. At an appropriate time in the order of business, the chamber clerk (or secretary) will read the title of the bill to all members present. Depending on the legislature, more than one reading may take place.

Referral An introduced bill or resolution is then subject to *referral.* The presiding officer assigns the bill to an appropriate committee for study and recommendation. The actual task is usually carried out by parliamentarians of each chamber, and the vast majority of referrals are routine. For example, in the U.S. Senate, a proposed law dealing with management of forests within National Parks would typically be referred to the Committee on Energy and Natural Resources, while a proposed law designed to address nonindustrial private forests would generally be referred to the Committee on Agriculture, Nutrition and Forestry. In some cases the legislation (e.g., water-pollution control laws) may be so complex and have so many potential consequences that referral is made to more than one committee, adding an additional point of approval—another delay—to the lawmaking process

Committee Action A referred forestry bill is subject to *committee action.* The responsible committee has several options:

- It may refuse consideration of the bill, which is equivalent to voting against the proposed law.
- It may consider the bill and recommend approval with or without amendments; that is, it may report the bill.
- It may consider the bill and recommend that it be rejected, which allows the full senate or house to act on the bill.

Most often, a committee chair will assign the bill to a subcommittee for study and hearings, after which the subcommittee marks up the bill (i.e., it meets to decide on the bill's specific language). If the language is approved by the full committee, a committee report on the bill is prepared and referred to the full chamber for action. Committee reports become an important source of information in that they describe the purpose and scope of the proposed law, the committee amendments and their rationale, the impact of the bill on existing statutes, and the estimated government cost of implementing the bill. They also include the texts of communications from department or agency heads whose views on the bill were solicited. Such information can prove especially valuable in later stages of the policy process, especially implementation.

Placement on a Calendar Once a bill has been reported by a committee, it must be placed on a *calendar,* an institutional agenda of business awaiting action. The process is largely controlled by the prevailing political party in each chamber. Numerous factors determine where and on what calendar a bill is placed, including:

- The rules of the chamber
- The timetable set forth by important laws (the Congressional Budget and Impoundment Control Act of 1974, for example)
- The pressure of national and international events
- The executive officer's expression of interest in a bill
- The policy and political preferences of the political leaders in the legislature
- The Committee on Rules (in the U.S. House of Representatives)

In the U.S. House of Representatives, four important calendars exist, namely (Congressional Quarterly 1983):

- *Union calendar* Bills directly or indirectly appropriating money or raising revenue
- *House calendar* Bills not directly or indirectly appropriating money or raising revenue
- *Consent calendar* Bills considered to be noncontroversial in nature
- *Private calendar* Bills addressing concerns of individuals (e.g., immigration, land titles, relief of claims against the United States)

A bill is calendared by the U.S. Senate in a far less rule-bound fashion. Reported bills are brought to the floor for action by either unanimous consent or motion of one member.

Floor Action After a bill has been placed on the agenda (or calendar), it may proceed to *floor action.* In the U.S. House of Representatives, for example, the process is fourfold, including (1) adoption of a "rule" (rules state the conditions under which the bill will be debated: length of debate, allowance for amendments, etc.), (2) general debate among all members of the chamber (known as the *Committee of the Whole*), (3) consideration of and vote on amendments (if allowed), and (4) final vote for passage of the entire bill. The U.S. Senate is less rigid in its floor procedures, in that amendments are freely offered and unlimited debate (filibuster) occurs on certain topics (Oleszek 1984). In both chambers, however, a bill is overseen by a floor manager (often the chair of the committee reporting the bill), whose responsibilities include developing strategy for parliamentary maneuvers, responding to member inquiries about the substance of a bill, advising colleagues on the meaning and importance of amendments, and controlling the amount of time spent on general debate on the bill and any proposed amendments. Floor managers are extremely important persons in the legislative process, often determining the fate of a proposed law.

Second Chamber Consideration After being agreed to by one chamber, a bill is transmitted for *consideration by the second chamber,* where it proceeds through the latter's legislative rules, process, and procedures for proposed laws. Because the second chamber may be already considering a very similar bill, the time required to act on the originating chamber's bill may be lessened. The second chamber may reject the bill, pass it with no changes (when this happens, no further legislative action is required), pass it with amendments, or nullify it by failing to act on its behalf. If the bill is rejected, or if the second chamber fails to act on the bill, legislative action ends: the bill will not become law during the session. If, however, the bill is passed in a form different from that passed by the first chamber, the differences must be reconciled. The originating chamber is given an opportunity to respond, and either to agree to the changes or to reject them.

Concurrence Differences in the house and senate versions of the same bill must be reconciled before proceeding with the legislative process. *Concurrence* is involved—that is, reconciling different versions by mutual agreement. If the differences are minor, the two chambers may informally agree to changes that will enable the bill to be adopted by both bodies. If, however, the two versions are significantly different, a formal conference must take place. This conference takes the form of an ad hoc joint committee composed of members selected by the presiding officers of the two chambers. Conference committee members are usually members of the standing committee that originally handled the bill. Conferees are expected to support their chamber's position and are often given party guidance on voting and limits of authority, such as specified limits on a proposed program's funding. Some of the legislature's most strenuous bargaining takes place in conference committees, where attempts are made to reconcile differences without granting changes that will be unacceptable to the chamber the conferees represent. If all goes well, the conferees will prepare and sign (signatures of a majority of the ad hoc committee members are usually required) a conference report describing the manner in which differences are handled. The agreed-to version of the bill is then acted on by both chambers. If approved, it is signed by their presiding officers. If agreement is not reached by the conference committee, or if one chamber turns down the conferees' report, the bill is defeated.

Approval The final step is *approval* by the executive officer (the President or governor). If the bill is signed, or if it is not signed within a specified period, it becomes law. If the executive vetoes the bill, it is returned to the legislature with an explanation of why the action was taken. A vetoed bill may still become law if both houses vote to override the executive's veto by a necessary margin.

Overtly, the lawmaking process may appear to be well-organized and efficient. In truth, however, there are a number of points at which the process may stumble—points at which individual members can stop or at least significantly delay the passage of a bill. Negative subcommittee, committee, or floor votes are obvious defeats for a bill. Subcommittee and committee inaction, motions to strike particular clauses, or motions by an entire chamber to recommit a bill to a committee can also defeat the bill. Delay can occur when committees postpone action on a bill, say, by delaying or prolonging hearings, by delaying preparation of reports, by delaying scheduling a bill, or by imposing restrictive rules of debate. Similarly, delays can occur when chamber actions on the floor prolong debates by requesting quorums and calling various points of order. The process of making law is an uneven one which is best served by those with great patience and significant tolerance.

Oversight and Review

Oversight and review are major legislative functions. They involve examination of executive agency policies and programs and are a legislature's primary means of exercising control over the implementing and evaluating stages of the policy process. For the most part, oversight is backward-looking, in that it attempts to assess how agencies and programs have been working and whether changes are needed to make them work better. Seldom is oversight a systematic, comprehensive activity; most often it is performed intermittently. Executive agencies responsible for implementing forest resource policies are simply too vast for unqualified legislative control. Choices have to be made about how often oversight will be exercised; who within the legislature will exercise such responsibility; which oversight techniques will be used; which agencies, programs, and personnel will be the subjects of oversight; and how much intensity will be applied to the oversight process. Because decisions about such questions arise from individual legislators and from numerous committees and their chairs, the accident of luck, politics, personality, and seniority weigh heavily in the oversight process. Accordingly, oversight inevitably becomes partial and selective.

A legislator's motivation for taking an active interest in oversight activities has much to do with politics and political survival. More specifically, legislators tend to actively engage in oversight activities when (Ogul 1981):

- Executive initiatives call for large new expenditures or new authorizations in controversial areas.
- Crises have occurred which have not been handled well by executive departments or agencies.
- Political opponents are in control of the executive branch of government.
- Confidence is modest in the administrative capacity of agency or departmental leaders.
- Constituents view an issue as being of direct and major concern.
- Political visibility or organizational support can be gained by addressing an issue.

Formal, congressional recognition of oversight responsibilities was granted by the Legislative Reorganization Act of 1946, which ruled that committees should exercise continuous watchfulness over administrative agencies. The act was expanded upon by 1970 amendments which stated:

> Each standing committee shall review and study, on a continuing basis, the application, administration, and execution of laws, or parts of laws, the subject of which is within the jurisdiction of that committee.

Oversight was largely ignored by state legislatures through the late 1960s. In the 1970s, however, the activity came of age; it has been a vigorously exercised state activity ever since (Rosenthal 1981).

Regulation Review Oversight actions can involve a number of subject areas. When they focus on *regulations,* the concern is with the administrative rules and regulations of executive agencies. When forest laws are enacted, legislatures often delegate broad rule-making power to agencies. Thus executive agencies are enabled to carry out in detail an often broadly stated legislative intent. Via oversight, legislatures are able to determine whether established rules are within the intent and scope of the enabling legislation, have been adopted via proper procedures, are correctly stated and widely communicated (published), and are necessary for accomplishing the intent of a law.

Agency Review Likewise, oversight can focus on *agencies,* raising questions of conduct and legitimacy (need for continuance). Does an agency have a legitimate mission? Should an agency be combined with other agencies having similar functions?

Policy and Program Review Oversight can also focus on *policies and programs.* Consideration is given to how well they are being implemented, to their effectiveness in meeting legislative objectives, and to opportunities for modification which would improve their effectiveness. Legislative oversight of policies and programs may also be carried out during legislative appropriation processes.

Personnel Review Legislative oversight can also be carried out by the exercise of control over *personnel,* especially in regard to selection, performance, and, as warranted, removal. The U.S. Senate, for example, is called upon each year to give advice and consent to thousands of executive nominations, many of which carry responsibility for major forest resource agencies and programs. Besides looking for knowledge and general honesty in appointees, legislatures may scrutinize them for unwarranted political participation and for general loyalty to the views of the political party in control. The results of such scrutiny often becomes the spice of the legislative diet.

Executive Decision Review A common means of implementing oversight control over executive decision making is to require annual or periodic reports from executive agencies. For example, the U.S. Congress requires the USDA-Forest Service to submit a resource assessment every 10 years and a resource program every 5 years (Forest and Rangeland Renewable Resources Planning Act of 1974). Annually the agency must inform Congress on progress toward accomplishment of the forestry objectives in the resource program. Legislative veto is also a decision-making control mechanism, wherein failure of a legislature to act on an executive proposal (often required by law) within a specific period of time (for example, 60 days) constitutes legislative disapproval, or veto, of the proposal; the proposal cannot be implemented. Alternatively, a law may provide that if the legislature fails to approve or disapprove of a proposal within a specified period of time, the agency may proceed to implement the proposal.

Approaches to Oversight The specific techniques for accomplishing legislative oversight are many and highly variable. A legislature may establish a specialized staff, such as a legislative auditors office to do the job, or may use the time-tested method of conducting committee hearings and carrying out special committee investigations. Some legislators have formed special groups and caucuses to monitor agency actions concerning specific issues and programs—for example, the Congressional Forestry 2000 Task Force. Some techniques are very informal, including such matters as the daily interactions between executive agency personnel and legislators (or staff); the processing of casework, wherein staff review agency performance while responding to constituent questions about executive actions; and the individual legislator's reviews of agency actions which may lead to charges of government waste and inefficiency. Most informal are the nuances of legislative language found in hearings, floor debates, and committee reports. In the latter, for example, the words ''expects,'' ''urges,'' ''recommends,'' ''desires,'' and ''feels'' convey (in roughly descending order) how obligatory a committee comment or viewpoint is intended to be (Oleszek 1984). The administrator and staff of a forestry agency had best be aware of such nuances.

Problems with Oversight The ostensible goals of legislative oversight are to promote efficiency and responsibility in government. The impact of oversight actions on such goals is far from clear. Despite demonstrable increases in legislative oversight activities in recent years, concern has been expressed about the process, and what is gained has been questioned. Here is an example (Keefe and Ogul 1989, p. 357).

> Although some legislators have high regard for [oversight] goals, they are also concerned with promoting their own careers and causes. Legislative oversight does not function in the abstract; rather it occurs in concrete situations where personal motives and broader goals are hopelessly intermingled. . . . Legislative scrutiny may serve merely to frustrate conscientious officials, to seek special favors, to promote political careers and to disrupt carefully conceived executive programs.

The dilemma can be attributed to a number of causes, including the limited time available to legislators for responsible oversight activities, the shortage of funding required to employ staff necessary to ferret out agency inadequacies, and legislators' frustration with the reality of overseeing large, complex programs and organizations. As stated by one legislator, oversight of an agency "is almost like trying to fight a pillow. You can hit it—knock it over in the corner—and it just lies there and regroups. You feel as though you are trying to wrestle an octopus. No sooner do you get a hammerlock on one tentacle than the other seven are strangling you" (Oleszek 1984, p. 237).

Whether effective oversight can occur or not depends on the approach to executive-legislative relations. Legislatures might choose to focus on the specific details of a forestry agency's operations—for example, scrutinizing specific harvest limits for forest wildlife, or assessing the operational efficiency of a specific forest ranger station. Alternatively, they may choose to oversee only broad forest policy and program directions, ignoring the details of program implementation and administration. Such an approach would be consistent with the perspective that legislatures have an obligation to represent broad interests of society, and that neither their structure, their organization, nor their available resources enable them to effectively oversee the details of policy and program implementation. In actuality, a combination of such approaches prevails. Circumstances ultimately determine the intensity of legislative oversight.

ORGANIZATION AND PARTICIPANTS

Legislatures are vast, complex collections of organizations and individuals. Their most basic structural feature is *bicameralism,* meaning that they almost always have two separate chambers for conducting business. (Nebraska is unique in having a unicameral legislature.) The *senate* is called "the upper body" by the senators who occupy it and "the other body" by other legislators. The second chamber is typically known as the *house of representatives,* although some states use other names, such as *house of delegates* or *general assembly.* Even though the two chambers are organizationally separate, the task of enacting laws frequently necessitates the building of alliances between them. Intense bargaining may be involved in getting favorable "other-chamber" treatment for a proposed law.

Characteristics of Chambers

Senates and houses of representatives differ in many respects—most commonly the result of differences in the number of members occupying them. The U.S. Congress is composed of 100 senators and 435 representatives, while state houses of representatives commonly have three times as many members as do state senates (Rosenthal 1981). Difference in chamber size has a considerable effect on legislature structure and on approaches to carrying out legislative business focused on a forest resource. The smaller chamber typically has an informal atmosphere and friendly, relaxed relationships, whereas the larger chamber is characterized by more confusion and comparatively impersonal relationships. The smaller chamber has an informal hierarchy and an air of collegial authority, as compared to more elaborate rules and a stronger sense of leadership in the larger chamber. The conduct of business is characterized by relaxed deliberations in the smaller chamber, contrasted with shorter but more structured debate in the larger chamber. The distribution of power is somewhat dispersed in the smaller chamber and more concentrated in the large chamber (Rosenthal 1981).

Leadership and Rules of Procedure

Every legislature is composed of individuals from a variety of backgrounds, and every legislature has innumerable agendas. An important activity of any legislature is organizing to carry out the business of government. Foremost among organizing activities is selection of legislative leadership (presiding officers) by vote of chamber members who usually vote along strict party lines. Once elected, the leaders select legislators to be committee chairs.

Rules to guide the legislative process must also be adopted. At the state level, provision is made for incorporating a parliamentary manual, usually Mason's Manual, Robert's Rules of Order, or Rules of the U.S. House of Representatives. Because of increasing work loads, many state legislatures have adopted rules that impose deadlines on the legislative process, including deadlines for the introduction of bills, for committee action on bills, and for submission of conference committee reports. Without rules of such a nature, the legislative process would in all likelihood become unmanageable.

Legislatures meet for varying periods of time and are not averse to organizing special sessions to deal with especially significant issues. By formal (constitutional) or informal (rule-adoption) arrangements, nearly all state legislatures meet annually, although constitutional limits are often placed on the lengths of sessions (e.g., 30, 40, or 60 days). The U.S. Congress organizes every 2 years to form a ''Congress'' for purposes of carrying out the nation's public business. The 102nd Congress, for example, meets during 1991 (first session) and 1992 (second session). Any bill that fails to become law by the time the 102nd Congress adjourns in 1992 must be reintroduced during a subsequent Congress in order to have a chance of passage, and must proceed through the lawmaking process from the beginning.

Legislators

Motivations and Characteristics Legislators are a focal point of attention within legislatures. Their desire to serve in such a capacity is multifaceted. For many the prestige of being elected and being one of a relative few is appealing, while to others the chance to serve the public, to accomplish something in the public interest, is of major importance. Socially and economically, legislators are seldom representative of the population as a whole. They are likely to be lawyers and businessmen, and are apt to be Protestants and Caucasians. By substantial margins, they are more likely to be male than female. Similarly, legislators are unlike the general population in terms of wealth: over a third of U.S. Senators in the late 1970s claimed a net worth of $1 million or more (Hinckley 1988).

Serving in a legislature can be an especially educational activity for a citizen. The opportunity to serve assumes a student-teacher perspective, with a focus on the art and craft of politics. Consider (Muir 1982, p. xii):

> A legislature is like a school. It educates its members in the science of public policy and the art of politics. Through its subject-matter committees it exposes legislators to a wealth of knowledge about human affairs. Through the bill-carrying responsibilities it imposes on authors of legislation, it teaches the art of negotiation. And through the division of labor, it invites its members to specialize and create a network of key persons who share an interest in a single field of public policy.

Work Loads Legislators are extremely busy. During a legislator's average day (11 hours and 20 minutes) in the U.S. House of Representatives, activities of the following nature take place (House of Representatives 1977):

- Attending sessions in the House chamber—26 percent of day
- Participating in committee work (hearings, markups)—13 percent of day

- Meeting with constituents and organized groups—8 percent of day
- Carrying out office work (answering mail and telephone, preparing speeches, reading, consulting staff)—23 percent of day
- Consulting with legislative leaders and other representatives—3 percent of day
- Engaging in other activities (socializing, speech making, drafting legislation, consulting executive agencies)—27 percent of day

More generally, a legislator's work load takes the form of acting on matters of policy including activities such as the following (Eulau 1959):

- Reacting to the policy preferences of constituents and organized interests, or sponsoring and securing passage of bills
- Providing nonlegislative services requested by constituents, such as providing information about forest resources or acting as an intermediary between an aggrieved constituent and a government natural resource agency
- Seeking allocation of specific government goods, services, and programs for a state or legislative district, such as securing special forestry grants or unique tax advantages for timber production
- Pursuing symbolic activities (with an emphasis on form and appearance) that demonstrate determination to keep in touch with clients, such as advertising via newsletters and local speech making, claiming credit for district projects, and taking positions on noncontroversial issues

Similarly, legislators expend considerable effort on legitimizing their actions to constituents (e.g., "I voted for the massive tax increase so that money would be available to build roads in 'your' state forest") and educating constituents (by describing, interpreting, and persuading) in the nature of complex issues and the policies available for addressing them ("The proposed wilderness land allocations would have the following major consequences. . . .") (Fenno 1978).

Relationships between Legislators The manner in which a legislator pursues legislative interests is very much dependent upon the development of relationships with fellow legislators. Over the years, legislators have established a variety of norms, folkways, and unwritten rules for standardizing such relations (Keefe and Ogul 1989). For example, legislators generally treat colleagues with courtesy and respect—taking care never to publicly embarrass or offend a colleague. They also refrain from personalizing disagreements over policy and procedures, which means they must lose graciously and avoid holding grudges. They are meticulous in these interpersonal matters because they are aware that opponents on one forest resource issue may be necessary supporters on the next.

Similarly, legislators are alert to the need to be honest, to keep their word, and to respect confidences. Refraining from obscuring a bill's real purpose, being reliable, and maintaining confidences expressed in private are considered important virtues. Legislators are also expected to be mutually helpful, to accommodate themselves to each other's needs, and to bargain in a way that leaves room for compromise.

Norms of Conduct Legislators tend to defer to the policy wishes of other legislators who face difficult issues in their home states or districts. If additions to the National Wilderness Preservation System are being proposed in a particular state, the wishes of the legislators from that state are often given extraordinary consideration by legislators from other states. They also specialize or seek to become experts on subjects within their committee assignments. Colleagues who attempt to become authorities on everything are often held in disdain. Very often, new legislators are obliged to serve an apprenticeship before actively taking the legislative stage—to be patient and to curb an inclination to speak too often, debate too vigorously, or discuss too many subjects.

Legislators do not always adhere to legislative norms. Among those who purposely deviate from norms are legislators who have had previous political experience (such as past governors), or who have great political ambition (a wish to be a governor, for example) or constituency problems (such as the problems faced by senators from large states that need many services very soon). Legislators whose political ideologies are liberal perceive norms of conduct as supportive of the status quo and may purposely deviate from such norms as may those who have a compulsive preoccupation with a special cause. Depending on the circumstances, legislators who violate norms run the risk of incurring sanctions such as the obstruction of their favored bills or assignment to undesirable office space.

Understanding and respecting legislative norms or standards of conduct is especially important for nonlegislators who are attempting to influence the course of legislation. Asking a legislator to violate a norm in order to secure a matter of self-interest may result in a brisk and pointed rebuff.

Types of Issues Addressed Legislators' decisions about policies, programs, and political circumstances are as numerous as they are complex. They range from decisions about legislative agendas to decisions about policy options, and from voting decisions on the senate floor to decisions about specific language to be inserted in a bill during a committee markup session. The nature of such decisions has much to do with the issues faced by legislators. Consider the following issue types, none of which are mutually exclusive (Kozak and Macartney 1987):

- *Controversial issues* The issue profile is very high, emotions run deep, and positions are strongly held. Information flow to a legislator is extremely broad and voluminous, involving lots of press coverage and constituent mail. Legislators do little to gather additional information, relying instead on experience gained during campaigns or previous sessions. Voting will be consistent with the legislator's personal ideology, campaign promises, or constituency interest. In natural resources, examples of such issues are public-land wilderness allocations and public regulation of private forest practices.

- *Low-visibility issues* The issue profile is very low, not widely publicized, and of concern to very few organized interests. Information flow to the legislator is almost nonexistent, little additional information will be gathered, and the legislator's vote on the issue will flow with the trend established by other legislators. In natural resources, examples of such issues are management of forest residues and minor adjustment of public forest boundaries.
- *Complicated issues* The issue profile is high to individuals or organizations immediately affected. The subject is complex, confusing, and hard to handle. Legislators engage in an extended search for information from staff, interest groups, executive agencies, and legislative support agencies, although their ultimate knowledge about the issue will be limited. The legislator's vote will be consistent with that of legislators who have specialized in the issue. In natural resources, examples of such issues are reauthorization of a major government agency (such as the U.S. Environmental Protection Agency) and approval of agency-proposed regulations on strip mining.
- *Recurring issues* The issue profile is modest, because it has been seen before and is considered normal or usual. Information gathering is minimal, and legislators' votes are most often based on ideology—on a standing position of voting for or against certain programs. In natural resources, examples of such issues are recurring noncontroversial authorizations and appropriations.

In addition to the influence of issue characteristics on voting patterns, a variety of actors on the political landscape also affect legislators' decisions about the use and management of forests. Among the major influences are those (Kozak and Macartney 1987):

- *External to legislature* Constituents, bureaucrats, organized interests (lobbyists), President or governor, media, public opinion, potential electoral outcome
- *Internal to legislature* Committee system, fellow members, party leaders, personal staff, legislative procedures, legislator's ideology

The relative importance of influences on legislators' decisions varies with the issue. Largely unknown is the significance of each source of influence on matters concerning forest resources. For issues in general, however, a significant proportion of U.S. House of Representative members cite constituents and fellow Congress members as having a major or determining influence on their decisions, while over half indicate that party leaders, staff, and the administration are unimportant (Table 10-1). In very few cases do members indicate any that one actor influenced their decision to the exclusion of all other actors (Kingdon 1989).

TABLE 10-1
Importance of various actors on legislative voting decisions of members of the U.S. House of Representatives (percentages)

Importance	Constit-uency	Fellow Congress members	Party leadership	Interest groups	Adminis-tration	Staff
Determinative	7	5	0	1	4	1
Major importance	31	42	5	25	14	8
Minor importance	51	28	32	40	21	26
Not important	12	25	63	35	61	66
Total	101	100	100	101	100	101

Note: Some totals exceed 100 percent because of rounding errors.
Source: Congressmen's Voting Decisions by John W. Kingdon. Reprinted with permission of The University of Michigan Press, Ann Arbor, 1989, p. 19.

Constituencies Constituents weigh heavily in legislators' decisions, and yet a legislator's perception of constituent claims is more complex than one might think (Fenno 1978). If constituencies are perceived as a series of concentric circles, the largest group of concern to a legislator is the *geographic constituency,* namely persons or organizations within geographic boundaries that have been fixed by legislative or judicial action—the district or state represented by the legislation. Next are *reelection constituencies,* composed of voting supporters, and *primary constituencies,* composed of the strongest voting supporters—loyalists, or those willing to actively engage in campaign activities such as fund-raising and recruitment of other supporters. The innermost circle is the legislator's *personal constituency:* the closest supporters, confidants, and advisers. These are the persons from whom a legislator may draw personal sustenance for political activities, and may seek candid advice about forest resources and related issues. To some degree, each of these constituencies influences a legislator's decision about which issues to address and which policies should be used to address such issues.

Representation Styles Constituencies are typically not monolithic in opinion nor alike in the intensity of the opinions they attach to important matters of forest resource policy. Hence, legislators are frequently torn between contradictory constituency claims. To handle such difficulties, legislators adopt various representation styles.

Some see their role as that of a *delegate*. They seek out and follow the wishes of their constituents, even when such wishes are not in accord with their personal views or convictions. In such a mode, a legislator may argue, "Because my constituents desire an increase in timber harvesting on public forests [a view the legislator disagrees with], I will work to have the harvest rates increased."

Other legislators adopt the role of *trustee*. They act in accord with their own judgment and conscience, largely ignoring the wishes of constituents. A legislator who adopts the trustee role is frequently politically secure and has little need to fear significant challenges in the home district. Interest groups that are intent on changing the voting pattern of a trustee legislator by encouraging constituents to pressure the legislator to vote more consistently with the group's views will probably be unsuccessful.

In reality, legislators vary the roles they assume, according to the forestry issue and the political conditions. Members of the U.S. House of Representatives, for example, allocate their own proportions as follows: 28 percent trustee, 23 percent delegate, and 46 percent combined role (Hinckley 1988, p. 79).

Decision-Making Process Explaining how legislators reach decisions on matters of forest resource policy is difficult. Few if any factors have been found to be consistently preeminent (Kingdon 1989). Often all that can be hoped for is an understanding of the decision-making process employed by legislators.

Consider the process shown in Figure 10-2. When a legislator is faced with the prospect of deciding how to vote on a proposed law, an overriding initial concern may be whether the issue to which the law is directed is controversial. Is, for example, a proposal to establish a new state park engulfed in controversy? If not, the legislator may well vote with legislative colleagues in general. If, however, the issue is controversial, consideration must be given to the extent of conflict among persons and organizations that are normally considered part of a legislator's field: the primary constituency, trusted legislative colleagues, and interest groups normally in contact. Are forest landowners, resource agency administrators, and environmental groups in disagreement over the proposed state park? If not, a legislator may vote with legislative colleagues in the field. If those in the field are in conflict, the goals of the legislator must be assessed: are they threatened? Will a vote in favor of establishing a park violate a legislator's interest in reducing government involvement in private affairs (including land ownership?) If not, the legislator may vote consistently with close and trusted legislative colleagues who agree with the legislator's general philosopohies. If a legislator's goals are at stake and are conflicting (for example, if a desire to reduce government involvement conflicts with an interest in enhancing the availability of quality environments via expansion of park systems), concern must be focused on constituents: Are they involved in the conflict? If they are, is the issue highly visible and will constituents disapprove of a certain vote? If the issue is highly salient, the legislator may vote according to the constituents' wishes.

In sum, legislators begin their decision-making process by making a significant distinction: Is the subject to be addressed controversial or not? From there they proceed to assess their environments in terms of agreement that exists in such environments. If they find consensus that conflict does not exist, they vote

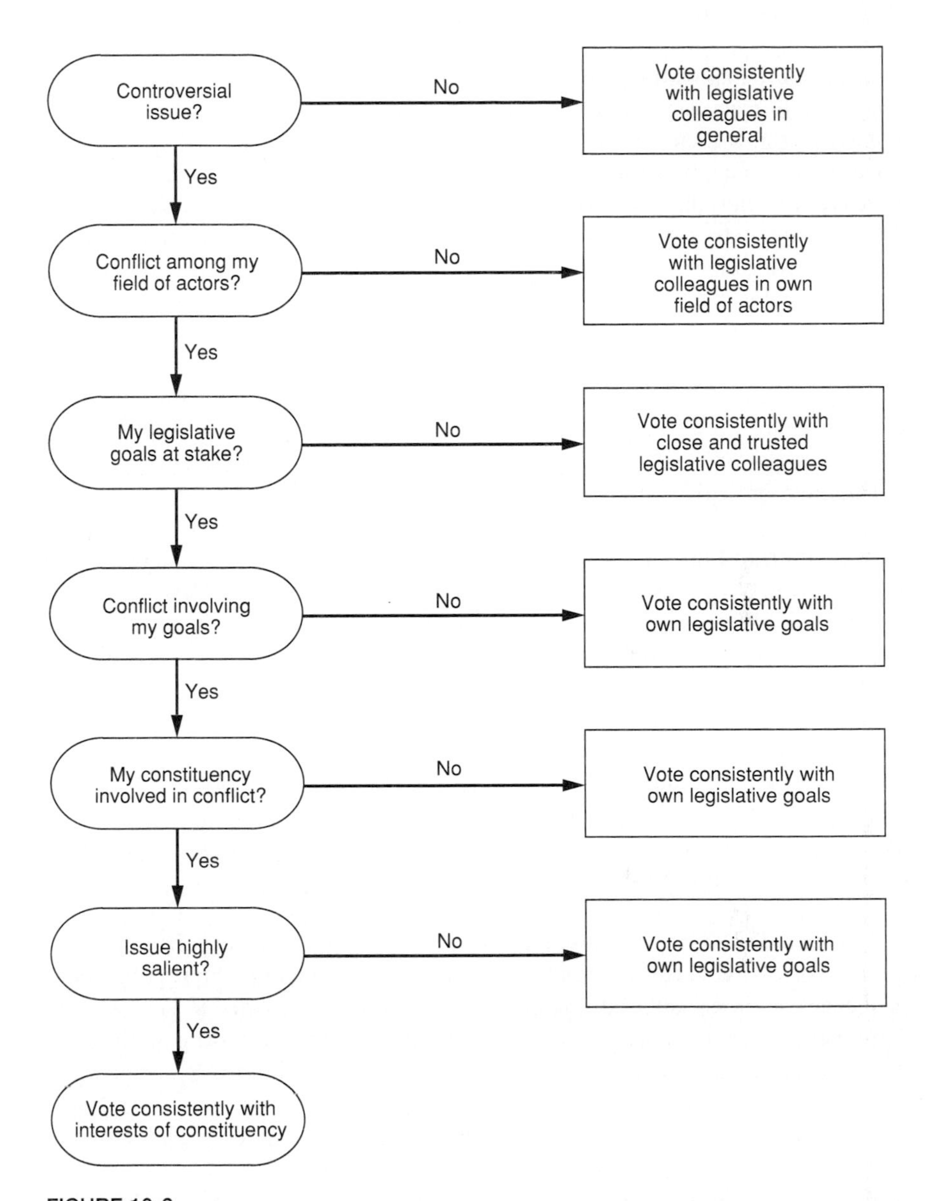

FIGURE 10-2
The sequence of events involved in the legislative voting decisions of legislators. (*Adapted from* Congressmen's Voting Decisions *by John W. Kingdon. Reprinted by permission of The University of Michigan Press, Ann Arbor, 1989, p. 244.*)

the direction of the consensus. If not, they proceed to search for a position that they consider comfortable.

Limits on Legislators' Discretion Although legislators often enjoy ample discretion to make legislative decisions on matters of forest and natural resource policy, at times there are factors that intrude on such freedom. For example, commitments may be wrung from reluctant legislators by groups that have electoral clout in their districts. The timber industry, for example, may seek to obtain from a candidate a commitment on favorable capital gains tax treatment of income from timber. If such a commitment is obtained, it will in all likelihood be honored by the legislator when policies addressing the issue are voted on. Similarly, a legislator may be obligated to support a certain legislative policy because of a previous logroll with a fellow legislator. The colleague may have voted for increased funding for a forest resource program, and in return, the legislator must now vote against proposed standards for air quality. Such logrolls are an integral part of the political environment of legislative systems, and they do impinge on legislators' voting discretion.

Information Flow and Quality Legislative decisions rely on a healthy flow of information. Commonly, legislators complain that they cannot make an intelligent choice because the available information is insufficient. In reality, however, information supplies are rarely insufficient; rather, they are usually overwhelming in terms of quantity. The form in which the information is presented to legislators (Kingdon 1989) is the problem. Legislators need forestry information that is condensed and focused, so that it will conform to their time and their cognitive abilities. Of little value are 800-page reports and 1000-page hearing records on, for example, policy options for public management of pesticides used in forestry. Such a glut of information is simply too much to handle; it must be condensed into a usable form.

Political Relevance of Information Information must be politically relevant if it is to be useful to legislators. Although information about policy consequences is important, information about political consequences is equally important, or maybe more so. Will a vote to increase the harvesting of elk bring forth the ire of constituents interested in nongame forest wildlife? Will a decision to vote against an administration proposal to ban the use of a pesticide jeopardize administrative interest in implementing other programs favored by the legislator?

Biased Character of Information Information that is explicitly biased is useful to legislators—information that takes a position and that is buttressed by arguments using selected facts. A legislator who faces mounds of neutral information must sift through it for clues to what position would be appropriate. Sometimes it is not even apparent that the information is neutral in the first place, which further complicates the matter. Biased information, in contrast, focuses on the issue and thus is largely void of irrelevant arguments and facts. If additional information is needed, the legislator can confront the adversaries and probe deeper for insight into which position should be favored.

In sum, legislators do not always lack information about forest resource issues. What they do lack is information that is brief, politically relevant, and to the point.

Forest Resource Information Though the specific nature of forest and related information demanded by legislators has only begun to be studied, some clues are available. Legislative committees in the South, for example, apparently request and prefer information concerning the economic implications of forest resource programs, including economc contributions, employment created, and costs and revenues associated with specific programs (Council of State Governments 1985).

A 1990 nationwide study of forestry information flows to the staff of state legislative committees suggests that legislative staff communicate very often with a state's forestry agency and with the staff of the other legislative chamber (Lewis and Ellefson, 1991). They very seldom, however, contract state budget offices, local forestry agencies or tourism interests (Table 10-2).

As to staff-perceived adequacy of forestry and related information, general information about the nature, extent, and condition of forestlands is apparently acceptable, whereas information concerning conversion of forestland to other uses, forest products markets, and the impacts of development on forest conditions (e.g., water pollutants) leaves much to be desired (Table 10–3) (Lewis and Ellefson 1991).

Some organizations have made an effort to inform legislators about forestry. For example, *A Legislator's Guide to Forest Resource Management* is available from the National Conference of State Legislatures (Meeks 1982).

Leadership

Legislators who attain positions of legislative leadership are decisive forces in matters of legislative policy. In a house of representatives, the principal leader is the speaker, followed by the majority leader (a member of the dominant party) and the minority leader (a member of the minority party). Leadership in a senate varies considerably. In the U.S. Senate, for example, the Vice President is constitutionally designated as the President of the Senate. When the Vice President is absent, leadership responsibility is delegated to the President pro Tempore. Majority and minority leaders also prevail in the Senate. In the U.S. Congress, each majority or minority leader has a whip (and may have numerous assistant whips) who aid in implementation of the respective political party's legislative program by gathering the necessary votes, checking attendence before critical votes, and building coalitions.

TABLE 10-2
State legislative policy committee (forestry and natural resources) staff communication with organizations active in state forest policy matters, by type of organization and frequency of communication, 1990 (percentages)

Group or organization	*N*	Frequency of communication: Very seldom	Fairly seldom	Fairly often	Very often
Personal staff of committee members	33	21	24	33	21
Appropriations/finance committee staff	38	21	37	29	13
Other committee staffs: same chamber	37	24	27	38	11
Counterpart committee staff: other chamber	34	21	18	32	29
Legislative analyst (budget/finance)	37	27	38	32	3
House/senate office of research	23	35	17	35	13
Leadership/caucus staff	30	27	30	27	17
Noncommittee legislators/personal staff	35	37	40	17	6
Governor's office staff	33	30	33	30	6
State budget office/department	29	66	14	17	3
State forestry agency	40	15	13	43	30
Other state natural resource agencies/ departments	39	21	23	41	15
State forestry boards/commissions	28	46	25	25	4
Federal forestry/natural resource agencies	32	47	38	13	3
County-local forestry/natural resource agencies	28	57	32	7	4
University/college departments: forestry/natural resources	38	29	37	24	11
Professional societies/journals: forestry	30	47	37	13	3
Wood products corporations/trade associations	31	20	39	29	13
Environmental groups	41	20	27	39	15
Recreation advocates: user groups	32	31	41	25	3
Tourism business interests	29	55	27	17	0
Mineral/energy interests	33	30	42	24	3
Real estate interests	28	54	32	14	0
Press/media	33	36	30	21	12
Contacts in other states	35	43	40	14	3

Note: N = number of staff persons. Because of rounding errors, communication percentages may not total 100.

Source: Staff-Mediated Information Flows to Forest Policy Committees in State Legislatures, by Bernard J. Lewis and Paul V. Ellefson. Agricultural Experiment Station Bulletin (forthcoming). University of Minnesota, St. Paul, 1991.

TABLE 10-3

State legislative policy committee (forestry and natural resources) staff assessment of the adequacy of forestry and related information available for legislative use, by type of information and adequacy rating, 1990

Type of forestry and related information	*N*	Percentage of staff rating information as:				
		Not adequate	Marginally adequate	Fairly adequate	Very adequate	Don't know/ don't obtain
State forest resource: all ownerships						
Nature, extent, and condition of forestlands in state	45	7	13	38	36	7
Public forestland management						
Forest protection (fire, insects, and disease): methods and impacts	45	2	7	56	22	13
Forestland ownership and jurisdictional matters: acquisition, sale, transfer, access	44	2	18	43	27	9
Timber management: outputs and impacts	45	2	31	36	20	11
Wildlife and ecological land management: outputs and impacts	45	7	22	38	31	2
Forest-based recreation management: outputs and impacts	45	7	27	33	29	4
Tradeoffs in forestland uses	45	11	29	33	11	16
State wood-based industry						
Structure and potential for economic development	44	5	34	30	18	14
Economic impacts: employment and income	44	11	11	52	21	5
Products, markets, and technology	44	14	16	41	21	9
Regulatory mechanisms: forest practices and wood processing	44	11	23	36	23	7

TABLE 10-3 (*Continued*)

Type of forestry and related information	*N*	Percentage of staff rating information as:				
		Not adequate	Marginally adequate	Fairly adequate	Very adequate	Don't know/ don't obtain
Nonindustrial private forestry						
Ownership patterns, outputs, and management practices	44	7	27	43	14	9
Intergovernmental forestry						
State and federal cooperative needs and opportunities	45	9	24	38	11	18
State, county, and local cooperative needs and opportunities	45	7	31	36	9	18
Forestry research and education						
Forestry research: program content and priorities	45	7	24	33	18	18
Forestry education and training: programs and needs	45	7	22	38	16	18
State and regional economic development: impacts on forestry						
Forestland conversion: pattern and extent	45	22	24	31	14	9
Impacts of development on forest resource condition (e.g., pollution)	45	13	27	36	16	9
Regional ecological issues (e.g., acid rain): impacts on forest resource base	45	9	36	24	20	11

Note: *N* = number of staff persons. Because of rounding errors, rating percentages may not total 100

Source: *Staff-Mediated Information Flows to Forest Policy Committees in State Legislatures,* by Bernard J. Lewis and Paul V. Ellefson. Agricultural Experiment Station Bulletin (forthcoming). University of Minnesota, St. Paul, 1991.

Selection Legislators compete for top leadership positions. When a legislature meets to organize for business after a general election, the candidate of the majority party is selected (by straight party-line vote) to be speaker of the house or president pro tempore of the senate. The candidate of the minority party becomes minority leader. Speakers are voted into office because of their personal prestige, their knowledge of legislative processes, and the general support they receive from party members. They are masterful politicians, known for their stamina, openness, legislative competence, and (when necessary) firmness. Their skill in the art of persuasion is often unsurpassed. They are keenly aware of the need to be associated with assured legislative victories and to avoid association with pending legislative defeats.

Responsibilities and Influence Legislators in positions of leadership bear great responsibility and can exert considerable influence over the establishment of forest laws (Rosenthal 1981). They are responsible for organizing the legislature, especially its committee structure. They exert substantial influence over the creation (and demise) of committees, the jurisdiction of each committee, and the number of legislators allowed to serve on a committee. Most important, however, leaders control committee appointments, including committee chairs. Leaders are also responsible for the smooth processing of legislation, including referral of forestry bills to committees, scheduling committee recommendations for floor action, promoting attendance on the house or senate floor, managing chamber debates, and convincing the President or governor to sign a forestry bill into law.

Legislative leaders are active negotiators, seeking to resolve disagreement within their political party and searching for means of securing favorable outcomes of conflicts with the opposing party or the administration. They also have major responsibility for handling the press, gathering and distributing reliable information on support or opposition to legislative matters, and acting as principal liaison with leaders in the executive branch of government. Legislative leaders also have the honor of dispensing certain benefits to legislators.

In addition to being in a position of favoring a member with the chair of an important committee, a leader can cosponsor a member's forestry bill, thereby giving it greater visibility, and can facilitate enactment of a proposed forest law by assigning it to an easy committee and scheduling it on a good day. Leaders can also persuade a reluctant President or governor to sign members' forestry bill and can help members secure important benefits for their home districts. In general, legislative leaders can make life easier (or more difficult) for a legislator by providing (or withholding) such amenities as extra staff, a comfortable office, and convenient parking. They play a decisive role in the policy process and must be recognized as important actors in the process of developing forest and related natural resource policy.

Committees

Committees are an extremely important organizational feature of a legislature. Their number and jurisdiction are important because of their power to advance or dispose of proposed laws, to conduct investigations, and to oversee the working of the executive branch of government. There are four types of legislative committees: (1) *Standing committees* are more or less permanent, continuing bodies with specified legislative responsibilities. (2) *Select committees* are appointed for a specific purpose and serve for a limited period of time. (3) *Joint committees* have as members both senators and representatives. (4) *Conference committees* are made up of representatives and senators appointed to resolve differences in versions of the same bill before final passage.

The number and nature of committees varies according to the legislature and also according to the chamber. During the 101st Congress, there were 40 standing committees—17 in the Senate and 23 in the House. U.S. congressional committees and subcommittees especially germane to forests and related natural resources are listed below (Government Publication Office 1989).

SENATE COMMITTEES

- *Agriculture, Nutrition, and Forestry* Subcommittees: Conservation and Forests; Agricultural Research and General Legislation
- *Appropriations* Subcommittees: Agriculture and Related Agencies; Energy and Water Development; Interior and Related Agencies
- *Budget*
- *Energy and Natural Resources* Subcommittees: Energy Regulation and Conservation; Energy Research and Development; Mineral Resources Development and Production; Public Lands, National Parks and Forests; Water and Power
- *Environment and Public Works* Subcommittees: Environmental Protection; Toxic Substances, Environmental Oversight, Research and Development
- *Government Affairs*
- *Judiciary*
- *Small Business* Subcommittees: Competition and Antitrust Enforcement; Government Contracting and Paperwork

HOUSE OF REPRESENTATIVES COMMITTEES

- *Agriculture* Subcommittees: Conservation, Credit and Rural Development; Department Operations, Research and Foreign Agriculture; Forests, Family Farms and Energy
- *Appropriations* Subcommittees: Energy and Water Development; Interior and Related Agencies; Rural Development, Agriculture, and Related Agencies
- *Budget*
- *Energy and Commerce* Subcommittees: Energy and Power; Health and Environment

• *Government Operations* Subcommittees: Environment, Energy and Natural Resources

• *Interior and Insular Affairs* Subcommittees: Energy and Environment; Mining and Natural Resources; National Parks and Public Lands

• *Merchant Marine and Fisheries* Subcommittees: Fisheries and Wildlife Conservation and the Environment

• *Public Works and Transportation* Subcommittees: Economic Development; Water Resources

• *Rules* Subcommittees: Rules of the House

• *Science, Space and Technology* Subcommittees: Energy Research and Development; Natural Resources, Agriculture Research and Environment; Science, Research and Technology

• *Small Business* Subcommittee: Regulation of Business Opportunities

• *Ways and Means* Subcommittee: Select Revenue Measures

Legislator Assignments Assignment of legislators to committees is an especially important event. Party leaders in each chamber submit their preferences to an appropriate mechanism of the majority or minority political party—for example, the Republican Committee on Committees in the U.S. House of Representatives. Depending on the party's rules, assignments may be made with the following goals in mind:

• Promoting regional representation on a committee
• Accommodation of members according to seniority
• Matching occupational backgrounds of legislators with committee assignments
• Rewarding especially loyal party activists
• Facilitating appointment to a committee having responsibilities of great concern to a legislator's home district or state

The distribution of Democrats and Republicans on a committee usually approximates the partisan composition of the chamber. Legislators are usually limited to two standing committee appointments. Some committees are more attractive to legislators than others. For example, of 18 U.S. House of Representatives committees ranked by legislators according to attractiveness, the Committee on Rules and the Committee on Veterans Affairs were ranked first and last, respectively. The Committee on Agriculture was ranked eighth, while the Committee on Interior and Insular Affairs was ranked fourteenth (Hinckley 1988).

Screening Legislation Legislative committees perform a number of tasks. The most imposing is the screening of legislation, and it can be an awesome responsibility, given the number of bills annually introduced by legislators. Over 10,000 bills are introduced annually in the U.S. Congress, but fewer than 2,000 are assigned into law every 2 years. More than 130 bills addressing forestry topics in general (such as forest fires, forest roads, lumber and trade, national forests, research and development, pest control, and appropriations) were introduced during the first session of the 100th Congress; an additional 190 bills focused on wilderness area designations were introduced (Congressional Research Service 1985).

Controlled largely by committee and subcommittee chairs, the legislative screening of bills involves the setting of an agenda (what is to be taken up and in what order), the carrying out of public hearings (if warranted), the markup of a proposed law (inserting and deleting specific language), voting (a sharply divided committee vote may portend trouble on the chamber floor), and the reporting of a bill to the chamber (in some cases a written report). Committees also have a history of studying issues and reporting findings and recommendations to a legislature in general. Proposed laws may result from such reviews and recommendations (Smith and Deering 1990).

Committee Hearings Hearings, often the most visible activity of legislative committees, are frequently a focal point for debate over proposals for natural resource policy (Ellefson 1985). They are historically associated with the right of citizens to petition democratic governments, and they serve multiple purposes. They provide, for example, a permanent public record of positions taken by committee members and interest groups on a proposed law. They also serve as a means for gauging the intensity of political support for or opposition to a bill. They are a channel through which both technical and political information can be transmitted to committee members from various interests. They can serve as a propaganda conduit through which committee members and staff can reach larger audiences through the news media, especially television and the press. It has also been suggested that hearings are a ritualistic means of relieving the intensity of conflict among interest groups. Acting as a safety valve, they enable irate citizens and frustrated interest groups to state their cases in a controlled environment (Keefe and Ogul 1989; Mater 1984). In addition, hearings provide a forum for criticism of public agencies, which sensitizes government to the wishes of individuals and the groups it is supposed to serve.

Committee or subcommittee hearings on matters of forestry are scheduled by committee or subcommittee chairs. Among the more important factors influencing decisions to conduct hearings are requests from agencies, interest groups, key legislators, and the legislative calendars of the other chamber. Of special importance is the chair's judgment that a timely public hearing will create momentum and support for a bill—that the intensity of interest is high and the appropriate mix of witnesses is available.

Once scheduled, committee hearings are supported by a significant amount of staff work; witnesses are contacted, research is undertaken, briefing books are prepared, and questions to be asked by members are prepared. The tone of a hearing depends on the issue and on whether the bill has appeared before the committee in previous sessions. Some hearings are conducted in a routine, almost disinterested fashion. Witnesses often read from prepared statements, while committee members either feign interest or discuss unrelated business with colleagues and staff. A surprising amount of testimony is taken with only a few committee members present.

Committee hearings attempt to blend the legislature's requirements for information with a member's (and a party's) requirements for influence and advantage. By no means are they models of efficiency and objectivity. Among the more serious charges leveled against hearings are that they provide a forum in which technically incorrect information may be given to a committee and that individuals and groups use them also as an opportunity to present opinions that are not representative of a larger public. A third charge concerns staging, and the allegation is that staff and committee members use them for such purposes as (Keefe and Ogul 1989, p. 187):

> slanting . . . the hearing, . . . manipulating the hearing machinery in such a way that a previous commitment on a bill would be made to appear as a decision reached by rational, detached deliberation. [Such is accomplished by] inviting the strongest, most persuasive witnesses to testify on behalf of the measure, by endeavoring to limit the number of strong opposition witnesses to a minimum; by asking "proper" questions of "friendly" witnesses and embarrassing ones of [unfriendly] witnesses; by arranging to close the hearing at a propitious time; by writing a report which brings out the best features of the testimony in favor of a bill; by subduing or discarding facts which do not fit the pattern of preconceived notions, opinions and prejudices.

Staff

Legislative staff have become an increasingly important component of legislative processes. At the turn of the century, U.S. representatives had no staff and U.S. senators had a total of 39 personal staff members, while Congressional Committee staff consisted of fewer than 12 clerks. By 1983, 11,655 personal staff members and over 3,000 committee staff members were on the congressional payroll (Kozak and Macartney 1987). Similar increases have occurred among state legislatures. Staff increases of this nature reflect the expanding role of government in American society, the enlarged work load assumed by legislators, and a legislative desire to lessen reliance on information and research and produced by executive agencies. As a result, legislative staff members whose training and speciality are in matters concerning forest resources are not uncommon in today's legislatures.

Legislative staff members are of various types, and distinguishing among them is not always easy. They range from security and maintenance workers to staff employed by legislative reference libraries, public information offices, and legislative audit bureaus. Among the most prominent staff, however, are the professional staff members associated with individual legislators or with committees, who serve at the pleasure of the legislator or a committee or subcommittee chair. They have no guarantee of continued employment. If a legislator is defeated in a general election or chooses to arbitrarily reduce employee levels, staff members quickly find themselves unemployed.

Personal Staff *Personal staff members* are responsible to an individual legislator who, on the congressional scene, may have as many as 30 personal staff members. U.S. Senate expenses for staff support are based on a scale linked to a state's population, whereas U.S. Representatives are given a set amount for staff support. Titles assigned to such staff members include: ''administrative assistant,'' ''legislative assistant,'' ''correspondence supervisor,'' ''press secretary,'' ''research assistant,'' ''personal secretary,'' and ''district director.'' Personal staff members tend to be young, predominantly male, highly educated, and strongly partisan. Their responsibilities are to provide:

- *Clerical and bureaucratic support* Answering mail, greeting visitors, and pursuing casework
- *Legislative support* Obtaining ideas for laws and assisting in bill drafting
- *Political support* Writing speeches, communicating with the press, managing fund-raisers, and assessing political conditions in the legislator's home district or state

Personal staff members in district offices (of which every member of Congress has at least one) increasingly handle casework and are the first contact of a group that wishes to invite a legislator to speak at a meeting or other gathering.

At the state level, fewer than 12 states provide for full-time personal staff support of a professional nature (Rosenthal 1981).

Committee Staff Congressional *committee staff members* are selected primarily by committee chairs, subject to nominal approval by the full committee. Legislators who belong to the minority party are often given an opportunity to appoint minority staff members, whereas majority staff members are from the prevailing political party. More than 200 professionals may be employed by a committee, and may hold titles such as ''chief of staff,'' ''press secretary,'' ''staff consultant,'' ''staff assistant,'' ''subcommittee consultant,'' ''financial clerk.'' The functions of committee staff are significant. They include:

- *Organizing hearings* Selecting witnesses, undertaking background research, preparing questions, informing the press, briefing members
- *Markup of bills and drafting amendments* Researching and explaining consequences of technical provisions
- *Preparing standing and conference committee reports* Writing reports to accompany bills, after consultation with committee chairs

- *Assisting during floor debate* Accompanying and providing support to the floor manager of a bill
- *Maintaining liaison with executive agencies and organized interest groups*
- *Handling selected press relations activities* Preparing press releases and background information

Congressional committee staff members tend to be male and older than personal staff members, and they are usually even better-educated. They have often had work experience in an executive agency, and they have strong ties to the ideological views of the committee chair who hires them. During the course of dispatching their responsibilities, they adhere to certain norms, including (Patterson 1970):

- *Limit policy advocacy* Refrain from publicly advocating their own policies and proposals
- *Maintain loyalty* Be loyal to committee chair or minority employer
- *Defer to legislators* Give legislators credit for staff ideas and initiatives
- *Assert anonymity* Limit personal visibility in public settings
- *Specialize* Focus and become an expert on a policy subject
- *Limit partisan activities* Refrain from involvement in campaign activities

Even while conforming to such norms, committee staff members have substantial opportunities to influence the outcome of legislative processes. Forest policies may be significant beneficiaries of the involvement of staff members if their involvement leads to their acquiring information that will enable legislators to make better judgments about competing policy options. However, if the judgments of staff members are substituted for the judgments of legislators, neither the legislative process nor the resulting policies are well served.

The organization of committee staffs of state legislatures is far more complex than that of U.S. Congress committee staffs. In some states, committee staffs are under the control of party caucuses or partisan legislative leaders. The dominant pattern, however, is a pool of legislative committee staff assigned to a bipartisan central-service office—for example, the Bureau of Legislative Research in Arkansas, the Legislative Counsel in Colorado, and the Office of Legislative Assistants in Maine. Although the responsibilities of such offices may include fiscal analysis, data processing, and statutory revisions, their principal task is to provide staff assistance to standing committees (Hansen 1989, Rosenthal 1981).

Support Agencies

Legislatures have responded to increased work loads and to the need for technical advice about important matters of public policy by establishing a number of legislative support agencies. The staff members employed by such agencies are in addition to personal and committee staff. Bills are drafted by special agencies often staffed exclusively by attorneys; an example is the U.S. Congress Office of Legislative Counsel. Reference libraries are maintained as a source of information for legislators and legislative staff, and public information offices are organized to facilitate the flow of legislative news to the general public and to members of the press. Nearly all legislatures have an audit office or a special evaluation office which is responsible for providing legislators with information about program performance and fiscal responsibility.

The best-known legislative support agencies of the U.S. Congress are as follows:

- *Congressional Research Service* (*CRS*) Part of the Library of Congress, the CRS serves as a politically neutral research arm of Congress. Staffed by over 900 persons, the CRS prepares legislative histories on measures taken up by committees, nonpartisan analyses of legislative proposals, and comprehensive digests of bills and resolutions introduced in both chambers. Natural resource topics addressed by the CRS in recent years include wild horses and burros, federal grazing fees, and USDA-Forest Service road construction.
- *Congressional Budget Office* (*CBO*) The CBO serves as a politically neutral organization (with over 250 employees) charged with providing Congress with budget-related information, including the fiscal ramifications of legislative proposals. It prepares federal revenue and expenditure estimates for future years, assists in the preparation of ceiling or target expenditure levels, and monitors revenue bills against agreed-to expenditure levels.
- *General Accounting Office* (*GAO*) The GAO (with over 4,000 employees) advises Congress on fiscal responsibility and legal compliance with laws enacted. It examines the efficiency of agency operations and probes into whether program results are consistent with legislatively prescribed objectives. Natural resource topics addressed by the GAO in recent years include Alaska land exchanges, USDI-Park Service maintenance funding, and endangered species of wildlife.
- *Office of Technology Assessment* (*OTA*) The OTA (with over 130 employees) advises Congress on complex issues involving science and technology. It clarifies the potential physical, biological, economic, and social impacts of a policy's technical elements. Natural resource topics addresses by the OTA in recent years include tropical forest resources, energy from biological resources, and evaluation of federal forest resource planning programs.

LEGISLATURE-BUREAUCRACY INTERACTION

Legislatures do not act in a vacuum on matters of forest and related natural resource policy. They actively engage in reciprocal relationships with client groups and various bureaus and agencies. By so doing they often find support for their interests. Adoption and implementation of policies become a tripartite product. Terms used to describe such relationships include *iron triangles* and *subgovernments*. For example, the Congressional Affairs Office on the USDI-Bureau of Land Management, working closely on matters of forest wildlife policy with the National Wildlife Federation and the staff of the Subcommittee on Public Lands, National Parks and Forests of the U.S. Senate's Committee on Energy and Natural Resources, plays a vital third-party role in an iron-triangle relationship.

Agencies and legislative units responsible for forest resource policies are inclined to interact for a variety of reasons. An agency's concern is with its own health and welfare—with how to expand its programs and resources and how to protect existing programs and resources from encroachment. Legislative concerns are the political survival of its members and the pursuit of certain policy interests.

Legislatures have numerous ways of drawing agencies into desired relationships (Ripley and Franklin 1984). Through covert means, they can threaten a program or an agency's existence; severely curtail an agency's jurisdiction and programmatic scope; and limit the agency's ability to maintain or increase the size of its staff by withholding funding, statutory authority, or confirmation of key appointments.

An agency can draw a legislature into the web of interaction by its approach to decisions that are considered important to legislators—for example, decisions about regulations guiding implementation of a forest resource program, about geographical and hierarchical location of professional personnel, about both the timing and location of program spending patterns (including facilities), and about the substance and timing of agency handling of client grievances. Agencies also can command legislators' attention because they are able to enhance a legislator's personal reputation by practicing the norm of deference and by providing timely information about forest resource programs.

Legislature and agencies interact on numerous occasions. Significant interaction may occur during the development of substantive forestry legislation. Legislative staff review of draft laws prepared by an agency may be involved, as may an agency's formal reaction to a law proposed during a legislative hearing. Interaction may also occur during the development of appropriate bills. Forest resource agencies attempt to reduce the uncertainty of the funding process by obtaining confidential information from appropriation subcommittee members and staff. Daily interaction can occur during the implementation of programs, the development of rules and regulations, the evaluation of programs, and the preparation of legislatively called for reports. Interactions between specific agencies and legislatures are highly variable. Some may be very formal, such as appearance before hearings or written reports, while others are quite casual, including telephone conversations, personal visits, and business luncheons.

LEGISLATIVE CHALLENGES

Legislatures are a democratic society's principal means of focusing widespread debate on major issues of forest resource policy and a significant means of legitimizing agreements for handling such issues. They perform a number of functions important to forestry and related communities, of which the most notable are as follows:

- *Constituency representation* They provide a means by which forestry interests can express their preferences for the use of management of forests.
- *Education* They provide a means by which various publics can be informed and educated about forests and the opportunities represented therein.
- *Lawmaking* They provide a means by which agreed-to forest resource policies can be given credence and legitimacy.
- *Oversight* They provide a means by which specific interests can make sure that policies and programs are being carried out according to legislative wishes.

These noble functions are worthy of the fullest attainment. In reality, however, legislatures frequently stumble and often falter in the performance of their duties. When this occurs, the consequences for forestry can be serious. In this section, we shall consider some common concerns about legislatures (Keefe and Ogul 1989).

Unresponsiveness

A frequent complaint is that legislatures are not responsive to the preferences of the majority. They are said to fail to represent the unorganized public and instead to be dominated by the wishes of organized special interests. Seldom are legislators attentive to the forest resource interests of the unorganized public. Being an audience to all people is nearly impossible. Besides, the noisy clamor of organized interests can be deafening, with the result that the weak voices of individual members of the general public cannot be heard. To the extent that the clamor represents the elusive public interest in forest resources, the unorganized segments of society may be represented in legislatures. However, there remain lingering concerns about whether the unorganized public is receiving its due.

Parochialism

Legislatures are also criticized for being parochial. Specifically, legislators are accused of looking only to their home districts for identification of important forest resource issues and for guidance on how to solve them. The concern is that forestry interests of a national, or even an international, nature are being treated with indifference—that legislators defer to the forestry claims of local individuals (say, loggers' interest in a favorable timber sale) and local organizations (say, a group that seeks an increase in forest wilderness designation), and that these claims are unlikely to be representative of broader geographic interests in forestland management. The obvious product of such parochialism is laws and appropriations tailored to fit local and regional concerns. Some critics speak harshly about such parochialism—for example, ''the legislature is populated by insecure and timorous individuals whose principle aim is to stay in office'' (Keefe and Ogul 1989, p. 7).

Lack of Innovation

Criticism is also leveled at legislatures for failing to innovate on matters of policy. Legislatures' cautious and conservative approach and unwillingness to experiment and cast off conventional perspectives are sometimes viewed as a deterrent to achievement of public interests in the management of forest resources. An example is the dilemma of how to balance forestry's long-term production cycles with needs for long-term sustained funding from legislative systems that operate on annual or biannual cycles. Similarly, there is concern that legislatures focus on trivial matters of forestry, such as a minor adjustment of state forest boundaries, or legislating the price of public timber. Legislative machinery becomes clogged with forest issues of little importance and with routine chores. Meanwhile, major opportunities for legitimizing innovative and far-reaching policies and programs languish.

Cumbersome Processes

The institutional arrangements employed by legislatures are frequently of concern because they obscure the legislative decision-making process and make it difficult to fix responsibility for legislative actions. Though a legislature may be operating according to detailed rules and agreed-to norms, even the most astute observers may have difficulty in following the course of a forestry bill through a legislature. Though overt processes and decisions are usually obvious, covert maneuvering and wheeling-and-dealing are difficult to follow and all but impossible to influence. Once a forest resource bill is enacted, who in the legislature will be held accountable if it fails, or if it succeeds? Which legislator? Which political party? Which committee? These questions are well-nigh unanswerable.

Minority Power

The inordinate power often exerted by minority views and positions in the legislative process is another matter of concern. Although minority viewpoints deserve to be protected from the politically strong, and although they certainly deserve to be heard, democratic theory is strained when a minority is able to resist, or even to veto, a forest policy preferred by the majority. Even though a persistent majority can eventually prevail, opportunities for minority-exercised delay in a legislature are mind-boggling. Some examples are as follows:

- A committee chair who opposes a bill and does not bring it up for consideration
- A committee that is unrepresentative of the chamber as a whole and that refuses to allow floor action on a bill
- A committee that prolongs hearings on a bill so long that a reporting deadline is missed
- Deference by many legislators to the opposition of a few legislators who have an immediate district interest in preventing passage of a bill

Although legislatures can be legitimately criticized, they are society's primary means of expressing the public's wishes on matters of forest resource policy. Over the years, they have legitimized a plethora of policies and programs that have served forestry interests very well. Major decisions of legislatures are typically temporary accommodations of groups that hold differing views on the use and management of forest resources. As circumstances change, old alliances are upset and the past consensus becomes damaged. Legislatures stand ready to legitimize the next consensus that appears on the political scene.

REFERENCES

Congressional Quarterly. *How Congress Works.* Congressional Quarterly Inc., Washington, 1983.

Congressional Research Service. *Bill Digest: Digest of Public General Bills and Resolutions.* 100th Congress, 1st Session. Library of Congress, Government Printing Office, Washington, 1985.

Council of State Governments. *Forestry in the South: A Survey of State Legislative and Executive Roles and Information Needs.* Lexington, Ky., 1985.

Davies, Jack. *Legislative Law and Process in a Nutshell.* West Publishing Co., St. Paul, Minn., 1986.

Ellefson, Paul V. Congressional Testimony: Who Presents Industrial Wood-Based Interests? *Journal of Forestry* 83(5):300–301, 1985.

Eulau, H. The Role of the Representative. *American Political Science Review.* September 1959.

Fenno, Richard. *Home Style: House Members in Their Districts.* Little, Brown and Company, Boston, Mass., 1978.

Government Publication Office. *Congressional Directory: 1989–1990.* 101st Congress, Washington, 1989.

Hansen, Royce. *Tribune of the People: The Minnesota Legislature and Its Leadership.* University of Minnesota Press, Minneapolis, 1989.

Hinckley, B. *Stability and Change in Congress.* Harper and Row Publishers, Inc., New York, 1988.

House of Representatives. *Administrative Reorganization and Legislative Management.* Commission on Administrative Review. H. Doc. 95-232. 95th Congress. U.S. Congress, Washington, 1977.

Keefe, William J., and M. S. Ogul. *The American Legislative Process: Congress and the States.* Prentice Hall, Inc., Englewood Cliffs, N.J., 1989.

Kingdon, John W. *Congressmen's Voting Decisions.* The University of Michigan Press, Ann Arbor, 1989.

Kozak, David C., and J. D. Macartney. *Congress and Public Policy.* The Dorsey Press, Chicago, Ill., 1987.

Lewis, Bernard J., and Paul V. Ellefson. *Staff-Mediated Information Flows to Forest Policy Committees in State Legislatures.* Agricultural Experiment Station Bulletin (forthcoming). University of Minnesota, St. Paul, 1991.

Mater, J. *Public Hearings Procedures and Strategies: A Guide to Influencing Public Decisions.* Prentice Hall, Inc., Englewood Cliffs, N.J., 1984.

Meeks, Gordon. *A Legislator's Guide to Forest Resource Management.* National Conference of State Legislatures, Denver, Colo., 1982.

Muir, William K. *Legislature: California's School for Politics.* The University of Chicago Press, Chicago, Ill., 1982.

Ogul, M. S. Congressional Oversight: Structures and Incentives, in *Congress Reconsidered,* by L. C. Didd and B. I. Oppenheimer. Congressional Quarterly Press, Washington, 1981.

Oleszek, Walter J. *Congressional Procedures and the Policy Process.* Congressional Quarterly Inc., Washington, 1984.

Patterson, Samuel B. The Professional Staffs of Congressional Committees. *Administrative Science Quarterly* 15(1970):22–37, 1970.

Ripley, R. B., and G. A. Franklin. *Congress, the Bureaucracy and Public Policy.* The Dorsey Press, Homewood, Ill., 1984.

Rosenthal, Alan. *Legislative Life: People, Process and Performance in the States.* Harper and Row Publishers, Inc., New York, 1981.

Smith, S. S., and C. J. Deering. *Committees in Congress.* CQ Press, Washington, 1990.

USDA-Forest Service. *A Case Study of Forest Service Legislation in the Legislative Process: The Small Tracts Act.* Legislative Affairs. U.S. Department of Agriculture, Washington, 1986.

USDA-Forest Service. *A Case Study of Forest Service Legislation in the Legislative Process: The Federal Lands Facilitation Act.* U.S. Department of Agriculture, Washington, 1990.

Willet, Edward F. *How Our Laws Are Made.* U.S. House of Representatives. 99th Congress, 2d Session. Government Printing Office, Washington, 1986.

11

JUDICIAL SYSTEMS AND PROCESSES

Judicial systems are the nation's primary means of furthering the cause of justice as prescribed by the U.S. Constitution. Such systems are most often viewed as being occupied with process, precedent, and the meaning of law—and as being remote from the push and pull of the politics that so often characterizes the legislative and executive branches of government. However, the need to grapple with sensitive constitutional issues, pour meaning into ambiguously worded forest resource laws, and resolve pointed forestry disputes between quarreling parties often places the judiciary squarely in the middle of the policy process.

In recent years, courts have been asked to rule on far-reaching issues—for example, the meaning of "major federal action significantly affecting the environment," as stated in the National Environmental Policy Act of 1970, a ruling which did much to sensitize government agencies to environmental concerns. Likewise, courts have been asked to rule on the circumstances under which an agency can identify, sell, and remove trees from certain public lands, a ruling which interpreted USDA-Forest Service authority as granted by the Organic Act of 1897. The ruling ultimately precipitated a small-scale revolution in the planning and management of 191 million acres of the nation's public forestland, rangeland, and recreation land. Initiated by disputes among public and private parties and ultimately addressed by the nation's courts, actions of this and a similar nature have generated broad public debate over the use and management of forests. In a democratic society in which judicial systems are very important, the situation is both inevitable and healthy.

In this chapter, we shall consider the prevalence of legal action concerning forestry and environmental issues, the character of law and legal proceedings, the organization and administration of the judiciary, the role and position of judicial systems in policy development, and challenges facing today's judiciary system.

OCCURRENCE OF LEGAL ACTIONS

A significant number of forestry and environmental disputes are brought to the judicial system for resolution. From 1970 through 1980, for example, 2,178 environmental law cases were adjudicated in federal courts—54 percent of them in trial courts, 43 percent in appeals courts, and 3 percent in Supreme Court (Wenner 1983). Of these, 1,001 were initiated by environmental organizations (with 926 focused on the actions of government), while the business community initiated 588 such cases. As for forestry and closely related subjects, 218 cases were decided by federal courts from 1975 through 1988; a substantial number of these cases (71) addressed the policies and programs of forest reservations and forest preserves such as the USDA-Forest Service's National Forest System (Table 11-1). Since many of the 218 cases involved dispute over more than one point of law (some more than 10 points of law), the actual number of forestry disputes addressed by federal courts during the period in question exceeded 700.

Case numbers per se are not always good indicators of the significance of judicial rulings, since some rulings are far more important than others in terms of the legal principles addressed and the scope of the geographic area to which they are applied. For example, the effects of a ruling addressing a two-party contract dispute may pale in comparison to the social consequences of a state forest practice law declared unconstitutional by a state supreme court.

The nature of the subjects addressed by federal courts and the segment of the policy process involved is far-ranging. Below are some examples selected for the sake of variety, not significance of impact of rulings by federal courts (West Publishing Company 1987).

- County land-use regulations (adopted under various provisions of Colorado law) apply to federal lands situated in a county to the extent consistent with federal law. *City and County of Denver By and Through Board of Water Commissioners v. Bergland* (District Court, Colo., 1981).
- Rules and regulations promulgated by the Secretary of the Interior and the Secretary of Agriculture are not exempt from notice and commitment requirements of the Administrative Procedures Act. *Conservation Law Foundation of New England v. Harper* (District Court, Mass., 1984).
- The USDA-Forest Service, which permitted development of a government-owned ski area on Coconino National forest, must comply with the Endangered Species Act and take appropriate measures to minimize danger to alpine plants proposed for listing as a threatened species. *Wilson, Hopi Indian Tribe, Navajo*

TABLE 11-1
U.S. federal court case decisions, by forestry subject area and by type of court, 1975–1988

	U.S. Courts			
Subjects	District courts	Courts of appeal	Supreme Court	Other courts
Woods and forests				
Power to protect and regulate	4	2		1
Statutory provisions	2	2		
Officers and commissions	3	3		
Forest reservations or forest preserves	34	37		7
Offenses		6		
Criminal prosecutions	3	3		
Public lands (government ownership of land and property)				
Government authority and control	11	1		
Trespass	1	1		
Cutting and removing timber	2	1		2
Statutory provisions	2	3		
Criminal prosecutions	1			
Parks				
General		27		
U.S. Property, contracts, and liabilities				
Sale of realty and timber in general	4	2	1	14
Health and environment (environmental impact statements)				
Lobbing and public land generally	13	6		
Logs and logging				
Sale and conveyance of timberlands	1			
Sale and conveyance of standing timber	10	4		3
Licenses	1			

Note: Case decisions included are those judged as directly involving or having a direct impacting upon the use and management of forests. Any one case may involve multiple decisions concerning more than one subject. Other courts are U.S. Claims Court and U.S. Bankruptcy Court.

Source: Reprinted with permission from *West's Federal Practice Digest 3d,* copyright © 1988 by West Publishing Company, St. Paul, Minn., 1988.

Medicinemen's Association v. Block (Court of Appeals, Washington, D.C., 1983).

• Under the public trust doctrine (common-law concept), all public lands of the nation are held in trust by the government, which has the duty under the doctrine to protect and preserve the lands for the public's common heritage. *Sierra Club v. Block* (District Court, Colo., 1985).

• As stated in the Organic Act of 1897 and to be adhered to by the USDA-Forest Service for purposes of selling timber from the National Forests, "mature tree" implies physiological maturity (growth tapers off and health and vigor decline) rather than economic or management maturity, and "designate" and "mark" mean indicating the area of cutting and identifying each tree to be cut by a blaze, paint, or marking hammer. *West Virginia Division of Izaak Walton League v. Butz* (Court of Appeals, W. Va., 1975).

• Numerical estimates in a contract for salvage and logging of timber were for informational purposes only and were not a guarantee of quantity. *LGD Timber Enterprises, Inc. v. U.S.* (U.S. Claims Court, 1985).

• So long as the USDI-Bureau of Land Management's decisions with regard to use of public lands are not irrational or contrary to law, it may manage public lands as it sees fit. *Northwest Coalition for Alternatives to Pesticides v. Lying* (District Court, Oreg., 1987).

• Provisions of the USDA-Forest Service's Manual do not preclude the agency from imposing new scaling provisions as a condition of timber sale contract extension. Timber company was not adversely affected by the new provisions. *Rough and Ready Timber Company v. U.S.* (Court of Appeals, Federal Circuit, 1983).

• Congress intended to allow land under USDI-National Park Service jurisdiction to be exchanged under section of Land and Water Conservation Act (accept title for nonfederal property received in exchange for federal property). *Committee of 100 on Federal City v. Hodel* (District Court, Washington, D.C., 1985).

• In preparing environmental impact statement for proposed categorization of National Forest Lands as wilderness, nonwilderness, and further planning, the USDA-Forest Service failed to adequately consider the site-specific impact of its decisions. *State of California v. Block* (Court of Appeals, Calif., 1982).

• The USDA-Forest Service failed to analyze the cumulative impact of all timber harvesting in and around a National Forest district in which timber harvesting was to take place. *National Wildlife Federation v. USDA-Forest Service* (District Court, Oreg., 1984).

• The USDI-Bureau of Land Management violated the spirit and letter of Council on Environmental Quality scoping regulations by failing to invite regional environmental organizations to participate in the scoping process prior to preparing final environmental impact statement on the use of herbicides. *Northwest Coalition for Alternatives to Pesticides v. Lying* (Court of Appeals, Oreg., 1988).

• In determining the value of standing timber, a real interest rate (or a rate net of inflation) must be used for discounting purposes. *York v. Georgia-Pacific Corporation* (District Court, Miss., 1984).

• A lumber company is entitled to recover damages from a landowner's breach of timber sale contract. *Lubecki v. Omega Logging Company* (District Court, Pa., 1987).

• Regarding the Secretary of Agriculture's forestry permitting procedure requiring "public meeting, assembly, or special event," the words "meeting," "assembly," and "special event" do not denote three wholly discrete events or activities. *U.S. v. Beam* (Courts of Appeals, Tex., 1982).

• USDA-Forest Service's denial of special-use permits to Native Americans for use of 800 forested acres of the Black Hills for a religious ceremony was "agency action," to be reviewed in accordance with the Administrative Procedures Act. *U.S. v. Means* (District Court, S. Dak., 1985).

• Site preparation and artificial regeneration in the USDA-Forest Service's reforestation program for a site devastated by southern pine beetle infestation would not have significant adverse environmental consequences; assessment procedures were in compliance with the National Environmental Policy Act of 1970. *State of Texas v. USDA-Forest Service* (District Court, Tex., 1987).

LAW AND LEGAL PROCEEDINGS

An ordered society seeks to ensure that all citizens are subject to uniform enforcement of agreed-to social principles or norms that a community values highly, such as rights to property, religion, and individual liberty. *Law* is the formal means by which social standards of conduct pertaining to society are set forth. Backed by the organized force of communities, law can take many forms (Abraham 1986; Arbuckle et al., 1983; Gellhorn and Boyer 1981; Van Horn et al. 1989) including:

Private Law Law governing the relationship between private citizens, namely

• *Contract law* Law governing formal and legally binding agreements among individuals and corporations. For example, contract between owner of forest land and hunters wishing to hunt on said land.

• *Torts* Law governing legal duty to avoid causing harm to the person or property of others (nuisance, trespass, negligence). Carelessness leading to harm may give injured party grounds for seeking restitution.

Public law Law governing relationships between citizens and government, namely

• *Constitutional law* Law governing fundamental power and organization of government based on a constitutional document (police power, due process, equal protection).

• *Statutory law* Law governing public and private actions and originating from specially designated, authoritative lawmaking bodies. Found within the body of laws and resolutions enacted by legislatures (for example, the National Environmental Policy Act of 1970).

- *Administrative law* Law governing the operations of administering agencies. Originates from administrative agencies of government which have been delegated rule-making authority by legislatures. Found in the collection of rules and regulations prepared by government agencies (for example, Code of Federal Regulations).

Categories

There is considerable overlap among categories of law. For example, torts may become public when a private party sues a government forestry agency for malfeasance or nonfeasance, or administrative law may move into the realm of statutory law when agency-promulgated forestry regulations are challenged and must be assessed in light of legislative intent. Other useful categories of law include *common law,* which originates from judicial decisions guided by previous court rulings in similar cases; *civil law;* and *criminal law.* In civil disputes, courts are called upon to define, establish, or adjust relationships between litigants (for instance, a community alleging air pollution caused by a local pulp and paper mill), while in criminal disputes the judiciary is called upon to adjust relationships between individuals and society as a whole (for instance, theft of property or physical harm inflicted upon on individual by another).

A major function of courts in the policy process is to settle disputes over the application of various types of law (legitimized policy) by an acceptable and authoritative means (McLauchlan 1977). When the Secretary of the Interior's authority to administratively establish rules to regulate off-road vehicle use on public lands is challenged, the dispute may require resolution by a court of law. Similarly, if the USDA-Forest Service is alleged to have failed to provide site-specific information in a statutorily prescribed environmental impact statement, the allegation may have to be examined by a judge and a jury. If toxic air pollutants from an electric power-generating facility are alleged to be adversely affecting the growth of forests owned by an industrial wood-based company, the courts may be asked to make a judgment.

Although the function of courts is to settle disputes, not all disputes require the services of a court. In fact, most disputes never reach the point of litigation, even though laws may be involved and legal rights affected. For example, a minor timber trespass case may be resolved by mutually agreeable payment and a friendly handshake. The disputes brought to a court of law for resolution generally involve situations in which the stakes are large, as when the dispute is over a large, 50-year or longer public timber sale, not a single small timber sale. Usually the attitudes and emotions of the parties involved are sensitive and strongly held; inflexible ideologies, for example, involving either timber or wilderness as a public-land use may be in question. Furthermore, usually the social pressure to resolve a dispute is great, as when a community's interest in domestic water supply is jeopardized by poorly designed and inadequately maintained logging roads in a forested area.

Adversary Process

Historically, the process by which American courts resolve disputes has evolved from adversary procedures followed by English law. The modern version of the proceeding is based on the disputing parties' initiating the process and subsequently presenting evidence and arguments that attempt to establish the rightness of their contentions. Through a representative (a lawyer), each side presents evidence which is ultimately judged (by judge or jury) against a legal standard (such as a statute). The outcome is not precisely predictable, although the parties involved can expect a court's decision to be made objectively, to be based on the evidence presented, and to be consistent with previous decisions (McLauchlan 1977). Key characteristics of the adversary process are:

• *Disputing parties (not courts) initiate the process* The burden is on the disputing parties to bring the case to trial and to come forward with evidence. Judges are passive enforcers of procedures. Courts do not create their own business as do legislators when they introduce bills.

• *Two parties and a trier of facts are the main actors in process* Views and opinions of only the two disputants are involved in the process. The trier of facts (evidence) is not one of the disputing parties. Not being party to a dispute, judges (or juries) do not sit in judgment of their own views and evidence as do legislators when voting on bills they introduce.

• *Decisions are based on evidence (facts) presented* The process is designed to yield an objective judgment based on evidence submitted, on procedural rules, and on a legal standard specified in law.

• *Outcomes are all or nothing* The process is a zero-sum game in that one party wins what the other party loses. The disagreement is over the actions and resources of the parties involved, not the actions and resources of others, as is frequently the case when legislators debate use of a third party's (e.g., taxpayers) resources.

• *Decisions (judgments) are authoritative* Court decisions carry the force of law and are accorded significant obedience.

The adversary process is not without problems. Because it is limited to two disputants, it may ignore or neglect important interests of parties that are not formally involved in the process. For example, a lawsuit involving a single timber company versus a single environmental group may exclude the relevant interests of other companies and other groups. Similarly, the antagonists in the process may define the dispute so narrowly that other more appropriate solutions to a forest resource issue cannot be considered. For example, a dispute over government regulation versus fiscal incentives may ignore the appropriateness of educational programs as a means of encouraging management of nonindustrial private forests. Since the process is a zero-sum game, compromises giving each party a large part, but not all, of their desired outcomes are not possible. Furthermore, in practice, judges (and juries) may be biased, some lawyers may be better than others in representing a party's interests, evidence needed to make informed decisions may not exist or may not be readily attainable, and the cost of the proceeding may be so great that certain disputants are excluded. Alternatives to the adversary process include dispute resolution by force, by binding arbitration, and by mediation (in which a third party seeks to persuade two parties to reach agreement).

Forms of Illegal Conduct

Courts are capable of responding to various forms of illegal conduct, of which three principal forms are most common (Sax 1970). Courts may find that there has been a *violation of legal standard specified in law*. If, for example, statutory law requires a landowner to demonstrate the existence of at least 300 vigorously growing seedlings per acre 3 years after timber harvesting, the landowner's failure to comply becomes grounds for a finding of guilt and subsequent imposition of a penalty (if such is prescribed).

Courts may also find that *procedures required by law have not been followed* (a procedural failing) or that a reasoned decision-making process has not been employed. For example, a law may require public agencies to conduct public hearings, make study reports, or consult with other agencies. Failure to do so may be judged illegal by a court of law. The great bulk of environmental litigation involves procedural failings, a situation fostered by a plethora of legislatively prescribed agency procedures (in which greater opportunity exists for plaintiffs to claim that proper procedures were not followed) and by a judicial tendency to avoid the substance of a dispute (e.g., Does clear-cutting lead to polluted water?) and instead to impose rulings that more hearings are necessary, additional information is required, or further consultations are needed.

Courts may also determine illegality by finding a defendant's *actions to be arbitrary, capricious, or unsupported by evidence*. For example, a forestry agency that has constructed a logging road without benefit of a transportation plan nor access to soil stability information may be judged as having acted in an arbitrary and capricious manner. Proving the existence of arbitrary and capricious actions in a court of law is very difficult, especially if defendants are able to bring forth substantial technical information supporting their actions as being consistent with conventional practices. For example, a forestry agency might be accused of acting in an arbitrary and capricious manner because it practiced even-aged management of a certain forest species; however, if it could present evidence that this practice was conventional, a finding of arbitrary and capricious actions would be unlikely.

Virtues of Courts

The ability of courts to settle disputes over the use and management of forests is significant. Courts are often believed to have virtues which are lacking in legislative and executive systems of government (Sax 1970). For example, courts can provide opportunity for significant independent action on a dispute. Legal norms of conduct make courts comparatively immune from political pressure to select one policy alternative at the expense of another. (In contrast, legislators and administrators are continually pressured by citizens and interest groups.)

Similarly, impartiality toward issues is fostered by the way judges are selected. Judges are not selected for cases on the basis of their attitude toward particular issues. In contrast, a governor may choose members of a forest practice board on the basis of their attitudes toward the issues. Nor do judges need to decide issues in such a way as to maintain a political balance among constituencies. In contrast, administrators may select a forest policy alternative which accommodates the views of all interested parties.

Judicial processes are also praiseworthy because they can provide opportunity for speedy access to an institutional agenda without prior political screening. In contrast, aggrieved parties often experience great difficulty in securing a place for a dispute on the action agenda of an agency head, a resources commission, or a legislative committee—and doing so without having to redefine the issue in a manner politically acceptable to a gatekeeper.

Courts also afford opportunity for quickly reducing disputes to concrete, well-defined concerns that can be the focus of a specific judicial ruling or order. In contrast, issues in legislative and executive systems may be allowed to float in the generality that so often accompanies public debate. For example, an issue may become lost among ambiguous allegations such as "Jobs will be lost," "The environment will be damaged," and "Prices will be increased." Parties that appear before courts must shape controversies into precise, manageable issues that can be the subjects of specific court orders.

ORGANIZATION AND PROCESS

Structure

The American judicial system is a large and complex structure. Employing a very decentralized organizational pattern, two systems (federal and state) operate side by side. The federal system is composed of the U.S. Supreme Court, 12 courts of appeals, and 90 district courts located throughout the nation. The Supreme Court has nine justices, while the federal courts of appeals have 168 permanent judgeships with between 6 and 28 judges in each court (depending on the appellate court's work load). Each state has at least one federal district court (populous states have up to four), and between 1 and 27 judgeships are assigned to each district court. In addition, there are a number of specialized federal courts, including the court of international trade, claims court, tax court, court of customs, and patent appeals. The only federal court that is constitutionally indispensable is the Supreme Court; authority to establish or abolish all other federal courts is vested with Congress.

Structurally, the federal court system is pyramidal. The highest level is the Supreme Court, the next or intermediate appellate level is the courts of appeals, and the lowest or trial level is the district courts. This organization enables the Supreme Court and appellate courts to correct legal errors made at the district court level and enables higher courts to assure uniformity of decisions when two or more lower courts have reached different decisions on the same subject (Jacob 1984).

State courts have a similar pyramidal structure, although the names, numbers, and functions of the state courts often differ markedly among states and from the federal system. In addition to supreme, intermediate appellate, and trial courts of general jurisdiction, states have established a number of courts of limited jurisdiction, such as probate courts, small-claims courts, juvenile courts, and municipal courts (Glick and Vines 1973).

Jurisdiction

Jurisdiction of courts is determined constitutionally or by law. In general, however, before a court will accept a forest resource dispute for resolution, the litigants must establish that the court has jurisdiction over the case, namely, that the subject at issue is within the subject area the court is charged by statute with considering. They must also establish that the defendant is physically within the court's geographic control. In addition, there must be a grounds to presume that the dispute is capable of being resolved or remedied by deciding the case, and that the plaintiff is entitled to bring the dispute to court (has standing) (Lawrence 1989).

The U.S. Constitution formally describes cases over which federal courts have jurisdiction (Article III and the Eleventh Amendment); by implication, all other matters are left to the states. Federal courts address all cases involving the U.S. Constitution (regardless of the parties involved), treaties, maritime matters, and laws enacted by Congress. For example, an alleged violation of the Wild and Scenic Rivers Act of 1968 would be handled by a federal court. In addition, a federal court will hear a case if any of the following parties are involved:

- *An agency of the United States* For example, the USDI-Bureau of Land Management or the USDI-National Park Service
- *Two or more states* For example, a disagreement over appropriate remedies for air pollutants originating in one state but damaging forests in an adjacent state
- *Citizens in different states in a dispute involving more than $10,000 (an amount that is subject to change by statute)* For example, a disagreement over

payment for timber purchased by a person in one state from a person in a second state

- *Foreign diplomats and foreign nations*

Trial Courts

Federal and state court systems include a variety of trial (civil and criminal) and appellate courts. Trial courts settle most cases with finality either by promotion of out-of-court settlements or by judgments resulting from a formal trial. They are usually the courts of original jurisdiction; they try cases in the first instance, either before a judge and a jury or before a judge. Trial courts deal with questions of a fact—such as whether a forestry event occurred and whether it was in conflict with a standard set forth in law. They also make judgments without issuance of a reasoned opinion and produce decisions which are not precedents for future trials (McLauchlan 1977). The legal procedures followed by a trial court are intricate (Abraham 1986). Forestry professionals involved in trials can benefit from learning to understand trial court procedures (Appler 1986).

In recent years, pretrial conferences have been used to expedite the settlement of disputes, wherein the parties involved may be required to specify in advance the disputed facts of the case. On the basis of such information, a judge may suggest negotiation and may even propose a financial sum for a settlement. Only exceptional cases are carried forward to full trial. In many instances, simply filing a case with a court provides sufficient incentive for the disputing parties to settle without engaging in the proceedings of a full trial. The caseload of federal trial, or district, courts is substantial, with, for example, over 295,000 pending cases in 1986, of which 86 percent were civil cases (Office of United States Courts 1986).

Civil cases brought before trial courts may involve claims that a party has suffered damages and injuries, the extent of which can be clearly specified in monetary terms (e.g., the value of lost business). In such cases, the court will order the losing party to pay the plaintiff the monetarily defined damages, based on a jury verdict. Many cases, however, involve injury that cannot be remedied by payment of monetary damages even though a judge or jury has found them to have occurred. For example, a defendant, such as a public transportation department, may be about to construct a road, under power of eminent domain, through a colony of rare and endangered forest plants. A plaintiff, such as a landowner or a class of society having standing, may allege permanent loss or injury—such as loss of scenic beauty, genetic material, or recreation opportunity—which cannot be repaired by the award of money. In such a case, where the loss is permanent and not subject to repair, either equity relief or relief which will prohibit construction of the road or will specify an alternative route will be sought.

Initially, equity relief can take the form of a *temporary restraining order* which is designed to preserve the status quo until a more detailed examination of the case can be made by the court. With supporting evidence from the plaintiff, a judge may issue such an order for a period up to 10 days. The defendant need not be present at the proceedings. Subsequently, a *preliminary injunction,* which can be issued only after the defendant has had an opportunity to present contrary information to the judge, may be sought from the court. Such an order may be issued for any period of time the judge chooses and is designed to prevent irreparable damage until a full trial can be held. The ultimate remedy for awarding nonmonetary relief, equity relief, is a *permanent injunction,* which is established after a complete trial by judge, or judge and jury, is held on the matter.

Appellate Courts

Litigants involved in civil and criminal disputes at the trial court level may wish to appeal the results of their trial, believing that the judge made prejudicial errors or that a forest resource law was improperly interpreted. For such appeals there exist appellate courts. Some appellate courts, such as the U.S. Court of Appeals, must accept all cases properly presented to them, whereas others have significant discretionary power to accept or decline trial (and some appellate) court decisions. Among the ways in which appellate courts differ from trial courts are that they:

- *Address questions of law (not fact) and procedure that might be raised by trial courts* Questions of law include interpretation of words included in a statute enacted by a legislature (e.g., what is the meaning of "mature or large-growth trees"). As such, appellate courts can play a major role in the development and implementation of policies.
- *Consist of a group of judges who must reach a decision as a group* Judges meet in conference, discuss a case's merits, engage in bargaining, and reach a consensus decision. A judge, for example, who has certain objectives in mind may have to modify an opinion in order to obtain another judge's support for those objectives (McLauchlan 1977). Since only questions of law are addressed, appellate courts never use juries.
- *Prepare reasoned essays justifying decisions* Decisions and essays become part of a jurisdiction's common law.
- *Base decisions on information contained in written briefs, trial transcripts, and oral arguments (often limited to 20 minutes) by a litigant's lawyer.*
- *Reach decisions that in most cases are final* If the appellate court affirms a trial court's decision, the trial court is ordered to execute it; a reversal may result in an order for a new trial. In the federal court system, however, some decisions of the U.S. Circuit Court of Appeals may be carried forward to the U.S. Supreme Court.

Participants in Judicial Processes

There are numerous participants in the judicial process, and their responsibilities are high variable. They range from lawyers (of whom there were over 750,000 in the United States in the late 1980s) employed by various private concerns (including interest groups) to attorneys engaged in the business of government, and from judges in various federal, state, and local courts to the general public which is the source of most ligation and which exerts substantial political pressure on courts.

Attorneys Attorneys for the federal government (of whom there are more than 20,000) are often employed by the U.S. Department of Justice but are frequently assigned to agencies responsible for the management of forest resources. For example, the USDA-Forest Service has an Office of General Counsel which is staffed by more than 15 lawyers. In addition, the agency has lawyers assigned to various regional offices and in some cases to specific national forests. Similarly, forest and related land-management agencies within the U.S. Department of the Interior—such as the Bureau of Land Management, the National Park Service, and the Fish and Wildlife Service—seek legal counsel from attorneys located in various divisions of the department's Office of the Solicitor.

Judges The most prominent and prestigious participants in judicial processes are judges. A court's reputation for honesty and impartiality is in large measure dependent upon the character and reputation of the judge in charge. As an administrator, a judge is involved in developing budgets, securing appropriate physical facilities, and appointing assistants. Judges are also responsible for maintaining an orderly flow of cases, enforcing legal procedures, encouraging pretrial settlements, handing down appropriate remedies, and—where called upon to do so by statute—using important discretionary power to decide the merits and the law of specific forestry cases. Federal judges are appointed; state judges may be appointed by a governor or may receive their position by legislative action or by election. Appointment of judges by the President and governors entails consideration of a number of factors, including legal competency, career experiences, ideological perspectives, geographic representation, religious and ethnic considerations, and political participation favorable to the executive making the appointment (Goldman and Jahnige 1985). Nearly all judges, however, have formal legal training and prior political experience, say as legislators or elected executives.

Decisions made by judges are influenced by a variety of conditions (Figure 11-1). For example, the extent to which a judge can be viewed as an activist on forest resource matters depends on the seriousness with which the judge distinguishes between making law and interpreting law. Making law is suppos-

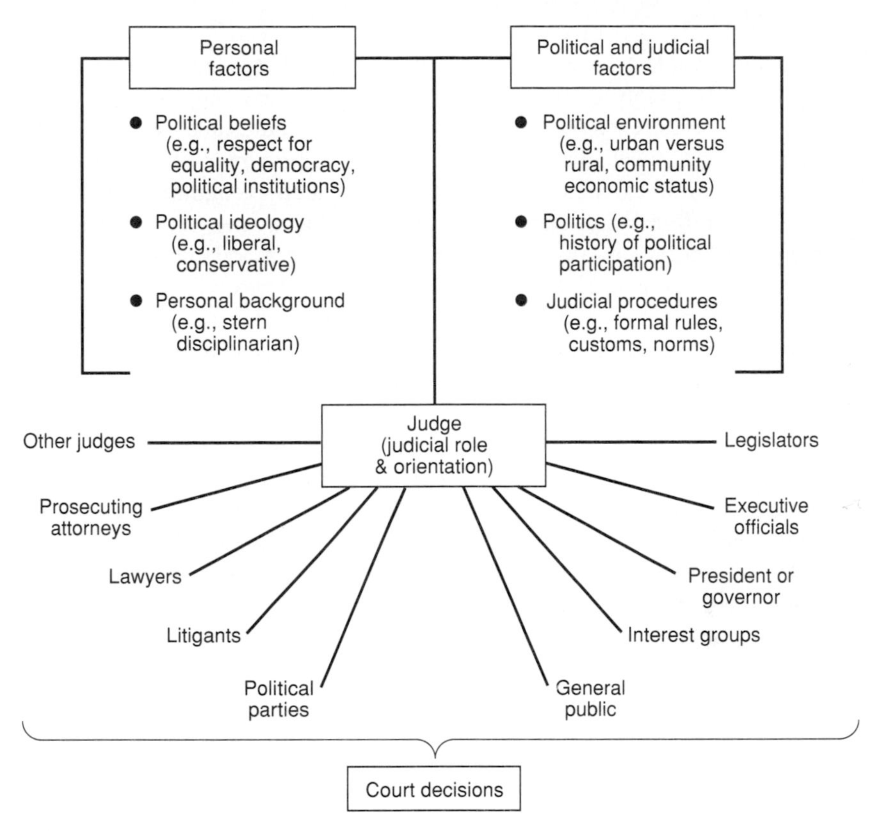

FIGURE 11-1
Factors influencing the role and decisions of judges in a judicial system. (*Source: Henry Robert Glick/Kenneth N. Vines,* State Court Systems, *© 1973, p. 55. Adapted by permission of Prentice Hall, Englewood Cliffs, New Jersey.*)

edly a legislative task, whereas interpreting law is a judicial task. Judges are also influenced by various norms of the judicial profession. Regarding partisan activity, for example, the Canons of Judicial Ethics (Canon 21) warns that "justice should not be molded by the individual idiosyncracies of those who administer it. A judge should adopt the usual and expected method of doing justice and not seek to be extreme or peculiar in judgments, or spectacular or sensational in the conduct of the court" (Jacob 1984, p. 209). The decisions of a judge may also be influenced by the extent to which a judge relates to local community norms. For example, a judge's residence in a community which is dependent upon forests for economic and social stability may have a significant influence upon the outcome of a forestry dispute (Goldman and Jahnige, 1985).

POLICY MAKING

The decisions reached by courts can have impacts that reach far beyond the litigants directly involved in a dispute. Although courts' approach to policy making may be different from the approaches of legislatures and executive agencies, courts must nonetheless be recognized as important participants in the policy process. For example, a judicial ruling which recognizes government imposition of reforestation standards upon a private owner of forestland can become part of common law within an appellate system and can serve as precedent for government intervention in similar private forests throughout a system's district. Similarly, a judicial decision to broaden statutory language (intent) concerning the management of endangered species can lead to significant changes in agency interpretation of law as carried out during the implementation phase of the policy process.

Debates over judicial intrusion into the policy-making responsibilities of legislatures and executive agencies are found throughout the literature of forestry and natural resources. Court involvement in the establishment of social and natural resource policy is widely thought to be the natural result of citizens' proclivity for thinking of social problems in legal terms—for people do tend to sue on all sorts of subjects, from wage rates to the allocation of airline routes. Equally popular is the view that such court involvement is often the consequence of widespread citizen disenchantment with the ability of other institutions, such as legislatures, to address perplexing public issues with workable public policies (Horowitz 1977).

The high visibility of a court that makes an unpopular decision is often the result of the willingness of a judge to tackle a difficult social problem. Decisions on such problems do indeed receive widespread public attention. Van Horn et al. have described the situation well (1989, p. 204):

> Judges, like other policy makers, are attentive to appearances, but are less likely . . . to substitute symbolic action for substantive action. Very few court decisions are purely symbolic. Unlike other public officials, judges do not avoid difficult problems or unpopular solutions. Instead, they confront many of society's most vexing problems.

Characteristics

Judicial policy making has a number of characteristics which make it different from the policy making of legislatures and executive agencies. For example, courts cannot, in a programmatic sense, seek out forest resource issues and subsequently establish forest policies and create extensive programs needed to implement them. A forest resource dispute must be presented to the courts before the machinery of judicial policy making can be set in motion; parties such as private citizens, interest groups, and government agencies must initiate the process. Because judicial agenda setting is passive, the forest resource policies which emanate from the courtroom may be neither comprehensive nor programmatic. They are likely to be disjointed, coming in fits and starts. Important social issues may never appear for action (Glick and Vines 1973; Jacob 1984). Even when an issue is placed on a judicial agenda, there is no guarantee that a range of innovative policies will be available for the judiciary to choose from. The distinctive presence of the very rigid adversary process may well narrow the scope of available policy options (Glick and Vines 1973, p. 90):

> Judicial decision making is dominated by the adversary process in which judges are presented with fairly rigid alternatives by opposing sides in a case. The decision reached by the court almost always reflects support for one of the two alternatives, thus limiting the range of judicial policies emanating from individual court cases.

Courts also differ from other policy-making institutions in that they cannot effectively implement the forest resource policies they establish. They cannot levy taxes and appropriate money to carry out their wishes. They must depend on the goodwill and resources of executive agencies, which may have sound reasons for circumventing judicial decisions. In addition they must face the often hostile attitudes of citizens who disagree with the policies that have been judicially established (Abraham 1986).

The source and type of information available to courts are also distinctive features of judicial policy making. The information in briefs and oral arguments is neither unbiased nor necessarily complete; it is designed to persuade the court to make a decision in favor of one of only two opposing interests. Having access to only two sources of information (both of which are biased and probably incomplete), the courtroom may lack access to a rich variety of information about the physical and economic characteristics of forests—information which could enhance the substance of a judicial ruling. This situation is in sharp contrast to that encountered in legislative policy making, in which various mechanisms, including legislative hearings, are used to access a variety of relevant information from various points of views.

The policy making of courts also differs from executive and legislative branches of government with regard to communication of agreed-to policy. For example, since the reasoning behind a judicially arrived at forest policy is not given at the trial court level, broader audiences are given no reasons as to why they should support a policy. At the appellate court level, broader audiences' understanding of reasons for compliance with a policy depends upon the extent to which the reasoning is free of legal terms and void of reference to complicated judicial precedents. Ambiguous judicial policies may also complicate communication efforts. At the appellate level, ambiguity may be the result of necessary bargaining and compromise over language; adherence to the policy product may consequently be difficulty to obtain. A flurry of dissenting opinions at the appellate level may also generate confusion about the convictions of the judiciary that established the policy (Van Horn et al. 1989).

Judicially established policy is also characterized by a special respect for precedent—a perspective which can confine the scope of searches for innovative forest resource policies. Legislative systems are not restrained by prior policies; in their political interest, they may trash old forest policies and probe in dramatically different directions for new solutions to especially vexing forest resource issues. Fortunately, the weight of precedent in legal decisions slows change but does not prevent it. Most courts (the exception is the U.S. Supreme Court) are also cognizant of the power of the legislative systems which created them. Judicial ability to innovate is certainly influenced by the knowledge that legislatures can reverse the ruling of courts or otherwise show displeasure with judicially established policies—for example, by withholding pay increases or changing jurisdictions (Patterson et al. 1989).

Appropriate Scope

Horowitz (1977) has offered suggestions about the appropriate scope of judicial action on matters involving policy, as follows:

- Courts should refrain from becoming involved in forestry issues that are rapidly changing, that have yet to be addressed statutorily, and that could place a court in the position of managing resources on a day-to-day basis.
- Courts should refrain from involvement in issues that are very narrow in scope. If a ruling would have little application elsewhere—as in a dispute over the habitat requirements of a single species of forest wildlife in a very limited geographic area—court action may be inappropriate.
- Likewise, courts should avoid cases in which there is insufficient incentive for the parties involved to adopt and implement a court's ruling (such as a dispute over widespread air pollution caused by technically and financially bankrupt wood-processing facilities), or in which there is a strong likelihood that a court's ruling will be distorted by the interaction of several groups (as when a ruling is expected to be implemented by numerous competing agencies).
- Courts might well avoid cases in which it would be very difficult for them to determine what would happen after a ruling was made. An example would be a dispute over water pollutants originating from private forestland over which no public agency has jurisdiction.

JUDICIAL CHALLENGES

Judicial systems have increased their participation in various segments of the policy process. The effects of this increase have been well characterized by the chief of the USDA-Forest Service , who in 1989 stated, ''The 3,300-plus lawsuits in our General Counsel's office sap our energy, requiring more attorneys. And we're having a tougher time making decisions stick'' (Robertson 1989).

The intensified role of judicial systems in forest policy making has received mixed reaction from the forest resource community. Many professionals fear that judicial actions will usurp their prerogative to guide forestland use and forest practice decisions in ways that will respect both the diversity of forest ecosystms and the many objectives of the owners of the ecosystems. Other professionals are deeply concerned about their ability to enter judicial processes and politically convince the courts of the validity of their particular point of view. Still other professionals view courts as a means of securing policy reactions from legislatures and executive agencies, institutions they regard as strangled by procedure and fearful of the political consequences of needed changes in forest policies.

A major challenge in the forestry-judicial interface is to determine the proper role of unelected judges in the policy process of a representative democracy. Some relevant questions are:

- Should judges interpret obscure segments of legislatively established forest law so as to force wholesale review of a natural resource agency's reason for existence?
- Should judges interpret the scope of a forest law's application so broadly that literally thousands of additional forest owners suddenly become subject to the law's provisions?
- Should judges declare federal interest in forests to be paramount and thereby negate the interests of state and local governments?

The relative merits of judicial activism and judicial restraint have been debated throughout much of the nation's history.

Supporters of judicial involvement in forestry and natural resources matters argue that highly educated, experienced judges should set limits for ambitious legislators who are interested in gathering votes and for aggressive nonelected bureaucrats who are interested in building bureaucratic empires (Patterson et al. 1989). They hold that judicial activism is an effective means of defending the rights of minority interests in public forest resource programs. Further, they maintain that judicial policy making is frequently the only means by which unpopular though needed forest resource policies can be adopted, because resistance in popular assemblies such as legislatures is too great.

Supporters of judicial restraint, on the other hand, argue that courts are elitist, nondemocratic parts of the governing apparatus and therefore should defer to elected officials (legislatures and executives). They believe that the views of voters, as embodied in the decisions of elected officials, should not be frustrated by the courts. They also claim that courts should not be entrusted with responsibility for resolving broad, socially involved forestry matters in a democratic society—that decisions affecting large numbers of people should be made by the most democratic method possible.

This long-standing debate continues. The usual rhetorical alternatives are (1) that all control of forests should be handed over to the courts and (2) that all court review of important matters of forest policy adopted legislatively or bureaucratically should be barred. Neither alternative occurs in reality. Debate that focuses on such extremes is seldom productive, and indeed, support for either court activism or court restraint is often a reaction to a specific decision that an observer or participant finds especially undesirable or distasteful, rather than a well-thought-out philosophical position.

Courts are also periodically challenged for taking action on issues involving highly technical forestry matters such as the application of even-aged silvicultural systems, the use of herbicides and pesticides for protection purposes, and the prescription of habitats for specific species of wildlife. Technical matters, it is argued, not only should be the exclusive domain of experts in administering agencies but are, in fact, the reason for establishment of such agencies.

Opponents of judicial involvement in technical matters argue (Dolgin and Guilbert 1974, p. 221):

> It is doubtful whether the federal courts have the technical competency or institutional capacity to deal with broad, relatively undefined environmental problems. . . . Environmental problems frequently involve a diverse mix of unrelated disciplines such as chemistry, physic, ecology, and medicine, in addition to economics. . . . In contrast to the courts, agencies have a considerable advantage in their ability to use internal structuring and staffing policies in dealing with complex, multifaceted problems.

Others, however, argue that judicial involvement in technical environmental issues, including forestry, is not a problem. Sax, for example, wrote (1970, p. 150):

> Courts are never asked to resolve technical questions—they are only asked to determine whether a party appearing before them has effectively borne the burden of proving that which is being asserted. . . . the question is not one of substituting judicial knowledge for that of experts, but whether a judge is sufficiently capable of understanding the evidence put forward by expert witnesses to decide whether the party who has the burden of proof has adduced evidence adequate to support his conclusions.

Proponents further contend that courts have significance experience in handling technically complex matters (including patents, product safety, industrial accidents, and medical malpractice) and also have procedures for obtaining technical information, if needed, including oral arguments, written briefs, amicus curiae briefs, the testimony of expert witnesses and independent technical consultants, and fuller or more concise explanations from plaintiffs and defendants. Furthermore, it is held, courts do not have to educate themselves to the level of the skilled practitioner of forestry science; instead, they need only learn that portion of the science which is necessary to judge the merits of the case before them.

More appropriate, however, may be judicial efforts to assure that decision-making procedures require careful attention to technical matters (Dolgin and Guilbert 1974, p. 213):

> In cases of great technological complexity, the best way for courts to guard against unreasonable or erroneous administrative decisions is not for the judges themselves to scrutinize the technical merits of each decision. Rather, it is to establish a decision-making process which assures a reasoned decision that can be held up to the scrutiny of the scientific community.

REFERENCES

Abraham, Henry J. *The Judicial Process: An Introductory Analysis of the Courts of the United States, England and France.* Oxford University Press, New York, 1986.

Appler, Charles I. The Forester as Expert Witness: A Litigation Attorney Offers Tips on Testifying. *Journal of Forestry* 84(3):21–24, 1986.

Arbuckle, J. G., G. W. Frick, R. M. Hall, and others. *Environmental Law Handbook.* Government Institutes, Inc., Publishers, Rockville, Md., 1983.

Dolgin, E. L., and T. G. P. Guilbert (eds). *Federal Environmental Law.* West Publishing Company, St. Paul, Minn., 1974.

Gellhorn, E., and B. B. Boyer. *Administrative Law and Process.* West Publishing Company, St. Paul, Minn., 1981.

Glick, H. R., and K. N. Vines. *State Court Systems.* Prentice Hall, Inc., Englewood Cliffs, N.J., 1973.

Goldman S., and T. P. Jahnige. *The Federal Courts as a Political System.* Harper & Row Publishers, Inc., New York, 1985.

Horowitz, Donald L. *The Courts and Social Policy.* The Brookings Institute, Washington, 1977.

Jacob, H. *Justice in America: Courts, Lawyers and the Judicial Process.* Little, Brown and Company, Boston, Mass., 1984.

Lawrence, B. M. Standing for Environmental Groups: A Review and Recent Developments. *Environmental Law Reporter: News and Analysis* 19(7):10289–10297. 1989.

McLauchlan, William P. *American Legal Processes.* John Wiley & Sons, Inc., New York, 1977.

Office of United States Courts. *Annual Report of the Director of the Administrative Office of the United States Courts.* Government Printing Office, Washington, 1986.

Patterson, S. C., R. H. Davidson, and R. B. Ripley. *A More Perfect Union: Introduction to American Government.* Brooks/Cole Publishing Company, Monterey, Calif., 1989.

Robertson, F. Dale. *The Six-Point Working Agenda.* Remarks at the Summer Regional Foresters and Directors Meeting in Santa Barbara, Calif., USDA-Forest Service, Washington, August 1989.

Sax, Joseph L. *Defending the Environment: A Strategy for Citizen Action.* Alfred A. Knopf, Inc., New York, 1970.

Van Horn, C. E., D. C. Baumer, and W. T. Gormley. *Politics and Public Policy.* Congressional Quarterly Press, Washington, 1989.

Wenner, L. M. Interest Group Litigation and Environmental Policy. *Policy Studies Review* 11(4):671–683. 1983.

West Publishing Company. *Federal Practice Digest: 1975–1987* (1988 Supplement). St. Paul, Minn., 1987.

12

BUREAUCRATIC SYSTEMS AND PROCESSES

Highly visible struggles over the selection and implementation of forest resource policies are often a reality among the media, interest groups, and legislative and judicial systems. Often less visible in the policy process, however, is the participation of myriad departments, agencies, commissions, and offices of government—in short, the *bureaucracy.* Lack of visibility in this instance is no measure of importance. The actions of bureaucracies are typically the concrete expressions of generalized statements of forest resource policy. For example, the lofty platitudes of multiple-use land management as legitimized by the Multiple Use Act of 1960 are given concrete expression in USDA-Forest Service actions which implement specific National Forest land-management plans. Similarly, noteworthy expressions of environmental policy are given meaning by the Council on Environmental Quality via rules and regulations which guide the preparation of environmental impact statements. The often uncertain definition of public interest in the management of private forests is given on-the-ground meaning by the state forestry agency that establishes standards for reforestation and rigorously enforces them. These bureaucracies—and hundreds more—are often the cornerstone of the policy process.

In this chapter, we shall consider the general character and organization of modern bureaucracies, the manner in which they are managed, and the nature of their policy- and decision-making activities.

CHARACTER AND ORGANIZATION

Structure

Modern bureaucratic organizations entrusted with programs focused on the use and management of forests have a number of common structural characteristics, namely (Rosenbloom 1986, pp. 120–121):

- *Specialization* Authority and work are divided according to specialities, such as offices of policy analysis, divisions of wildlife management, and branches of telecommunications.
- *Hierarchy of authority* At each level of a bureaucracy, superiors direct and coordinate activities of numerous employees. For example, the flow of authority goes from the director to assistant directors, to deputy assistant directors, to officers, to chiefs, and to branch chiefs. In most designs, an organization is headed by a single individual authority, such as the director of the USDI-Bureau of Land Management or the commissioner of the MI Department of Natural Resources.
- *Specialized career structure* Employee career paths move through various specializations and ranks according to merit and/or seniority. For example, a silviculturalist may first become district ranger, then forest staff specialist, then forest supervisor, then regional specialist, and then regional forester.
- *Permanence* The organization tends to remain stable regardless of the flow (entry or exit) of employees. Society becomes dependent upon the functioning of the bureaucracy.
- *Large size* Bureaucracies tend to be very large organizations in terms of employees, budgets, and the number of clients served.

Procedure

Given such structural characteristics, bureaucracies tend to assume certain procedural characteristics that influence the manner in which they engage in the policy process, including the following (Weber 1977):

- A tendency to become *impersonal,* with a special emphasis on highly rational approaches to problems and policy implementation. Individuals must fit an organization's slot; the slot is not adapted to fit an individual's personal, mental, or physical character.
- Assumption of a *formalized style.* Communication occurs in proper channels and usually takes the form of written documents that are stored in files. Access to these documents is limited and is frequently a source of power.
- Adoption of *rules* which specify proper procedures and ensure regularity in dealing with outsiders.
- Placement of high value on *discipline.* Employees are bounded by rules

and regulations and may be disciplined for violating them. The power of inflicting discipline is a means of bolstering hierarchical authority.

The results of these structural and procedural characteristics often produce bureaucracies that are (Rosenbloom 1986, pp. 121–122):

- *Efficient* They act with significant continuity, rationality, expertise, speed, and discipline. Use of discretion is fairly predictable since bureaucracies are structurally and procedurally constrained by rules and well-established administrative processes.
- *Powerful* Respect for and adherence to decisions is achieved as a result of deference to authority granted by legislatures and courts; admiration of extensive expertise and the use of rational approaches to the policy process; and support from political alliances, affiliations, and connections with outsiders that are dependent on bureaucracies for provision of services and the application of constraints.
- *Ever-expanding* Tasks perceived as needing attention grow over time; growing and increasingly complex clients demand additional services. Some energy is devoted to nurturing the health and longevity of the bureaucracy per se.

Notions of efficiency, power, and growth should not give the impression that bureaucracies are meritorious organizations in all circumstances. Real-world bureaucracies can be plagued by arbitrary and zany rules, confusion and conflict among employees and superiors, leadership void of significant technical competence, and informal communications which subvert or even replace formal policy processes.

Organization

Managerial Perspective Bureaucracies involved in the process by which forest resource policies are developed and implemented can be organized in distinctly different manners (Rosenbloom 1986). For example, they may be heavily influenced by a need to emphasize productivity, efficiency, and effectiveness. In so doing they assume a *managerial perspective* in which labor is specialized, formality prevails, a hierarchy manages and coordinates, and techniques such as management by objective (MBO) and program evaluation and review technique (PERT) become common. (MBO involves setting goals, developing plans, allocating resources, implementing plans, and monitoring progress, whereas PERT involves mapping the sequence and timing of steps needed to accomplish an activity.) The product of a managerial perspective can be bureaucracies organized by purpose (e.g., research, fire management, or environmental protection), by process (e.g., accounting, policy analysis, planning, or budgeting), by clientele (e.g., timber, recreation, or wildlife), or by place (e.g., state or region, rural or urban). The interest is in accomplishing such an activities or focus in an efficient and productive manner.

Political Perspective Bureaucratic organization can also assume a *political perspective,* in which stress is placed on representation, political responsiveness, and accountability to citizens. Here are some examples:

- A state division of wildlife management may be established to accommodate the interests of sportsmen.
- An office of wilderness resources may be established to accommodate the interests of wilderness recreation enthusiasts.
- A bureau of lands and minerals may be organized to accommodate the concerns of mining industries operating in forested areas.
- An environmental quality board may be established to act as a focal point for the environmental interests of citizens.

Rather than emphasizing clear lines of functional responsibility, bureaucracy is a microcosm of the political forces at work in society in general (including friction and competition). The organizational landscape is one of many settings in which interest groups, political officials, and interested citizens express preferences for missions and programs. In such an environment, bureaucracies, and their agencies or divisions, tend to have significant autonomy. Their legislative connections are important and intense, and they are likely to be highly decentralized. Interested clients can express their program preferences through many regional or field offices.

Legal Perspective Bureaucratic organization can also be influenced by a *legal perspective,* in which the adversarial process is the organizing principle. Examples are the Federal Trade Commission, the Interstate Commerce Commission, and state pollution-control agencies. Such a bureaucracy is usually headed by a commission or board which exercises quasi-judicial functions via hearing examiners or administrative law judges. It usually has significant concern for procedural due process and the rights of individual citizens, as well as a special interest in the equitable distribution of a policy or program's costs and benefits.

Specialization Managerial, political, and legal considerations have a strong influence on the manner in which forestry bureaucracies are organized. The specific decisions or actions leading to a bureaucracy's structure, however, are far more complex. Of initial concern is *specialization* or division of responsibilities into smaller sets of related tasks and activities, such as watershed management, timber sales, environmental protection, fiscal management, and program planning.

Authority Once specialization has been determined, concern focuses on the need to appropriately distribute *authority* among managers and supervisors. Examples include authority to decide budgets, forestland uses, employment of personnel, and forest-management prescriptions.

Spans of Control With tasks determined and authority delegated, tasks must be properly grouped into *spans of control* which are comfortable to those responsible for supervision—for example, control over aviation and fire management; range and watershed management; and recreation, wilderness, and cultural resources.

The nature of control, authority, and specialization decisions ultimately determines:

- *The degree of formality of a forest bureaucracy* A highly formal bureaucracy requires issuance of numerous, detailed directives, and has substantial and specific requirements for the format of written reports.
- *The degree of centralization of a forest bureaucracy* A highly centralized bureaucracy requires that decision-making authority be centrally located.
- *The complexity of a forest bureaucracy* A highly complex bureaucracy has numerous departments, occupational groupings, and authority levels.

Inevitably, control, authority, and specialization decisions determine the manner in which a bureaucracy participates in the policy process (Gibson et al. 1991).

Participative Organization Forestry bureaucracies of the future may look considerably different from those currently in existence. Today's bureaucracies, it has been argued, are unable to adapt to the rapid pace of change in contemporary life (Rosenbloom 1986, p. 164):

> Agencies today are authoritarian, rigid, defensive, unable to utilize effectively their human resources, alienating and repressive. In short, they simply lack the flexibility to keep up with the constantly changing technological, political, and social environments with which they must interact. Hierarchy, in particular, is unable to tap the full talents and utilize the perspective of employees in the lower and even middle ranks . . . [and] tends to over emphasize the authority and overstate the ability of those at the top. . . . The use of static representational devices, such as advisory committees, creates structures of privilege that are too resistant to change. . . . Adjudication emphasizes adversary relationships rather than seeking to promote cooperation in complex areas of public policy [and] also tends to overemphasize procedure over substance.

The harshness with which today's bureaucracies are criticized suggests a need for an alternative concept for guiding the design of a bureaucracy's organization and approach to management. Most often suggested is a *participative organization,* wherein communication is abundant and unbridled (regardless of rank and power), consensus is the decision mode (rather than authority and coercion), influence is based on knowledge and technical competence (rather than on prerogatives of power), emotional expressions are encouraged, and the inevitability of conflict between individuals and organizations is recognized and willingly mediated (Bennis 1978). Furthermore (Rosenbloom 1986, p. 165),

> A scientific attitude of inquiry and experimentation will prevail, loyalty to organizations per se will decline as those who are less committed are likely to be more able to take advantage of change, and structural arrangements will be task-oriented rather than based on fixed specializations, rigid jurisdictions, and sharply defined levels of hierarchical authority.

A participative organization may work differently in the policy process than does an organization heavily influenced by managerial considerations. For example, abundant communication is likely to involve a substantial amount of advocacy from within as well as from outside an organization. Employees with professional skills and official standing may vigorously advocate particular forestland-use options, and may even oppose certain agency policies and programs without being fearful of charges of insubordination. Advocacy from outside the organization may entail greater citizen participation in the policy processes managed by an agency. Advisory or citizen boards may flourish, open hearings may become more common, and mechanisms may be established whereby citizens, interest groups, or specific agency clients actually become part of an organization's structure. For example, an interest group may manage a public wilderness area, with the help of technical advice from public employees, or a timber company may directly invest in timber production on public forestlands. Greater citizen participation will be based on the belief that a valuable opinion on the desirability of pursuing a particular land-use or land-management option need not be predicated on expertise in technical forestry matters (USDA-Forest Service 1983).

MANAGEMENT AND DECISION MAKING

Bureaucracies are heavily involved in decisions involving various stages of the policy process. In carrying out their responsibilities, they actively engage in agenda-setting activities designed to foster or discourage the placement of forest resource issues on their or another organization's agenda. They pursue formulation activities, frequently expanding or constraining the scope of policies suggested for addressing an issue or problem. In most cases, bureaucracies are the primary implementors of forest resource policies that have been legitimized elsewhere. Because bureaucratic personnel are especially knowledgeable about the details of a policy's implementation, they often have a special perspective on the evaluation of forest resource policies and programs. Even though the bureau cratic culture is often opposed to termination, bureaucracies can also be the deciding force in the termination of forestry programs.

Bureaucratic Power

Bureaucracies are powerful entities. They can make major decisions about forests that affect many individuals and result in significant financial and human investments. When a forest law is enacted by a legislature, for example, bureaucracies exercise considerable discretion in interpreting and administering the law. They secure their power from various sources, including access to and control over specialized information and professional talent. Because they have technical know-how, they can act authoritatively and command respect. They gain additional power from authority that is conferred upon them. Legislatures, for example, customarily grant forestry agencies the power to promulgate rules, and they frequently delegate to agency heads special authority to act and to impose their wills upon their organizations. Outside political alliances, affiliations, and connections can also be a source of power. Forest resource agencies frequently have strong constituencies which are willing to come to an agency's defense, thereby giving the agency greater ability to resist hostile actions, including proposed budget reductions.

Power Struggles

Struggles over power to decide matters of forest resource policy are not uncommon within agencies. Struggles, for example, may occur between a state forester and a state regional forester whose will and direction the former is attempting to guide. Struggles also take place between forestry agency heads and political appointees with unclassified government positions, when the appointees attempt to promote partisan or ideological agendas. When internal bargaining and accommodation procedures break down, intense conflict may spill outside a forestry agency into a larger political arena, where interest groups, media, and legislators often lie in wait. Just about anything can happen, including firings, resignations, and dramatic policy changes. In general, however, the normal state of a forestry agency is one in which low-visibility politics, unchallenged authority over certain policy and program territory, and support from outside groups and certain legislative committees prevail. Bureaucracies try to avoid letting outsiders seize control of power.

Policy Decisions

Policy-level decision making within bureaucracies can be viewed from various perspectives.

Managerial Perspective From a managerial perspective, policy decisions are made so as to promote rationality. The *rational comprehensive approach* to decisions is favored; objectives are clearly specified, alternatives designated, and efficiency criteria imposed.

Political Perspective From a political perspective, policy decisions involve pluralistic give-and-take activities. The approach endorsed is *rational incrementalism,* in which goals remain fuzzy, objectives are continually redefined, and arriving at a consensus is the measure of a good policy. In such a mode of operation, agencies react to feedback about the success or failure of a very limited number of forest policy options, choosing that option which seems to satisfy all or most of the major clients in the agency's political environment. Past policies serve as highly valued bases for subsequent decisions.

Legal Perspective From a legal perspective, decision making relies on adversarial processes which are bounded by highly formalized rules that are intended to surface the facts regarding an issue, to set forth the interests of opposing parties, and to balance facts and interests against a legal standard or a measure of the public interest. Such a process is commonly employed by independent hearing examiners and administrative law judges (Gellhorn and Boyer 1981).

Shared Responsibility Agency decisions about matters of forest resource policy are seldom made by one agency acting alone. Decisions are most often made in the context of sharing concerns and responsibilities with other organizations and other sources of power. For example, major policy decisions by the USDI-Bureau of Land Management are the product of interactions with higher officers within the U.S. Department of the Interior, examiners within the Office of Management and Budget, and various committees of the U.S. Congress. Consultations with related federal agencies such as the USDA-Forest Service may even play a part in such decisions. Any of these interactions may be either informal (as when policymakers have private consultations) or very formal, as in legislative oversight. They take place as a result of mutual interests in budgeting activities, long-range planning proposals, agency reorganization, personnel matters, and the location and incidence of benefits resulting from a policy decision. Within government, interaction is often required. The federal Office of Management and Budget, for example, rules on legislative proposals suggested by the agency and oversees the preparation of budget requests for future years.

Outside Influences. Agency policy decisions are also affected by influences from outside government, such as clients, persons being regulated, or adversaries in a lawsuit or an appeals process. A major wood-based company that purchases timber from a public forest may interact forcefully with forest administrators on matters of allowable harvest levels and appropriate uses of public forestland. Similarly, private landowners subject to public regulation of forestry practices may attempt to influence policy decisions concerning the nature of the forestry standards they will be asked to meet. Some forestry agencies are required by law to provide appropriate channels through which the general public can interact with a forestry agency. Such channels are typically in the form of advisory committees or requirements for general public involvement in the development and selection of policies and programs. Interaction between agencies, their clients, and legislative committees can be such that each reinforces the interests and goals of the other. The result is an *iron triangle* of alliances.

Leadership

Critical to an agency's success in the policy process is leadership. Agency heads and supporting staff with the ability to influence people and motivate them to a common purpose must be available. The history of the natural resources community is replete with references to leaders and their roles in guiding or influencing forestry bureaucracies. Although the qualities and skills most often associated with leadership are difficult to specify, they include:

- *Belief in the possibility of success* Good leaders assume that their actions will make a difference.
- *Significant communication skills* They are able to express clearly to followers what is expected.
- *A measure of empathy* They can understand the aspirations, fears, and limitations of their followers.
- *Boundless energy* They are willing to work long hours; they are workaholics.
- *Sound judgment* They are able to make reasoned judgments, and to avoid emotional or arbitrary actions.

A prerequisite for effective leadership of large forestry organizations is a positive viewpoint on the significant potential for good that is represented by complex bureaucracies—not a pessimistic belief that bureaucracies are inefficient, ineffectual, and resistant to change (Rosenbloom 1986).

Legitimizing Policies

The forest and related policies selected by a bureaucracy can be legitimized in various ways. Some policies are informally legitimized via speeches and advisory opinions given by agency heads or their representatives. Others are legitimized through mountains of rules and regulations formally established by an agency. Such rules can be found in the *Federal Register,* the *Code of Federal Regulations,* and comparable documents at the state level. Examples of the kinds of rules that guide the USDA-Forest Service are published in the *Code of Federal Regulations* (Title 36) are:

- Organization, functions, and procedures
- Involving the public in formulation of forest service directives
- Planning—National Forest System land and resource management planning
- Law enforcement and support activities
- Landownership adjustments
- Sale and disposal of National Forest timber
- Range management
- Minerals
- Protection of archeological research

Statements of policy produced by a bureaucracy can also be found in manuals and handbooks issued by forestry agencies. The USDA-Forest Service Handbook is a good example. Typical of the subjects contained therein are:

- *Organization and management* Laws, regulations, and orders; Forest Service mission; directive system; management; controls; external relations; information services; civil rights; human resources programs; planning
- *National Forest Resource Management* Environmental management; range management; recreation; wilderness and related resource management; timber management; watershed management; wildlife, fish, and sensitive plant habitat management; special-uses management; minerals and geology
- *State and Private Forestry* Authority, objectives, policies, and responsibilities; cooperative fire protection; rural and urban forestry assistance; forest pest management; cooperative watershed management; rural resource conservation and development; organizational management assistance; statewide forest resource planning
- *Research* Objectives and policies; timber management; wildlife, range, and fish habitat; watershed management; forest fire and atmospheric sciences; forest insects and diseases; forest products and harvesting; forest resources economics; forest recreation
- *Protection and development* Fire management, law enforcement, landownership, land classification, aviation management, personnel and procurement management, property management, finance and accounting, systems management safety and health
- *Engineering* Engineering operations, communications and electronics, buildings and structures, public health and pollution-control facilities, water storage and transmission, transportation system

BUREAUCRATIC LANDSCAPE

Bureaucracies with jurisdiction or influence over the development and implementation of forest resource policies and programs are vast and complex. At the federal level in the United States, some departments, agencies, and bureaus are directly involved via the ownership and management of forests; an example is the USDI-Bureau of Land Management. Others, such as the Internal Revenue Service and the Appalachian Regional Commission, are involved via the indirect impact of their programs (Backiel et al. 1987; Montanic 1990). Among the federal bureaucracies that had significant involvement in forestry and related activities in 1988 were the ones shown in Table 12-1 (General Services Administration 1988; Office of Management and Budget 1989). Keep in mind that not all an agency's employees nor all an agency's budget are focused on forestry. Of the Tennessee Valley Authority's budget, for example, only $3.7 million was spent on land and forest resources.

TABLE 12-1
Selected federal agencies responsible for the use and management of forest and related resources, by agency title, number of employees, and annual budget, 1988

	Approximate number of full-time permanent employees	Budget (millions of dollars)
U.S. Department of Agriculture		
Forest Service	31,700	2,513.9
Soil Conservation Service	11,400	686.6
Agricultural Stabilization and Conservation Service	2,800	1,289.2
Cooperative State Research Service	200	352.0
Extension Service	200	358.0
U.S. Department of the Interior		
Fish and Wildlife Service	5,000	645.9
National Park Service	12,000	956.5
Geological Survey	6,300	447.8
Bureau of Indian Affairs	9,400	1,572.4
Bureau of Land Management	6,900	842.5
U.S. Department of the Army		
U.S. Army Corp of Engineers	26,900	3,226.4
Independent Agencies		
Environmental Protection Agency	11,200	4,968.4
Tennessee Valley Authority	21,500	1,123.7
Executive Office of President		
Council on Environmental Quality	12	0.8
Office of Management and Budget	500	41.3

Source: The United States Government Manual 1987–1988, by the Office of the Federal Register, National Archives and Records Service, General Services Administration, Washington, DC, 1988; and *Budget of the United States: Fiscal Year 1990,* by the Office of Management and Budget, Executive Office of the President, Washington, DC, 1989.

State governments in the United States also have major responsibility for the development and implementation of forest policies and programs. Every state has a lead organization that is directly responsible for the protection and management of forests within the state. As might be expected, the state agencies and departments vary in title and are organizationally quite different. Examples of titles are Division of Forestry, Office of Forestry, Bureau of Forestry, Forestry Service, Lands and Forestry Division, and Division of Parks and Forestry.

During the early and mid-1980s, responsibility for forestry in 24 states resided within a unit of a state's department of conservation or natural resources. Under this most common arrangement, the administrative responsibility for managing various state natural resources (including forestry, parks, minerals, and fish and wildlife) was combined under a single executive-level department. In 5 other states, the lead forestry agency was located within an environmental protection unit which had broad responsibility for natural resource management and environmental protection. Under both of these arrangements, policy-level guidance usually originates with a governor-appointed board or commission charged with overseeing each unit within the larger department. A notable exception is Minnesota, where major policy authority rests with department officials.

Seven states (mostly in the South) have independent forestry commissions consisting of representatives from various public and private sectors. In Oregon and California, forestry responsibilities reside in a single-purpose, executive-level forestry department. Forestry agencies in the remaining states are located in a variety of nonforestry departments that are closely allied with forestry interests, such as departments of parks, lands, energy, agriculture, game and fish, economic development, and a university school of forestry (McCann and Ellefson 1982; Council of State Governments 1985, 1988).

The organization and programmatic interests in the development and implementation of forest resource policies can best be appreciated by examination of specific agencies. We shall consider a major federal land-management agency (the USDA-Forest Service), a major federal environmental protection agency (the U.S. Environmental Protection Agency, or EPA), three state forestry agencies (two of the board-commission type, in Oregon and Georgia; the other a departmental type, the Minnesota Division of Forestry), forestry agencies in selected foreign countries, and an international agency (the Forestry Department of the United Nations' Food and Agriculture Organization, or FAO).

Federal Agencies

USDA-Forest Service The USDA-Forest Service was established in 1905 by Congressional action which transferred the federal forest reserves from the Department of the Interior to the Department of Agriculture. Assuming major responsibility for national leadership in forestry, the USDA-Forest Service has adopted the following objectives to guide its actions (General Services Administration 1988):

• Promote patterns of natural resource use that best meet the needs of people now and in the future.

• Protect and improve the quality of the nation's air, water, soil, and natural beauty.

• Create forestry opportunities that are able to accelerate growth in rural communities.

• Encourage growth and development of forest-based enterprises that readily respond to consumers' changing needs.

• Encourage optimum forestland ownership patterns.

• Improve the welfare of underprivileged members of society via various forestry programs and activities.

• Involve the public in the formulation of forest policies and programs.

• Encourage development of forestry in other countries of the world.

• Expand public understanding of environmental conservation.

• Protect and improve the quality of open-space environments in urban and community areas.

• Develop and make available a firm scientific base for the advancement of forestry.

The USDA-Forest Service carries out activities in three major areas:

• Protection and management of resources on 191 million acres of National Forest System lands

• Research on all aspects of forestry, rangeland management, and forest resources utilization

• Cooperation with state and local governments, forest industries, and private landowners

In addition, the agency has significant responsibility for the development and wise use of human resources through the Youth Conservation Corp's, the Job Corp's Civilian Conservation Centers, and the Forest Service Volunteers Program. To carry out activities focused on agencywide objectives, the USDA-Forest Service is administratively organized into the National Forest System, State and Private Forestry, and Research (Figure 12-1). The agency has been the object of numerous administrative studies, some of which have become classics in the field of administration (Kaufman 1967; Robinson 1975). Others have become well known for defining or advocating change in the agency's approach to the use and management of forests (O'Toole 1988; Frome 1971; Leman 1981).

The National Forest System is a line organization of the USDA-Forest Service, decentralized from the Washington Office to nine regional offices. Under the regional offices are grouped 156 National Forests, 19 National Grasslands, and 16 Land Utilization Projects. A National Forest ranges in area from 400,000 to 3,000,000 acres and is composed of from four to nine ranger districts. Line

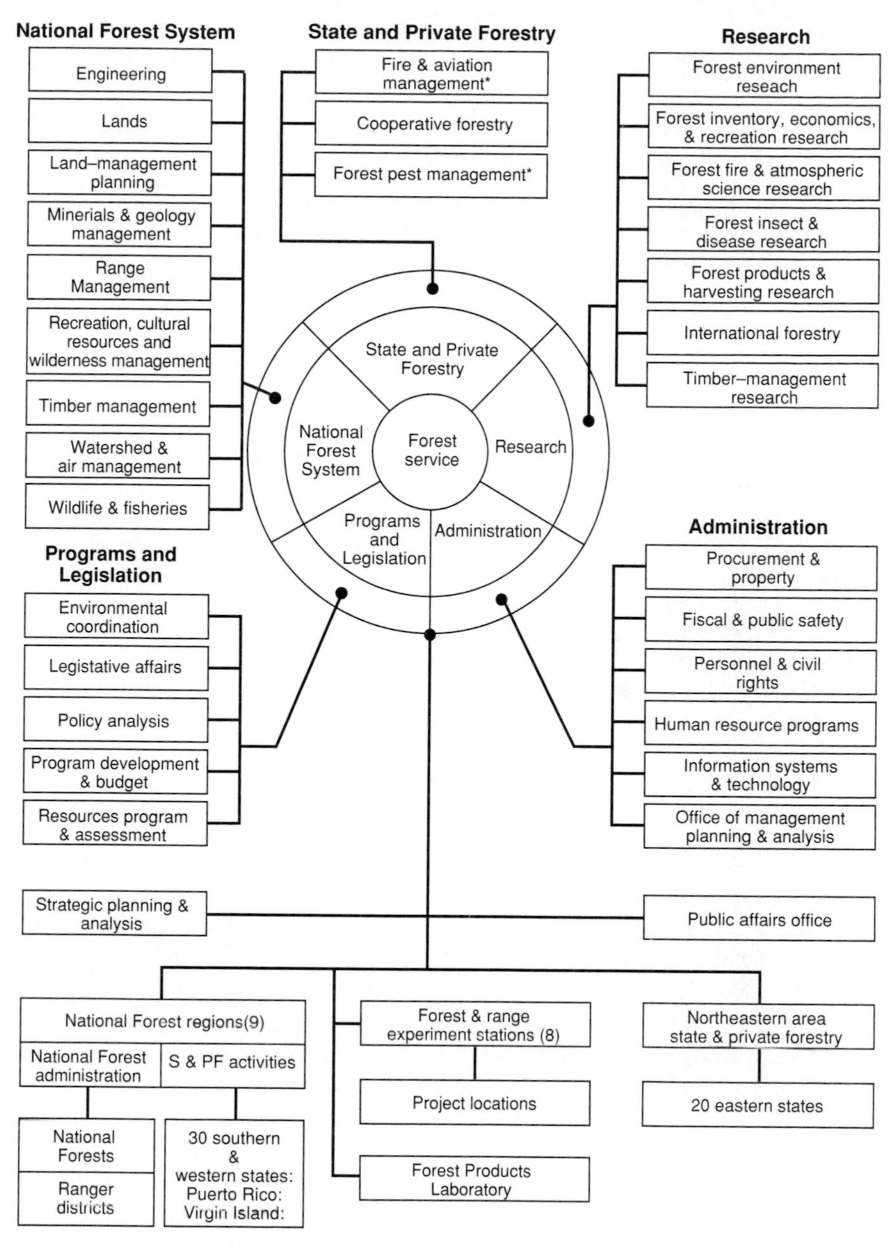

* Also provides program support to National Forest System.

FIGURE 12-1
Administrative organization of the USDA-Forest Service, 1989. (*Source: Organizational Directory: Forest Service. FS-65. U.S. Department of Agriculture, Washington, 1989.*)

officers (including the chief, regional foresters, forest supervisors, and district rangers) are responsible for the development of programs for their units and the management of allocated funds. In 1988, funding of the National Forest System totaled $2.28 billion. Among the many important functions performed in 1988 (USDA-Forest Service 1989b and 1990):

- Provided habitat for 60 percent of wild animal species in the nation, including habitat for 30 percent of nation's endangered species
- Supplied more outdoor recreation than lands under any other federal jurisdiction, namely a quarter billion visitor days
- Provided 32 million acres of wilderness to the National Wilderness Preservation System
- Provided half of the West's water supply
- Provided 13 percent of the wood harvested annually in the nation, and served as the storehouse for nearly half the nation's softwood sawtimber
- Contained 25 percent of the nations's potential energy reserves

The USDA-Forest Service's State and Private Forestry performs the following functions:

- Provides financial and technical assistance needed to improve fire control and to protect forests from insects and diseases
- Assists in the planning of state and private forestry programs
- Advises on practices to improve harvesting, processing, and marketing of forest products
- Encourages development of state forestry agencies and their administration
- Assists in the transfer of forestry research to potential users

Administratively, such programs are grouped under Cooperative Forestry, Forest Pest Management, and Fire and Aviation Management. Programs are administered through the Northeastern Area regional office and through offices associated with the regional offices of the National Forest System. Cooperative programs are very often carried out through state foresters or equivalent state officials. In 1988, funding of State and Private Forestry totaled $90.6 million.

The Research unit of the USDA-Forest Service provides new knowledge and technology that can reduce costs, improve productivity, and enhance the efficiency of forest management in environmentally responsible manners. Scientific and technical knowledge is sought about a number of subjects, including forest fire and atmospheric conditions; forest insects and diseases; forested watersheds; wildlife, range, and fish habitats; economics and inventory problems; forest recreation; and timber-harvesting and forest products. Research activities are conducted at eight forest and range experiment stations and the Forest Products Laboratory, plus more than 70 research laboratories located throughout the United States. USDA-Forest Service research funding totaled $140 million in 1988 (USDA-Forest Service 1989b).

Employees of the USDA-Forest Service exceeded 31,700 persons in 1988, of whom 33 percent were classified as professional, 12 percent as administrative, 42 percent as technical, and 13 percent as clerical or wage earners (USDA-Forest Service 1988). The orientations of employees classified as professional included forestry (56 percent); civil engineering (11 percent); wildlife biology (6 percent); soil science (3 percent); and hydrology, landscape architecture, fisheries biology, and entomology (each approximately 2 percent). Other professional orientations include geology, accounting, and archeology.

Policy development at the national level in the USDA-Forest Service is centered in the agency's Washington office. This office is organized in a hierarchial fashion, headed by a chief to whom report an associate chief, five deputy chiefs (Programs and Legislation, National Forest System, Research, Administration, and State and Private Forestry), and two assistant chiefs (civil rights and strategic planning). Policy development is a continuing process involving the flow of ideas from within and outside the agency; legislative staff members are involved, along with interest groups, the Office of Management and Budget, and the U.S. Department of Agriculture. Primary responsibility for liaison with the U.S. Congress is assigned to the deputy chief for Programs and Legislation.

Environmental Protection Agency EPA was established in 1970 for the express purpose of systematically abating various forms of pollution via a variety of research, monitoring, standard-setting, and enforcement activities. As a complement to such a purpose, the agency is charged with making public its comments through environmental impact statements. It also has a responsibility to discourage other federal agencies from undertaking activities which would adversely impact the environment. The agency is guided by a number of federal environmental laws, including the Federal Water Pollution Control Act; the Clean Air Act; the Federal Insecticide, Fungicide, and Rodenticide Act; and the Resource Conservation and Recovery Act. In 1988 the agency, which has its headquarters in Washington, employed over 11,200 persons and had a budget outlay of $4.8 billion.

The activities of the EPA are determined in large measure by the federal environmental laws which have been assigned to the agency for implementation. For purposes of administration, these legal mandates have been grouped into five major activity categories, as follows (Environmental Protection Agency 1988; Office of Federal Register 1988):

- *Water resources* Establishment of water-quality standards and effluent guidelines, issuance of permits, groundwater protection, municipal source pollution control, and drinking-water quality control
- *Pesticides and toxic substances* Evaluation of hazards, establishment of tolerance levels, registration of pesticides, and enforcement and compliance
- *Air and radiation* Establishment of air-quality standards and emission guidelines, and enforcement and compliance

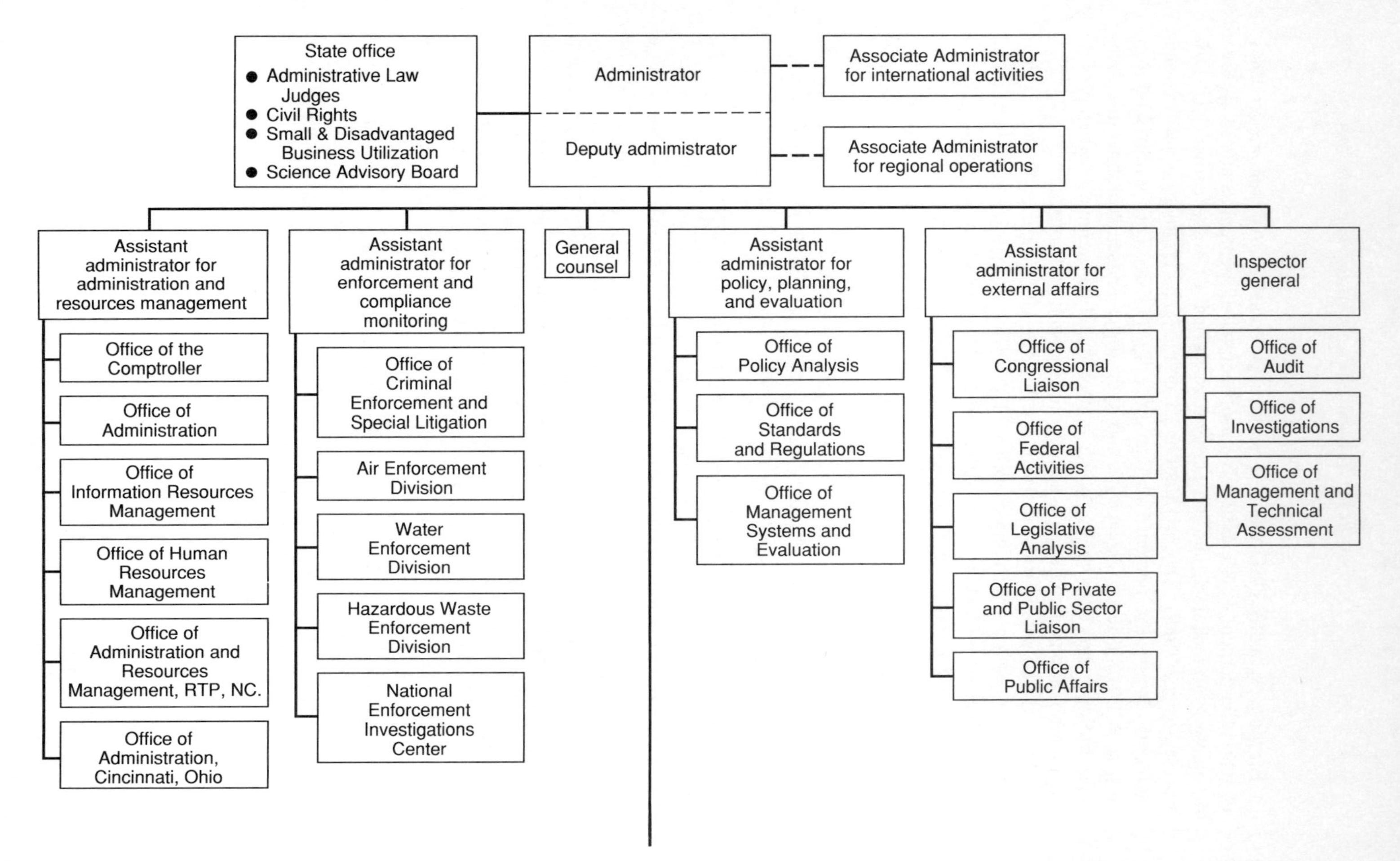

State office
● Administrative Law Judges
● Civil Rights
● Small & Disadvantaged Business Utilization
● Science Advisory Board
Administrator
Deputy admimistrator
Associate Administrator for international activities
Associate Administrator for regional operations
Assistant administrator for administration and resources management
Office of the Comptroller
Office of Administration
Office of Information Resources Management
Office of Human Resources Management
Office of Administration and Resources Management, RTP, NC.
Office of Administration, Cincinnati, Ohio
Assistant administrator for enforcement and compliance monitoring
Office of Criminal Enforcement and Special Litigation
Air Enforcement Division
Water Enforcement Division
Hazardous Waste Enforcement Division
National Enforcement Investigations Center
General counsel
Assistant administrator for policy, planning, and evaluation
Office of Policy Analysis
Office of Standards and Regulations
Office of Management Systems and Evaluation
Assistant administrator for external affairs
Office of Congressional Liaison
Office of Federal Activities
Office of Legislative Analysis
Office of Private and Public Sector Liaison
Office of Public Affairs
Inspector general
Office of Audit
Office of Investigations
Office of Management and Technical Assessment

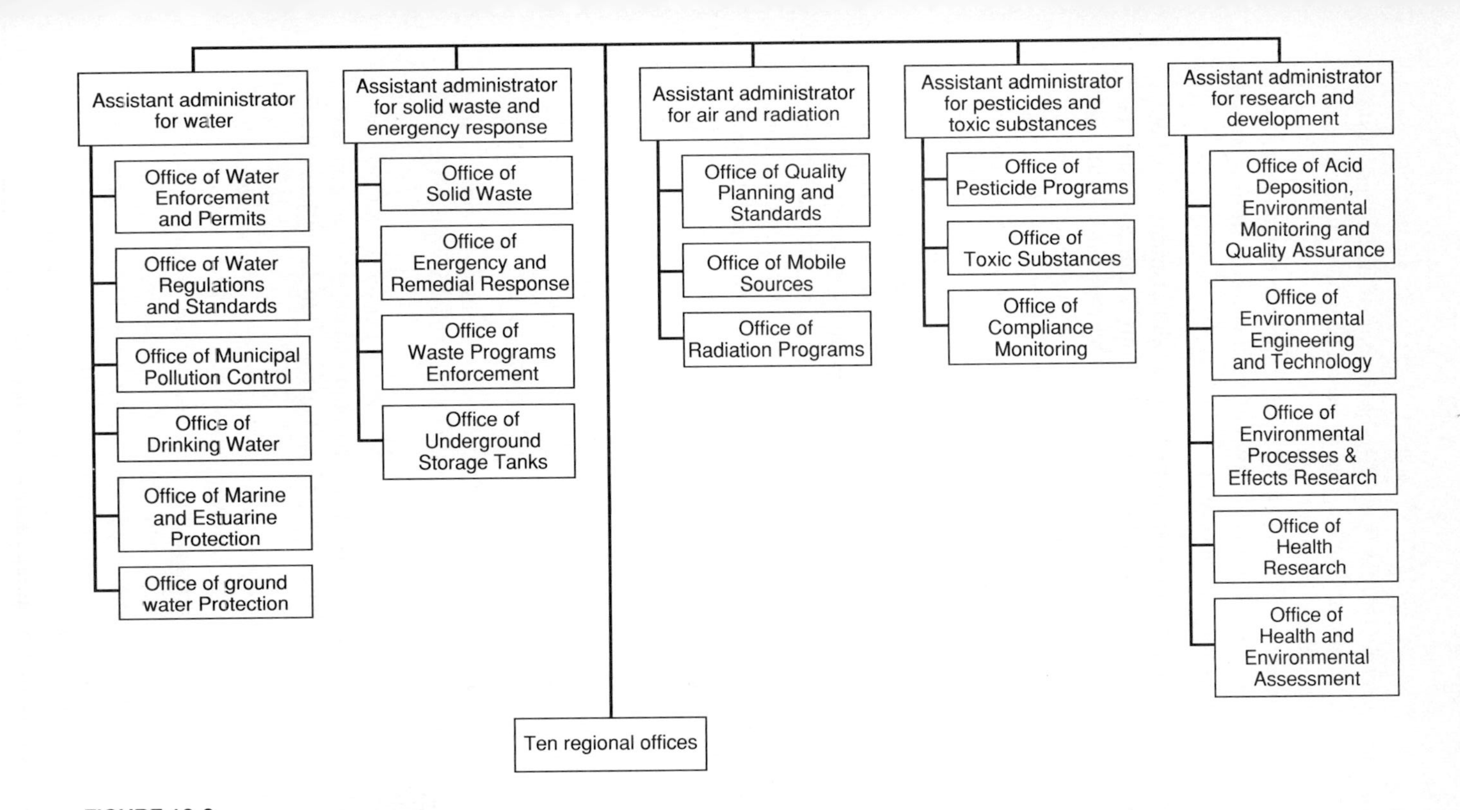

FIGURE 12-2
Administrative organization of the U.S. Environmental Protection Agency, 1988. (*Source: Adapted from* The United States Government Manual, 1987–1988. *Office of the Federal Register. National Archives and Records Service, Washington, 1988.*)

• *Solid waste and emergency response* Establishment of hazardous waste standards; guidelines for treatment, storage, and disposal; hazardous waste clean-up programs; compliance and enforcement

• *Research and develoment* Nationwide research in pursuit of technological controls of all forms of pollution, and evaluation of such controls from biological, physical, and social science perspectives

The EPA is a hierarchal organization headed by an administrator, to whom report nine assistant administrators (plus a general counsel and an inspector general) and 10 regional administrators (Figure 12-2). The regional administrators are the administrator's principal regional representatives for contacts involving other government agencies (federal, state, and local), various industries, and other private groups interested in environmental subjects. Ten research laboratories are responsible to the administrator, reporting through the assistant administrator for research and development. The Office of Assistant Administrator for Policy, Planning and Evaluation is the agency's principal focal point for oversight and coordination of all policy, program guidance, and evaluation functions. The office represents the administrator in communications with Congress, the Office of Management and Budget, and other federal agencies with an interest in the programs of the EPA. Ultimate responsibility for agency policy, however, rests with authority granted to the administrator (National Research Council 1977).

State Agencies

Oregon Department of Forestry The Oregon Department of Forestry is a state executive-level department responsible for policies and programs involving, according to the 1987–1989 budget (Oregon Department of Forestry 1987):

• *Fire protection ($36.5 million, or 42 percent)* Planning, prevention, detection, mobilization, initial attack, and fuel management on 15.8 million acres of forest.

• *Insect and disease management ($0.8 million, or 1 percent)* Surveys to determine presence of insect and disease pests, recommendations of appropriate pest control measures, and coordination of pest management on 12 million acres of forestland.

• *State forest management ($25.0 million, or 29 percent)* Management of nearly 800,000 acres of state-owned forestland. Sixty-four percent of the Board of Forestry Land's timber sale receipts are returned to the county in which timber is harvested; the remainder is retained by the Department of Forestry. Except for reimbursed departmental expenses, timber receipts from Common School Lands are deposited in the Common School Fund.

• *Service forestry ($2.6 million, or 3 percent)* Providing forestry services to 25,000 private owners of 3.5 million acres of forestland.

• *Forest nursery and tree improvement ($7.2 million, or 8 percent)* Providing tree seedings for reforestation of state-owned forests and to meet private reforestation requirements called for by forest practices regulations; improving the genetic stock of trees used in reforestation.

• *Forest practices regulation ($5.2 million, or 6 percent)* Implementing reforestation and related rules as called for by forest practices law on 10.1 million acres of private, state, and county forestland.

• *Forest products market development ($0.4 million, or less than 1 percent)* Assisting in the development of new markets or expansion of existing domestic and foreign markets for Oregon's forest products.

• *Forest resource planning ($0.7 million, or 1 percent)* Providing background information and recommendations on major issues concerning the use and management of forests in Oregon.

• *Administrative services [$8.5 million (excluding equipment pool budget), or 10 percent]* Providing administrative support necessary for the conduct of department programs.

Overall development of policy and direction for the Department of Forestry is the responsibility of the Oregon State Board of Forestry, a seven-member board appointed by the governor with confirmation by the state senate (Figure 12-3). The state forester (who is also secretary to the board) is responsible for day-to-day administration of the department, including appointment of four assistant state foresters to head the divisions of administrative services, resource policy, forest management, and forest protection. A deputy state forester is responsible for the department's field operations, which are administratively divided into three geographic areas (northwest, eastern, and southern). Area directors and district foresters coordinate all field activities of the department in their respective sections of the state. Each region has a regional forest practices advisory committee which advises on the implementation of the state's forest practices regulations in its region. The department's forestry activities are guided by a statewide forestry plan or program entitled "Forestry Program for Oregon."

Georgia Forestry Commission The Georgia Forestry Commission is responsible for guiding the investment of nearly $40.6 million into a variety of forestry programs, including (Georgia Forestry Commission 1988):

• *Forest protection* Detection and suppression of wildfire on 23.7 million acres of forestland within the state. Responsibilities include fire defense involving vehicle and structural fires in rural Georgia.

• *Forest management* Various cooperative forestry assistance programs focused on private landowners (e.g., forestry incentives program, conservation reserve program, service forestry assistance), urban forestry programs, water-quality-improvement programs, insect and disease management and assistance, and management of the state's 36,000-acre state forest.

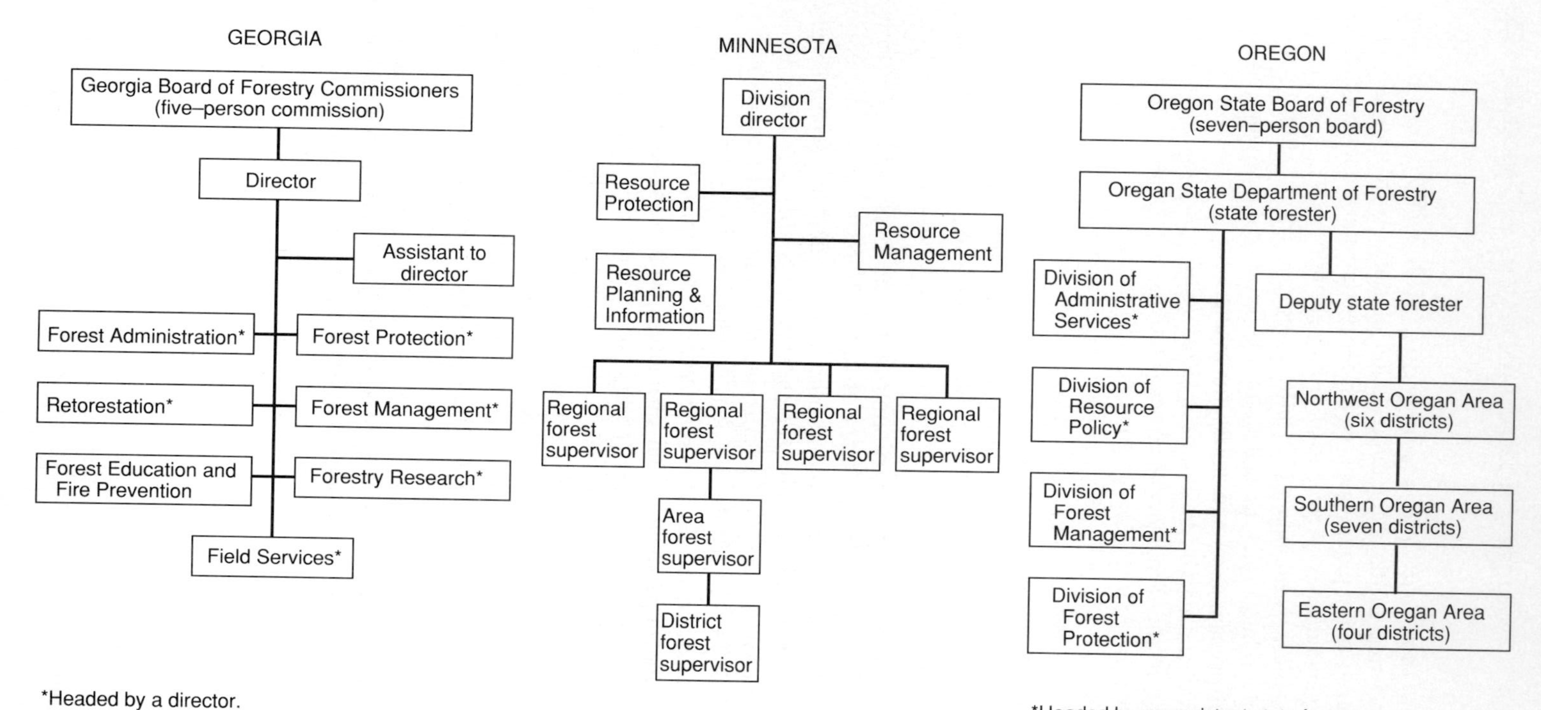

FIGURE 12-3
Administrative organization of public forestry organizations in Georgia, Minnesota, and Oregon, 1989. (*Source: Georgia Forestry Commission, Macon; Minnesota Division of Forestry, St. Paul; Oregon State Department of Forestry, Salem, 1989.*)

• *Reforestation* Management of state tree nurseries (sale of more than 175 million seedlings); development of genetically superior stock through tree-improvement programs and management of seed orchards.
• *Information and fire prevention* Development of publications, special exhibits, field demonstration, and promotional events concerning the use and management of forests; publicity regarding the prevention of wildfire in forested areas.
• *Research* Sponsoring forestry research (25 projects in 1988 involving commission investments of $250,000) involving topics such as market development, acid rain assessment, and wood energy.
• *Administration* Monitoring and coordinating commission programs.

The Georgia Forestry Commission has five members, who are appointed by the governor and confirmed by the state senate. By law, commission members' terms are 7 years; three members are to be owners (or representatives of owners) of 50 or more acres of forestland within the state, while two members are to be manufacturers or processors of forest products (or representatives thereof). With the advice and consent of the governor, the commission appoints a director who is responsible for day-to-day conduct of commission business within the policy constraints established by commission members. The director has administratively organized the commission's work into seven programmatic areas: the field organization, forest administration, reforestation, forest protection, forest management, forestry research, and forest education and fire prevention. Each area is headed by a director. The field organization is headed by a director of field services (Figure 12-3). The state is divided into districts, each headed by a district forester who reports to the director of field services. The commission's forestry activities are guided by the statewide Georgia Forest Resource Plan: 1986–1990.

Minnesota Division of Forestry The Minnesota Division of Forestry is responsible for the management of 2.6 million acres of state-owned forest and holds in trust for county governments an additional 2.3 million acres of forest. The division's budget in 1989 totaled approximately $27.5 million; 435 persons were employed by the division. The responsibilities of the division can be characterized by the goals set forth in the state's forest resources plan (prepared by the division), which include (Minnesota Division of Forestry 1983):

• Encourage expansion of the forest-products sector of the state's economy.
• Facilitate meeting of the state's energy demands through increased use of energy-related forest resources.
• Provide efficient forest resources protection.
• Support improved multiple-use management of nonindustrial private forests.

- Encourage intensified multiple-use and sustained-yield management of county forestlands.
- Promote optimum patterns of forestland ownership for multiple-use management purposes.
- Provide developed and dispersed recreation opportunities.
- Maintain and improve forest-related wildlife habitat for game and nongame species.
- Develop a forest road system that provides for protection, management, and use of forest resources.

The Minnesota Division of Forestry is one of eight divisions in the Minnesota Department of Natural Resources, which is headed by a governor-appointed commissioner. The commissioner in turn appoints the directors of the various divisions (waters, minerals, fish and wildlife, and parks and recreation). The Division of Forestry is organized into three functional staff groups, each of which is headed by an assistant to the director (Figure 12-3). In 1985, the emphasis (in terms of full-time equivalent employees) among the groups was: resource protection, 9 percent; resource management, 64 percent; resource planning and information, 15 percent; fiscal and personnel administration, 12 percent. The field units, from smallest to largest, consist of 80 districts, 18 areas, and 5 regions; all ultimately report to the division director.

Agencies Worldwide

The organization of public forestry agencies in countries other than the United States varies considerably. One common thread, however, has been a tendency to place forestry responsibilities in a country's ministry of agriculture. The disadvantages of this arrangement have been bluntly described by Italy: "There are two reasons for no longer including forestry in the National Plan for Agriculture: first, forestry differs markedly from agriculture in its characteristics and functions, and secondly [sic], forestry tends to be treated as a residual appendage when combined with agriculture" (Hummel and Hilmi 1989, p. 11).

For such reasons (and others), many countries have recently established autonomous or semiautonomous forestry units. For example, in the U.S.S.R., the supreme state forestry organization is the State Committee for Forestry, while forestry responsibility in Ireland rests with the Ministry of Energy, in Switzerland with the Ministry of the Interior, and in Israel with the autonomous Land Development Authority. In Sweden, forestry generally is directed by the Ministry of Agriculture, but the State Forestry Enterprise, which manages the state forests, is responsible to the Ministry of Industry.

Debates over the organizational location of forestry agencies aside, available evidence suggests that the ministerial location of a forestry unit is less important than the existence of forest policies which view the forestry sector as a whole and which provide a clear definition of agency responsibilities and authority to implement them (Hummel and Hilmi 1989).

Forestry administrative structures in countries other than the United States are best understood by examples. consider the following (Forestry Department 1988; Hummel and Hilmi 1989).

In Norway, the Department of Forestry, under the Ministry of Agriculture, has direct responsibility for matters concerning private forestry, including implementation of strict regulatory actions focused on private forests as called for by the Forestry and Forest Protection Act. The Department of Forestry is organized into a national level, a county level (there are 19 counties, each of which has a county land board and a county forest service), and local municipal governments which are serviced by 184 units of the District forest Service. Publicly owned forests in Norway are managed by the Directorate of State Forests and Land, which has a certain amount of autonomy outside the Ministry of Agriculture.

In the United Kingdom, the federal Forestry Commission, governed by the ten-member Board of Commissioners, is organized into a three-tiered administrative structure, consisting of a headquarters; seven conservancies, or regions (three in Scotland, three in England, and one in Wales); and 65 local forest districts. The commission is directly linked to the Ministry of Agriculture, Fisheries and Food in England and to the Secretaries of State in Scotland and Wales. The Forestry Act of 1967 requires the commission to achieve a reasonable balance between the management of forests for the production of timber and their conservation for purposes of flora, fauna, and natural beauty. The Forest Service in Northern Ireland is a division of the Department of Agriculture for Northern Ireland. Its powers and responsibilities are set forth in the Forestry Act (Northern Ireland) of 1953.

The forestry administration of Spain is situated within the Ministry of Agriculture, Fisheries and Food. Specific administrative units within the Ministry and their primary responsibilities are as follows:

- *National Institute for the Conservation of Natural Resources* Prevention of forest fires, reforestation activities, management of protected areas
- *Directorate General of Agrarian Research and Training* Forestry research and foresters' training schools
- *Directorate General of Agrarian Production* Forestry production and grants to private forest owners
- *Directorate General for Agro-Industries* Encouragement of primary wood-processing industries

United Nations' FAO Forestry Department

In addition to forestry administrations that are responsible to a single country, there are a number of agencies and organization that have missions to improve forestry conditions in countries throughout the world. one such organization is the Forestry Department of the United Nations' Food and Agriculture Organization (FAO). Although the word "forestry" does not appear in the latter's title, FAO is responsible for forestry development in the United Nations system. This responsibility is met in several ways. For example:

- There are sixty-one professional forestry staff members at FAO headquarters in Rome and 499 professional project staff members in various forestry field posts. This makes FAO the largest international organization in terms of forestry staff (in 1990). Technical forestry information and assistance is provided to 158 member countries on request.
- FAO forestry field programs provide technical and management services for more than 177 field projects in 81 countries, ranging from nursery operations in Kenya to forest-fire protection in India, and from watershed-management training in Iran to community forestry in the Peruvian Andes.
- The FAO Committee on Forestry, established in 1972, and the FAO Committee on Forestry Development in the Tropics, established in 1965, meet together in alternate years for purposes of allowing key forestry officials from around the world to discuss technical and policy subjects.
- As one of the initiators of the Tropical Forestry Action Plan, FAO facilitates processes within the international forestry community to more effectively focus assistance on priority areas of tropical forestry concern.

The FAO Forestry Department in Rome is one of FAO's smaller departments, with 160 staff persons out of a total FAO Rome staff of 3,300. Department funding comes from two major sources: regular program monies obtained from countries making up FAO's membership ($10.9 million in the 1990–1991 biennium) and extrabudgetary funding obtained from a variety of sources, such as the United Nations Development Program and trust funds. The trust funds are composed of monies given by donor countries for specific projects. The department has two divisions, Forest Resources and Forest Products, and two service units, Operations and Policy and Planning. Policy and Planning provides assistance in the areas of economics, planning, policy, law, and statistics as related to forestry.

Forest Resources Division The FAO Forestry Department's Forest Resources Division provides assistance in the development and conservation of forest and wildlife resources, with special emphasis on implementation of the Tropical Forestry Action Plan (Committee on Forest Development in the Tropics 1985). Other division activities include efforts to develop and conserve forest genetic resources, management of wildlife and national parks, and integration of forests and trees in land-use systems designed to produce food, fodder, and fuelwood.

Forest Products Division The FAO Forestry Department's Division of Forest Products focuses on development of forest industries, logging and transportation, wood-based energy systems, and development of nonwood fuel products. Special emphasis is put on training via preparation of information material and sponsorship of workshops. The division also undertakes to reduce waste during the course of harvesting, transporting, marketing, and processing wood fiber, and provides assistance needed to improve the efficient use of forest resources in cottage, village, and rural industries.

Forest Field Program The forestry field program of the department is administered by the Operations Service unit, and projects are carried out by field staff. The largest number of projects is situated in Africa, where there are 70 projects; this is followed by 68 projects in the Caribbean Region and 22 projects in the Near East Region. The largest source of project funds (58 percent) for the Operations Service unit is the United Nations Development Program. Trust funds provide an additional 36 percent of the unit's finances, while the remaining 6 percent comes from FAO's regular program budget.

BUREAUCRATIC CHALLENGES

Organization and management of bureaucracies that can effectively meet the public's wishes for the use and management of forests is a challenging social task. It is also an especially important task, in light of a bureaucracy's ability to access the policy process at all stages, thus influencing the sum and substance of forest policies and forest resource programs. Bureaucracies are not mindless organizations that carry out policies established by some distant group of elected or appointed officials. They can and do have significant influence over policy development and often have substantial leeway to impose their own meaning on forest resource policies which are selected and legitimized elsewhere. In such a context, bureaucracies face a number of challenges.

Destructive Ideologies

As large social organizations, bureaucracies often develop their own ideologies, which are frequently reinforced by intense employee loyalty to the organization. Such a situation can be especially difficult to an agency's leadership. Among the more common ideologies attributed to large bureaucracies are the following tendencies (Downs 1967):

- To emphasize the positive benefits of policies and programs while deemphasizing the costs
- To encourage expansion of policies and programs while claiming that curtailment would be undesirable

- To emphasize the benefits of policies to society in general rather than to specific interests or clients
- To stress high levels of efficiency in achieving policy goals
- To emphasize policy achievements while ignoring failures and inabilities

A bureaucracy that is sensitive to the public's demands for its services and sincere in its attempts to meet such demands in an efficient and effective manner will strive to ensure that ideologies such as the ones listed above are confronted and dealt with effectively. Such ideologies should not be allowed to detract from the development and implementation of effective forest policies.

Confusion of Interests

Also challenging to public forestry bureaucracies are situations in which broader public policy interests become confused with the interests of a narrow constituency or clientele group—in which the reactions of parties who are intensely interested in a narrowly defined policy or program are viewed as reactions of the general public. Narrow outlooks and close relationships with particular interest groups, for example, can easily distort a policymaker's view of a broader public interest in a forest resource policy—however difficult that interest may be to define. When bureaucratic politics encourages the formation of iron triangles among agencies, legislative committees, and organized interest groups, the problem can become especially acute.

Ineffective Policy Selection

Another challenge to bureaucratic involvement in the policy process is the reality that selection of policies by an agency is not always a smooth and easy process. As Rosenbloom aptly stated (1986, p. 305):

> There are many sources of pressure: time, interest groups, members of the legislature and their staffs, the media, chief executives and their staff, personal advancement and personal goals are among the more common. Specialization may limit the public administrator's view and definition of reality, and administrative jargon may obscure matters. Furthermore, group decision making carries within it a tendency toward conformity, the stifling of dissent, and constant reinforcement of the agency's traditional view of matters. It is also difficult to know precisely when to decide and when to await further developments before adopting new policies and new procedures.

One of the more commonly identified obstacles to effective agency decision making in a policy context is the lack of clear policy goals presented to an agency, a situation which often reflects the political price that has been paid to have a policy at all. Compounding the problem is the assignment of vague policy goals to multiple agencies without specifying appropriate means of coordinating agency efforts to achieve such goals (Kilgore and Ellefson 1991). Other obstacles are:

- *Rigid conservatism in the sense of strict adherence to rules, procedures, and practices* Aversion to taking risks; going by the book
- *Specialization that causes administrators to oversimplify reality* Confining vision to one of a few policy options; a narrow view of an issue's causes and effects
- *Excessive reliance on quantification which deemphasizes qualitative factors in decision making* Judgments based on quantification of unimportant factors versus judgments based on subjective assessment of important qualitative factors
- *Reluctance to engage in serious policy and program evaluation* Perfunctory and superficial evaluations

Political versus Technical Skills

Significant gaps in understanding between agency policymakers, who are adept in general management and policy skills, and their subordinates, who are often highly trained in specialized subject areas, can also be troublesome. The policymaker who is responsible for programs involving complex and highly technical principles of forest resource management often finds it difficult to be a technical expert (or at least to have the confidence of experts) while at the same time exercising the political and managerial skills that are needed to guide an agency through the innumerable political hurdles that are posed by the policy process. Technical experts lose confidence in policymakers, and policymakers lose patience with experts. Tension that is destructive to sound policy making can be the result.

Accountability, Advocacy, and Information Sharing

Still more challenges to bureaucracies involved in the development and implementation of forest policies can stem from factors such as those listed below (Rosenbloom 1986):

- *Recognition of the growing complexity of policy development and implementation* This complexity is less likely to be construed as a matter of technical forestry abilities and more likely to be seen as an ability to coordinate and direct relationships among political, economic, social, organizational, managerial, legal, and technological systems.
- *Increasing focus on personal accountability and responsibility for the success or failure of a policy or program* The potential for harm to humanity and the environment from mismanagement is becoming so great that policymakers and managers can no longer afford to plead that they are ''just following orders.''

• *Growing public interest in participation and representation in the development and implementation of policies and programs* There is now less emphasis on political neutrality of a bureaucracy and more emphasis on representing the public in administrative decisions and on challenging technical forestry experts. There is also now less inclination toward secrecy in the development of policies and programs and more commitment to sharing information as a basis for informed participation by policymakers and clients alike.

REFERENCES

Backiel, A., M. L. Corn, R. Gorte, and others. *The Major Federal Land Management Agencies: Management of Our Nation's Land and Resources*. CSRS Report for Congress 87-232 ENR. Congressional Research Service. Library of Congress, Washington, 1987.

Bennis, W. Organizations of the Future, in *Classics of Public Administration,* by J. M. Shafritz and A. C. Hyde (eds), pp. 276–288. Moore Publishing Company, Oak Park, Ill., 1978.

Committee on Forest Development in the Tropics. *Tropical Forestry Action Plan*. Forestry Department, Food and Agriculture Organization of the United Nations, Rome, Italy, 1985.

Council of State Governments. *State Administrative Officials Classified by Function 1985–86*. Lexington, Ky., 1985.

Council of State Governments. *Resource Guide to State Environmental Management*. Lexington, Ky., 1988.

Downs, Anthony. *Inside Bureaucracy*. Little, Brown and Company, Boston, Mass., 1967.

Environmental Protection Agency. *Organization and Functions Manual: 1985 Edition* (updated through 1988). Management and Organization Division, Office of Administration, Washington, 1988.

Forestry Department. *Forestry Policies in Europe*. FAO Forestry Paper 86. Food and Agriculture Organization of the United Nations, Rome, Italy, 1988.

Frome, Michael. *The Forest Service*. Praeger Publishers, New York, 1971.

Gellhorn, E., and B. B. Boyer. *Administrative Law and Process*. West Publishing Company, St. Paul, Minn., 1981.

Georgia Forestry Commission. Annual Report 1988. Macon, Ga., 1988.

General Services Administration. *The United States Government Manual 1987–88*. Office of the Federal Register, National Archives and Records Service, Washington, 1988.

Gibson, J. L., J. M. Ivancevich, and J. H. Donnelly. *Organizations: Behavior, Structure and Process*. Richard D. Irwin, Inc., Homewood, Ill., 1991.

Hummel, F. C., and H. A. Hilmi. *Forestry Policies in Europe: An Analysis*. FAO Forestry Paper 92. Food and Agriculture Organization of the United Nations, Rome, Italy, 1989.

Kaufman, Herbert. *The Forest Ranger: A Study in Administrative Behavior*. The Johns Hopkins University Press, Baltimore, Md., 1967.

Kilgore, Michael A., and Paul V. Ellefson. *Coordinating State Natural Resource and Environmental Policies and Programs*. Station Bulletin (forthcoming). Agricultural Experiment Station, University of Minnesota, St. Paul, Minn., 1991.

Leman, Christopher K. *The Forest Ranger Revisited: Administrative Behavior in the U.S. Forest Service in the 1980s.* Paper delivered at the Annual Meeting of the American Political Science Association: 1981. American Political Science Association, Washington, 1981.

McCann, Brian D., and Paul V. Ellefson. *Organizational Patterns and Administrative Procedures for State Forest Resources Planning.* Staff Paper Series No. 31. Department of Forest Resources, University of Minnesota, St. Paul, 1982.

Minnesota Division of Forestry. *Minnesota Forest Resources Plan.* Department of Natural Resources, St. Paul, 1983.

Montanic, D. A. *Bureau of Indian Affairs' Policies, Field Implementation and Issues Affecting Its Forestry Program,* pp. 304–307, in Forestry on the Frontier: Proceedings of the 1989 National Convention. Society of American Foresters, Bethesda, Md., 1990.

National Research Council. *Decision Making in the Environmental Protection Agency.* Commission on Natural Resources, National Academy of Sciences, Washington, 1977.

Office of the Federal Register. *Protection of Environment: Environmental Protection Agency* (Statement of Organization). Number 40 (Part One). Code of Federal Regulations. National Archives and Records Administration, Washington, 1988.

Office of Management and Budget. *Budget of the United States Government: Fiscal Year 1990.* Executive Office of the President, U.S. Government Printing Office, Washington, 1989.

Oregon Department of Forestry. *Goals, Objectives and Issues: 1987–1989 Biennium.* Salem, Oreg., 1987.

O'Toole, Randal. *Reforming The Forest Service.* Island Press, Covelo, Calif., 1988.

Robinson, Glen O. *The Forest Service: A Study in Public Land Management.* The Johns Hopkins University Press, Baltimore, Md., 1975.

Rosenbloom, David H. *Public Administration: Understanding Management, Politics and Law in the Public Sector.* Random House, New York, 1986.

USDA-Forest Service. *Participative Management.* Management Notes 27(1). Administrative Management Staff, U.S. Department of Agriculture, 1983.

———. *Equal Opportunity Is for Everyone: 1988 Civil Rights Report.* Personal and Civil Rights, U.S. Department of Agriculture, Washington, 1988.

———. *Organizational Directory.* FS-65. U.S. Department of Agriculture, Washington, 1989a.

———. *Report of the Forest Service: Fiscal Year 1988.* U.S. Department of Agriculture, Washington, 1989b.

———. *Report of the Forest Service: Fiscal Year 1989.* U.S. Department of Agriculture, Washington, 1990.

Weber, M. Bureaucracy, in *Perspectives on Public Bureaucracy,* by F. A. Kramer (ed.), pp. 6–20. Winthrop Publishers, Cambridge, Mass., 1977.

13

INTEREST-GROUP SYSTEMS AND PROCESSES

The prevalence of interest groups and their influential role in the process by which forest resource policies are established and implemented is nearly legendary. By implementing various strategies and tactics on behalf of the constituents they represent, interest groups are often in a position to influence the direction, magnitude, and timing of virtually all aspects of forestry. *Interest groups* are formally organized collections of people or organizations. They are established for purposes of promoting the common interest of group members by various methods, including intervention in the policy process. Their pursuit of specialized goals (such as promoting labor, business, or environmental causes) in many arenas, including the legislative, judicial, and bureaucratic systems, distinguishes them from political parties, which are policy generalists that focus primarily on the electoral system with an interest in securing control of government in general.

People are attracted to interest groups for various reasons—to share ideas, to promote common interests, to seek intellectual and professional betterment, to spread their ideas, and to press demands on government. Attraction to and establishment of interest groups is facilitated by the economic, social, and cultural diversity which exists in the nation's political economy. The multitude of activities and relationships that are spawned by such diversity promotes development of specialized organizations that can link diverse concerns to government and to one another. If diversity stimulates the formulation of interest groups, then legal guarantees of freedom of speech, freedom of association, and freedom to petition government work to legitimize their role in society. Because the nation's governments are highly decentralized (into states, counties, cities, and special districts) and divided (into executive, judicial, and legislative entities), individuals are encouraged to organize on a great many levels and to wage public policy battles on a variety of fronts. An interest group that has little chance of success at the national level may be in a position to greatly influence or even control forest policy formation at a local community level. Similarly, a group that fails in the legislative arena may seek success in the courts.

The role of interest groups in the development of forest resource policies cannot be overemphasized. In this chapter, therefore, we shall consider the function and purpose of interest groups in general, their major types and number, the manner in which they are organized and governed, and the nature of the power and influence they so often exercise, especially their decisions to enter the policy process and the strategies and tactics they subsequently employ.

FUNCTION AND PURPOSE

Interest groups perform a variety of functions as they seek to influence the direction and substance of forest resources policies From a general perspective, such functions can be categorized as follows (Berry 1984, pp. 6–8):

- *Representing constituents* Interest groups provide a formal link (organization) through which citizens can voice opinions to those who govern them. Noble democratic intentions are embodied in institutions through which citizens can effectively speak to government about specific policy preferences.
- *Educating the public* Interest groups provide citizens with a variety of often-contested information about the nature of public issues and the policy alternatives available for addressing them. Spirited interest-group debate over reported facts leads to development of a more informed citizenry.
- *Building agendas* Interest groups bring important yet unrecognized issues to the attention of government. Vigorous agenda-building activities turn important problems into political issues that need prompt attention by government.
- *Monitoring policies and programs* Interest groups follow the status of government policies and programs viewed as important by their constituents. They draw attention to shortcomings and work toward their correction.
- *Fostering opportunities to participate* Interest groups provide significant opportunities for interested citizens to participate in political processes. By directing their attention to lobbying, organized protests, letter writing, and campaign contributions, citizens come to feel politically more gratified than if they simply voted in general elections.

Conflicting Views

However admirable and noteworthy such functions may be, interest groups are very often viewed with significant disfavor. Not uncommon is the charge that they are geniuses at deception, agents of greedy demands, and organizations virtually untouched by ethical standards. A frequent argument is that the relationship between government and interest groups becomes too cozy—that legitimate political actors in government simply defer to the wishes of interest groups. As a result, agencies and programs proliferate, conflicting regulations expand, programs are multiplied in number, and budgets and expenditures skyrocket. What is implied is the existence of a network of subgovernments or *iron triangles.* It is thought that the policy and program interests of interest groups, key legislators, and government agencies become very similar and that such actors develop an ability to negotiate their differences with much ease. The supposed result is that difficult policy choices are seldom actually made—that the actors simply accommodate each others's wishes.

Although negative charges are commonly made about interest groups, they can be viewed from a more positive and understanding perspective. Arguments can be made that interest groups are (Lineberry 1981, p. 267):

- *An important link between people and government* Almost all organized interests can secure a hearing with government.
- *Constantly competing and making claims on one another's interests.*
- *Unlikely to be dominated by a single group for any great period of time* Opponents intensify their activities and bring balance to the system.
- *Inclined to play by publicly acceptable rules* Political skirmishes are fair; cheating, lying, stealing, or engaging in violence are not typically tolerated.
- *Capable of substituting one resource for another* Groups weak in one resource, such as financing, compete by using another resource, such as a large, active membership.
- *A rough approximation of the public interest is the result of their political battles and aggressive competition.*

Policies through Conflict

The political conflicts among interest groups can be fierce. In many respects, the task of the established institutions of government—the legislatures, courts, and executive agencies—is to manage such conflict. They accomplish this task by:

- *Establishing rules for the conduct of political contests* For example, enacting campaign finance laws and laws requiring registration of lobbyists.
- *Arranging compromises* Bringing groups together at public hearings and negotiating sessions
- *Legitimizing compromises* Enacting laws or establishing rules and regulations that speak of the policies agreed to by groups

• *Enforcing legitimized compromises* Requiring compliance with the substance of policies stated in laws, rules, and regulations

The character of an implemented policy will favor the group that has the most influence, as depicted in Figure 13-1.

Variety of Focal Points

Interest groups are most widely known for their efforts to influence policy development in legislatures. They do, however, focus substantial resources on policy development in executive agencies and the courts. Successful interest groups recognize, for example, that forest resource laws enacted by a legislature require continuing attention during the implementation and evaluation stages of the policy process. Interest-group staff may be ceaseless visitors to the administrative offices of forestry agencies, where they prod professionals to adopt rules and propose budgets that will facilitate the group's purposes as expressed in a forest law.

Similarly, interest groups may be willing participants in judicial processes, often because they advocate unpopular or minority forestry causes for which support cannot be obtained in other branches of government. Groups may enter the judicial arena in a number of ways, the most common of which are direct participation as a litigant and the filing of *amicus curiae (friend of the court) briefs* which demonstrate concern and provide specific views on a forest resource issue.

In sum, to presume that interest groups attempt to influence only legislative activities is to be woefully in error. They also have vital interests in the activities of the executive and judicial branches of government.

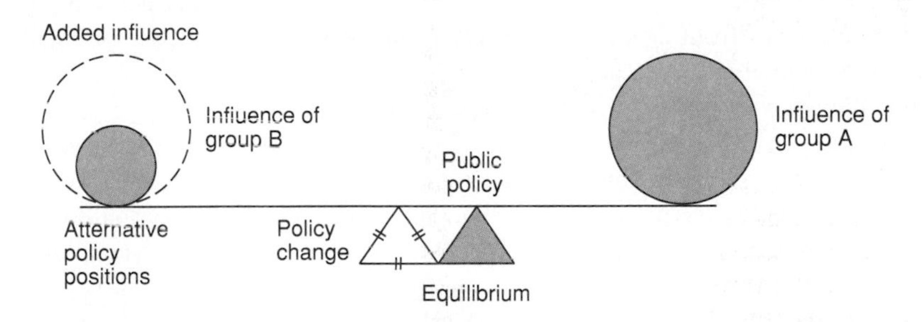

FIGURE 13-1
Interest group competition model of policy development. (*Source:* Understanding Public Policy, *by Thomas R. Dye. Reprinted by permission of Prentice Hall, Inc., Englewood Cliffs, N.J., 1981,* pg. 27.)

UNIVERSE AND CHARACTER

Growth in Numbers

Interest groups of various types and persuasions dot the political landscapes of most nations throughout the world. Although the absolute number of interest groups is unknown, the *Encyclopedia of Associations* identifies nearly 21,000 nonprofit interest groups in the United States (Gale Research Company 1989). The number of such groups has increased markedly in recent years; there were 40 percent more interest groups in the 1980s than in 1968. The increase has been especially notable among groups with social, ethical, and ideological goals that are not of a business or labor orientation. This is hardly surprising in light of the citizenry's propensity to join organizations. Nationwide, 62 percent of the population belong to at least one voluntary organization (such as a labor union, a religious group, or an environmental organization), and four out of ten persons consider themselves actively involved in their chosen group (Patterson et al. 1989).

The proliferation of interest groups in recent years is the product of many factors. Four major factors are as follows:

First, waves of reaction to social (civil rights), economic (poor and homeless), and environmental injustices have fired citizens' interest in securing a more forthright and organized way of obtaining government responses to society's ills.

Second, government has nurtured interest-group proliferation by making political and government structures more open and accessible. Especially noteworthy is the proliferation of laws requiring citizen participation (including organized participation) in policy making—for example, certain provisions of the National Environmental Policy Act of 1970, the Forest and Rangeland Renewable Forest Resources Planning Act of 1974, and the National Forest management Act of 1976.

Third, the general decline of political parties has also facilitated growth in the number of interest groups. Very few of today's public affairs issues sharply divide Republicans from Democrats; activists must often move outside political parties to effectively further their causes.

Fourth, citizen involvement with interest groups has grown because groups have become more effective in conveying their messages and concerns to government. Groups have been able to enhance their influence by using more effective leaders, employing more advanced technologies (such as computerized mailings), and using more sophisticated management techniques (including organized fund raising).

Categories of Groups

Interest groups important to the development of forest resource policies can be categorized in a number of ways. One method (which excludes labor unions) is as follows:

- *Conservation and environmental groups* Composed primarily of lay citizens who are interested in the quality and quantity of goods and services provided by forest and related environments. In general, there are no requirements for membership other than payment of dues.
- *Business and manufacturing groups* Composed primarily of individuals, companies, or other associations interested in the position of wood and closely related products in the marketplace and in the availability of resources needed to sustain the manufacture of such products. In general, membership is predicated on an economic interest in wood products and on the payment of dues or fees.
- *Professional groups* Composed primarily of natural resource professionals. In general, membership is predicated on attainment of a professional degree in natural resources (or closely related field) and payment of dues. Membership is independent of a professional's employer.

There are also some groups that participate with some frequency in the development of forest resource policies but that do not fit neatly in the above categories. Examples are the National Association of State Foresters, the International Association of Fish and Wildlife Agencies, and the National Association of Professional Forestry Schools and Colleges. Nonprofit research and public education groups are also to be noted, including The Conservation Foundation and Resources for the Future. Some organizations act as umbrellas for other interest groups—for example the National Resources Council of America, an organization of 50 national groups that are concerned about the use and management of natural resources, and the Renewable Natural Resources Foundation, an organization of 13 groups concerned about the use and management of renewable resources. There is also a significant number of organizations that have primary concerns other than natural resources but that may at times become involved in certain issues pertaining to forest resource policy. Examples include associations such as the Business Roundtable, the Chamber of Commerce of the United States, and the National Association of Home Builders of the United States.

Interest groups that focus all or a portion of their resources on forest resource policies are numerous in the United States (Table 13-1). The *Conservation Directory,* for example, identifies over 430 private organizations that are concerned with natural resources (including forests) while the number of business and manufacturing groups with a direct interest in forests probably exceeds 500 (National Wildlife Federation 1989; Ellefson 1984). As a conservative estimate, the number of national (or at least regionally oriented) interest groups in the United States that have forestry or closely related subjects as a major concern probably exceeds 1,000. If state and local organizations were included, the number would in all likelihood double.

If interest groups in a worldwide context are considered, the number expands considerably. Listed below are selected examples of interest groups (and their

TABLE 13-1
Organized interest groups having interests in forest and related resources, 1987–1989

Name	Membership	Staff	Operating budget (thousands of dollars)
Conservation and environmental groups			
American Forestry Association	27,000	20	2,000
American Rivers	14,000	20	1,500
Conservation International	55,000	65	4,600
Defenders of Wildlife	80,000	22	4,100
Ducks Unlimited	600,000	210	62,000
Earth First	15,000	4	210
Earth Island Institute (San Francisco, Calif.)	32,000	16	1,100
Environmental Action	20,000	23	1,200
Environmental Defense Fund	150,000	109	12,900
Friends of the Earth	30,000	40	3,100
Greenpeace USA	2,300,000	175	50,200
Izaak Walton League of America	50,000	22	1,400
League of Conservation Voters	55,000	70	1,400
National Association of Conservation Districts	3,000	16	1,300
National Audubon Society (New York)	600,000	335	35,000
National Parks and Conservation Association	100,000	30	3,400
National Wildlife Federation	5,100,000	860	87,200
Natural Resources Defense Council	168,000	140	16,000
The Nature Conservancy	600,000	1,000	156,100
Rainforest Action Network (San Francisco)	30,000	8	900
Rainforest Alliance (New York)	18,000	9	800
Ruffed Grouse Society (Coraopolis, Pa.)	23,000	14	1,200
Save-the-Redwoods League (San Francisco)	45,000	10	700
Sierra Club (San Francisco)	560,000	290	35,200
Sierra Club Legal Defense Fund (San Francisco)	120,000	67	6,700
Sport Fishing Institute	18,000	13	820
Trout Unlimited	58,000	27	2,300

TABLE 13-1 (*Continued*)

Name	Membership	Staff	Operating budget (thousands of dollars)
The Wilderness Society	370,000	140	17,300
Wildlife Management Institute	1,000	9	720
World Wildlife Fund	940,000	250	35,500
Business and manufacturing groups			
American Forest Council	135	27	2,850
American Furniture Manufacturers Association (High Point, N.C.)	336	14	
American Paper Institute (New York)	180	125	14,000
American Plywood Association	136	180	12,000
Cedar Shake and Shingle Bureau (Bellevue, Wash.)	350	22	4,000
Hardwood Plywood Manufacturers Association	200	24	1,000
International Snowmobile Association	29	4	500
National Forest Products Association	650	85	6,000
National Hardwood Lumber Association (Memphis, Tenn.)	1,250	37	2,200
Northwest Forestry Association (Portland, Oreg.)	70	12	736
Northeastern Retail Lumber Association (Rochester, N.Y.)	1,532	27	11,000
Recreation Vehicle Industry Association	625	46	6,500
Southern Forest Products Association (New Orleans, La.)	220	31	2,300
Timber Operators Council	495	38	3,300
Timber Association of California (Sacramento, Calif.)	50	15	1,500
Western Wood Products Association, (Portland, Oreg.)	250	100	8,000
Professional groups			
American Fisheries Society	8,500	19	1,500
Association of University Fisheries and Wildlife Program Administrators (Raleigh, N.C.)	75		25
Forest Products Research Society (Madison, Wisc.)	2,910	10	633

(*continued*)

TABLE 13-1 (*Continued*)

Name	Membership	Staff	Operating budget (thousands of dollars)
International Society of Tropical Foresters	1,800		30
National Association of State Park Directors	50		30
National Recreation and Parks Association	20,000	40	4,300
Renewable Natural Resources Foundation	13	5	800
Society for Range Movement (Denver, Colo.)	5,700	6	500
Society of American Foresters	20,000	35	2,500
Soil and Water Conservation Society of America (Ankeny, Ia.)	14,000	16	1,000
The Wildlife Society (Bethesda, Md.)	8,600	10	830
Others			
Council of State Governments (Lexington, Ky.)	50	100	4,500
Forest Farmers Association (Atlanta, Ga.)	3,400	3	300
Forest History Society (Durham, N.C.)	1,750	6	300
International Association of Fish and Wildlife Agencies	450	6	400
National Association of State Foresters	54	1	
National Conference of State Legislatures (Denver Colo.)	10,000	145	8,000
Natural Resources Council of America	50	1	150
Outdoor Writers Association of America	1,840	4	200

Note: Unless noted in parentheses, all organizations have headquarters in Washington or the immediate vicinity. Membership may reflect individual citizens (as is typically the case for conservation and environmental groups and professional groups) or companies or other associations (as is typically the case for business and manufacturing groups). Blank spaces in this table indicate that the information is not readily available or, in the case of staffing, that members or part-time workers serve as staff.

Sources: U.S. Wood-based Industry: Industrial Organization and Performance, by Paul V. Ellefson. Praeger Publishers, New York, 1984. *Encyclopedia of Associations 1990* by Gale Research Company, Detroit, Mich., 1989. *Conservation Directory: 1989,* by National Wildlife Federation. Washington, 1989. Inside the Environmental Groups, by B. Gifford, *Outside* 15(9): 69–84, 1990. Inquiries to specific interest groups.

headquarters) that have international interests or that focus primarily on forestry and related matters in countries other than the United States (Union of International Associations 1985).

- International Union of Forestry Research Organizations (Wein, Austria)
- Greenpeace International (Lewes, United Kingdom)
- International Union of Societies of Foresters (Edinburgh, United Kingdom)
- World Wildlife Fund (Gland, Switzerland)
- International Society of Tropical Foresters (Bethesda, Md.)
- International Council of Environmental Law (Bonn, Germany)
- Scandinavian Society of Forest Economists (Umea, Sweden)
- International Association of Fish and Wildlife Agencies (Washington, D.C.)
- International Popular Commission (Rome, Italy)
- Earthwatch (Nairobi, Kenya)
- Asian-Pacific Peoples Environmental Network (Penang, Malaysia)
- International Professional Association for Environmental Affairs (Bruxelles, Belgium)

The character and function of interest groups concerned with forests can be better understood by a focused look at a representative of each major category of interest group. The Sierra Club (representing conservation and environmental interests), the National Forest Products Association, or NFPA, (representing business and manufacturing groups), and the Society of American Foresters, or SAF, (representing forestry professionals) are discussed below.

Conservation and Environmental: The Sierra Club Founded in 1982 by John Muir, the Sierra Club is one of the most widely known and most vocal conservation and environmental groups in the nation. The far-ranging interests of the club, which has its headquarters in San Francisco, include (Gale Research Company 1989):

- Protecting and conserving natural resources worldwide
- Undertaking and publishing scientific and educational studies concerning the environment and natural ecosystems
- Educating others about the need to preserve and restore the quality of natural and made-made environments
- Working on urgent campaigns to save threatened environmental areas and species
- Addressing issues involving wilderness, forestry, clean air, coastal protection, energy conservation, and land use
- Lobbying for environmental legislation at all levels of government
- Sponsoring and scheduling wilderness outings

To accomplish these interests, the club is guided by the 17-person, member-elected Board of Directors. Each member serves a three-year staggered term, and the board is supported by an executive director and five vice-presidents responsible for legal affairs, planning, political affairs, international affairs, and parks and protected areas. The board receives advice from a number of national committees focused on topics ranging from air quality and energy to public lands and human population levels. Fifteen regional vice-presidents (appointed by the board) are charged with involvement in regional environmental matters. They are supported by 12 regional club offices, each staffed by regional representatives. Examples are the Appalachia Regional Office located in Harpers Ferry, West Virginia, and the Northwest Regional Office located in Seattle, Washington. The office in Washington, D.C., is staffed by 8 to 10 persons, including a legislative director. The club has nearly 60 formally organized state groups and is allied with over 330 local Sierra Club chapters.

The Sierra Club publishes the bimonthly magazine *Sierra* and the biweekly newsletter *Sierra Club National News Report*. In addition, the club is widely known for publication of books and calendars focused on a variety of environmental topics. It is also responsible for a political action committee (PAC) the Sierra Club Committee on Political Education (SCOPE). It is affiliated with the Sierra Club Foundation and the Sierra Club Legal Defense Fund, which has a staff of 67 with offices in five different cities.

Sierra Club membership has been growing—for example, it was 15,000 in 1960, 83,000 in 1969, 136,000 in 1972, 346,000 in 1983, 426,000 in 1987, and 560,000 in 1989. Nearly 300 staff persons are employed by the club; they are organized into six departments, including the Department of Public Affairs, the Department of Outings, and the Department of Conservation, which is a focal point for club environmental policy activities. Operating revenues in 1987 exceeded $26 million, 35 per cent of which came from membership dues; 19 percent from sales (for example, of books); 13 percent from grants and contributions; and the remainder from royalties, investments, outing fees, advertising, and the like. Revenues in 1987 were programmatically distributed to information and education (32 percent), administration and member services (27 percent), influencing public policy (22 percent), outdoor adventure activities (7 percent), chapter allocations (7 percent), and other activities such as fund raising (Sierra Club 1988).

Environmental and natural resource issues of concern to the Sierra Club in recent years are great in number and far-reaching in scope The club recognizes, however, that to be effective it must prioritize its concerns and focus its staff and financial resources on selected issues. To this end, the Sierra Club Board of Directors designates national conservation campaigns that correspond to the 2-year cycles of the U.S. Congress. In early 1987, the board selected the following issues (along with specific club goals for each issue) for 1987–1988 campaign attention:

- Clean Air Act Reauthorization
- Arctic National Wildlife Refuge protection

- USDI-Bureau of Land Management wilderness-national park designations
- USDA-Forest Service National Forest planning and management
- USDI-National Park Service park management and park designations
- Implementation of toxic materials control
- Onshore oil and gas leasing
- Disposal of high-level nuclear waste

In addition to these issues, continuing efforts were focused on environmental impacts of international development aid and lending, as well as on population growth, with emphasis on family planning (Sierra Club 1987).

Business and Manufacturing: The National Forest Products Association Founded in 1902, the NFPA has its headquarters in Washington and is among the nation's most influential wood-based interest groups. ''As the national association for all producers of hardwood and softwood products, the Association represents industry at the national level on public policy, legislative and regulatory matters, and maintains and increases the availability, acceptance and use of all types of forest products both within the United States and in international markets'' (National Forest Products Association 1988, p. 1).

The association also provides a lawful forum for discussion of matters of concern to the industry and its associations. Consistent with such interests, its organization includes four councils: the American Wood Council, the Public Timber Council, the International Trade Council, and the American Forest Council (which is responsible for guiding the American Tree Farm System). The program of the International Trade Council emphasizes:

- Strengthening government trade programs (such as the USDA-Foreign Agricultural Service)
- Achieving reductions in foreign market trade barriers
- Supporting laws that expand trade
- Gathering and distributing information about foreign market opportunities

Jointly sponsored by NFPA and the American Paper Institute, the American Forest Council's mission is to:

- Maintain and improve long-term growth, quality, and availability of wood from private lands
- Foster public acceptance of productive forestry on all forest ownerships, with special focus on assured timber supplies from public lands, and profitable forestry (without undue public regulation) on industrial and nonindustrial private forestlands. In part, such is accomplished by Project Learning Tree, a conservation education program focused on young people.

Each council's activities is guided by an elected board and a staff. Besides the councils, the Association's organization includes the Department of Congressional Relations and the Department of Environment and Health. The Association also directs the Forest Industries Political Action Committee.

The NFPA is supported by direct company members (162 in 1987), supporting members (422 in 1987), member associations (40 in 1987), allied members (56 in 1987) and contributors (4 in 1987) (National Forest Products Association 1988, 1989). Membership in 1988 represented 50 percent of the nation's volume of softwood lumber production. The association's 1988 operating budget was approximately $6 million, which supported programs guided by a staff of about 70. The American Forest Council's operating budget for 1989 was over $2.8 million. Membership dues are based on the volume of wood fiber produced and vary from one council to another. For example, 1988 dues rates for basic association programs were 6 cents per thousand board feet (or equivalent) of timber produced (or sold), while dues for the American Forest Council were 5 cents per thousand board feet. Each member company pays the association's basic rate as well as the rate of the specific council which it wishes to support financially and in which it wishes to participate most actively. This arrangement enables members to target dues and representation to their own areas of greatest concern. The association and its various councils and departments distribute a number of publications, including the biweekly *NFPA in Focus* and the *International Trade Report* and the *Environmental Report.*

Forestry resource issues addressed by the National Forest Products Associations in recent years are many. They include forest health research, timber sale appeals, spotted-owl habitats, capital gains taxation rates, forest chemicals, forested wetlands, forest plan challenges, nonpoint pollution standards, design specification for wood construction, implementation of National Forest System plans, and disposal of treated wood.

Professional: The Society of American Foresters An organization of forestry professionals, the SAF was founded in 1900 by Gifford Pinchot. Its objectives are to advance the science, technology, education, and practice of professional forestry and to use the knowledge and skills of the profession to benefit society. Membership in the society (nearly 20,000 in 1989) is predicated on graduation from a society-accredited forestry curriculum or on being a scientist or practitioner in a field closely allied to forestry who holds a bachelor's degree and has 3 or more years of forestry-related experience. The society has its headquarters in Bethesda, Maryland, and is governed by elected officers (president, past president, and vice-president) and an 11-member regionally elected council. A council-appointed executive vice-president directs four major program areas: finance and administration, resource policy, science and education, and publications. The society's Forest Science and Technology Board supports the activities of 27 subject-oriented working groups. More than 20 national committees focus on a variety of topics, including membership, professional recognition, forest policy, continuing education, and world forestry. In addition, there are a varying number of task forces charged with special, short-term assignments (for example, the Task Force on Wilderness Management). Thirty-four state societies (each with an elected chair and related officers), 12 divisions within state societies, and 244 chapters are also geographically and constitutionally designated units of the society. With a staff of about 35, the society has an annual national office budget of approximately $2.5 million. Its publications include the *Journal of Forestry* and *Forest Science,* as well as three regional journals of applied forestry.

The SAF has an elaborate procedure for selecting forest resource issues, studying issues, and preparing positions focused on such issues. A member-approved document, the SAF Forest Policies, addresses various forestry principles and procedures, and guides the substance of the positions to be taken. In recent years, the society has chosen to address issues of the following nature: below-cost timber sales on National Forests, multiple use of forestlands, off-road vehicles in forestlands, appropriations for the USDA-Forest Service, forest wilderness, and federal tax treatment of income from timber investments.

GOVERNANCE AND ORGANIZATION

Leadership

Leadership and governance of interest groups is highly variable. Although seemingly very democratic, interest groups are in reality almost always dominated by staff, elected officials, and a small cadre of very active members (Berry 1984). The most common democratic connections between a group's members and its leadership are member-elected governing boards and a group's national meeting or conference (usually held annually). The effectiveness of these avenues is continually called into question; for example (Patterson et al. 1989, p. 296):

> Few groups regularly poll their members (Common Cause is a conspicuous exception). Some groups have only the most haphazard ways of finding out what their members want. Many groups are guided by tiny clinques of activist; the largely inactive members may not know or care what their leaders are doing. In some groups, dissent is bullied into silence.

Governance of interest groups by relatively few individuals is not surprising, since rank-and-file members of interest groups seldom have the time or financial resources for active daily participation in a group's business and advocacy decisions. Full-time staff, who have greater command over information and more management expertise, can easily gain preeminent influence within a group. If group members are to be more than passive in group activities, they must demonstrate to staff and elected officers that they will actively participate in group business, and that they will not acquiesce at important junctures in a group's policy advocacy process. Individuals or companies who desire direct involvement in an interest group's governance are apt to join small groups in which opportunity exists for face-to-face contact with staff and group leaders.

Although the dominance of many interest groups by staff or influential members may be disconcerting, what must also be recognized is that many individuals, companies, and associations join interest groups because they strongly agree with the goals espoused by the groups. They do not seek active involvement in important group decisions because they generally agree with the values and perspectives of the group's leadership, as those values and perspectives are ultimately reflected in the group's actions. Whenever members become deeply troubled by the manner in which a group is governed, they cast their votes by leaving the group.

Administrative Structure

Interest groups invariably have an administrative structure, although they may often appear to be loosely grouped affiliations of citizens or companies, void of any organizational order. In some groups, the members communicate directly with the central office—as in the American Forestry Association, for example. Other groups are federations in which the members deal with geographic units, which in turn communicate with the central office; the National Wildlife Federation has this type of organization. Yet others assume a peak-like structure; the interest group is composed of other interest groups. The Natural Resources Council of American and to some extent the National Forest Products Association are examples of organizations with a peak-like structure.

Governing Boards, Officers, and Staffs

Internally, most moderate to large-sized interest groups organize around an elected or appointed governing board or council, a group of elected or appointed officers, and a staff which reports to the board and the officers. Boards or councils that vary in name (the members may be called trustees, directors, advisers, or officers) are usually elected by the membership, serve for terms varying from 1 year to a lifetime, and can assume an active or a largely ceremonial role in governing the activities of the group. The officers may also vary in name and function (president, secretary, treasurer). They are usually elected by the group's membership but may be elected by the governing board. They serve from 1 to several years, and they often form (in concert with staff) the core leadership of a group.

An interest group's staff contributes continuity and efficiency to a group's activities. The staff is usually organized by functional specialization (such as business management, publications, or general policy advocacy) or by issue specialization (perhaps forest wilderness, pesticides, or international forestry). As the size of an interest group grows, the degree of staff specialization tends to increase. An especially important staff person is the group's executive director (or executive vice-president, managing director, or chief executive officer). Usually appointed by the governing board, the executive director is responsible for the general financial and political health of the group. An executive director is selected with great care; the person must not only be an administrative wizard and a flawless operator in public relations, but also a staunch supporter of the group's values, biases, and prejudices. In some groups, the executive director is the reason for the group's existence; if the director leaves, the reason for belonging may no longer exist, and the group's membership may decline.

Government Regulation

The structure and conduct of interest groups is often limited by government efforts to regulate and monitor their activities. The Regulation of Lobbying Act of 1946, for example, requires lobbyists to file (with the clerk or secretary of the appropriate chamber) a quarterly lobbying report of all money expended, a list of the people or organizations to whom it was paid, and the legislation it supported or opposed. A *lobbyist* is defined as any person who directly or indirectly solicits, collects, or receives money for purposes of aiding the passage or defeat of legislation before the U.S. Congress. The U.S. Supreme Court limited the law's application to persons who personally solicit money for purposes of influencing legislation, who have as their main purpose the passage or defeat of legislation, and who attempt to accomplish such goals through direct communication with members of Congress. In 1980, Friends of the Earth, Environmental Policy Center, and Zero Population Growth each claimed one registered lobbyist, and the Sierra Club claimed two. In 1981 the NFPA claimed five (Ellefson 1984; Wootton 1985).

The Federal Election Campaign Act of 1971 also limits interest-group organization and administration by authorizing and regulating PACs. Though the statute allows such committees to contribute money to candidates for elected office, the contributions must be made in a limited and highly regulated fashion; the regulations place limits on spending, reporting requirements, and organizational affiliation. Similarly, the Ethics in Government Act of 1978 regulates group activities (especially staffing) by prohibiting former government employees from representing private interests before agencies in which they were previously employed. A 1-year waiting period is required before a past high-level official many contact a former government employer. Such regulations are designed to prevent conflicts of interest.

Requirements of the Internal Revenue code of the Internal Revenue Service (IRS) also pose administrative and organizational challenges to interest groups. Groups like to qualify under section 501(c)(3) in order to be exempt from federal income tax, to have the privilege of using favorable postage rates, and to receive the prestige of being known as a charitable or scientific organization. Another advantage is being able to solicit contributions from individuals more effectively, since contributions are tax-deductible for donors. Such privileges are not granted without restrictions. The most critical restriction is that a substantial portion of the group's budget may not be used to directly lobby a member or employee of a legislative body or to indirectly influence legislation through efforts to change public opinion about a proposed law. "Substantial" is strictly defined according to the proportion of a group's budget that is used for such purposes; when expenditures for such purposes reach certain thresholds, they are considered substantial. The proportion decreases as the budget of a group increases, and a maximum dollar amount is imposed regardless of budget size.

Certain interest-group activities are excluded from consideration by the Internal Revenue Service, in determining 501(c)(3) status—that is, such activities can be engaged in without limit. These activities are:

- Making available the results of nonpartisan research and analysis
- Providing technical advice to legislatures in response to written requests
- Communicating with the executive branch of government (except when the principal purpose is to influence legislation)
- Communicating with legislatures on matters of direct concern to the life of an organization (such as a proposal to outlaw a group's existence)
- Communication between a group and its members on matters of legislation, if such communication does not directly encourage members to lobby for or against a proposed law

Although most interest groups covet the benefits which result from qualifying under section 503(c)(3) of the Internal Revenue Code, some have chosen to forgo such a classification. Examples are the Sierra Club, Environmental Action, and Friends of the Earth. For such groups, the advantages of being free to act without fetters in the political arena apparently outweigh the benefits of favored status in the eyes of the IRS (Gifford 1990).

POWER AND INFLUENCE

Interest groups that operate as part of the nation's forestry community have a deep-seated concern about how forest resources are used and managed. Their concern is expressed at various points in the policy process and in many ways—ranging from testimony presented at legislative hearings to funding of research focused on forest wildlife problems, and from litigation about contentious matters of forest policy to financial contributions to candidates for public office. Such obvious expressions of concern about the use and management of forests are preceded by important group decisions, which may be either explicit or implicit. Among the matters that must be decided are which issues should be addressed, what strategies and tactics should be used to influence an issue, and under what circumstances a group's efforts to influence an issue should be considered a success. The strategies and tactics of interest groups are well known, but little is commonly known about the decision processes which precede an interest group's use of such strategies and tactics.

Decision to Influence

Interest-group involvement in a forest resource issue is often the product of a series of decisions made by individual or combined actions of a group's staff, governing board, or influential members. Although the events that lead to selection of an issue varies among groups, they are likely to involve judgments about the following:

- *The issue's relationship to purpose and goals* Is the issue consistent with overall goals and purposes expressed by the group? Is the issue included among the subjects previously judged worthy of the group's attention?
- *Adequacy of resources* Are sufficient time and financial resources available for addressing the issue? Are the staff or important group committees knowledgeable about the issue? Is the opportunity cost of addressing the issue significant (i.e., must work on other issues be delayed)?
- *Chances of success* Is there evidence that the group can influence the outcome of the issue? Are gatekeepers and decision makers sympathetic to the group's views on the issue?
- *Strategy selection* If the issue is to be addressed, what overall strategy should be used to ensure that the group's interests will prevail (e.g., direct staff actions, coalition building, activation of group membership, or indirect influence on the actions of important actors)? Does the group have the knowledge and skills required to implement a particular strategy?
- *Tactic selection* Given a strategy, what tactics should be used to make the group's interests prevail (e.g., testimony at hearings, or meetings with key administrators)? Does the group have the knowledge and skills to implement a particular tactic?
- *Effectiveness* Are implemented strategies and tactics accomplishing group goals? Do strategies and tactics need to be changed? Should the issue be dropped as a concern of the group?

Priority Assignment Interest groups are invariably confronted with an enormous number of forest resource issues. Since not all such issues can be addressed in a responsible fashion, priorities must be established—a challenging exercise for most interest groups. Although few would admit it, interest groups seldom control their own destinies on matters of issue selection and issue timing. Their long-range planning of issue priorities very often becomes ''it-all-depends'' planning. When a public agency, for example, concludes that the time is ripe for a hearing on a forestland plan, the substance and timing of the plan's issuance are usually firmly established. The interest group must react accordingly. Similarly, when an adversary decides that the time has come to litigate a forest resource issue, the group's response must be consistent with court-imposed schedules and issue scoping.

Selection of issues is also influenced by the reality that some issues so closely match an interest group's purposes that questions about the group's involvement in an issue cannot in good judgment be raised; the issue must receive high-priority attention, and if necessary, other issues must be deferred. For example, laws proposing additions to the National Wilderness Preservation System are virtually guaranteed the attention of The Wilderness Society, and the National Parks and Conservation Association is sure to respond to a proposed reduction in the budget of the USDI-National Park Service. Similarly, the National Forest Products Association would be unlikely to have second thoughts about involvement in issues concerning capital gains treatment of income form the sale of timber.

Selection of issues for interest-group attention is also influenced by a group's often deep-seated wish to present a united front. Leaders and staff will often reject involvement in an issue (however important it may be to society in general) if there is evidence that the issue would be divisive among the membership. The Society of American Foresters, for example, would have an especially difficult time securing membership approval for advocating legislative designation of a specific forest wilderness on public forestland or for promoting public regulation of forest practices prescribed by private forestland owners. Opinions among the SAF's membership are much too diverse to allow agreement on these issues, and consensus would thus be very unlikely.

Issue selection is also adversely affected by rapid changeover amongst interest-group staff. New staff may bring new issues to the foreground, meaning that long-term, consistent focus on any one issue is unlikely. Newly elected officers likewise may have personal agendas. In addition, undue consideration may be given to the wishes of a few members who constantly bring topically and geographically narrow issues to the attention of an interest group's leaders. When this happens, the group's beleaguered staff may simply give in and direct limited group resources to a very minor issue of public forest policy or to a forest resource issue that has already been decided in the public area.

Being faced with innumerable obstacles to prioritizing issues does not mean, however, that interest groups abandon systematic thinking about which issues they will address. They most certainly do attempt to prioritize issues. In the late 1980s, for example, the American Forestry Association chose to focus on urban forestry, on atmospheric deposition (including global warming), and on planning the use and management of forests within the National Forest System. The Sierra Club's 1987–1988 conservation campaign spoke to, among other issues, Clean Air Act reauthorization, Arctic Wildlife Refuge protection, and USDI-Bureau of Land Management wilderness designations. The 1989 Annual Report of the

National Forest Products Association stated that in 1989 attention would be devoted to USDA-Forest Service appropriations and National Forest plans (especially their implementation, including simplification of appeals procedures).

Interest-group issue selection is more often than not an informal activity precipitated by discussions among staff members and group leaders (usually driven by notions of consensus) (Henley and Kelly 1988). The process is very often void of rules designed to guide issue selection. An exception is the SAF, in which each of the following questions must be answered affirmatively if the society is to become involved in a forest resource issue (Society of American Foresters 1988):

- Is the issue of major importance to the public?
- Is the issue covered by *Forest Policies* (a member-approved document outlining broad statements of forest resource policy)?
- Is the issue within the knowledge and skills of the forestry profession?
- Is the issue of general interest to the level of the society that is considering it?
- Is there time for the society to act responsibly on the issue?
- Does the society have the resources to act responsibility on the issue?

The SAF's issue-selection criteria have been expanded upon by its national Committee on Forest Policy (Ellefson 1988a, 1988b). Among the committee's suggestions to the society and its state and chapter units are the following:

- *Address a small number of issues* Limited financial and professional resources implies a need to a address one or two issues well rather than many issues poorly.
- *Address issues whose outcome can be influenced by the involvement of the Society* Ignore issues for which a policy option has already been agreed to or for which a policy option differing from the Society's recommendation is destined for adoption.
- *Address fundamental forest resource issues* Ignore current budget allocations, and instead focus on options for long-term sustained funding of forest resource programs. Ignore site-specific public forestland use issues, and instead focus on processes for the allocation of public forestlands to various uses.
- *Address issues for which a natural and informed membership consensus is obtained* Refrain from issues involving limits on human population levels, on specific public land wilderness allocations, and on reorganizing major public forestry agencies.
- *Avoid issues implying a vested professional interest* Abstain from issues proposing increased forestry staffing or increased staff salaries.
- *Address issues for which policy consequences can be clearly stated* Avoid "we're against" or "we favor" statements. Present consequences of budget reductions and specific land allocations.

Selection Procedures Specific procedures followed by interest groups intent on influencing various stages of the policy process are as variable as interest groups themselves. Some, such as those of the American Bar Association, are supersophisticated, while others boarder on chaos. Generally, however, issues are selected and studied; policy options are set forth; one or more options are legitimized by staff, a governing board, or member referendum; and a product (a position or policy statement) describing the group's views is made available for use in a variety of ways, including testimony at hearings. To be effective, a group's procedures should (at a minimum) provide for (Henly and Kelly 1988):

- *Efficiency* Products are developed without waste of professional, financial, and organizational resources.
- *Representation* Products are a representation of group consensus on policies to be used to address an issue.
- *Substantive competence* Products are thorough and factual, and represent state-of-the-art knowledge about an issue.
- *Simplicity and sensibility* Products are understandable by members and various external audiences.
- *Swift action* Products are developed in a timely fashion consistent with the state of the policy process being addressed.

A group's governing board usually has final authority on issue selection and on the content of policy statements the group chooses to use when addressing a selected issue. Procedurally, a statement acted upon by a governing board may be the product of a number of individuals or units operating within the group (staff, committees, task forces); the more units, the more likely a statement is to reflect a broad consensus of the group. The involvement of an excessively large number of individuals, units, or committees can slow the group's policy process considerably and can lead to adoption of positions or policy statements which are very generalized (required in order to achieve consensus). The latter are often of limited value when used to influence the adoption of specific policies needed to address a particular forest resource issue. The technical correctness of a group's views on an issue is very often secured via the involvement of special task forces of ad hoc committees with members who have special technical training or related technical experiences. Nearly all groups recognize and provide procedures for addressing fast-breaking issues that are in need of a group's immediate response. Very few groups employ member referendum to approve specific responses to an issue.

A generalized example of the SAF procedures for developing national positions is presented in Figure 13-2.

Policy Products The products of a group's policy procedures can be very diverse. They may be *fundamental policy statements* that reflect a group's most

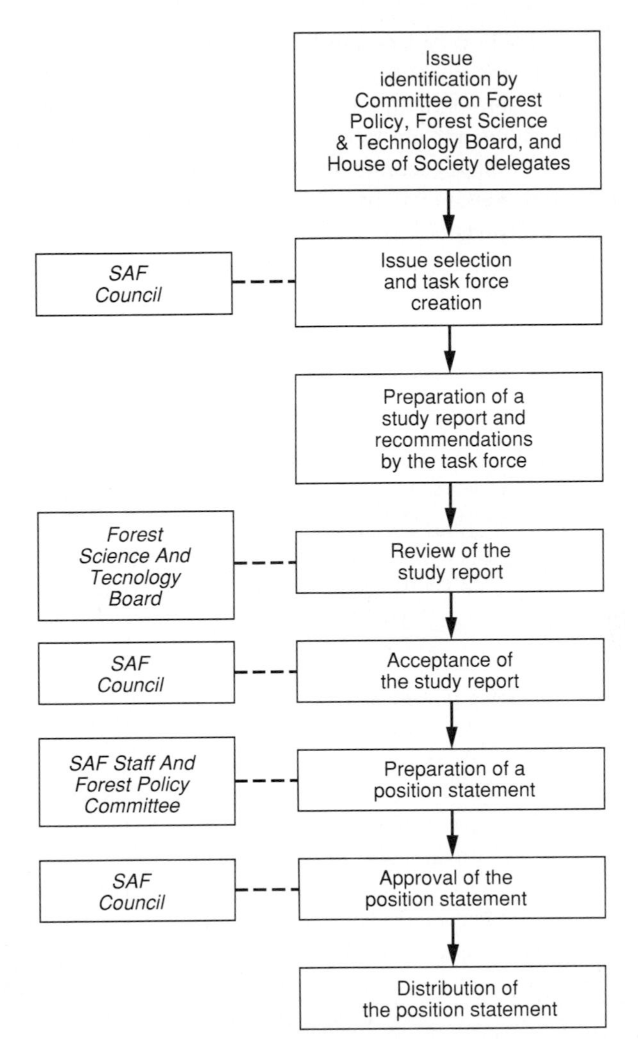

FIGURE 13-2
Society of American Foresters procedures for developing national positions on forest resource issues, 1988. (*Source: SAF Forest Policies and Positions. Reprinted by permission of Society of American Foresters, Bethesda, Md., 1988.*)

basic views or principles about the use and management of forest resources. Acting much like a constitution, such statements are only occasionally reviewed and are reflective of a broad-based member consensus (for example, the SAF Forest Policies). Some products of a group's policy procedures may be *background papers and reports* that result from careful research designed to ascertain factual information about a forest resource issue (for example, the American

Forestry Association's White Papers and Focus Papers). They may not contain policy or action recommendations. Other products are *position statements,* which are brief declarations summarizing a group's view on an important forest resource issue. Such statements are frequently presented at legislative hearings; they may or may not contain recommendations for action. Similar to position statements, but usually containing little or no factual background, are *resolutions.* They are generally statements of recommended action. Last, a group's policy product may be a short *letter of comment* which typically is void of substantial prior analysis or laborious group consensus building.

Strategies and Tactics

An interest-group decision to address a forest resource issue leads quickly to a need to select appropriate strategies and tactics. *Strategy* selection implies decisions about broad or very general approaches to influencing the outcome of an issue, such as building coalitions, supplying information, taking legal action, confronting authorities, and staging protests. *Tactics* are specific actions to be taken to gain a favorable policy response, such as letter writing, testifying at hearings, and meeting with policymakers. Which of many possible strategies and tactics should be selected is dependent upon several factors:

- *Where an issue is in the policy process* A membership letter-writing campaign against a proposed forest law may be of little value if the proposed law is about to be signed by a governor.
- *The resources available to the group* A staff that is experienced in legislative affairs may emphasize personal contacts with legislative staff, while a legal staff with substantial finances may encourage litigation.
- *The nature of a policymaker's expected response* If policy development is going against a group's preferred strategies and tactics (if setbacks have occurred), the group may implement other strategic and tactical approaches, or may continue in the same mode of operation but search for a more sympathetic audience, such as a different legislator or agency policymaker.

The number of strategies and tactics employed by interest groups is mind-boggling (Table 13-2). Consider in summary form the following examples.

Lobbying Influencing policy decisions by direct contact with persons responsible for such decisions; meeting with legislators, agency heads, forest supervisors, wildlife superintendents, and the like. Efforts are made to persuade officials of the merits of the group's views. The term "lobbyist," which often conjures up a negative stereotype (the smooth-talking arm-twister) is disliked by many persons professionally engaged in the practice. Interest groups consider personal presentations among the most effective means of influencing policy development (Berry 1977; Evans 1984; Wolfe 1990).

TABLE 13-2
Strategies and tactics used by organized interest groups to influence development of public policies, 1982

Strategy or tactic	Proportion of groups using (percent)
Testifying at hearings	99
Contacting government officials directly to present a viewpoint	98
Engaging in informal contacts with government officials at conventions, over lunch, or at social gatherings	95
Entering into coalitions with other interest groups	90
Discussing policy options with the press and the media	86
Consulting with government officials to plan legislative strategy	85
Helping government officials to draft legislation	85
Sponsoring letter-writing or telegram campaigns	84
Mounting grass-roots lobbying efforts	80
Having influential constituents contact a legislator's office	80
Helping government officials to draft regulations, rules, or guidelines	78
Serving on advisory commissions and boards	76
Alerting legislators to the effects that a bill may have on their districts	75
Filing a lawsuit or otherwise engaging in litigation	72
Making financial contributions to electoral campaigns	58
Doing favors for government officials who need assistance	56
Influencing appointments to public office	53
Publicizing legislative candidates' voting records	44
Running advertisements in the media about the group's positions on issues	31
Contributing work or personnel to electoral campaigns	24
Making public endorsements of legislative candidates for office	22
Engaging in protests or demonstrations	20

Note: Strategies and tactics used by 174 interest groups based in Washington, D.C. An average of 19 strategies and tactics were used per interest group.
Source: More of the Same: Washington Pressure Group Activity in a Decade of Change, by K. L. Schlozman and J. T. Tierney. Reprinted by permission of *Journal of Politics* 45:37, May 1983.

Public hearings Influencing policy decisions by presenting information at agency or legislative hearings. This is one of the most common activities of interest groups, although it is usually considered an ineffective means of influencing policy being developed by legislatures (Berry 1977; Rosener 1984). Appearance at hearings has much symbolic value for interest groups. It legitimizes a groups's presence in the policy process, provides means of getting views on an official record, and demonstrates to group membership the existence of an action-oriented staff (Ellefson 1985).

Protest and confrontation Influencing policy decisions by open defiance of persons or organizations responsible for legitimizing or implementing policies. Examples range from peaceful marches (logging trucks encircling a state capital building) to violent acts of vandalism (iron spikes driven into trees). Protest and confrontation are important means by which groups make officials more sensitive to and cognizant of a particular point of view. They are also an often-used method of accessing wider audiences via the media which are attracted to a protest scene.

Legal action Influencing policy decisions by challenging the legality of laws, rules, and certain private actions. In addition to lawsuits, groups may be involved in commenting on proposed rules, filing complaints, appealing administrative decisions, and when allowed, intervening in adjudicative proceedings (Wasby 1983). Since litigation tends to be expensive, it is often used only as a last resort. Much litigation initiated by interest groups results from previously unsuccessful efforts to influence policy decisions made by legislatures or administering agencies. Between 1970 and 1980, environmental organizations initiated 1,001 environmental law cases in the federal courts. Of the 416 cases initiated by national groups, the Sierra Club originated 102, while the National Resources Defense Council originated 83 (Wenner 1983). An example of a group specializing in legal strategies and tactics is the Sierra Club Legal Defense Fund.

Coalition building Influencing policy decisions by demonstrating strength through large numbers of like-minded groups. Although coalitions are usually formed on an ad hoc basis and last only until an issue is resolved, they provide a means by which groups can combine resources (staff, experience, communication networks) in order to have greater influence over policy development. Over three-quarters of the interest groups in Washington, D.C., participate in coalitions; the average group belongs to at least two coalitions (Berry 1977). A successful coalition is one which is temporary, limited to a specific issue, involved in an issue soon to be resolved, staffed by staff members from each group (not a separate coalition staff), inclined to rotate leadership, and willing to credit success to individual group members of the coalition (Berry 1984). Staff members of an interest group are often reluctant to fully support a coalition; their concern is to avoid diluting their own group's image and influence.

Grass-roots lobbying Influencing policy decisions by organizing and encouraging group members (or selected portions thereof) to contact and persuade policymakers. Mass telephone, telegram, or letter-writing campaigns may be involved. Grass-roots campaigns organized by interest-group staff assume that member attitudes will sway policymakers' opinions on an issue. Grass-roots mobilization demands good organization, effective communication, and precise timing. Seasoned interest groups reserve grass-roots lobbying for especially important issues, knowing how difficult it is to arouse members to action. If members are not directly affected by an issue, they are apt to ignore staff efforts to mobilize. A version of grass-roots lobbying is the use of visible and very influential group members to sway policymakers' opinions through such means as visiting agency heads and testifying at legislative hearings.

Endorsements and electioneering Influencing policy decisions by facilitating the placement of favored (elected or appointed) officials in positions of power. Actions may involve recommending a candidate to voters, seeking voter turnout for elections, and stimulating debate on a candidates voting record.

Public relations and advocacy advertising Influencing policy decisions by attempting to educate citizens or influential government officials. The tactics involved may be carefully crafted, advocacy advertisements in prestigious magazines; soft-sell public-service announcements in newspapers or on television; widespread distribution of legislator voting records; and dissemination of research results and special analyses. Such research and analysis projects are undertaken, and the results distributed, in the belief that if previously unknown facts about an issue are identified, policymakers will be more receptive to an interest group's views. In this respect, The Wilderness Society has recently become especially active on a variety of issues concerning the National Forests, including reforestation, timberland suitability, biological diversity, water quality, and management of old-growth timber.

Political action committees An interest-group strategy which has gained notoriety in recent years is the financing of electoral campaigns via PACs. Although campaign funding as a general strategy for influencing the development of policy is strictly limited by complex federal laws and rules, PACs are one tactical means by which interest groups can provide money to favored candidates for public office. PACs were authorized by the Federal Election Campaign Act of 1971 (and subsequent legal interpretations). The number of PACs has grown from a modest 600 in 1974 to over 4,100 in 1988. Virtually anyone (including an interest group) can establish a PAC; the only requirements are compliance with certain reporting requirements and with limitations established by the Federal Elections Commission. Individuals, for example, may contribute up to $1,000 per PAC per election, to a maximum of $5,000. In turn, PACs may contribute up to $5,000 per candidate per election and up to $15,000 per political party per election—and a candidate or political party may be involved in more than one election per year (e.g., primary election, final election). With a $5,000 per-candidate, per-election ceiling, the contribution of a single PAC is not likely to be a major influence on the election of a candidate. In the early 1980s, for example, nearly $300,000 was needed to win a U.S. House of Representatives seat, and over $2.3 million to win a U.S. Senate seat.

PAC money, however, is also used for other purposes, including educational campaigns and administrative operation of PAC offices (e.g., salaries, rent, postage). Most PACs are administered by the staff or elected officials of an interest group.

An interest group may form a PAC for any number of reasons, including retention or placement of people in elected office who express views similar to those of the interest group, assuring later access to successful candidates by creating a sense of indebtedness, and providing members with a means of pooling campaign gifts in order to be more effective (Sabato 1985). More detailed insight is provided by the Sierra Club's PAC, the Sierra Club Committee on Political Education, which cites the following objectives (Ellefson 1986):

- To encourage the election of public officials who will listen to conservationists
- To directly influence legislators who will cast votes on environmental issues
- To organize and promote electoral activities that help to build the club's grass-roots political structure
- To provide a mechanism that enables club volunteers to become involved in political campaigns with important environmental consequences
- To make sure that environmental issues are widely recognized as important campaign issues
- To make sure, by monitoring, that winning candidates who campaigned as friends of the environment carry out their campaign promises

Funds raised by PACs are used for various purposes, although donations to candidates for elected office are dominant. How are candidates chosen to receive PAC monies? PACs have varying approaches. Business association PACs often go by the following rules: Give to challengers who might unseat nonbusiness-oriented candidates, who have favorable voting records on issues important to business, and who need to retire campaign debts (on the theory that good candidates usually run again); and give to only one party in a campaign.

On the other hand, the League of Conservation Voters, the 99th largest distributor of PAC funds in the early 1980s, has a different set of guidelines. Its monies are distributed primarily to persons who have key congressional committee assignments (such as committee chairs); who already have a positive record on environmental issues (the league is reluctant to give to unproven candidates); and who have a good chance of winning their elections (so that league contributions will really make a difference). Furthermore, the contributions must be consistent with the endorsement of environmentalists in the candidate's home district or state.

Although cash contributions to candidates are made by the Sierra Club Committee on Political Education, the Club's preferred strategy is to use PAC money to promote campaign activities by its members, such as distribution of literature and sponsorship of forums (Ellefson 1986).

PACs affiliated with forestry or related natural resource interest groups numbered nearly 20 in 1983–1984; they contributed a modest $600,000 to the campaigns of candidates for federal elected office—although if PACs of wood-based corporations are included, the number jumps to more than 70, with contributions totaling nearly $4 million (Table 13-3). [PACs with names that defy identification of sponsorship (e.g., Filthy Five Campaign PAC) are excluded.] Although a number of nonforestry professional societies sponsor PACs (for example, the American Institute of Architects and the American Veterinary Medical Association), none having forestry or closely related interests have seen fit to do so.

TABLE 13-3
Financial contributions to candidates for federal office made by PACs with interests in forest and related natural resources, 1983–1984

PAC	Contribution (dollars)
Conservation and environmental committees	
California League of Conservation Voters	1,109
Campaign for Clean Air	444
Environmental Action's PAC	24,653
Friends of the Earth PAC	56,051
Georgia League of Conservation Voters	0
League of Conservation Voters	185,880
New Jersey Environmental Voters Alliance	0
New Mexico Conservation Voters Alliance	0
Oregon League of Conservation Voters	0
Sierra Club Committee on Political Education	253,927
Solar Lobby PAC	4,556
Virginia Conservation Voters	0
Business and manufacturing committees	
American Wood Preservers Institute PAC	0
Forest Industries PAC	64,814
International Hardwood Products INPAC	725
Lumber Dealers PAC	100
National Association of Furniture Manufacturers PAC	1,850
Society of American Wood Preservers, Wood Preserving PAC	2,000
Southeastern Lumber Manufacturers Association PAC	11,050

Note: Although a PAC may not have contributed to political candidates, a PAC may have used solicited funds for other purposes (e.g., PAC administration, educational advertising).

Source: Reports on Financial Activity: 1983–1984. Final Report on Party and Non-Party Political Committees, vols. 1–4. Federal Election Commission, Washington, D.C. 1985.

Measures of Success

Interest groups are a reality in the process by which forest resource policies are developed, but whether they are capable of influencing the process in their favor (and thus the policies resulting from the process) is a much-debated subject. For example, is the NFPA able to significantly influence congressional appropriations for forest roads built by the USDA-Forest Service in the National Forests—and, if so, what strategies and tactics have proved to be most effective? Similarly, has The Wilderness Society been able to increase wilderness allocations on lands administered by the USDI-Bureau of Land Management—and again, if so, what strategies and tactics have proved to be most effective? Questions of this nature are difficult to answer. The interest groups involved would probably answer, "Yes, we had a very significant impact on the outcome, but we hesitate to discuss our strategies and tactics publicly."

Complementary Interests The ability of interest groups to successfully influence the development of public policy has much to do with political mood of the citizenry and the complementarity of interest-group and policymaker concerns. As Ziegler stated well, the impact of interest groups "depends more on the harmony of values between groups and legislators than it does on the ability of a group to wield its power either through skillful techniques or presumed electorial influence" (1964, p. 274). For example, the temperament of the 1970s was one of widespread national concern over the condition of the nation's natural and man-made environments. The environmental concerns of citizens, elected officials, agency heads, and interest groups were in large measure very similar. However modest or large their influence, the work of interest groups in securing enactment of far-reaching environmental laws was most definitely facilitated by this similarity of concerns. Doors were opened, ideas acknowledged, and support extended—experiences that interest groups seldom had prior to the wave of 1970s and 1980s environmentalism.

Opposing Groups The influence of an interest group is also a function of the number and the tenacity of opposing interest groups. The use and management of forests is determined by an incredibly complex, multilevel political system which is followed by national, regional, and local interest groups. In one segment of the system (for example, in an area in which the use of a state forest is being planned), interest groups with conflicting concerns may abound, while in other segments, competitors may be few or even nonexistent. When competition for a decision maker's attention is keen, an interest group may find its ability to influence a policy outcome nearly thwarted. An interest group that is the sole representative of an important clientele is more likely to be effective in the legislative arena than is a group which must share with other groups the representation of a single client. For example, a solitary group representing big-game hunting interests in a specific district of a state forest is more effective than each of numerous groups representing the same client would be. In such a situation, the solitary group is able to demonstrate the primacy of its concerns more forcefully and thus is better able to influence the views of legislators. Other factors working in favor of interest-group influence are, first, support for (or, at least, lack of opposition to) its aims by important legislators or agency heads, and second, a focus on issues which are not very visible publicly. As public attention is drawn to an issue, the impact of interest groups in the legislative process apparently diminishes (Ripley 1975).

Available Resources Interest groups are successful in political settings to the extent that they have access to various resources. For example, access to resources such as a large membership and significant funds are especially important. Organizational resources are also meaningful—especially leadership, membership unity, effective policy-making procedures, and access to technical information. Also important are political resources such as knowledge of the policy process (for example, lawmaking and rule-making processes) and political reputation (an interest group that is known as an honest and reputable source of information has an obvious advantage). Prestige and status can also be important group resources, and a large measure of highly motivated, ideological commitment to ensuring that a group's forestry views become public policy is another valuable resource.

Organization and Administration The success of interest groups is also dependent upon how well they are organized and administered. Although groups may vary in size, structure, and purpose, interest group leaders have suggested the following as common elements of a well managed interest group (Langton and Foote-Smith 1984):

- *Establish clear statements of purpose* Establish a planning process that leads to a clear statement of mission, clearly stated long-range goals, and well-defined operational goals
- *Encourage sound administrative practices* Place emphasis on good management skills focused on personnel, accounting, publications, conference management, and membership record-keeping
- *Stress good financial management* Emphasize development and adherence to realistic budgets via planning, income and expenditure monitoring, and cost control measures
- *Promote fund-raising skills* Build diversified income sources and espouse accomplishment of groups' mission
- *Define governing board responsibilities* Recruit good leaders and separate board responsibility for policy making from staff responsibility for day-to-day management and administration

- *Carefully select staff* Select talented staff with the aid of search committees, journal advertisements, and thorough interviews; look for team players and persons interested in producing results
- *Utilize volunteers* Accomplish major portion of group's work load with volunteers and encourage good staff management of volunteers
- *Develop effective publications* Offer high-quality publications that have well-defined purposes
- *Organize to act effectively on issues* Monitor issues, select important few, research issues, develop consensus on group response, and develop strategy to implement group response
- *Access technical information* Secure the best available technical information about an issue via scientists on groups' staff, technical advisory committees, and scientific contacts in government
- *Utilize coalitions* Build alliances and share information and resources with groups having common interests

Strategy and Tactic Success The effectiveness of specific strategies and tactics used by interest groups is very much dependent upon the group in question and the issue being addressed. Of 83 interest groups (including 21 environmental groups) based in Washington, D.C., 53 percent indicated that personal presentations to legislators, legislative aides, or agency personnel were either an "effective" or a "very effective" means of influence, whereas 7 percent indicated that they were not "effective." When asked about testimony at legislative hearings, 20 percent of the groups thought that it was "effective" or "very effective," while 42 percent rated it as "not effective" in influencing policy development (Berry 1977).

In another study, when nearly 600 legislators in four states were asked whether they respond to the lobbying activities of interest groups, 34 percent replied that they question previously held opinions, 33 percent said that they lean more toward the views of group lobbyists, and 33 percent said that they change to the positions suggested by group lobbyists (Zeigler 1976). However, even though they are influenced by the views of interest-group lobbyists, legislators apparently accord less trust to technical information provided by lobbyists than to information provided by other sources, such as a government agency (Pierce and Lovrich 1983).

In a study of litigation as a tactic, 617 environmental cases in federal courts involved the government and an interest group; the government was victorious or partially victorious in over half. The proportion of cases won by the government increased as the cases moved from district court to appellate court to the Supreme Court. At all levels, national interest groups (such as the Natural Resources Defense Council and the Sierra Club Legal Defense Fund) were consistently more victorious than local or ad hoc interest groups. The organizational strength and judicial experience of national groups is an obvious advantage in the judicial arena (Wenner 1983).

Effective lobbying is dependent upon a number of factors, of which access is often paramount. The ability to identify important points in the policy process, to establish contacts with persons involved with decisions at such points, and subsequently to influence the decisions is predicated on access. Interest-group staff members prize access, and attempts to achieve and exploit access use up much of their time and energy. In communicating with the members of the interest groups, staff members frequently stress their contacts with government officials—for example, at meetings, formal hearings, and social events. Without access, an interest group is unable to use the innumerable strategies and tactics that are at its disposal—which means that its ability to influence the outcome of important matters of forest resource policy becomes doubtful.

Interest-group lobbyists often have strong opinions about what makes for success in the highly charged political world in which they operate. Their task is to convince decision makers that the group's constituency has an intense preference for the viewpoint being presented. Lobbying can be done gracefully or crudely; sometimes it is heavy-handed or even threatening. Full-time professionals often have a smooth approach, while the technique of part-time amateurs is likely to be rough around the edges.

Success in lobbying generally requires respect for the following principles (Berry 1984, pp. 118–123):

- *Credibility comes first* Honesty and sincerity are fundamental. Misleading statements and deceitful tactics will close doors and alienate policymakers far into the future.
- *Never burn bridges* Cordiality and friendliness are critical, no matter how intense the political struggle. The luxury of venting anger toward a policymaker who acts contrary to a group's wishes cannot be afforded; enemies on one issue may be needed friends on the next issue.
- *Success equates with compromise* Flexible and negotiable attitudes toward positions are important. That which is least valuable to a group must be given up willingly in order to secure more important priorities.
- *Create a dependency* Reciprocal relationships in which policymakers rely on groups for expertise and information are essential. Repeated opportunities to interact and display group knowledge and understanding are important elements in creation of such dependency.
- *Persuasive communication* Clear and concise messages in carefully packaged verbal or written modes are essential. Lofty platitudes about public interest usually fall on deaf ears, as does repetition of well-worn, unworkable approaches for solving issues.

INTEREST GROUP CHALLENGES

The growth of interest-group politics in forestry and related natural resource fields may well be applauded as a powerful step toward broadened representation in decisions concerning the use and management of forests. Yet it would be unfair to say that such representation has been obtained without difficulty. In truth, both the groups themselves and various other actors involved in the policy process have encountered many difficulties. It is often alleged, for example, that competing interest groups get all (or part) of their opposing preferences included in a law, with the result that forest resource policy contains built-in contradictions requiring simultaneous attainment of opposing goals. Another frequent allegation is that interest groups participate in cozy arrangements with powerful legislators and agency administrators, making forest policy without the benefit of other important actors. Interest groups are also said to provide policymakers with large amounts of issue-related information that naturally tends to be biased—to supports their own views. It is sometimes said that interest groups fail to represent broader legitimate interests in forests. Finally, it is claimed that interest groups simply work to maintain the status quo, thus buttressing policymakers' preferences for familiar, nonthreatening solutions to familiar forest resource issues (Browne 1991; Ripley 1975).

Whether such accusations are true and whether they are relevant to the forestry scene are not well understood. It may be that the accusers are simply using interest groups as scapegoats for other, more fundamental ills of society. However differing interests in the use and management of forests are expressed, they will always exist. Interest groups, in spite of their imperfections, offer a means by which a particular constituency's interest can be linked to government. Interest groups are fundamental expressions of a democratic society.

Interest groups often experience internal problems which affect their ability to express policy preferences and to interact favorably with policymakers. For example, interest groups may have difficulty in determining which of a myriad of issues should receive the benefit of their often-meager resources. Symptoms of such a difficulty may be manifest in public statements which obviously denote little or no expertise on an issue, or in a focus on issues that are narrow in scope (either in subject or geographically, or both) and of concern to only a few members of the group. Such symptoms are likely to appear when too little attention has been paid to development of a group's agenda-setting criteria.

Another internal difficulty is the propensity of unauthorized individuals (members as well as nonmembers) to speak on behalf of a group. Geographic and subject units within a group may also speak on behalf of an entire group, presenting conflicting views that should be reconciled before official statements are made.

Interest groups are also challenged by the need to reply responsibly to the multitude of requests made of them by policymakers. For example, they are commonly asked by policymakers to provide opinions on voluminous and highly complex environmental impact statements. In order to effectively do so, interest groups have often added more staff, or established short-term task forces composed of members that have special knowledge about the issue of concern. In some cases, the requests are simply ignored.

Interest groups also have ongoing concerns about maintaining adequate operating funds (money to pay staff, rent, and operating expenses). They must also be careful to maintain an appropriate balance between revenue from member dues and from outside grants, so that, for example, a granting organization will not begin trying to set the group's agenda of forest policy issues. Nondues sources of revenue, such as book sales and life insurance sales, and revenue provided by special sponsors (including foundations, trust funds, and large private donors) are common means of securing a stable flow of funds (Mitchel 1980). Financial difficulties are compounded by the so-called free-rider problem, in that individuals who fail to join a group nevertheless often benefit from the policy successes achieved by the group.

The composition of an interest group's membership can make it exceedingly difficult, or sometimes impossible, for the group to speak out on sensitive issues. Diversity within a group—in terms of employment, expertise, and geographic distribution—may be so intense that consensus on which issue to address and the nature of a substantive response is impossible to attain. The dilemma has been well stated by David Stahl, past president of the NFPA (1987).

> If the [forest products] industry is to effectively represent itself nationally and to the American people, it must do so with a united voice. . . . Some companies have a tendency to resign over, or to withhold support for, activities which are not consistent with the views of their company. Many more use their disagreements for not supporting industry efforts. . . . I have concluded that the forest products industry is an industry frequently at war with itself. As a result of this, it often has little time, energy, or resources to fight its common enemies. Virtually every federal agency, the Congress and outside groups that NFPA deals with have exploited this disunity to achieve objectives which harmed the entire industry.

Policymakers should respect such conditions and refrain from prodding groups into public positions that may force them over the brink into internal chaos.

Management and administration of interest groups can also pose special challenges. As important as it is for a group to address the substance of a forest resource issue, doing so effectively may be dependent upon how well a group is organized and administered. To remain effective in the years ahead, suggestions have been made (Langton 1984) that group leaders should be cognizant on the need for:

- *Managerial skills* Seek employees with skills in accounting, planning, public affairs, budgeting and program evaluation; resist casual or sloppy management practices
- *Collaboration* Encourage meaningful and cooperative relations with other interest groups; defy the inefficiencies often associated with fragmented efforts
- *Political competence* Develop effective political action skills, especially in the areas of lobbying, citizen involvement, and the legislative processes
- *Scientific and technical capacity* Acquire a capacity to assess the scientific

and technical substance of issues, matters which will grow in the future as sources of dispute

Suggested also is that interest groups "should be as concerned with advocating realistic solutions to environmental problems as in warning about environmental hazards . . . [groups must] gear educational efforts toward constructive discovery and advocacy as well as toward dedicated opposition. Without the former, the credibility of the latter will be minimized" (Langton 1984, p. 10).

REFERENCES

Berry, Jeffrey M. *Lobbying for the People: The Political Behavior of Public Interest Groups.* Princeton University Press, Princeton, N.J., 1977.

______. *The Interest Group Society.* Little, Brown and Company, Boston, Mass., 1984.

Browne, W. P. Issue Niches and the Limits of Interest Group Influence, in *Interest Group Politics* by A. J. Cigler and B. A. Loomis (ed.). CQ Press, Washington, 1991.

Dye, Thomas R. *Understanding Public Policy.* Prentice Hall, Inc., Englewood Cliffs, N.J., 1981.

Ellefson, Paul V. *U.S. Wood-based Industry: Industrial Organization and Performance.* Praeger Publishers, New York, 1984.

______. Congressional Testimony: Who Presents Wood-Based Industrial Interests? *Journal of Forestry* 83(5):300–301, 1989.

______. Political Action Committees: Forestry PACs. *Journal of Forestry* 84(5):20–26, 1986.

______. Professional Associations Address Resource Policy Issues: What Criteria Should Guide Issue Selection? *Renewable Resources Journal* 6(4):15–17, 1988a.

______. SAF Selects Issues for Policy Action: A Perspective for Improvement. *Journal of Forestry* 86(8):47–48, 1988b.

Evans, Brock. Lobbying for the Environment, in *Environmental Leadership* by S. Langton (ed.). D.C. Health and Company, Lexington, 1984.

Gale Research Company. *Encyclopedia of Associations: 1990.* Detroit, Mich., 1989.

Gifford, B. Inside the Environmental Groups. *Outside* 15(9):69–84, 1990.

Henly, R. A., and R. Kelly. *Policy and Procedures in the Society of American Foresters and Other Professional Organizations.* Working Paper. Society of American Foresters, Bethesda, Md., 1988.

Langton, S. The Future of the Environmental Movement, in *Environmental Leadership* by S. Langton (ed.). D.C. Health and Company, Lexington, 1984.

Langton, S. and C. Foote-Smith. One Hundred Ideas for Successfully Managing and Leading Environmental Groups, in *Environmental Leadership* by S. Langton (ed.). D. C. Heath and Company, Lexington, 1984.

Lineberry, Robert L. *Government in America: People, Politics and Policy.* Little, Brown and Company, Boston, Mass., 1981.

Mitchel, Robert C. National Environmental Lobbies and the Apparent Illogic of Collective Action, in *Collective Decision Making: Applications from Public Choice Theory,* by C. S. Russell (ed). The Johns Hopkins University Press, Baltimore, Md., 1980, pp. 87–121.

National Forest Products Association. *1987 Annual Report.* Washington, 1988.

______. *1988 Annual Report.* Washington, 1989.

National Wildlife Federation. *Conservation Directory: 1989.* Washington, 1989.

Patterson, S. C., R. H. Davidson, and R. B. Ripley. *A More Perfect Union: Introduction to American Government.* Brooks-Cole Publishing Company, Pacific Grove, Calif., 1989.

Pierce, J. P., and N. P. Lovrich. Trust in the Technical Information Provided by Interest Groups: The Views of Legislators, Activists, Experts and the General Public. *Policy Studies Review* 11(4):627–639, 1983.

Ripley, R. B. *Congress: Process and Policy.* W. W. Norton and Company, New York, 1975.

Rosener, Judy B. Environmental Testimony and Public Hearings: Some Questionable Assumptions, in *Environmental Leadership* by S. Langton (ed.). D. C. Heath and Company, Lexington, 1984.

Sabato, L. J. *PAC Power: Inside the World of Political Action Committees.* W. W. Norton and Company, New York, 1985.

Stahl, David E. Public Letter to Members of NFPA. National Forest Products Association, Washington, Feb. 13, 1987.

Society of American Foresters. *SAF Forest Policies and Positions.* Bethesda, Md., 1988.

Sierra Club. *Sierra Club Conservation Campaign.* 1987–88 National Conservation Campaigns, San Francisco, Calif., 1987.

______. Annual Report: 1987. *Sierra* 73(2):77–79, 1988.

Union of International Associations. *Yearbook of International Organizations: 1985/86.* Brussels, Belgium, 1985.

Wasby, S. L. Interest Groups and Litigation. *Policy Studies Journal* 11(4):657–670, 1983.

Wenner, L. M. Interest Group Litigation and Environmental Policy. *Policy Studies Review* 11(4):671–683, 1983.

Wolpe, Bruce C. *Lobbying Congress: How the System Works.* Congressional Quarterly, Inc., Washington, 1990.

Wootton, G. *Interest Groups: Policy and Politics in America.* Prentice Hall, Inc., Englewood Cliffs, N.J., 1985.

Zeigler, Harmon. *Interest Groups in American Society.* Prentice Hall, Inc., Englewood Cliffs, N.J., 1964.

______. The Effects of Lobbying: A Comparative Assessment, in *Public Opinion and Public Policy,* by N. R. Luttbeg (ed.). The Dorsey Press, Homewood, Il., 1976, pp. 225–251.

14

THE GENERAL PUBLIC, POLITICAL PARTIES, AND THE MASS MEDIA

A variety of political influences shape the substance of forest resource policies as they proceed through the policy process. Such influences are exerted by legislatures, the judiciary, bureaucratic agencies, and interest groups, as described in earlier chapters. Also important to the development of forest policies are influences exerted by the general public, political parties, and the mass media. Each is briefly examined in this chapter.

GENERAL PUBLIC

Participation by all citizens in the politics of government is supposedly the hallmark of democratic societies. In reality, however, direct involvement of the general public in the establishment and implementation of policies and programs is a rarity. To act on the thousands of laws being considered by legislatures or the hundreds of rulings pending before executive or judicial systems, or to become involved in the millions of administrative decisions required each year of government managers, would be simply overwhelming for the average citizen. Indeed, the public as a whole has an informed opinion on only a small fraction of the policy matters facing government at any one time. As well stated by Van Horn et al., ''Public opinions are involved in virtually everything that government institutions do, yet most citizens are typically little more than bystanders. The public policy enterprise occurs in the background of their lives; they hear the noise, but seldom listen'' (1989, p. 228).

In general, citizen participation in matters of politics is probably as follows (Hennessy 1985):

- Officeholders, candidates, interest group leaders, media specialists, and political party leaders that fully participate in politics (1 percent of the population or less)
- Members of issue-specific organizations that would mobilize quickly to act on an issue (3 to 5 percent of the population)
- People who occasionally seek out political information and periodically discuss politics with friends and acquaintances (60 percent of the population)
- Persons that are apolitical (20 percent of the population)

The views and concerns of the general public may be represented in the policy process in a number of fashions. In most cases, political officials and important social leaders simply decide what public opinion is. Government officials, media representatives, and interest-group leaders construct an idea of the general public's desires in relation to the nation's forest resources after listening to a variety of knowledgeable persons and to reports and polls on public sentiment. The policy process is implemented within the boundaries of such ideas about public opinion. Public officials use these interpretations of public sentiment in their day-to-day struggles for control of various segments of the policy process—liberally citing public support for or opposition to the forest resource policies they favor. To lend credence to their cause, officials may formally check out public sentiments toward a particular policy proposal or implementing action; they may, for instance, conduct a legislative hearing, sponsor a public involvement session, or carry out a formal public opinion poll. From this perspective, the general public's involvement in the policy process is largely passive and is interpreted from the top down.

The general public's passive involvement in the policy process can at times be awakened by deep-seated concern over certain issues. Serious economic hardship in the wood-based industry, wanton destruction of rare and endangered forest ecosystems, and public health concerns stemming from misuse of pesticides can so stir the passions of ordinary citizens that they become active in the policy process. When this occurs, the general public may exert significant pressure on public officials. This pressure may take a variety of forms. Citizens may exercise their right (provided for in more than 20 states) to place issues on the ballot for approval or disapproval, or they may become involved in grass-roots politics. Aroused citizens may be transformed into political activists who have a fierce desire to change the substance of a particular forest policy. They may even engage in electoral politics or agency-structured public hearings. The degree to which the general public engages in such actions is frequently a function of their impatience with the pace of decision making, their frustration with the unwillingness of public officials to address their forestry concerns, or their anger over policies and programs that infringe upon their basic rights or fundamental interests.

When members of the general public are aroused, their perspective on the policy process turns into a bottom-up involvement. Under such circumstances, citizens are no longer a passive audience, watching as the political elites engage in contests over forest policies and work their way through various stages of the policy process. Drawn by various political ideologies and influenced by complex cultural backgrounds, citizens enter the fracas and attempt to directly influence policy development (Fortmann 1987).

POLITICAL PARTIES

The primary purpose of political parties is to capture control of government by securing the election of candidates whose views on important social and political issues are in accord with the views of the parties. These organizations differ from interest groups in terms of the scope of their interests (their concerns are very broad) and their primary goal (which is, as stated above, gaining control of government). Political parties are also the primary means of formally contesting elections in the United States. A political party pursues its interests by various means, including nominating candidates, mobilizing voters, influencing public opinion, organizing branches of government, and building the coalitions necessary for the establishment of public policies. Today's political parties are vigorous and active. Ironically, however, voter loyalty to political parties has never been thinner; even elected officials are often willing to forsake party loyalty in favor of their own policy preferences (Patterson et al. 1989).

Political parties are active in many segments of the policy process. They are visibly involved in the agenda-setting process via the often widespread publicity given to the establishment and implementation of party platforms. The nature of Democratic and Republican national platform pronouncements on matters of forestry and related subjects in 1987 is represented by (Congressional Quarterly 1988a, 1969; Congressional Quarterly 1988b, 2387–2388):

DEMOCRATIC

- National parks, forests, wildlife refuges and coastal zones must be protected and used only in an environmentally sound manner.
- Nation must redouble its efforts to provide clean water-ways, sound water management, and safe drinkable ground water.
- Address the depletion of the ozone layer, the greenhouse effect, the destruction of tropical forests and other global threats, and create a global action plan for environmental restoration.

REPUBLICAN

- Work for further reductions in air and water pollution [to] be achieved without harmful economic dislocation.

- Upgrade recreation, fisheries, and wildlife programs in parks, wildlife refuges, forests, and other public lands.
- Support efforts, including innovative public-private partnerships, to restore declining waterfowl populations and enhance recreational fisheries.
- Fight to protect endangered species and to sustain biological diversity worldwide.
- Protect the productive capacity of lands by minimizing erosion.
- Public lands should not be transferred to any special group.
- Keep public lands open and accessible.
- Require that federal departments and agencies meet or exceed the environmental standards set for citizens in the private sector.
- Develop international agreements to solve complex problems such as tropical forest destruction.

Once elected, partisan public officials engage the policy process at various points to advance the principles and purposes of their party platforms. The platforms become blueprints for legislators and executives and are the means by which like-minded officials work together to propose, legitimize, and implement policies and programs that they generally favor. Interest groups also use platforms, to remind partisan officials of the need to keep promises made during political campaigns.

MASS MEDIA

The process by which forest resource policies are developed and subsequently implemented is sensitive to and often influenced by the mass media—newspapers and magazines, radio and television. Indeed, citizens, policymakers, interest groups, and others involved in the policy process would probably be at a severe disadvantage without the opportunity to communicate via the mass media. Many important publics would not be well-informed about major forestry decisions, and decision makers would not be well-advised of the will and the wishes of the various publics they are charged with serving. However, communication via the mass media does have shortcomings. The media can distort reality, they do magnify certain facts or events, and they also sometimes ignore or minimize certain phenomena which ultimately prove to be of major importance in the development of effective forest resource policies. From a policy perspective, these characteristics of the mass media complicate the task of managing forest resources, or at least managing them effectively.

The mass media are of various types, and each publication, radio station, or television station has its own degree of interest in forestry, depending upon its location (its proximity to forests and to officials who are responsible for establishing forest policies) and upon the public-affairs interest of its owners and editorial boards. In the late 1980s, the print media in the U.S. included over 11,000 newspapers (of which 1,900 were dailies) and more than 12,300 periodicals. Widely read daily newspapers such as the *Washington Post, The New York Times,* the *Denver Post,* and the *San Francisco Chronicle* frequently report on major forest resource issues, as well as periodically publishing investigative reports on major matters of forest policy. Many of the nation's more local, weekly or semiweekly newspapers (of which there were over 9,100) report forestry news on a continuing basis.

The electronic media also are important carriers of forestry news. Television has the unique ability to carry especially graphic representations of forestry activities (timber harvesting, sport hunting, forest recreation) into the living rooms of nearly all the nation's citizens. There are more than 10,000 radio stations in the country, and there are nearly 1,300 television stations and about 7,800 cable television systems (Patterson et al. 1989, p. 321).

The mass media are fundamentally concerned with communication—a five-fold process involving a *source* (for example, a district ranger, an interest-group executive, or a legislative staff member), a *message* (perhaps a proposed policy or a major policy decision), a *medium* (a newspaper, a radio or television station, a speech, or a press release, for example), an *audience* (the general public or, on occasion, a specific policymaker, such as a legislator), and an *effect* (public outrage, say, or a policy change).

Exactly what messages do (or should) flow through the communication process is debatable, involving differing views over what constitutes news. In general, *news* is information about events that are recent (or at least recently disclosed), can be quickly described, and are relevant to an audience. Timeliness and immediacy are special aspects of *newsworthiness.* Even though a forestry event is momentous, it may not be considered newsworthy if the event is remote from the interests and lives of the public. Central to the debate over newsworthiness is whether an issue's surface events or its underlying conditions should be stressed by the media. Should the regeneration clear-cut (which can be seen and experienced directly) in the forest environment be stressed, or should the basic ecological and economic principles underlying even-aged silvicultural systems (concepts that can be understood but not necessarily experienced) be emphasized? If surface events are given prominence, the media need only describe them as correctly as possible; if underlying conditions are important, explanation and interpretation are involved.

The following general ideas have been suggested about news (Patterson et al. 1989, p. 329):

- *Surface events tend to be the focus of news* Examples are major forest wildfires, destruction of endangered forest wildlife, massive forest insect infestation, and the harvesting of timber.
- *Highly visible people tend to be highlighted in news* Examples include an agency head, a legislative leader, a prominent conservation leader, and a public celebrity.

- *Surprising, strange, or unusual events are especially newsworthy* A major unexpected change in forest policy is an example.
- *Positive or fortunate events are usually less newsworthy than dangerous or unfortunate events* For example, a disastrous forest wildfire is more newsworthy than a large-scale reforestation program.
- *Conflict among individuals or organizations is usually very newsworthy* For example, two forest resource agencies, one that is opposed to a proposed policy and one that is in favor of it, may be attempting to influence the fate of the policy.

News stories about forestry do not just happen. Reporters must interact with a source—someone who has special information, such as an agency head or the chair of a legislative forestry committee. Reporters and their sources are often locked into a love-hate relationship. Reporters have the power to create publicity—and positive publicity is usually of vital interest to leaders. Sources have information—and reporters must have information. Most such relationships are formal, in that they involve interviews, press conferences, and press releases. However, there are usually opportunities for off-the-record contacts between reporters and sources (e.g., news leaks and off-the-record briefings).

The relationship between reporters and news sources is not a one-way street. Reporters do not simply communicate forestry news that is supplied by their sources. Officials can use reporters to shape news about important forestry topics. They provide reporters with previously prepared press releases at appropriate times. They may demonstrate favoritism toward reporters who communicate news in a favorable manner, perhaps giving them helpful background material or special access to important policymakers. Nearly all public forestry agencies have public affairs offices or offices of information and education which serve as initial contact points for reporters (General Accounting Office 1986).

The news media can become involved in virtually every segment of the policy process. Especially significant is the media's influence on agenda-setting activities. Interest-group leaders, government policymakers, and citizen activists are well aware of the role the media can play in keeping the attentive public informed and in arousing the apathetic general public to support or oppose a particular cause (Anderson 1984; Cook 1989). Interest groups that are intent upon securing changes in a public forest policy frequently seek out the media for help in expanding the scope of public interest in and support for their cause. Group members may distribute petitions, hold press conferences, make speeches, appear at editorial board meetings, give interviews to reporters, or appear on televised public affairs programs. The media communicate the substance of such events and may report that public interest in the concerns expressed by groups in building. Eventually, policymakers (sensing a groundswell of public concern) may respond by acknowledging the need to address the issue at hand—by setting the agenda. The gatekeeping role of the media should not be overlooked, for they have a significant ability to decide what news will (or will not) be passed on to the public.

The media are often in a position to shape the substance of forest resource policies and the manner in which they are implemented. It has been argued that the media (Ranney 1983):

• *Compress the time available for decision makers to carefully develop policies and implement programs* The media are eager for stories and action; they desire quick responses to problems and favor simple, uncluttered solutions to complicated issues.

• *May reduce policy options to be considered by highlighting and publicizing proposed solutions prematurely* Inviting criticism of policy options early in the formulation stage may eliminate effective options from serious consideration.

• *Highlight leaders who simplify policy options, take extreme stands on solutions to issues, and fight courageously with little interest in compromise* The media may weaken the less glamorous—but necessary—process of building coalitions and securing agreement on a particular policy or program.

It has also been suggested that the media constitute the fourth sector of government, "watching over [government] officials and keeping them responsible." The disadvantage of such a role for the media is that it may cause public officials to become somewhat reluctant to develop and propose the bold and imaginative policies that are sometimes required to address important forest resource issues of concern to the public (Patterson et al. 1989).

REFERENCES

Anderson, N. W. Mastering the Media, in *Environmental Leadership* by S. Langton (ed.). D. C. Heath and Company, Lexington, 1984.

Congressional Quarterly. Democratic National Platform: 1988. *Congressional Quarterly Weekly Report* 46(29):1967–1970, 1988a.

______. Republican National Platform: 1988. *Congressional Quarterly Weekly Report* 46(34):2369–2399, 1988b.

Cook, T. C. *Making Laws and Making News.* The Brookings Institution, Washington, 1989.

Fortmann, L. People and Process in Forest Protest. *American Forests* 93(3,4):12, 13, 56, 57, 1987.

General Accounting Office. *Public Affairs: Public Affairs and Congressional Affairs Activities of Federal Agencies.* GAO/GGD-86-24. Washington, 1986.

Hennessy, B. *Public Opinion.* Brooks-Cole Publishing Company, Pacific Grove, Calif., 1985.

Patterson, S. C., R. H. Davidson, and R. B. Ripley. *A More Perfect Union: Introduction to American Government.* Brooks-Cole Publishing Company, Pacific Grove, Calif., 1989.

Ranney, A. *Channels of Power.* Basic Books, New York, 1983.

Van Horn, C. E., D. C. Baumer, and W. T. Gormley. *Politics and Public Policy.* Congressional Quarterly Press, Washington, 1989.

FOUR

POLICIES AND PROGRAMS

Charged with an issue or opportunity and then vigorously activated by interested individuals and organizations, the policy process can produce a variety of noteworthy policies and programs which focus on forest resources. From a broad social perspective, such policies can be viewed as lying on a continuum. At one extreme society trusts the private sector to provide the economically and socially desirable forestry benefits it desires, while at the other extreme society pursues such benefits by retaining for itself (through government) exclusive ownership of forests. In between such extremes, society can choose from a variety of progressively more intrusive government-initiated means for securing important forestry benefits, including information and service initiatives, fiscal and tax initiatives, and regulatory initiatives.

15

PLANNING AND BUDGETING USE AND MANAGEMENT OF FORESTS

Forests provide a remarkable array of useful benefits to society. Careful consideration must be given to planning the use and management of forests and to deciding on the appropriate levels of investment to be made in implementing the designs set forth in such plans. The policy process and its various participants have produced a number of policies and programs to guide planning and investment. Although these policies and programs do not necessarily belong in discrete subject areas, they can be roughly categorized as follows: strategic program planning, land use and management planning, and fiscal and budgetary planning. Each of these categories entails policies which have been legitimized and implemented by forestry or broader communities. Examples are discussed below.

STRATEGIC PROGRAM PLANNING: THE RPA

Strategic Planning

Strategic planning is a disciplined effort to produce fundamental decisions and actions that guide what an organization is, what it does, and why it does it. Such planning typically requires broad-scale information gathering, an exploration of far-reaching alternatives, an emphasis on future implications of present decisions, and an ability to accommodate divergent interests and values (Bryson 1988). In a forestry program context, strategic planning focuses on matters such as the role of the federal government versus the roles of state governments in forest protection, the degree of emphasis to be placed on threatened and endangered species recovery in forested environments, and the role of public timber-production programs in maintaining healthy and diversified local economies. Such planning is usually agencywide and tends to avoid consideration of project-level or site-specific topics.

Forest resource policies addressing strategic program planning have been legitimized by laws enacted by the federal government and by a number of states (for example, California and Minnesota). Examples of federal laws are the Soil and Water Resources Conservation Act of 1977, which the USDA-Soil Conservation Service is responsible for implementing, and the Forest and Rangeland Renewable Resources Planning Act of 1974 (RPA), as amended.

Implemented by the USDA-Forest Service, the RPA is designed to address political and administrative topics concerning forests in a national setting, especially:

- The availability of facts concerning current and potential condition of forests
- The nature of long-term strategic goals involving the use and management of forests
- The financial commitments required to achieve long-term societal interests in forests

In such a context, the provisions of the RPA call for preparation of an assessment, a program, a presidential statement of policy, and annual statements identifying progress to date and the implications of proposed budgets (Shands 1981a; Wilkinson and Anderson 1985). These provisions are considered in detail in what follows.

Resource Assessment

Prepared at 10-year intervals, the assessment presents the factual basis for the development of a nationwide strategic plan (called the ''Program'') for forests and related resources. As a status report on the condition and capability of the nation's forestlands, the assessment addresses the broad range of benefits (including water, wildlife, timber, recreation, and wilderness) that forests are capable of producing or influencing—regardless of ownership. Included in the assessment are:

- An analysis of present and anticipated uses of, demand for, and supply of forest and range resources on all forestland ownerships
- An inventory of present and potential forest resources and an evaluation

of opportunities for improving their yield of tangible and intangible goods and services, together with estimates of investment costs and direct and indirect returns to the federal government

- A description of USDA-Forest Service programs and responsibilities in research, cooperative programs and the management of National Forest System, their interrelationships, and their relationships with other public and private activities
- A discussion of important policy considerations, laws, regulations, and other factors expected to significantly influence the use, ownership, and management of forests and related lands and resources

The 1989 assessment document addressed these requirements by presenting information on (USDA-Forest Service 1989a):

- Projected population and income levels
- Forest and rangeland area
- Location and type of vegetative cover
- Supply-and-demand condition of forest resources (including identification of acceptable, impaired, or uncertain resource conditions)
- Social, economic, and environmental implications of projected demands and supplies
- Opportunities for responding to the implications of such projections

Based on such information, major findings were set forth and opportunities identified (Table 15-1).

Resource Program

The program, which is prepared at 5-year intervals, looks ahead to conditions four decades in the future. It represents a response to conditions and opportunities identified in the assessment. The program, like the assessment, addresses the variety of benefits all forest ownerships are capable of providing. It does not, however, prescribe policy or program direction for agencies other than the USDA-Forest Service, nor does it prescribe directions for programs carried out by private organizations or state governments. The USDA-Forest Service emphasis is felt when the program is implemented by the agency's three major administrative units—Research, State and Private Forestry, and the National Forest System. The program includes (but is not limited to):

- An inventory of specific needs and opportunities for both public and private investments in forest resources.
- Identification of specific results (outputs) to be anticipated from various proposed investment opportunities. Costs must be related to program benefits.
- Discussion of priorities assigned to various investment opportunities.
- Identification of personnel needed to implement investment opportunities and ongoing programs.

TABLE 15-1
Selected findings and opportunities identified by the 1989 RPA assessment of U.S. forest and rangeland resources

Major findings (selected examples)

Timber Demands for all major timber products will increase over the next five decades (79 percent domestic increase for hardwood timber, 35 percent increase for softwood). Harvest on forest industry lands will increase 31 percent, on other private lands 70 percent, and on National Forests approximately 20 percent. It will remain constant on other public lands. Supplies will meet demands in U.S. markets, but prices will be higher.

Wildlife and fish Big- and small-game hunting will decline to the year 2000 and then increase. Increases in nonconsumptive uses of wildlife will grow twice as fast as population to 2000. National Forests and other public lands will become more important for unique wildlife and fish habitats. Access to remaining free and unrestricted National Forests and other public lands will become more important for big-game hunting and cold-water fishing. Hunting for a fee on private lands will become more important in the future.

Outdoor recreation Participation in all forms of outdoor recreation will increase over the next five decades. Total demand for recreation will grow in line with future population growth (proportion of population participating in recreation and the allocation of leisure time per capita have stabilized). Projected increases in public- and private-sector recreation opportunities will meet projected increases in demand. Most increases in demand for recreation will occur near urban centers far away from federal lands in the West. Shortfalls in the supply of specific types of recreation opportunities can be made up for by increased supplies from private lands, especially by supply responses to fees for recreational use of private lands.

Wilderness The National Wilderness System contains 89 million acres (1 of 6 acres in the National Forest System is designated wilderness). Major future growth of the system is not expected. Wilderness recreation accounts for less than 1 percent of all outdoor recreation. The total recreation time spent in designated wilderness in recent years is stable. Wilderness areas contribute to species diversity and habitat for threatened and endangered species.

Water Water demands will increase significantly. Limits on water supplies will force reallocation among users (from irrigation to municipal use). Ninety percent of inland surface waters are fishable and swimmable because of control over point sources of pollution. Nonpoint sources of pollutants are the principal existing sources of water pollutants. Market mechanisms may be useful for allocating quantities and qualities of water.

Forage Demand for domestic range forage is expected to increase 54 percent by 2040. The supply of range forage will increase 52 percent by 2040, mostly from private lands.

Minerals Demand for energy minerals, metallic minerals, and industrial minerals will increase in future. Although prices for minerals will increase, no physical shortage is expected.

Major opportunities (selected examples)

Timber Timber supplies may be extended by (1) increasing the useful life of wood products, improving the efficiency of harvesting and manufacturing, utilizing unused wood materials (harvest residue), and increasing recycling of paper and paperboard; (2) increasing the harvest from existing timber resources by expanding harvest on Eastern timberlands and accelerating harvest on public forests that have large quantities of old-growth softwood timber; and (3) increasing timber growth by planting trees, converting existing stands to more desirable species, applying intensive timber-management practices (especially to nonindustrial private forests), and protecting forests from wildlife, insects, diseases and poor logging practices.

TABLE 15-1 ***(Continued)***

Wildlife and fish Wildlife and fish supplies may be extended by (1) managing habitat, including increasing food supplies and size and diversity of cover; (2) directly managing species, including reintroducing species, removing competing species, and increasing fish hatchery capacity; (3) controlling users and people, including increasing access to private lands, and arranging for better distribution among available wildlife and fish populations; and (4) planning resource use and management, including increasing interagency coordination, integrating wildlife and fish needs with other forest resource uses, and increasing the availability of management information gained via research.

Outdoor recreation and wilderness Outdoor recreation may be extended by rehabilitating deteriorating recreation sites; encouraging partnerships between public and private recreation suppliers; improving coordination and integration of recreation uses with other forest resource uses; developing stable sources of revenue to cover costs of providing recreation opportunities; and expanding visitor information services, especially field interpretation and educational services.

Water Quantity and quality of water supply may be extended by improving vegetative management to enhance natural recharge of surface and ground water; expanding and improving reservoirs to increase storage, regulate flows, and reduce evaporation; improving timber-harvesting and road-building practices; and rehabilitating deteriorated watersheds.

Forage Range forage supplies may be extended by (1) managing range vegetation, including increasing seasonal availability via adjusting mix of plant and animal species, developing biological controls for noxious weeds, and restoring deteriorated range conditions; (2) managing grazers and browsers, including increasing predator control and mixing animal species in order to take full advantage of available forage; and (3) planning range forage use and management, including increasing multiresource planning across ownerships, agencies, and resources.

Minerals Mineral supplies may be extended by improving information on the location, quantity, and quality of mineral supplies; expanding availability through more efficient recovery in processing and use; and encouraging substitution of nonmineral materials for minerals and abundant minerals for scarce minerals.

Source: RPA Assessment of the Forest and Rangeland Situation in the United States: 1989. Forest Resource Report No. 26. U.S. Department of Agriculture, Washington, 1989.

The program is required to be cognizant of:

- Multiple-use and sustained-yield principles
- Opportunities for various types of forest landowners to participate in the program
- Interrelationships between various forest benefits (for example, timber and wildlife interactions)
- The need to protect fundamental soil, water, and air resources

The development of the 1990 program entailed a number of important actions, including (USDA-Forest Service 1990):

- Careful review of resource conditions identified by the assessment
- Identification of major natural resource issues of concern to the public

- Specification of appropriate USDA-Forest Service roles for responding to issues and resource conditions
- Identification of alternative long-term strategies to guide USDA-Forest Service management actions
- Solicitation of public comments regarding issues, roles, and alternative strategies
- Ultimate selection of a preferred long-term strategy

The strategy selected is a combination of complementary programs to be undertaken by the USDA-Forest Service's National Forest System, State and Private Forestry, and Research units. Major themes of the selected strategy are (USDA-Forest Service 1990):

- *Recreation, wildlife, and fisheries resources will be enhanced* The USDA-Forest Service will heighten production of outdoor recreation, wildlife, and fisheries on National Forest System land and on state and private lands through increased technical and financial assistance, and will increase research on how to enhance compatibility of resource use on all lands.
- *Commodity production will be more environmentally responsive* The USDA-Forest Service will produce commodities from the National Forests in a more environmentally sensitive manner (possibly adjusting commodity production downward). Through technical assistance, it will emphasis environmental sensitivity for the management of state and private forestlands. Research will be designed to develop a better understanding of the ecological consequences of managing forests.
- *Scientific knowledge about natural resources will be improved* USDA-Forest Service will expand research focused on securing a better understanding of resource production in the context of protecting the environmental integrity of forest resources.
- *Global resource issues will be responded to more intensively* The USDA-Forest Service will increase scientific exchange and technology transfer to other countries so they may more wisely use the benefits offered by forest resources while at the same time reducing the adverse impact of management activities on global ecosystems. Research will be undertaken to develop a better understanding of global ecological interactions involving forests.

The program reflects a continued commitment to multiple-use management in manners that will effectively meet the demands expected for various forest resources. The strategy is also considered to be environmentally sensitive and economically efficient. Details of the National Forest system portion of the long-term strategy are presented in Table 15-2.

Presidential Policy Statement

The assessment and the presidentially agreed to program are transmitted by the President to the U.S. Congress along with ''a detailed [Presidential] statement of policy intended to be used in framing budget requests by [the] Administration for Forest Service activities,'' which must be carried out in order to implement the program. Congress may accept, modify, or reject the Presidential Policy Statement; once agreed to, the statement becomes a document which Congress and the President feel obligated to observe. In the Presidential Policy Statement implementing the 1990 program, President George Bush stated that during the next 5 to 10 years the four aforementioned RPA themes ''are the cornerstones of my commitment to providing a proud natural resource heritage for future generations of Americans.'' Furthermore, he added, ''the RPA Program strongly renews my commitment to multiple-use management—conserving and wisely using our resources so our children and their children have the resources they need to build their homes, to enjoy the great outdoors, and to satisfy other needs as well'' (USDA-Forest Service 1990).

Statements of Intent and Accomplishment

When presenting annual USDA-Forest Service budget requests to Congress, the President is required to state in quantitative and qualitative terms the extent to which the proposed annual budget meets the substance of the program and the Presidential Policy Statement. If the proposed budget fails to meet the intent of these documents, ''the President shall specifically set forth the reason or reasons for requesting the Congress to approve [lesser budget levels].'' A further means of reinforcing presidential budget performance is the required submission (by the Secretary of Agriculture) to Congress of an annual report which evaluates accomplishment of the component parts of the program, information which can be used by Congress to carry out ''oversight responsibilities and improve the accountability of agency [USDA-Forest Service] expenditures and activities.'' State and Private Forestry accomplishments in 1987 are an example, as shown in Table 15-3 (USDA-Forest Service 1989b).

Design versus Reality

The strategic planning policy legitimized by the RPA were designed to provide a firmer basis for policy and budget selection. The assumption was that the policy process and the actors participating in it will produce better decisions if more ample information is available about the potential mix of outputs that can be produced by forests in a long-range context. The spirit of this intent is reflected in the assessment of forest and related resources, the national program which gives direction to investments in forests, the presidential commitment to

TABLE 15-2
Recommended renewable resources program (1990–2040) and long-term strategy for the National Forest System

Key components	Units	Year				
		1989	1995	2000	2005	2040
Recreation and wilderness						
Program or management activity						
Trail construction	Miles	1294.0	2396.0	2016.0	1869.0	1471.0
Outputs						
Camping, picnicking, swimming	MMRVD	72.3	83.5	92.6	101.8	133.7
Travel—motorized, nonmotorized	MMRVD	80.2	95.4	104.5	111.1	147.5
Hiking, riding horses, sailing	MMRVD	21.4	25.2	29.3	31.5	44.5
Winter sports	MMRVD	17.5	21.3	22.9	25.3	33.5
Resorts	MMRVD	14.6	17.8	19.1	20.5	25.0
Wildlife and fish use	MMWFUD	41.7	48.9	55.0	61.3	119.8
Recreation use—all other	MMRVD	12.2	15.9	17.5	19.2	27.4
Total recreation use	MMRVD/MMWFUD	259.9	308.0	340.9	370.7	531.4
Less-than-standard recreation use	Percent	42	18	6	2	0
Wilderness	MMRVD	12.2	15.3	17.7	19.8	28.7
Special-use receipts	MM$	24.0	30.6	33.2	36.7	49.5
Recreation user fees	MM$	14.0	16.1	17.9	19.3	26.1
Cost						
Total recreation and wilderness cost	MM$	187.8	294.2	326.1	334.0	407.5
Wildlife and fish						
Program or management activity						
Anadromous fish improvements	Structures	3827.0	4929.0	5104.0	5258.0	1850.0

Hunting capability	MMWFUD	17.4	19.1	19.9	20.7	27.7
Fishing capability	MMWFUD	18.7	21.8	25.1	28.7	52.7
Nonconsumptive capability	MMWFUD	10.8	13.0	15.6	18.2	47.1
Outputs						
Hunting use	MMWFUD	16.7	18.0	18.6	19.2	25.9
Fishing use	MMWFUD	17.3	20.4	23.0	26.0	50.8
Nonconsumptive use	MMWFUD	7.7	10.5	13.4	16.1	43.1
Total wildlife and fish recreation use	MMWFUD	41.7	48.9	55.0	61.3	119.8
Commercial anadromous catch	Mlbs	117556.0	125898.0	135377.0	141397.0	173759.0
T/E species objectives met	Number	16.0	75.0	121.0	153.0	202.0
Cost						
KV funds—wildlife, fish	MM$	17.0	16.7	18.0	18.0	18.4
Total wildlife and fish cost	MM$	99.0	144.9	189.8	198.1	220.1
		Range				
Program or management activity						
Noxious farm weeds	MAcres	24.0	52.0	57.0	55.0	44.0
Range vegetation management	MAcres	87785.0	88307.0	87966.0	87214.0	84938.0
Outputs						
Permitted grazing	MMAUMs	9.9	9.3	8.8	8.8	9.2
Grazing receipts	MM$	10.9	9.9	9.5	9.5	36.8
Cost						
Total range cost	MM$	36.9	56.9	62.3	63.7	64.8
		Timber				
Program or management activity						
Reforestation	MAcres	476.0	416.0	421.0	420.0	427.0
Timber stand improvement	MAcres	343.0	323.0	334.0	311.0	319.0

(continued)

TABLE 15-2 (*Continued*)

Key components	Units	Year				
		1989	1995	2000	2005	2040
Clear-cut harvests	MAcres	320.0	265.0	256.0	252.0	233.0
Partial-cut harvests	MAcres	590.0	602.0	696.0	733.0	900.0
Road construction for timber	Miles	1698.0	1602.0	1646.0	1470.0	365.0
Road reconstruction for timber	Miles	3232.0	5037.0	6014.0	6507.0	6859.0
Outputs						
Softwood sawtimber offered	MMBF	8008.3	8044.5	8317.8	8396.0	8780.4
Softwood roundwood offered	MMBF	1144.3	1225.6	1202.2	1264.5	1223.9
Hardwood sawtimber offered	MMBF	291.4	419.9	455.1	507.4	622.6
Hardwood roundwood offered	MMBF	550.0	423.1	444.0	436.0	488.1
Total fuelwood offered	MMBF	605.7	729.2	756.3	773.2	859.3
Total timber offered	MMBF	10599.7	10842.3	11175.4	11377.1	11974.3
Timber receipts	MM$	909.6	1432.0	1538.4	1902.0	1734.9
Total KV collections	MM$	209.5	217.3	216.7	224.6	234.4
Cost						
KV funds for timber	MM$	168.0	163.4	161.1	161.5	175.8
Total timber cost	MM$	777.3	779.2	800.2	806.8	810.2
Water, soil, air						
Program or management activity						
Soil inventory	MAcres	6012.0	6919.0	6680.0	6680.0	6665.0
Watershed improvements	Acres	39190.0	46001.0	40816.0	38705.0	38020.0
Air-quality monitoring	Number	222.0	362.0	455.0	530.0	636.0
Outputs						
Increased water yield	MAcre Feet	1346.0	1479.0	1486.0	1488.0	1520.0
Cost						
KV funds for soil, water, air	MM$	5.4	6.2	6.2	6.0	6.0
Total water, soil, air cost	MM$	50.7	69.7	69.3	68.4	63.0

Minerals						
Program or management activity						
Minerals operations processed	Number	15720.0	20198.0	19948.0	19913.0	19920.0
Minerals operations administration	Number	13432.0	17701.0	18158.0	18203.0	18206.0
Leased, nonenergy	MAcres	177.0	287.0	*	*	383.0
Leased energy	MAcres	14169.0	25573.0	*	*	25673.0
Geology resource inventory	MAcres	54782.0	19562.0	*	*	6461.0
Outputs						
Minerals receipts	MM$	266.0	298.1	329.9	362.2	689.2
Cost						
Total minerals cost	MM$	28.7	42.3	43.3	43.4	45.1
Protection and support						
Program or management activity						
General-purpose roads, construction	Miles	129.0	281.0	213.0	138.0	123.0
General-purpose roads, reconstruction	Miles	493.0	949.0	1249.0	1408.0	1516.0
Outputs						
Lands receipts	MM$	4.5	6.6	7.5	7.7	8.5
Cost						
Total support cost	MM$	512.3	675.2	700.9	716.4	721.9
Total protection cost	MM$	399.2	463.9	475.6	482.4	498.2
Total	MM$	1579.6	1851.1	1966.6	1996.8	2108.9

* Data not available.

Note: MM$ = million dollars; MAcres = thousand acres; MMAcres = million acres; MMRVD = million recreation visitor-days; MMWFUD = million fish and wildlife user-days; Mlbs = thousand pounds; MAUMs = thousand animal unit months; MMBF = million board feet; MAcre feet = thousand acre feet; KV = Knudsen-Vandenberg funds; T/E = threatened, endangered, and sensitive species.

Source: The Forest Service Program for Forest and Rangeland Resources: A Long-Term Strategic Plan. Forest Service, U.S. Department of Agriculture, Washington, 1990.

TABLE 15-3
USDA-Forest Service State and Private Forestry accomplishments in 1987 compared to RPA-recommended program levels for 1987

	Units	Funded	Accom-plished	RPA recom-mended level*	Percent of RPA accom-plished
Forest pest management					
Insect and disease management surveys	MM acres	547	640	170	376.47
Insect and disease suppression	MM acres	—	0.98	—	—
Insect and disease special projects	Projects	21	21	—	—
Forest management and utilization					
Forest resource management					
Forestland management plans	MM acres	3.5	4.3	2.2	195.45
Timber prepared for harvest	MM cubic ft	—	305.3	—	—
Reforestation	M acres	—	1,098.9	323	340.22
Timber-stand improvement	M acres	—	240.1	156	153.91
Woodland owners assisted	M owners	—	158.4	—	—
Wood utilization	MM cubic ft	—	—	64	0.00
Seedling, nursery, and tree improvement	MM seedlings	725	828	—	—
Urban forestry assistance	Areas assisted		4,633	—	—
Management improvement					
State forest resource planning	Person Years	—	28	44	63.64

* Low bound level.
Note: M = thousand; MM = million.
Source: Report of the Forest Service Fiscal Year 1987. Forest Service, U.S. Department of Agriculture, Washington, 1988.

the national program, and the oversight role assumed by Congress. These activities are not carried out in a vacuum. For example, drafts of the assessment and program, which are prepared by the USDA-Forest Service, are published and widely distributed for review and comment by any interested parties. Since the program is considered a major federal action that significantly affects the environment, it is prepared as an environmental impact statement; thus, it is subject to extensive requirements for public comment. In addition, the assessment and program are subject to considerable review by the U.S. Department of Agriculture and the President's Office of Management and Budget. Themes that permeate the processes established by the act include (Shands 1981a):

- Consideration is to be given to all forest benefits when determining the biologically and economically optimum mix of such benefits.
- Assessment and programs activities are to reflect long-range social interest in forests.
- Attention is to be given to all forest and related resources, regardless of ownership.
- Improvements are to be made in data collection and evaluation capabilities, especially economic evaluation capabilities.
- Federal programs expressed through the Program are to be coordinated with forestry programs of state and local governments.
- Decision-making processes leading to the assessment and program are to include views of the public (however defined).
- Accountability and commitment will exist between Congress and the executive branch of the federal government.

In addition, the planning processes set in motion by the RPA represent a sincere attempt to accomplish the following ends:

- *To address the long-term sustained financial needs required to produce many forest benefits* This means reconciling annual legislative appropriation process with long-term financial requirements.
- *To explicitly link public investments in forests with the specific nature and magnitude of outputs that flow from such investments* The goal is to foster program efficiency and effectiveness.
- *To present clear documentation of the reasoning behind emphases given to certain programs* The criteria used to judge the merits of alternatives are explained.

Implementation of the planning process legitimized by the RPA has not been without concern and difficulty. Reviews of the process have been numerous (for example, Behan 1990; Binkley 1989; Binkley et al. 1988; Irland Group 1989; Larsen et al. 1990; Office of Technology Assessment 1990; Sample 1989; Shands 1990; Shannon 1990). Some have been especially onerous (U.S. Congress 1986):

> The Committee does not agree to continue to spend millions of dollars on planning documents that are not provided in a timely manner, do not reflect reasonable and professional judgments and estimates when they are released, and are not of particular value to the Committee when finally available.

The concerns expressed over RPA planning activities often reflect the innumerable problems that must be faced during implementation of a nationally developed program strategy at local sites. Amazingly dissimilar physical, biological, and administrative conditions inevitably affect implementation. Such concerns also often reveal the forestry community's unrealistic expectations for clear policy direction from a Congress that must deal with many (often more important) societal issues other than forestry and that must reconcile forestry's demands on the federal treasury with those of many equally legitimate nonforestry demands (such as education and national defense).

Among the specific concerns expressed about RPA strategic planning are:

- *The need for greater program emphasis on strategic directions and on more focused, less comprehensive evaluations in the assessment* For instance, wilderness, timber, and water might each be given emphasis every third year.
- *The inability of the assessment to operate effectively as a distant early warning system for major pending problems regarding forests and related resources* Atmospheric pollutants, global warning, and economic recessions are examples.
- *The dominance of quantitative analytical methods, which often preclude strategic consideration of nonquantifiable or poorly understood issues and management concerns* Problems such as atmospheric pollutants and global warming may receive inadequate consideration.
- *The limited integration of program-established directives with annual budgeting processes* It has been suggested that the program has not been grounded in reasonable assumptions of future funding availability, nor in recognition of future legislatures' probable unwillingness to follow budget commitments made by current legislatures.
- *The lack of convincing integration of strategic RPA program planning with forestland management planning* Conflicts between requirements of the RPA and the National Forest Management Act of 1976 have not been adequately resolved.

Planning Principles

Critical reviews of the planning activities that are required by the RPA have identified numerous concerns. They also lead to suggestions for principles of good planning. For example (Larsen et al. 1990):

- *Integrate and balance resource allocations* Good planning integrates consideration of all forest resources. It does not pit one resource use against another.

• *Communicate a clear vision* Good planning generates a clear vision of a forest's unique contributions to meeting local, regional, and national needs.

• *Recognize limits* Good planning recognizes limits on a forest's ability to produce a mix of goods and services in perpetuity.

• *Seek informed consent* Good planning welcomes citizen involvement. Decisions should be made and explained openly. Dialogue among disparate interests should be facilitated.

• *Finish in a reasonable time* Good planning is completed in a reasonably short period of time. Short periods facilitate incremental planning and stability among key players. People can actually harvest the fruits of their labor.

• *Be people-oriented* Good planning recognizes that individuals, both inside and outside an agency, make the difference between good and bad plans.

• *Promote active administrative leadership* Good planning requires active involvement and personal leadership on the part of responsible administrators.

• *Match analysis to questions at hand* Good planning involves use of analytical tools for purposes of evaluating options. Such tools should not drive or dominate the process.

• *Be both locally oriented and nationally balanced* Good planning should be locally oriented and should also give ample consideration to national constituencies.

Concerns over legitimized strategic planning processes focused on forestry are many. Not to be overlooked, however, are the virtues of such processes. Systematic evaluation by key constituents (such as state foresters, administrative officials overseeing state forestry organizations, state budget directors, legislators, and forest industry and environmental interest groups) of state-developed strategic forest plans identified the following benefits of such plans (Gray and Ellefson 1987):

• Improved sense of long-term agency direction

• More effective decision making, with a greater focus on critical matters that need policy-type decisions

• Greater compatibility with and integration of forest resource programs

• Improved ability to anticipate and respond to major issues and opportunities

• Improved sense of authority, accountability, and control over programs

• Increased public awareness of forests and forest resource programs

• Greater awareness by decision makers of the policy needs of their agencies

• Increased political support for programs, often reflected by higher forestry budgets

• Improved communication and coordination among important forestry actors, such as federal, state, and local natural resources agencies

LAND USE AND MANAGEMENT PLANNING: THE NFMA

Strategic planning of the type legitimized by the RPA provides important national direction for many forestry and related activities. Equally important, however, are planning processes that provide greater detail about appropriate uses for specific tracts of forestland and the exact management practices to be applied thereon.

Such processes have, for example, been legitimized in the National Forest Management Act of 1976 (NFMA) and its associated regulations. They are to be implemented for each of the nation's 156 National Forests (and associated Grasslands). As an amendment to the RPA (all but one of the NFMA's first 12 sections are amendments to the RPA), the NFMA is designed to be an integral part of the RPA's strategic planning processes. For example, National Forest plans provide information for the assessment, particularly with regard to National Forest productive capability (Wilkinson and Anderson 1985). When viewed in total, the RPA and the NFMA have legitimized planning processes at three levels—national, including all forest ownerships; regional, covering National Forest System lands; and individual, specifically for National Forests. Detailed procedures set forth in rules which interpret the intent of the NFMA are the primary means by which the act is implemented (USDA-Forest Service 1979; Shands 1981b). Among the requirements established by the rules are "public participation in the development, review and revision of land and resource plans, and the coordination of such plans with state and local units of government and other federal agencies" and "integration of planning for National Forests and grasslands, [to include] timber, range, fish and wildlife, water, wilderness, and recreation resources."

Planning Process

The *planning process* established by the NFMA and its associated rules is designed to generate regional National Forest System plans as well as plans for each forest within the system. The process consists of 10 steps, which can be grouped into four activity categories: analysis, alternatives, decision, and implementation (Table 15-4). As an example, step 2 requires that planning criteria be set forth—both process criteria (used to judge type and amount of data to be collected, or type and amount of public involvement to be carried out) and criteria to be used for selection from among many formulated alternatives (decision or evaluation criteria). Criteria used in the process are to be based on, for example,

- Existing laws, executive orders, or agency policy
- The RPA program and, for forest plans, the appropriate regional plan
- Plans and programs of state and local governments and of other federal agencies
- Ecological, technical, and economic principles

Step 5 of the planning process requires that an interdisciplinary team formulate alternative courses of action to reflect a range of output levels for a forest or a region. These alternatives (Jameson et al. 1982):

TABLE 15-4
National Forest land-management planning process

Analysis
1. Identification of issues, concerns, and opportunities
2. Development of planning criteria
3. Collection of data and information
4. Analysis of management situation

Alternatives
5. Formulation of alternatives
6. Estimation of effects of alternatives
7. Evaluation of alternatives

Decision
8. Selection of an alternative

Implementation
9. Implementation of the alternative (the plan)
10. Monitoring and evaluation of plan implementation

Source: Regional and National Forest Planning, by William E. Shands, in A Citizen's Guide to the Forest and Rangeland Renewable Resources Planning Act, by W. E. Shands (ed.). FS-365. USDA-Forest Service, Washington, 1981.

- Must be achievable
- Must seek to be cost-effective
- Must address each previously defined issue with at least one alternative
- Must state the condition and uses of the forest in the long-term, including when goods and services will be produced
- Must provide for the elimination of backlogs (for example, reforestation) needed for resource restoration
- Must include a no-action alternative that describes conditions likely to occur with no change in management intensity

Once the alternatives have been formulated and their consequences determined, the forest supervisor (for a forest plan) or the regional forester (for a regional plan) must examine them and recommend a preferred alternative. A draft environmental impact statement setting forth the preferred alternative is made available for public review and comment. A final selection, based on the statement and reactions to it, is made and subsequently legitimized (becomes the forest or region plan), implemented, and monitored.

Administrative Procedures

The act and associated regulations also set forth a series of *administrative procedures* which must be adhered to, as follows:

- *Responsibilities are defined* The forest supervisor of a National Forest, for example, is responsible for a plan's development and implementation. Responsibility includes appointing an interdisciplinary planning team.
- *Points of approval are established* A regional forester, for instance, approves or disapproves a forest plan.

- *Recording of grounds for decisions is called for* A document which discusses the rationale for selecting an alternative is to be prepared.
- *Public appeals process is established* A procedure the public can follow if it seeks to change a forest plan is to be developed (USDA-Forest Service 1988).
- *Appropriate timing for plan revision is specified* A plan must be revised at least every 10 years.
- *Record-keeping procedures are defined*
- *Subject material to be included in a forest plan is prescribed.* At a minimum, each plan is to include:
  - A description of major issues and management concerns, and their disposition
  - A summary of analyses of the management situation, namely the ability of the forest or region in question to supply the goods and services demanded by society
  - A presentation of long-range policies, goals, objectives, and specific management prescriptions
  - Specification of management standards and guidelines, and identification of major sources of information
  - A list of interdisciplinary team members and their qualifications

Management Actions

The rules interpreting policies set forth in the NFMA also call for a number of specific *management actions*. For example, National Forest land that is suitable (or unsuitable) for timber production must be identified. The criteria to be used to determine suitability include:

- *Land not withdrawn from timber production by Congress* For example, legally designated wilderness areas
- *Land not withdrawn from timber production by administrative action* For example, natural areas designated for research
- *Land on which technology will permit harvesting without irreversible resource damage*
- *Land for which there is reasonable assurance that trees (or other vegetation) can be adequately restocked*

As a further example, long-term, sustained-yield timber capabilities of a forest must be set forth. These must be consistent with management intensities and timber unitization standards set in the RPA program, and must ensure a forest structure that will enable perpetual timber harvest at a sustained-yield capacity consistent with the multiple-use management objectives of a forest. Also action-oriented is the required specification of acceptable vegetation management, allowable timber sale quantities, conditions for departure from nondeclining even-flow harvest policies, and appropriate silvicultural guides for employing harvest rules based on mean annual increment of growth.

Standards and Guidelines

The NFMA's implementing rules also establish a number of *management standards and guidelines* for National Forests. For example, streams, stream banks, shorelines, lakes, wetlands, and other bodies of water are to be protected. Detrimental changes in water temperature or chemical composition, blockage of streams, and deposit of sediments which could adversely affect fish habitat are prohibited. Diversity of plant and animal communities is to be maintained. To the extent practical, practices that result in plant and animal diversity as great as that which would be expected in a natural forest and in a planning area generally are required. As for timber harvesting, maximum size limits are set forth for clear-cuts—for example, 60 acres for Douglas fir in California, Oregon, and Washington; 80 acres for southern yellow pine types in the South; 100 acres for hemlock-sitka spruce in Alaska; and 40 acres for all other forest types. Further, management prescriptions must ensure that harvested forestland can be adequately restocked. Cuttings are to be made in such a way that adequate restocking will occur within 5 years after harvest.

The development of forest and regional plans is a complex, time-consuming, and costly process. As might be expected, the substance of stated management directions varies from plan to plan. There is, however, very little variation in the procedures used to develop plans or the format used to present them, for these are fixed by law or rules.

Clearwater National Forest: An Example

An appreciation of consistency in process and variability in substance is best gained by reference to a specific forest plan. The plan for the Clearwater National Forest, located in northeastern Idaho, is an example. The forest consists of over 1.8 million acres divided into five administrative districts. A portion of the Selway-Bitteroot Wilderness (159,000 acres) and a portion of the Middle Fork-Lochsa Recreation River (26,000 acres) are within the forest's boundaries. Average annual timber harvest is 170 million board feet; sawmill capacity in the area adjacent to the forest is 478 million board feet.

As required, the documents for the plan include the plan itself (with its detailed maps), a final environmental impact statement, and a record of decision (USDA-Forest Service 1987). The plan is the preferred alternative selected by the regional forester from among 12 alternatives considered and set forth in the environmental impact statement. There are four major sections in the plan, as follows:

A. *Forestwide management direction* Composed of five subsections (goals, objectives, research needs, standards, and desired future conditions), focused on 17 or more forest benefits or supporting services (such as wildlife, cultural, range, timber, facilities, water quality, and recreation). For recreation, as an example,

1. One of three goals is to "Provide a range of quality outdoor recreation opportunities."
2. One of five objectives is to "Provide . . . dispersed recreational opportunities in a mix of approximately 60 percent roaded and 40 percent unroaded settings."
3. One of two research needs is to "Develop a regional recreational demand prediction model for North Idaho."
4. One of ten standards for recreation is to "Emphasize low impact recreational techniques in dispersed recreational areas."
5. A desired future condition is that, for instance, "The trail system will have been reduced to a stable mileage of about 1,260 miles—a reduction of 25 percent of the present system."

B. *Management area direction* Provides direction for specific areas (management areas) of land within the forest. Seventeen areas are described; for each, management goals are specified and standards set forth for the goods and services the area is capable of providing. Management Area C8S, for example, is composed of over 207,000 acres (mostly unroaded) of productive timberland, key big-game summer range, and valuable fisheries.

1. Area goals in general include management to "maintain high quality wildlife and fisheries objectives while producing timber from the productive forest land; . . . objective can be met by modifying standard timber practices and . . . prohibiting most public motorized uses."
2. Timber subject-area goals include planning for and distributing openings to achieve maximum elk use, while the standards accompanying such timber goals include designing "harvest areas within each elk analysis area that do not require re-entry within ten years."
3. Activities to take place annually in the area during the first decade include: timber harvested from 3,099 acres, 38.4 miles of road constructed, timber stand improvement on 476 acres, and soil and water improvements on 5 acres.

C. *Implementation* Specifies how the forest plan will be put into effect. Entailed is identification of site-specific management practices needed to achieve management area goals, organization and design of projects (including budget estimates), execution of designed projects, and monitoring and evaluation of results. Monitoring will be carried out to assure that goals are being met and that standards and guidelines are being adhered to. The proposed budget for implementing the plan was set at $31.1 million.

D. *Summary of analysis of management situation* Describes the ability of the

Clearwater National Forest to supply the goods and services demanded by society, and serves as the basis for determining whether a need to change current directions exists.

The Clearwater National Forest Plan contains 16 appendixes addressing topics ranging from management of insects and diseases to scheduled review of mineral withdrawals, and from management of visual travel corridors to vegetative management for timber production purposes. The plan's final environmental impact statement lists the individuals (99 persons are identified as having prepared the plan) responsible for preparation of the plan, including:

- *The forest supervisor* Responsible for overall plan direction
- *The core team which formulated plan alternatives* 6 persons
- *The management team which made decisions about formulated alternatives* 10 persons
- *Support teams and resource specialists who supplied information to the core team* 20 persons, including engineers, editors, geologists, foresters, and landscape architects
- *A task force responsible for formulating alternatives based on public comments* 19 persons

The preparation of the forest plan for the Clearwater National Forest was not without cost. The draft plan was developed during the period 1981–1985 (fiscal years) and required an investment of over $2.3 million (includes costs at the forest, regional, and headquarter levels). The $2.1 million invested at the forest level (2.3 percent of the forest's $93.7 million operating budget for the period 1981–1987) was distributed among the following actions: preparation of criteria to guide planning (5 percent), data and information collection (21 percent), analysis of management situation (29 percent), formulation of alternatives (21 percent), estimation of effects of alternatives (9 percent), evaluation of alternatives (8 percent), and recommendation of preferred alternative (7 percent). Since the plan at this stage was a draft, the costs of approving, monitoring, and evaluation are not included (General Accounting Office 1986).

Design versus Reality

Land use and management planning policies of the type legitimized by the NFMA have also received significant review and criticism (for example, Baltic et al. 1989). In 1986, 231 appeals were filed on 96 final or draft plans. The outward reflection of this criticism is continuing and often bitter conflicts over the use and management of National Forests—in spite of the fact that process is being implemented. The many reasons for the controversial nature of forestland use and management decisions resulting from such a process include the following accusations:

- Certain resource user groups have amassed enough power to control agency decisions
- Agency power is sufficient to ignore all (or most) user groups
- Agency discretion is constrained by budgetary limitations or legislative mandates
- Rational comprehensive planning processes are not effective in dealing with highly controversial forestland uses

Constructive criticism of National Forest planning processes may be warranted; excessive pessimism over planning process in general is not. As measured by the extent to which forest users are satisfied that agency actions accommodate their interests, land-management planning has been successful in several national forests, including the Monongahela National Forest and the Jefferson National Forest. Their experiences can offer guidance about how the land-management planning process can be designed to remedy problems that foster especially bitter conflicts over the use and management of forests. Noteworthy in this respect are the needs to (Wondollek 1988):

- Build trust among the various participants in the process, and in the process itself
- Build understanding of the process, the bounds of decision making, and the true concerns and stakes of those involved
- Incorporate the values held by different stakeholders in such a manner that agreement can eventually be reached
- Provide opportunity for joint fact finding by affected groups, allowing issues and questions to be raised early and providing all parties with equal information
- Provide incentives for cooperation and collaboration in a problem-solving rather than an adversarial manner

FISCAL AND BUDGETARY PLANNING

Society's demands for the benefits forests are capable of producing can be better met if the use and management of forests are planned. Equally important is the willingness of society to invest appropriate sums of financial and related resources in forest management. Money, in reality, is the critical ingredient which often determines whether abstract statements of forest resource policy are translated into tangible goods and services. The budgeting processes used to decide how much money will be invested and the policies toward which it will be directed have been legitimized by law and by various administrative actions. Federal examples of the former include the Budget and Accounting Act of 1921, which required preparation of a presidential budget; the Budget and Impoundment Control Act of 1974, which clarified presidential and congressional budget responsibilities; and the Gramm-Rudman-Hollings Act of 1985, which mandated planned reductions in federal budget deficits. Examples of administratively legitimized budget process can be found in the Code of Federal Regulations and the USDA-Forest Service Manual.

In nearly all cases, legitimized budgetary processes call for continual interaction between the legislative and executive branches of government; their design invites intervention by literally hundreds of individuals and organizations. Such processes can be generalized into four phases: executive preparation and submission, legislative approval, executive execution and control, and review and audit (Table 15-5).

Character of Budget Process

Complex Process Budgetary processes and the subjects they deal with are complex. At any one time, the process may be addressing a host of technical subjects ranging from the national defense implications of a newly proposed strategic weapon to the consequences of a mammoth tree-planting program suggested as a means of reducing atmospheric CO_2 in a worldwide context, or from the consequences of subsidizing the production of wheat in North America to the uncertainties surrounding proposed public investments in the production of timber to be harvested 150 years hence. Even if such investment opportunities are technically understood, policymakers must still confront the problems

TABLE 15-5
Development and execution of a federal budget, by stage, timing, and government participant

	Executive preparation and submission	Legislative approval	Executive execution and control	Review and audit
Stages	Requests, central review, and submittal	Authorization, appropriation, budget resolution	Obligation and outlay	Audit
Main participants	Agencies, departments, OMB, President	Congress (budget, authorizing, and appropriations committees)	Agencies and Office of Management and Budget	General Accounting Office
Timing	12–24 months (agencies), 9 months (President) before start of fiscal year	January to September 30 (9 months before and up to start of fiscal year)	October 1 to September 30 (12 months of fiscal year)	October 1 (up to 12 months after start of fiscal year)

Source: Adapted from *Budgetary Politics,* by Lance T. LeLoup. Reprinted by permission of Kings Court Publishers, Brunswick, Ohio, 1988.

involved in comparing the products of investments in programs that have different values to different people. Investments in forestry must always compete with investments in defense, school, and transportation, for example.

Faced with technically complex, value-laden budgetary matters, policymakers often seek refuge in certain traits of the budgetary process (Wildavsky 1988). For example, since forestry budgets are fragmented, and since they are handled by multiple layers of specialization in agencies and in legislatures, they never really need to be dealt with in their entirety. Since budgets are usually rooted in previous years' experiences, they can be comfortably addressed in an incremental fashion; the largest determinant of next year's budget may be this year's budget. Since budgets are dealt with at different times and in different places, budgetary decisions need not be made all at once. Further, since policymakers seek simplified rules to guide their decision making, judgments about forestry budgets can often be made on the basis of an administrator's competence and dependability rather than on the merits of the forestry program in question.

Lengthy Time Schedule Public forestry organizations are constantly involved in budgetary matters, and their activities can be multifaceted. In February 1989, for example, the USDA-Forest Service:

- Prepared reports and evaluations documenting the consequences of investments made with funds from the fiscal year 1988 budget
- Carried out programs made possible by funds available from the fiscal year 1989 budget
- Participated in congressional hearings for the fiscal year 1990 budget
- Began formulation of budget estimates for the fiscal year 1991 budget
- Undertook development of budget instructions and program direction for the fiscal year 1992 budget

The period during which a specific year's budget for a federal forestry agency is being developed can often span 5 years. Budgetary planning for programs to be implemented in fiscal year 1989 began in fall 1985. As an example, the USDA-Forest Service process involves:

- *Issuance of budget instructions* Instructions which guide agency units in the preparation of funding proposals. Suggested may be identification of projects to be funded at current budget levels, at levels 20 percent more (or less) than current budget levels, or at budget levels specified in the RPA Program or in individual National Forest plans.
- *Development of alternative budget proposals* Alternatives prepared by agency units are submitted to the agency's Washington office. The agency's national budget is developed by an agency task force on budgets.
- *Presentation of the budget to the Department of Agriculture* Explanation and justification of proposed budget. Negotiations occur and changes are made as appropriate.

• *Presentation of the budget to the Office of Management and Budget (OMB)* Explanation and justification (by the department) of the agency's proposed budget. The OMB agrees to, modifies, or rejects the proposal. After appeals are considered and resolved, the budget becomes the President's budget.

• *Allocation of the budget to agency units* Presidential targets, program directions, employee allocations, and fund distributions are made known to units within the agency. The budget is not yet law; budget authority to spend does not yet exist.

• *Presentation of the budget to Congress* The presidential budget is explained and justified to Congress. Hearings are held, information requests are met, and the agency's proposed budget is enacted into law.

• *Agency allocations updated and obligation authority granted to agency units.*

• *Reviews and reports* Midyear and end-of-year financial reports are prepared, and an annual agency report describing program accomplishments is prepared.

Proposed forestry budgets must eventually be presented to a legislative body for review and approval. The progress of legislative consideration of proposed budgets can be characterized by the following sequence of events, occurring over a period of 9 months, which is the procedure of the U.S. Congress (Collender 1990):

Submission: President submits budget (first Monday after January 3).

Review: Standing committees review proposed budget and submit their views and estimates of expenditures to the appropriate House and Senate committees on budget (February 15). Congressional Budget Office submits major economic report (revenue and budgetary information) to budget committees.

Resolution House and Senate committees on budget review reports of standing committees, assess expected revenues to be generated by existing tax policy, and seek advice and counsel from the Congressional Budget Office. Spending targets are established and reported to committee's appropriate house (April 1). Concurrent resolution on budget is adopted by House and Senate (April 15).

Reconciliation If congressionally approved budget resolution directs funding reductions for certain programs (e.g., reconciliation instructions), authorizing and appropriating committees must meet to establish law appropriate to such reductions. Revised budget resolution is confirmed by House and Senate (June 15).

Appropriation Within limits established by the budget resolution, House and Senate committees on appropriation proceed to determine appropriations for specific programs. The House Committee on Appropriations recommends an appropriations bill (June 15), on which the entire House of Representative acts (June 30). (By tradition, the House acts first on spending matters.) The

Senate then proceeds to act on the bill; differences between House and Senate versions are addressed by conference committees. House and Senate enact law (October 1).

Fiscal year Beginning of fiscal year (October 1).

Although the process appears very logical and amenable to implementation in a timely fashion, technical and political realities often make logic and efficiency elusive. Because of the massive size and scope of budgeting activities, conflict is inevitable and delays in budget formulation are common. The result is often government by continuing joint resolution, wherein Congress provides an agency with temporary (3- to 4-month) emergency authority to spend money, usually at the previous year's level.

Arbitrary Base Budget Common throughout the budgetary process is reference to the *base budget,* a term used to describe a program's (or an agency's) current budgetary level. The base budget is typically the lowest level at which an agency expects a program to be funded in the coming year. Inclusion of a program in the base budget is usually a sign that the program will continue, will be funded, and will not be subject to overly intense scrutiny as it proceeds through the budgetary process. Absence of a base budget makes budgetary decision making much more difficult; there is nothing against which to compare proposed changes.

Authorization versus Appropriation Fundamental to legislative action on a budget is the distinction between authorization and appropriation. *Authorization* is legislative action that establishes legal authority for a program, or an agency to exist, to operate, and to spend money. For example, the Renewable Resources Extension Act of 1978 states "There is hereby authorized . . . to implement this Act $15,000,000."

Authorization for expenditure of funds does not guarantee that monies will be forthcoming. An *appropriation* is needed, namely a legislatively agreed-to financial commitment that allows agencies to make financial obligations (or that gives them budget authority to spend money) up to a specified amount. Appropriation is one type of budget authority; others are authority to *borrow* and authority to *contract* (i.e., to obligate funds in advance of appropriations).

Legislatures typically organize so as to separate decisions to authorize from decisions to appropriate. In Congress, for example, certain authorizing committees (such as the Senate Committee on Energy and Natural Resources) legitimize policy that authorizes the spending of money for forestry programs while other committees (such as the Senate Appropriations Committee) actually appropriate, or provide, the funds for such programs.

Budget Development Strategies

The amount of program funding requested by a public forestry agency is influenced by both technical and political factors. The request may be the product of ponderously complex planning processes that involve the recommendations of several administrative units at various layers in an organization. Funding requests may also be the result of heavy-handed exercise of political power. For example, a powerful legislator may simply demand that funding for a new or expanded forestry program be included in an agency's budget recommendations. Consider the following comments on how agencies decide their requests for funding (Wildavsky 1988, pp. 85–86):

> The simplest approach would be to add up all the costs of all worthwhile projects and submit the total. This simple addition rarely is done; . . . the reason is strategic. If an agency continually submits requests far above what it actually gets, the [Office of Management and Budget] and the appropriations committees lose confidence in it and automatically cut large chunks before looking at the budget in detail. It becomes much harder to justify even items with top priority because no one trusts an agency that repeatedly comes in too high. Yet it is unrealistic for an administrator not to make some allowance for the inevitable cuts that others will make.

Whether agencies are disposed toward inflating their budget requests is subject for conjecture. What is certain is that they do carefully evaluate the political feasibility of achieving various budget levels. For example, agencies are continually on the alert for budgetary signals about what might be emphasized or dememphasized) from other executive agencies (such as the OMB), from legislative officials (committee staff), and from important client groups. They are also alert to public announcements about how stringent a chief executive officer intends to be on proposals for new expenditures. The likes and dislikes of influential legislators may be charted, and records of prior budget hearings may be perused for indications of attitudes (pro or con) toward specific programs. Emerging issues, such as global warming, may be carefully monitored for their budgetary significance; global warming, for instance, may offer opportunities to work toward expansion of reforestation programs.

Many budgetary strategies are employed by an agency to maintain or increase funding levels (Wildavsky 1988). Especially important is the cultivation of a clientele that will support an agency's program. Feedback from such clients can be invaluable. The development of trust and confidence between agency administrators and important budgetary officials is essential. Budgetary officials attempt to develop a reputation for honesty, reliability, and the avoidance of extreme claims.

Skill in politically exploiting specific opportunities can also be important. Here are some examples of ways in which budget reductions are resisted:

- *Proposing to eliminate especially popular programs for budgetary reasons.* Legislators are likely to restore the needed budget level.

- *Protesting that even modest reductions in a program's budget will jeopardize the entire program* All or nothing is required.
- *Imposing legal but politically unpopular means of raising funds* Charging user fees is an example.
- *Arguing that a program has a special virtue, in that it pays for itself* The program's cost is less than the revenue it generates.

Attempts to increase budgets can be facilitated by repackaging old programs in such a manner that they appear to fit currently popular themes or interests. Another noteworthy strategy is to secure seed money for new programs, with the expectation that, once modest funds are committed, funding for the program will continue, and may even expand. Finally, even though many political strategies are available, budget managers should not overlook the obvious strategy of simply pointing out that a program is effective—that it succeeds, produces tangible results, and satisfies important constituents.

Budgetary Challenges

Budgetary processes legitimized in law and in rules and regulations pose innumerable challenges to the funding of public forestry programs. For example, the *length of annual budget cycles* (often 5 years) requires almost continual agency attention to budgeting and related fiscal activities. Budget plans must be drawn up far in advance, without knowledge of either the revenue implications that might be posed by unpredictable downturns in the economy or the program implications that might surface when newly elected or appointed political leaders take office. Long *budgetary cycles are also very costly:* "The budget in the Forest Service is recognized throughout the agency as the single most costly administrative job in the organization" (USDA Forest Service n.d.).

Budgeting processes may also challenge the *administrative flexibility* needed by managers and agreed to by legislatures. The necessity for legally mandating the funding of narrow program categories or specific projects, along with the rigid accounting practices which soon follow, limits a manager's ability to invest in worthwhile management opportunities that often arise outside narrowly confined budgetary categories. All too often, excess funds in one account cannot be transferred to another which is seriously lacking funds, even though the two accounts have similar purposes. The difficulty has been partially addressed by lump-sum approaches to government budgeting (USDA Forest Service n.d.). Related are budgetary accounting procedures that do not allow for an accurate accounting of all costs and all benefits incurred from investments in a particular program area. For example, investments in roads used to harvest timber may also provide recreation opportunities which may not be directly assigned as a benefit to the road investment.

Especially discomforting challenges that are often encountered during the development of program budgets are the *political games* engaged in by the many participants in the budgeting process. Such games do not build public confidence in government nor in its many activities. Examples are:

- Padding funding requests
- Threatening to cut popular programs first
- Repackaging old programs to fit new political themes
- Questionable seeking of program seed money
- Reinstatement by legislative allies of budget cuts made by an executive officer

Political games often reflect the difficulties of attempting to apply a managerial approach to budgeting in a political setting—an approach that incorporates efficiency, economy, and managerial effectiveness.

The *long time periods often required to produce certain forest products,* such as timber, and the *lack of a marketplace in which to value some forestry services,* such as much forest recreation, also pose special challenges to budgeting. Production of timber requires long-term sustained financial commitments from legislative systems (which are not always prone to make such commitments), while a marketplace in which the value of certain services can be assessed is necessary (though seldom available) for making comparisons among public investment opportunities.

One means of securing budgetary commitments over long forestry production cycles is to use trust funds (special accounts). These are established by law to provide money for specific purposes; they are not available for general government use. Monies for trust funds usually come from specially designated receipts such as timber sales, recreational user fees, and arms and ammunition taxes. USDA-Forest Service special accounts and trust funds, for example, amounted to over 30 percent of the agency's 1987 appropriation. Included were forest improvement work with Knutson-Vandenberg Act receipts from the sale of timber, as well as reforestation work with Recreational Boating Safety and Facilities Improvement Act (1980) tariffs on the import of solid wood products (Gorte and Corn 1989). Trust funds may accommodate the long-term financial requirements of unique forestry programs; they do, however, become legislatively resistant to reform as their purposes become outdated or as the programs they support become ineffective.

REFERENCES

Baltic, T J. J. G. Hof, and B. M. Kent. *Review of Critiques of USDA Forest Service Land Management Planning Process.* General Technical Report RM-170. Rocky Mountain Forest and Range Experiment Station, Fort Collins, Colo., 1989.

Behan, R. W. The RPA/NFMA: Solution to a Nonexistent Problem. *Journal of Forestry* 88 (5):20–25, 1990.

Binkley, C. S. Economic Analysis in the Resource Planning Act and Program. Report (draft) to the Office of Technology Assessment. U.S. Congress, Washington, 1989.

______, G. D. Brewer, and A. A. Sample. *Redirecting the RPA: Proceedings of the 1987 Arilie House Conference on the Resources Planning Act.* Bulletin 95. Yale School of Forestry and Environmental Studies, New Haven, Conn., 1988.

Bryson, John M. *Strategic Planning for Public and Nonprofit Organizations.* Jossey-Bass Publishers, San Francisco, Calif., 1988.

Collender, S. E. *The Guide to the Federal Budget: Fiscal 1991.* The Urban Institute Press, Washington, 1990.

General Accounting Office. *Land Management Planning: Forest Planning Costs at the Boise and Clearwater National Forests in Idaho.* GAO/RCED-87-28FS. U.S. Congress, Washington, 1986.

Gorte, Ross W., and M. Lynne Corn. *The Forest Service Budget: Trust Funds and Special Accounts.* 89-75 ENR. Congressional Research Service, The Library of Congress, Washington, 1989.

Gray, Gerald J. and Paul V. Ellefson. *Statewide Forest Resource Planning: The Effectiveness of First-Generation Plans.* Miscellaneous Publication 20-1987. University of Minnesota Agricultural Forest Experiment Station, St. Paul, 1987.

Irland Group. *RPA as Strategic Thinking: Background, Comparative Experiences, and Some Implications.* Report (draft) to the Office of Technology Assessment. U.S. Congress, Washington, 1989.

Jameson, D. A., M. A. D. Moore, and P. J. Case. *Principles of Land and Resource Management Planning.* Land Management Planning Office. USDA-Forest Service, Washington, 1982.

Larsen, Gary, A. Holden, D. Kapaldo, and others. *Synthesis of Critique of Land Management Planning.* FS-452. Policy Analysis Staff, USDA-Forest Service, Washington, 1990.

Office of Technology Assessment. *Forest Service Planning: Setting Strategic Direction under RPA.* OTA-F-441. U.S. Congress, Washington, 1990.

Sample, V. Alaric. Improving the Integration of RPA Planning with National Forest Planning and the Budget and Appropriations Process. Report (draft) to the Office of Technology Assessment. U.S. Congress, Washington, 1989.

Shands, William E. (ed.). *A Citizen's Guide to the Forest and Rangeland Renewable Resources Planning Act.* FS-365. USDA-Forest Service, Washington, 1981a.

______. Regional and National Forest Planning, in *A Citizen's Guide to the Forest and Rangeland Renewable Resources Planning Act,* by W. E. Shands (ed.). FS-365. USDA-Forest Service, Washington, 1981b.

______, V. A. Sample, and D. Le Master. *National Forest Planning: Fulfilling the Promise.* The Conservation Foundation, Washington, 1990.

Shannon, M. A. Renewable Resources Planning Technologies for Public Lands. Report (draft) to the Office of Technology Assessment. U.S. Congress, Washington, 1990.

U.S. Congress. *Department of Interior and Related Agencies Appropriation Bill (1987) Report.* Senate Committee on Appropriations, Government Printing Office, Washington, 1986.

USDA-Forest Service. The Lump-Sum Approach to Government Budgeting, in *New Thinking for Managing in Government.* U.S. Department of Agriculture, Washington [n.d.].

______. *National Forest System Land and Resource Management Planning (Final Rule)*. 36 Code of Federal Regulations Part 219. U.S. Department of Agriculture, Washington, 1979.

______. *Forest Plan: Clearwater National Forest. Environmental Impact Statement Forest Plan: Clearwater National Forest. Record of Decision Forest Plan: Clearwater National Forest*. USDA-Forest Service, Orofine, Idaho, 1987.

______. *A Guide to the Forest Service Appeal Regulation (36 Code of Federal Regulations Part 211.18)*. FS-388. U.S. Department of Agriculture, Washington, 1988.

______. *RPA Assessment of the Forest and Rangeland Situation in the United States: 1989*. Forest Resource Report No. 26. U.S. Department of Agriculture, Washington, 1989a.

______. *Report of the Forest Service: Fiscal Year 1988*. U.S. Department of Agriculture, Washington, 1989b.

______. *The Forest Service Program for Forest and Rangeland Resources: A Long-Term Strategic Plan*. U.S. Department of Agriculture, Washington, 1990.

Wildavsky, Aaron. *The New Politics of the Budgetary Process*. Scott, Foresman and Company, Boston, Mass., 1988.

Wilkinson, C F., and H. M. Anderson. *Land and Resource Planning in the National Forests. Oregon Law Review* 64(1, 2):1–373, 1985.

Wondolleck, J. M. *Public Lands Conflict and Resolution: Managing National Forest Disputes*. Plenum Press, New York, 1988.

16

PRIVATE FORESTRY PROGRAM INITIATIVES

The United States has sustained a significant portion of its forested landscape in private ownership. The policies and programs of private owners of forests are, in essence, an expression of the manner in which society wishes such forests to be used and managed for the public good. Of the nation's 483 million acres of unreserved timberland, nearly three-quarters (72 percent) is owned by private individuals or organizations. Private ownership can take many forms, including forested campgrounds owned by corporate organizations and forested habitats owned and managed for pleasure by nonprofit sport hunting groups. Two commonly identified major categories of private ownership are industrial private forests (forest industry) and nonindustrial private forests. Each ownership category plays a major role in implementing a variety of forest policies which ultimately result in benefits of interest to society in general.

NONINDUSTRIAL PRIVATE FORESTS

Extent of Ownership

Society has secured a variety of forest benefits via an assortment of policies and programs that are implemented by nonindustrial private owners of forestland—a category that excludes only industrial forestry holdings and private owners who possess a means of processing the timber produced by their forests. Owners of nonindustrial private forests are a heterogeneous lot. They include commercial farmers, part-time farmers, rural nonfarm residents, retired persons, homemakers, wage earners, professionals, business persons, and a host of other owners with a variety of forest interests—or with a notable disinterest in their forests. By some estimates, 7.7 million individuals or corporate entities each own 500 or fewer acres of nonindustrial private forest; 5.5 million are known to own 1- to 9-acre parcels (Birch, Lewis, and Kaiser 1982). As a group, nonindustrial private owners hold over 276 million acres of timberland or 57 percent of the nation's total. Of this total, 89 million is concentrated in the eastern portion of the United States, where it accounts for 70 percent of the region's total timberland (Table 16–1). Nonindustrial private holdings account for only 12 percent of the Rocky Mountain and Pacific Coast regions' timberland base.

Benefits Produced

Nonindustrial private forests are an important part of the United State's timber-producing economy. They contain over 345 billion cubic feet of the nation's growing stock inventory (48 percent of the national total), of which 61 percent is in the hardwood species group. They are the nation's storehouse of hardwood timber, accounting for 70 percent of the national total. In the late 1980s, growing stock situated in such forests increased at an annual net rate of 12.4 million cubic feet, or 56 percent of all net annual timber growth on all ownerships nationwide, accounting for 72 percent of nation's hardwood growth. On a per-acre basis in the same period, timber in industrial private forests annually grew at a rate of 45 cubic feet—a rate nearly equal to the average for all timberland ownerships. As for timberland productivity, the distribution of nonindustrial private timberland acreage by productivity class is nearly identical to the national average for all timberland ownerships (USDA-Forest Service 1989).

Timber harvest from nonindustrial private forests totaled 9.2 billion cubic feet in 1986, representing over 51 percent of the volume harvested from all timberland ownership categories in the United States. These forests were an especially important source of hardwood timber, yielding 4.8 billion cubic feet, or three-quarters of the total hardwood harvested nationwide. Total net annual growth in 1986 (12.4 million cubic feet of hardwood and softwood) exceeded harvest by 3.2 billion cubic feet; for each 1 billion cubic feet harvested, 3.9 billion cubic feet grew to take its place (USDA-Forest Service 1989).

Nonindustrial private forests are also a rich source of benefits other than timber—water, forage, scenic beauty, recreation, and fish and wildlife. For some owners, the nontimber benefits are the major reason for forest ownership; timber is of distinctly secondary concern. Such owners often take great pleasure in personal use of recreational opportunities of their forests, such as hiking, picnicking, skiing, snowmobiling, and living in recreation homes, while others find their pleasure in providing a host of recreational opportunities, such as hunting, for consumption by others. Nonindustrial private forests are also a source of

TABLE 16-1

Timberland area in the United States, by region and ownership, 1987 (in thousands of acres)

Region	All ownerships	Public								Private				
			Federal									Nonindustrial private		
		Total public	Total federal	National forest	Bureau of Land Management	Other	Indian	State	County and municipal	Total private	Forest industry	Total	Farmer	Other private
Northeast	80,102	9,770	2,936	2,212	0	724	122	5,812	900	70,332	12,590	57,742	12,926	44,816
North Central	74,584	21,217	7,990	7,253	44	693	842	7,503	4,882	53,367	4,361	49,006	23,255	25,751
Great Plains	3,529	1,225	993	943	0	50	83	138	11	2,304	21	2,283	1,586	697
Southeast	84,594	8,772	6,983	4,871	0	2,112	60	1,434	295	75,822	16,793	59,029	20,552	38,477
South Central	110,790	10,911	8,917	6,896	11	2,010	57	1,485	452	99,879	21,438	78,441	28,157	50,284
Rocky Mountain	57,611	42,905	37,709	34,819	2,768	122	2,652	2,428	116	14,706	2,943	11,763	6,422	5,341
Pacific Northwest	54,697	31,958	22,424	19,487	2,677	260	1,713	7,474	347	22,739	9,702	13,037	2,579	10,458
Pacific Southwest	17,412	9,595	9,051	8,742	300	9	99	431	14	7,817	2,757	5,060	1,523	3,537
Total	483,319	136,353	97,003	85,223	5,800	5,980	5,628	26,705	7,017	346,966	70,605	276,361	97,000	179,361

Note: Totals may not add due to rounding.

Source: Forest Statistics of the United States: 1987, by K. L. Waddell, D. D. Oswald, and D. S. Powell (eds.). Resource Bulletin PNW-RB-168. Pacific Northwest Forest and Range Experiment Station. USDA-Forest Service, Portland, Oreg., 1989.

forest benefits which are consumed by society in general. Forest owners have little control over the outputs, from both the production and the consumption standpoints. Examples include scenic beauty and the flow of water, both of which are often consumed and enjoyed by larger audiences (Clawson 1979).

Deterents to Management

Public and private organizations are frequently committed to policies and programs designed to encourage the owners of nonindustrial private forests to manage their forests more actively. The assumption is that the owners face obstacles that discourage such investments, especially with regard to timber production and timber-harvesting decisions. Among the often-cited deterrents to investment in timber management by owners of nonindustrial private forests are (Clawson 1979; Department of Agriculture 1978; Ellefson et al. 1990; Fedkiw 1983; USDA-Forest Service 1989):

- *Nontimber goals and objectives of landowner* Some landowners perceive timber as secondary to or in conflict with forest benefits such as recreation, wildlife, or scenic beauty.
- *Divergent intensity of interest in investing* Even when timber is not ruled out as an objective, some owners—*custodial investors*—simply hold timberland pending some further disposition of the land. Others are *sideline investors,* who invest as an afterthought in forests which are part of other assets, such as farms. *Speculative investors* acquire forestland in hopes of a windfall appreciation in property value in excess of increases due to timber growth. Still others, *hobby investors,* own and manage forestland for the personal satisfaction of being involved in land stewardship.
- *Operational problems of growing, harvesting, and marketing timber* These include lack of technical information and assistance on timber production and timber harvesting; vague expectations as to financial returns from timber production; perception of high risk from fire, insects, and disease; lack of readily available capital; excessively long payback periods for investments; the uncertainty of future market conditions (especially timber prices); legally imposed environmental constraints; limited availability of labor and equipment; adverse consequences of income, inheritance, and property taxes; and inability to capture appropriate economies of scale.

Policy and Program Options

Public interest in the use and management of nonindustrial private forests can be expressed in numerous ways. Conceivably, it could even include transfer of such forests to an ownership category that is capable of more effectively meeting society's forestry wishes. Some policy alternatives are described below, in terms of degree of government involvement in the activities of nonindustrial private owners of forests (Sedjo and Ostermeir 1978):

- *Benign neglect* Gradually reduce government involvement in nonindustrial private forestry to the point where few or no public programs are focused on nonindustrial private forests. Maintain only public programs (such as protection from fire, insects, and diseases) focused on the private sector generally. Tax policies would be neutral. *Arguments in favor:* Relative to other forestland owner categories, timber removals, productivity and growth rates of nonindustrial private forests are acceptable. Low rates of owner investment in timber production truly reflect the value society places on timber produced by nonindustrial private forests. Timber shortage as measured by price increases is not expected. Production of nontimber benefits by nonindustrial forests is satisfactory in type and amount.
- *Status quo* Continue existing government programs and expenditure levels directed toward nonindustrial private forestry. *Arguments in favor:* Problems with existing programs focused on nonindustrial private forests are not serious enough to warrant abandonment. The amount of money invested via programs focused on nonindustrial private forests is relatively small. Continuation of programs and investments levels is necessary for the protection of timber-management investments made in the past. Although a timber shortage in the future is doubtful, the risk could be covered by continuation of existing programs.
- *Reform* Modify and improve the design and implementation of existing programs, with the intention of producing improved program results. *Arguments in favor:* Existing programs are often plagued by a multiplicity of administering agencies and are often ill-defined, poorly coordinated, and ambiguous in purpose. The cost effectiveness of programs is questionable. Improvements in design and delivery are warranted.
- *Market improvement* Improve market performance and information availability. *Argument in favor:* Lack of market information fosters poorly operating markets which owners of nonindustrial private forests have difficulty in responding because of lack of price information, lack of product standardization, lack of futures market, and lack of market opportunities.
- *Increased government involvement* Substantially increase government programs focused on nonindustrial private forests. *Arguments in favor:* Nonindustrial private forests are a large untapped resource which society ought to manage. The benefits would be increased outputs and more favorable species. Even though a shortage of timber is not perceived as likely, government involvement is warranted as a precaution against future uncertainties, because markets cannot be relied upon to bring forth necessary investments.

Nonindustrial private owners of forestland are frequently cited as lacking an interest in the management of their forests; when expressed, their interests are unplanned or irregular. From a social perspective, commentators often argue that such attitudes are not consistent with development of the potentials of nonindustrial private forests nor with society's interest in meeting expected demands for the goods and services that could flow from such forests.

Other commentators, however, are equally convinced that the condition of nonindustrial private forests is falsely perceived as poor by many public policymakers and administrators. One of their arguments is that nonindustrial private forests are being managed at levels consistent with the interests of society and of individual landowners (Clawson 1979). Another argument is that the problem is not poor management but rather that the private nonindustrial owner's management objectives are simply not the same as the objectives of advocates of a need for greater landowner investments in their forests (Lyons 1983; Zivnuska 1974).

These differences have inspired debates over nonindustrial private forests for decades, and the debates will likely continue in the future. Even so, it is undeniable that nonindustrial private forests provide a significant portion of the nation's supply of important forest benefits. The policies and programs implemented by their owners are an aspect of broader societywide interests in this extremely important landownership category.

INDUSTRIAL PRIVATE FORESTS

Extent of Ownership

Society has also secured a variety of forest benefits (especially timber) through an assortment of policies and programs implemented by industrial timberland owners. (*Timberland* is defined as forestland capable of producing at least 20 cubic feet of wood per acre per year and not reserved for other purposes). In 1987, timberland owned by operators of primary wood products manufacturing facilities added up to nearly 71 million acres, or 15 percent of the nation's total. Industrial timberland ownership has been increasing slightly since the early 1950s; in 1952, 60 million acres of timberland was owned by corporations; in 1962, the figure was 62 million acres, in 1970 it was 66 million acres, and in 1977 it was 69 million acres. The acreage is expected to stabilize at 71 million acres over the next 50 years. The South, with 38 million acres, contains the largest amount of industrially owned timberland—nearly 6 million acres more than industry owns in all other regions combined (Table 16-1).

The nation's largest 40 industrial timberland-owning companies in 1979 held 58 million of the nation's 69 million acres (estimated 1979 area) of industrially owned timberland—an average of 1.45 million aces each. The 90 sales-leading companies in the same year accounted for 91 percent of the nation's industrial timberland. Seven companies owned more than 2 million acres each, while 14 owned between 1 and 2 million acres each. International Paper Company was the largest corporate timberland owner, with more than 7.1 million acres, mostly in the South. The company's timberland was slightly larger than the forty-second largest state, Maryland. The other six were Weyerhaeuser (with 5.9 million acres), Georgia-Pacific (4.1 million), St. Regis (3.2 million), Champion International (3.0 million), Great Northern Nekoosa (2.7 million), and Boise Cascade (2.6 million) (Ellefson and Stone 1984). Company patterns of timberland ownership are not stable, however, and may be difficult to determine. Individual companies are continually purchasing and selling timberland. The merger of Champion International and St. Regis in the late 1980s, for example, moved the Champion International Corporation to the position of second largest owner of industrial timberland nationwide.

Timber Production

Timberland owned by forest industry tends on average to be more productive than the aggregate of other timberland ownership categories. Productivity is measured by the amount of wood that can annually be produced in fully stocked natural stands of timber and is generally arranged into four productivity classes (as measured by cubic feet per acre per year), namely:

Productivity class	Industrial timberland	All timberland
120+	20 percent	11 percent
85–120	25	23
50–85	39	39
20–50	16	27
	100	100

One-fifth of industrially owned timberland is in the highest productivity category; the proportion of the lowest-productivity timberland owned by industry is considerably less than the average for all forest ownerships. In addition to timberland, forest industry owns approximately 700,000 acres of forestland which is not capable of producing at least 20 cubic feet of wood fiber per acre per year (USDA-Forest Service 1989).

Timberland owned by industrial forestry concerns contains nearly 102 billion cubic feet (14 percent) of the nation's nearly 714 billion cubic feet of growing stock. Over 67 percent of this industrial volume is softwood timber; the remainder is hardwood timber volumes. Regionally, the 107 billion cubic of industrial growing stock is distributed as follows (Forest Service 1989):

North, 12 percent

South, 24 percent

Rocky Mountain, 39 percent

Pacific Coast, 25 percent

In 1979, the volume of timber owned by the 40 largest company timberland holdings totaled 80 billion cubic feet, or 75 percent of industrywide growing stock volume. The four leading companies in that year were (Ellefson and Stone 1984):

Weyerhaeuser—11,500 million cubic feet

International Paper—9,000 million cubic feet

Georgia-Pacific—6,000 million cubic feet

Crown Zellerbach—3,900 million cubic feet

Of the 40 largest companies, 15 relied on their own land for half or more of their wood supplies, but only three were close to being self-sufficient.

Industrially owned timber grows each year by approximately 4.2 billion cubic feet, 27 percent of which is hardwood growing-stock growth. This is 19 percent of total net growth nationwide, or approximately 60 cubic feet per acre per year. The growth rate for all timberland ownerships is 46 cubic feet per acre, which means that industrially owned timberland has a growth rate far above that experienced by the nation's three other major landownership categories—National Forests, with 39 cubic feet per acre per year; other public (state and federal) forests, with 46 cubic feet per acre per year; and nonindustrial private forests, with 45 cubic feet per acre per year (USDA-Forest Service 1989).

Timber harvested from growing stock owned by the forest industry in 1986 exceeded 5.4 billion feet of wood fiber. This volume equaled nearly one-third of the timber harvested from the growing stock controlled by all categories of forestland owners in the United States. It was 2.1 times as high as the National Forest harvest rate of 2.6 billion cubic feet, and it was equal to 59 percent of the nonindustrial private forest harvest rate of 9.2 billion cubic feet.

The vast majority—79 percent—of timber harvested from industrially owned timberland came from softwood growing stocks. The remainder was from hardwood timberlands. In 1986, overall harvests from industrially owned growing stock exceeded growth by 28 percent, and harvests of softwood species exceeded growth by 40 percent. Hardwood growth, in contrast, exceeded removals by 30 percent. These figures can be compared to the 4.3 billion cubic feet by which growth exceeded harvests for all categories of forestland ownership categories; growth was 24 percent more than harvested. Timber growth-harvest balances are unique to a specific year; they do not necessarily reflect harvest response to long-term market conditions, nor do they accurately portray long-term efforts to develop the sort of fully regulated forest that might be required for a sustained, even flow of timber (USDA-Forest Service 1989).

Timberland Management Policies

Company policies toward timberland ownership and management are as diverse as the corporate interests of the many companies in the industry. One executive in the mid-1970s stated, ''We are the stewards of the land under a public franchise,'' while another said, ''We own the land and we do with it what we want'' (Enk 1975). Company policies concerning ownership of timberland include the following (Ellefson and Stone 1984; O'Laughlin and Ellefson 1982):

- *Raw material* Assure a supply of raw material (timber) for highly capital-intense production facilities, both in a physical sense (to reduce the possibility of a physical shortage of timber) and in an economic sense (to dampen short-term price fluctuations).
- *Investment* Secure a respectable rate of return on investments, viewing timberland as a sound, low-risk opportunity for investing corporate finances. Considered to be a hedge against inflation.
- *Strategic importance* Maintain a sound corporate competitive position for raw materials in a specific region (i.e., exclude competitors from raw material supplies).
- *Integration* Provide a complete vertical integration of processing and marketing operations, of which an assured raw-material supply is an integral part.
- *Speculation* Achieve financial gains from appreciation in the value of timber and the land on which its grows.
- *Public relations* Assure creation among various publics of an image of the company as a land steward—an image of concern for the long-run health of land, forests, and the corporate entity.

Companies also own timberland for various other reasons, including the usefulness of timber as a source of revenue for paying off corporate debts and as collateral for securing loans needed to make improvements in a variety of company operations. Through the mid-1980s, timber harvested from company timberlands was afforded favorable capital gains tax rates, a condition which was often viewed as an incentive for timberland ownership. In some cases, however, timberland can be simply of incidental importance, having been acquired as a result of a merger designed to enhance a company's sales or manufacturing position.

Ownership Disadvantages

Timberland ownership by wood-based enterprises cannot always be viewed as a positive entry in the corporate ledger. Acquiring and holding large blocks of timberland has obvious disadvantages. As an asset, for example, timberland ties up large amounts of capital for long periods of time. Active timber management entails expenses for such expenses as administration, silvicultural practices, property taxes, and interest on borrowed capital. Positive returns from such investments are unlikely to result until after the timber has been sold or processed. Until then, depending on what expenses can be deducted from income tax payments, cash flows are likely to be negative. To compound the problem, the timber being grown is in danger of destruction by insects, disease, and fire. In addition, timber markets, like all commodity markets, may be depressed at the time a company decides to sell the timber it has produced, or the price of open-market timber may be significantly less than the cost of producing the company-grown timber. Under such circumstances, a company might have been better off purchasing timber on the open market instead of growing its own timber.

Professional Forestry Staff

The professional forestry staffing of wood-based companies can be a reflection of company policy toward timber production and land stewardship in general. In 1986, 70 companies owning 49.4 million acres of timberland (70 percent of the nation's industrially owned timberland) employed a total of 3,569 foresters. Excluding the 445 staff and administrative foresters located at central or regional headquarters, 71 percent of the foresters were located in the South, 12 percent in the North, and 17 percent in the West. The responsibilities of the 3,569 foresters were distributed as follows: 12.5 percent were administrators, 47.9 percent were managers of company land, 34.3 percent procured timber from noncompany land, and 5.3 percent provided services to nonindustrial private owners of forest. Of the foresters charged with company land management, each was on average responsible for 28,800 acres of forest. In general, the larger a company's timberland ownership, the fewer acres managed per forester; however, among companies that owned more than 1.8 million acres, per company, the area managed per forester increased (Ellefson and Irving 1989).

Public Interest in Industrial Forests

That nearly 71 million acres of forest is owned and managed by private industrial concerns is a reflection of society's interest in fostering an explicit commitment to policies and programs that result in the production of timber and related raw materials. Since timber is of major importance to the economic and social fabric of society, industrial ownership of timberland (and the privately implemented policies implied therein) can be viewed by society as the most direct means by which it can promote a continuing flow of this important raw material. Although society views other landownership categories as important sources of timber, industrially owned timberland is the only forest ownership category that is devoted primarily to the production of wood. Its importance as a continuing source of wood raw material cannot be overemphasized.

REFERENCES

Birch, T. W., D. Lewis, and H. F. Kaiser. *The Private Forest-Land Owners of the United States*. Resource Bulletin W0-1. USDA-Forest Service, Washington, 1982.

Clawson, M. C. *The Economics of U.S. Nonindustrial Private Forests.* Resources for the Future, Washington, 1979.

Department of Agriculture. *The Federal Role in the Conservation and Management of Private Nonindustrial Forest Lands.* USDA-Interagency Committee—Soil Conservation Service, Forest Service, and Extension Service, Washington, 1978.

Ellefson, Paul V., M. D. Bellinger, and B. J. Lewis. *Nonindustrial Private Forestry: An Agenda for Economic and Policy Research.* Station Bulletin 592–1990. Agricultural Experiment Station, University of Minnesota, St. Paul, 1990.

——— and Frank D. Irving. Industrial Forester Staffing of Leading Wood-Based Companies: An Examination of Forester Responsibilities. *Journal of Forestry* 87(3):42–44, 1989.

——— and Robert N. Stone. *U.S. Wood-Based Industry: Industrial Organization and Performance.* Praeger Publishers, New York, 1984.

Fedkiw, J. *Background Paper on Nonindustrial Private Forest Lands: Their Management and Related Public and Private Assistance.* USDA-Forest Service, Washington, 1983.

Enk, Gordon A. *A Description and Analysis of Strategic and Land Use Decision Making by Large Corporations in the Forest Products Industry.* Ph.D. Dissertation. Yale University, New Haven, Conn., 1975.

Lyons, James R. A National Perspective on [Nonindustrial Private Forestry] Issues, in *Nonindustrial Private Forests: A Review of Economic and Policy Studies,* by J. P. Royer and C. D. Risbrudt (eds.). School of Forestry and Environmental Studies, Duke University, Durham, N.C., 1983, pp. 23–29.

O'Laughlin, Jay, and Paul V. Ellefson. Strategies for Corporate Timberland Ownership and Management. *Journal of Forestry* 80(12):784–788, 1982.

Sedjo, Roger A., and David M. Ostermeier. *Policy Alternatives for Nonindustrial Private Forests.* Society of American Foresters, Washington, 1978.

USDA-Forest Service. *An Analysis of the Timber Situation in the United States: 1989–2040.* Parts I and II. A Technical Document Supporting the 1989 RPA Assessment. U.S. Department of Agriculture, Washington, 1989.

Zivnuska, John A. Forestry Investments for Multiple Uses among Multiple Ownership Types, in *Forest Policy for the Future: Conflict, Compromise, Consensus,* by M. C. Clawson (ed.). Resources for the Future, Washington, 1974, pp. 222–279.

17

INFORMATION AND SERVICE PROGRAM INITIATIVES

The self-motivated behavior of forest landowners (producers) and the consumers of goods and services produced by such landowners may lead to patterns of production and consumption that are efficient and are consistent with the expectations of society in general. Given such circumstances, changes in the patterns of production and consumption as expressed through markets would be unlikely to make anyone better off without making someone else worse off. Market systems are characterized by numerous producers and buyers, homogeneous goods and services, easy access to or exit from production activities, and unlimited availability of information about market potentials and market transactions.

Under other circumstances, however, the operation of market systems may be neither efficient nor effective. Forestry benefits and the manner in which they are produced may be inadequate from a broader social perspective. Such circumstances include the incidence of (Gregory 1987):

• *Public goods* Goods and services that are desired by the public but are not made available through markets because providing them to any one member of a group makes them simultaneously available to all other members of the group. There is no need for all members to express their demands through a market system; to a point, most will satisfy their desires without active involvement in a marketplace. As a result, reliance on a market system for the production of public goods will eventually reduce rather than increase the supply of such goods. Classic examples of public goods are national defense and certain environmental resources such as air.

• *Externalities* Goods or services that are produced and subsequently imposed on others without their permission or that are produced and subsequently consumed by others without their payment. Typical examples of negative externalities are the noise of jet airplanes and emissions from automobile exhausts. Positive externalities include the scenic landscapes often found in rural forested areas.

• *Monopolies and oligopolies* Goods or services produced by a single producer or relatively few producers, often resulting in discriminatory pricing and poor-quality products. Highly concentrated industries such as the automobile and steel-manufacturing industries are often alleged to be monopolistic or oligopolistic.

• *Distributional inequities* Goods and services produced for and consumed by preferred income groups, preferred geographic regions, or preferred generations, often at the expense of other groups, regions or generations. Examples are continued accumulation of wealth by those already wealthy and consumption of raw materials by a current generation at the expense of future generations.

• *Information inadequacies* Goods or services that are not produced because uncertainties exist about their production or potential for consumption, or goods and services that are overconsumed because of similar information-based uncertainties. An example is producer reluctance to invest in the production of products for markets that will not occur until many years in the future.

Market inefficiencies are very common in forestry. Nonindustrial private landowners may refrain from investing in timber-management practices that would result in financial returns at a far-distant period in the future (50 to 75 years hence). They may lack access to the financial resources required to carry out forestry practices that would foster habitats needed by rare and endangered species of flora or fauna. They may practice forms of forestry that lead to soil erosion, loss of site productivity, or production of a variety of environmental pollutants, including air pollutants from prescribed burns, water-polluting sediments from road construction, and destruction of wildlife as a result of use of certain chemicals. Similarly, industrial forestry interests focused on timber-producing activities may overlook opportunities to simultaneously provide a broader range of other benefits which are of interest to society in general, such as wildlife, quality water, and natural beauty. Likewise, private owners of forests in general may be unable to shoulder the consequences of especially risky investments. If, for example, wildfire or the widespread incidence of insects and diseases become deterrents to private investment in the management of forests, society (through government) may act to serve broader social interests by preventing or reducing the frequency of such events. Some form of public intervention may be warranted.

Broader societal involvement in the forestry activities of private owners of forests can take many forms. Among its least intrusive manifestations are the furnishing of technical assistance directly to landowners, the distribution of information (via extension) to groups of landowners, and the provision of technical and related support to landowners in an indirect fashion. In this chapter, we shall consider examples of these approaches.

DIRECT TECHNICAL ASSISTANCE

Trained professionals who provide advice and counsel on matters of technical forestry to individual owners of forestland form the nucleus of direct technical assistance programs.

Public Sponsorship

Probably the most widely known programmatic means of providing direct technical assistance are *publicly sponsored private forest-management assistance programs* (often referred to as *cooperative forestry assistance programs* or *rural forestry assistance programs*). Although such programs are capable of being independently sponsored by virtually any unit of government (federal, state, or local), the most visible means of sponsorship involves partnerships between federal and state forestry agencies.

Federal-State Programs The technical assistance program jointly sponsored by state forestry agencies and the USDA-Forest Service is highly visible. The USDA-Forest Service provides financial support and specialized technical support to state agencies, which in turn complement the federal funding and employ foresters, known variously as *service foresters, farm foresters, county foresters,* or *local foresters.* These foresters provide forestry assistance at the local, rural level to individual private landowners. As stated by the USDA-Forest Service, the objective of private forest-management assistance programs is to (1985, FSH 3210.2):

> Advance forest resource management by ensuring that both the individual landowner's objectives and the public need for goods and services for non-federal forest lands are met now and in the future.

Upon request by a nonindustrial private owner, service foresters examine the forest; prepare management plans and tree-planting recommendations; mark timber for improvement cuttings and firewood harvest; and provide advice on erosion and sediment control, wildlife habitat improvement, insect and disease management, forest recreation, and road construction. The specific services provided depend on the state in which the service forester is operating. Although the program is primarily focused on individual landowners, service foresters may also assume such responsibilities as certification of tree farms, administration of forest practice laws, certification of loggers and vendors, administration of cost-share programs, formation of landowner associations, and provision of technical advice needed to implement certain pollution-control laws (for example, the application of best forest-management practices).

The forest landowners served by technical assistance programs are generally concentrated among the smaller forest ownerships. In the nation's Central and Northeastern region, for example, 75 percent of all clients served in the early 1980s had holdings of less than 100 acres; assistance to owners of less than 10 acres has decreased significantly in recent years (Fedkiw 1983). Although fees are not charged for most of the technical advice provided by service foresters, a distinct trend toward fee schedules for certain practices has occurred. In 1980, for example, 12 states had fee schedules for one or more forestry assistance activities, including site preparation, prescribed burning, timber marking, and timber sale administration.

In the early 1980s, investment in direct technical assistance programs cooperatively administered and funded by the USDA-Forest Service and state governments exceeded $50 million, of which approximately 85 percent was state monies (Council of State Governments 1984; Fedkiw 1983). Such programs were the most common of the many private forestry assistance programs offered by state forestry agencies in 1985; they focused on forestry activities ranging from forest protection to reforestation and from wildlife management to protection of water quality (Table 17-1). As for the number of professionals involved in such programs nationwide, over 970 service foresters provided technical forestry assistance to individual landowners in 1983; an additional 734 aides, technicians, and related staff supported such professionals (Fedkiw 1983).

In 1989, service foresters made over 153,000 general assists to owners of forestland and furnished advice which led to the completion of nearly 71,000 forest-management plans (Table 17-2). General forestry assists averaged 152,000 per year during the period 1980–1989. Leading all other states in such assists in 1989 were Illinois (with 17,014 assists), Georgia (16,719 assists), and Mississippi (16,390 assists).

Including the 346 million cubic feet of timber harvested with the assistance of service foresters in 1989, service foresters also assisted in the marking for harvest of over 1,242 million board feet of timber. They also facilitated forest landowners' recreational objectives on nearly 171,000 acres of forestland. Over 527,000 acres of forestland was affected for wildlife purposes by the technical advice provided by service foresters (USDA-Forest Service 1990).

The range of advice given by such foresters in furtherance of landowner nontimber objectives is well characterized by the following examples from the case records of service foresters (Gansner and Herrick 1980).

- Helped choose trees to leave for development of new homesites
- Advised on the development of hiking and cross-country ski trails

- Helped to set up protection zones for rare and endangered species
- Gave instruction on appropriate use of cultural practices for Christmas trees
- Helped to lay out ponds and access roads for fire protection

Agencies other than state forestry organizations and the USDA-Forest Service also provide direct technical assistance to private owners of forest. In Vermont, for example, state wildlife managers of the Department of Fish and Game provide technical advice directly to landowners who are interested in promoting forest wildlife. The finances for the assistance program come from receipts obtained from the sale of antlerless deer hunting permits. Federally, the USDA-Soil Conservation Service, in cooperation with 13,000 Conservation Districts, offers landowners, especially those associated with agricultural operations, a variety of technical assistance. Included is assistance designed to prevent immediate problems with erosion and sediment control; a usual solution is to plant grasses or trees. In a longer perspective, assistance in preventing future erosion problems may be offered in the form of assessment of sensitive soil or drainage conditions and prescription of appropriate land-management practices. The agency will also assist landowners in the preparation of conservation plans, including plans for the installation and management of forested windbreaks in rural areas, and will provide advice on the most appropriate means of managing wildlife habitat and recreation amenities. Most technical requests for assistance with the management of timber are referred by the agency to state or other public or private service foresters.

Program Effectiveness The effectiveness of the private forest-management assistance provided by service foresters has generally been assessed as positive, although accurate measurement has often proved to be a major challenge to policy and program analysts. Assistance in the form of a forester-prepared management plan or a general consultation with a landowner is of little value unless the landowner responds by applying the recommended treatments. The most recent approach to determining effectiveness has been to compare the forestry consequences of assisted and unassisted forest landowners. In Georgia, for example, landowners assisted by a service forester generally had less pine removed, more softwood timber volume standing after harvest, and more pine seedlings after natural stand harvest. They also generally received higher than average prices when their timber was sold (Cubbage et al. 1985). Similar results have been found in Minnesota (Henly et al. 1988) and Illinois (Budelsky et al. 1989). In Minnesota, differences in most physical comparisons—residual volume, regeneration, soil compaction, and erosion—between assisted and unassisted sales were minimal for all forest types considered. There were, however, important differences in the sale prices received for aspen; assisted landowners received 40 percent more per cord of aspen than did unassisted landowners.

TABLE 17-1

State-administered forestry programs focused on major private forestry activities, by program type and region, 1985

Major forestry activity and program type	Frequency of states in region having program type									
	New England	Middle Atlantic	Lake	Central	South Atlantic	South	Pacific	Northern Rocky Mountain	Southern Rocky Mountain	Total
Water quality protection										
Tax incentives	0	0	1	0	0	0	0	0	0	1
Financial incentives	0	1	1	0	1	1	1	0	0	5
Educational programs	5	2	3	5	3	8	3	3	3	35
Technical assistance	6	5	3	6	3	6	2	4	5	40
Voluntary guidelines	3	4	1	3	3	9	2	3	2	30
Legal regulations	5	4	3	1	0	0	5	3	3	24
Reforestation and timber management										
Tax incentives	1	2	3	5	1	2	0	2	0	16
Financial incentives	1	3	3	4	3	4	2	1	1	22
Educational programs	5	4	3	6	3	8	3	3	2	37
Technical assistance	6	5	3	7	3	8	4	5	5	46
Voluntary guidelines	0	2	2	2	3	3	1	1	2	16
Legal regulations	1	3	1	1	0	0	4	1	3	14
Forest protection										
Tax incentives	0	1	0	0	0	0	0	0	0	1
Financial incentives	0	1	1	0	0	1	1	0	0	4
Educational programs	5	5	3	6	3	9	1	3	3	38
Technical assistance	6	5	3	7	3	9	4	4	5	46
Voluntary guidelines	1	1	1	2	3	3	1	3	2	17
Legal regulations	6	4	2	6	3	8	5	4	4	42

Wildlife and aesthetic management										
Tax incentives	0	1	1	1	0	0	0	0	0	3
Financial incentives	0	1	1	3	0	0	1	0	0	6
Educational programs	4	3	3	5	3	7	1	4	2	32
Technical assistance	5	5	3	6	3	7	4	4	4	41
Voluntary guidelines	1	1	1	2	2	3	1	1	1	13
Legal regulations	2	2	1	2	0	1	5	1	0	14

Note: Water quality protection focuses on nonpoint silvicultural sources of pollutants, vegetative buffer strips along water, and road and skid trail design and construction. Reforestation and timber management focuses on seed trees and other reforestation forms, timber-harvesting systems, and clear-cut size and design. Forest protection focuses on slash treatment, other wildfire-related treatments, prescribed burn smoke management, herbicide and pesticide application, and disease and insect management. Wildlife and aesthetic management focuses on wildlife habitat, scenic buffers along roadways, and coastal zone management requirements.

Source: State-Administered Forestry Programs: Current Status and Prospects for Expansion, by Russell K. Henly and Paul V. Ellefson. *Renewable Resources Journal* 5(4):19–23, 1987.

TABLE 17-2
Federal-state government cooperative private forest-management assistance activities, by region, 1989

Assistance activity	North	South	West	Total
Forest owners assisted (number)	72,352	64,091	17,412	153,855
Forest-management plans prepared (number)	17,246	51,856	1,871	70,973
Reforestation (acres)	109,585	1,002,662	50,476	1,162,723
Timber stand improvement (acres)	46,502	138,135	47,269	231,906
Outdoor recreation development (acres)	59,008	94,759	17,227	170,994
Wildlife habitat improvement (acres)	143,202	354,745	29,472	527,419
Forested range improvement (acres)	10,204	15,013	29,334	54,551
Timber sale harvest assistance (thousand cubic feet)	91,294	210,555	44,294	346,143
Urban forestry assistance (number of communities)	4,102	2,535	1,327	7,964
Referrals to consulting foresters (number of referrals)	8,377	6,644	1,527	16,548

Source: Report of the Forest Service: Fiscal Year 1989. U.S. Department of Agriculture, Washington, 1990.

Private Sponsorship

Although the major focus here is on publicly sponsored programs involving direct technical assistance to nonindustrial private owners of forests, the assistance offered by private forestry consultants and by industrial forestry firms should also be acknowledged. The nation's approximately 2,000 private forestry consultants provide services similar to those provided by public service foresters. In addition, they survey land, prepare tax returns, offer legal advice, and negotiate land and timber sales—services often considered inappropriate for public service foresters.

Among the landowners assistance programs sponsored by industrial forestry firms is the Tree Farm Program. Once the forestland of a nonindustrial private owner becomes a Certified Tree Farm, it is eligible for industrially sponsored technical assistance. Certified Tree Farms, which serve as examples of forests well-managed for timber-production purposes, number over 45,000 and encompass over 7 million acres of nonindustrial private forestland nationwide. There is an additional 27 million acres of industrially owned forest land within the system.

Industry also engages in the leasing of nonindustrial private forestland, an arrangement which means that a company often applies to a lessee's land the same forest-management practices that it applies to its own forestland. In 1982, nearly 4.7 million acres of nonindustrial private forestland was leased by industry in the South (Meyer and Klemperer 1984). Industrial concerns are also often willing to carry out management practices for nonindustrial private owners, if the owners are willing either to pay the costs of such practices or to provide the involved company with first refusal right (the right to meet or exceed any other firm's bid) when the mature timber is sold. Industry generally limits such arrangements to large-sized tracts of forestland located close to the company's mill. In the South, enrollment in formal industrially sponsored assistance programs of this nature exceeded 4.2 million acres in 1984 (Meyer and Klemperer 1985; Cleaves and O'Laughlin 1983).

Technical Assistance Challenges

Programs involving direct technical assistance to private owners of forestland face many economic and administrative challenges. Often of central concern is whether public intervention, via technical assistance programs, in the forestry activities of private landowners is warranted. If it is found that clearly identifiable, market-driven problems suggest a need for publicly supplied technical assistance, the next question is, what is the most suitable mix of public (service foresters, extension foresters) and private (forestry consultants, industrial assistance) suppliers of such assistance? Further, which bureaucracies are most appropriate for delivering technical assistance to landowners? At first glance, involvement of many agencies may seem appropriate; viewed more critically, however, a bureaucratic landscape composed of many agencies having intertwining responsibilities can be confusing to landowners as well as administrators. For example, multiple state agencies may be delivering services in the same field in which multiple federal agencies are providing funding and specialized assistance.

The targeting of publicly supplied technical assistance, both in an equity sense and in an efficiency sense, can be another major challenge in the provision of direct technical assistance to landowners. One concern is whether publicly supplied assistance should be available to all landowners regardless of income and wealth. Another is whether technical assistance should be targeted to just those landowners who can capture the economies of scale needed to assure highly efficient public investments. (This approach might entail a focus only on large land ownership.) Another challenging question is whether direct technical assistance should be focused on producing a limited range of forestry outputs, such as timber, or on fostering the multitude of benefits which forests are often capable of supplying, including quality water, scenic landscapes, recreational opportunities, and wildlife habitats. Even when direct technical assistance is effectively targeted to specific landowners categories that are willing to produce the range of forestry benefits of interest to society in general, owners of forestland may not seek out the assistance or may not respond to it. Of concern is the failure of many private owners of forests to use the assistance available, especially at crucial decision times such as immediately following timber harvest. Still another challenge to the establishment and administration of programs involving direct technical assistance is their general efficiency and effectiveness. High standards are a necessity, from both the public and the private perspectives.

EXTENSION OF INFORMATION

Private owners of forests are frequently faced with management problems that can be effectively resolved by access to appropriate and timely forestry information. One of the most widely used programmatic means of placing such information in the hands of forest owners and users is extension forestry programs. These programs are frequently considered to be the primary means by which the nation's interest in the transfer of forest technology is achieved, especially the transfer of products resulting from forestry research.

Educational Activities

The breadth of educational activities undertaken through extension forestry is well characterized by the following directives, stated in the Renewable Resources Extension Act of 1978 (USDA-Extension Service 1986):

- Provide educational programs that enable individuals to recognize, analyze, and resolve problems dealing with renewable resources.
- Disseminate the results of research on renewable resources, and help to identify problem areas in need of additional research.
- Implement educational programs that give special attention to the forestry education needs of nonindustrial private owners of forests.
- Undertake educational programs concerning the breadth of renewable resources, including fish and wildlife, forested watersheds, outdoor recreation, rangelands, and timber.
- Help forest and rangeland owners to secure appropriate technical and financial assistance.
- Assist in providing continuing education programs for professionally trained managers of renewable resources.

Aimed most often at group audiences, these interests are accomplished by extension forestry in a number of ways, including the development and distribution of interpretative publications; the distribution of information via radio, television, and other media; the use of short courses, workshops, conferences, meetings, and field tours; and the development of demonstration projects designed to display appropriate means of addressing problems and opportunities encountered by owners of forestland.

Federal, State and Local Cooperation

Extension forestry in the United States is a cooperative venture involving the U.S. Department of Agriculture, state governments, and local communities (often counties). At the federal level, programmatic focus is on natural resource program areas (forest management, wildlife and fisheries, and soil and water) as identified by the USDA-Extension Service. Extension forestry is administered by land-grant universities at the state level, and extension forestry professionals are members of academic institutions. The relationship between federal and state extension units is not hierarchical but is more one of negotiation between concerned equals. State extension organizations typically have a director, program leaders, subject matter specialists, and county agents. The role of specialists is to provide specialized technical assistance (such as forest taxation, forest regeneration, and forest policy) to county agents and others, while county agents are charged with providing for the direct delivery of information to owners and users of forests. In some states, agents are assigned to more than one county; that is, an area concept has evolved.

Program Planning and Funding

County extension offices and state specialist groups annually prepare work plans for the coming year. Such plans are reviewed at state and federal levels and are often developed with the help of citizen advisory committees. An example of extension forestry's program response at the national level is the extension programs developed in 1985 to address the following issue areas concerning the use and management of nonindustrial private forests (Baughman et al. 1985):

- *Public awareness* Inform forest owners, policymakers, and the general public about the economic and environmental role of private forestlands in contributing to local, state, and national needs.
- *Policy education* Inform forest owners about how they might more effectively participate in the development of local, state, and federal forest policies.
- *Program coordination* Improve coordination among organizations and professionals that provide educational assistance to forest owners and users.
- *Professional education* Provide continuing education opportunities for extension professionals, to enable them to better assist forest owners with the latest available technology.
- *Forestland management* Inform forest owners about the importance of professional forestry services, and develop and supply information that will enable them to more effectively achieve their forestry objectives.

Funding of natural resource extension programs in 1989 totaled \$9.5 million (including approximately \$2 million from the Renewable Resources Extension Act of 1978). In 1985 these programs had 573 staff members nationwide, accounting for 3.6 percent of cooperative extension personnel. The specialities of these staff members were: forest management, 258 staff members; range management, 79; fish and wildlife management, 131; outdoor recreation, 39; and environmental quality and public policy, 66. Forty-three states were sponsors of natural resource education programs in that year; the number of personnel per state ranged from 1 to 20. Program funding sources for all cooperative extension programs is approximately 20 percent federal, 40 percent state, and 20 percent county (Baumgartner and Deneke 1986; USDA-Extension Service 1986).

Extension Challenges

Challenges faced by extension forestry programs are similar to those encountered by policies and programs designed to provide direct technical assistance. They include questions of public versus private roles in the supplying of information, the targeting of information to appropriate groups of landowners, the meshing of extension programs with other types of programs (such as cost sharing and technical assistance), and concerns about program efficiency and effectiveness in general.

INDIRECT TECHNICAL ASSISTANCE

Private owners of forestland, both industrial and nonindustrial, frequently face forestry problems or discover management opportunities for which publicly supplied direct technical assistance is not sufficient. For example:

- *Landowners may defer forestry investments because they fear the adverse consequences of wildfire or the widespread damage that can be caused by insects and diseases* A government-sponsored protection program may be useful.
- *Owners of forestland may elect not to reforest their harvested timberland because investment in the production of the necessary tree seedlings via a forest nursery is simply beyond them economically* A government-operated nursery may be in order.
- *Forest owners may avoid investment in timber production because prospective timber markets are likely to be meager or nonexistent* Regional expansion of wood-products industries facilitated by government policies and programs could help.

Implied in such situations is the need for technical forestry assistance of a broader scope—assistance that is not focused directly on any single owner of forestland. Forest pest management, fire protection, and research, examples of indirect technical assistance, are discussed below.

Pest Management

State governments and the USDA-Forest Service cooperatively provide a variety of services designed to protect private forests from the ravages of destructive insects and diseases, as follows:

- Detecting pest outbreaks (often via surveys)
- Evaluating current damage and the likelihood of future outbreaks
- Providing technical assistance concerning suppression of damaging pest outbreaks, including counsel on the use of pesticides
- Dispensing advice on means of preventing insect and disease infestations

Functionally, federal funding for pest management was allocated as follows in the early 1980s: prevention, 2 percent; detection, 9 percent; evaluation, 24 percent; technical assistance, 26 percent; and suppression, 39 percent (Fedkiw 1983). In recent years, major suppression projects have been directed at gypsy moths, southern pine beetles, western spruce budworms, dwarf mistletoe, and mountain pine bark beetles. USDA-Forest Service funding of cooperative forest pest-management programs in 1988 was approximately $44 million. State pest-management investments focused on state and private lands in the early 1980s were 1.3 times the federal investment on the same lands; 28 percent of federal pest-management funding was focused on nonindustrial private forests (Fedkin 1983).

Fire Protection

Protection of privately owned forests from destructive wildfire is a form of indirect technical assistance in which complex program elements are involved. These include suppression, presuppression, fuels management, fire prevention, aviation management, personnel training, and rural fire prevention and control. Such activities are funded by state governments and by the USDA-Forest Service, although state governments have assumed the bulk of the funding responsibility for state and private lands in recent years. Federal funding was less than 10 percent of the $224 million total in the early 1980s (Fedkiw 1983). In 1987, over 1 billion acres of state and private land (forest and nonforest) was protected from fire by federal and state cooperative protection efforts. Federally appropriated funding of the USDA-Forest Service's State and Private fire protection program in 1988 was approximately $14 million (Forest Service 1989a).

Research

Owners and users of private forests frequently face problems which cannot be resolved on the basis of current information. Research is needed to develop or discover new scientific knowledge that can be used for economic, social, and environmental enhancement of forest management and utilization of the goods and services produced by forest. Among the subject areas addressed by forestry research are:

- *Fire and atmospheric conditions*
- *Forest insects and diseases*
- *Forest inventory and analysis*
- *Renewable resource economics*
- *Forest recreation*
- *Timber management*
- *Watershed management*
- *Wildlife, range, and fish habitat*
- *Timber harvesting and timber products*

In 1988, investments in forestry research programs by the USDA-Forest Service totaled $138 million (representing over 800 scientist-years)—approximately two-thirds of a nation's publicly supported forestry research (USDA-Forest Service 1989a). State agricultural experiment stations and forestry schools also conduct research of interest to private owners and users of forestland. In 1988, for example, the federal *Cooperative Forestry Assistance Program Act of 1962* (also known as the *McIntire-Stennis Act*) provided $17.5 million in funding for forestry research conducted by state experiment stations. Nonfederal research funds available to states for forestry research as part of this cooperative effort have been about five times the federal funding levels (USDA-Forest Service 1989b).

Private owners and users of forests also benefit from a variety of additional public programs which indirectly supply technical assistance concerning the use and management of forests (USDA-Forest Service 1983). For example, the USDA-Forest Service actively pursues nursery and tree improvement programs which, in cooperation with state governments, provide for the upgrading of seedlings and nursery facilities. The intent is to improve the genetic quality of forests on state and private lands and to assist states in the production and distribution of tree seeds and seedlings which can be used to establish forests, windbreaks, and shelterbelts and to facilitate erosion control and land reclamation. In addition, the agency contributes technical leadership for the forestry aspects of small watershed and flood-prevention programs (a major responsibility of the USDA-Soil Conservation Service), and assists the private sector in improving the efficiency of timber harvesting and the utilization and marketing of wood products.

The USDA-Forest Service also cooperates with the U.S. Environmental Protection Agency in the administration of long-term contracts designed to foster *best management practices.* Such practices are important to the control of non-point sources of water pollutants from private lands. States have often seen fit to make such practices part of their suggested voluntary guidelines for timber harvesting on private forestlands (Georgia Forestry Commission 1988; Minnesota Department of Natural Resources 1989). The guidelines themselves are still another example of the transfer of technical forestry information to private owners of forests.

REFERENCES

Baumgartner, D. M., and F. J. Deneke. Extension Forestry in the United States, in *Proceedings of the International Union of Forest Research Organization's 18th World Contress.* Division 4. Lubijuania, Yugoslavia, 1986, pp. 287–294.

Baughman, M. J., D. M. Baumgartner, et al. *Reaching the Potential of Private Forest in the United States through Education: The Extension Committee on Organization and Policies Response to the USDA National Form on Nonindustrial Forestlands.* Cooperative Extension Service. Michigan State University, East Lansing, 1985.

Budelsky, C. A., J. H. Burde, F. H. King, D. R. McCurdy, and P. L. Roth. *An Evaluation of State District Forester Timber Marketing Assistance on Non-Industrial Private Forest Lands in Illinois.* Department of Forestry, Southern Illinois University, Carbondale, 1989.

Council of State Governments. *The Book of the States: 1984–1985.* Lexington, Ky., 1984.

Cleaves, D. A., and J. O'Laughlin. Forest Industry's Management Assistance Programs for Nonindustrial Private Forests in Louisiana. *Southern Journal of Applied Forestry* 7(2):85–89, 1983.

Cubbage, F. W., T. M. Skinner, and C. D. Risbrudt. *An Economic Evaluation of Georgia Rural Forestry Assistance Program.* Research Bulletin 322. Agricultural Experiment Station, University of Georgia, Athens, 1985.

Fedkiw, John. *Background Paper on the Nonindustrial Private Forest Lands: Their Management and Related Public and Private Assistance.* USDA-Forest Service, Washington, 1983.

Gansner, David A., and Owen W. Herrick. *Cooperative Forestry Assistance in the Northeast.* Research Paper NE-464. Northeastern Forest Expedriment Station, USDA-Forest Service, Broomall, Pa., 1980.

Georgia Forestry Commission. *Recommended Best Management Practices for Forestry in Georgia.* Macon, 1988.

Gregory, G. Robinson. *Resource Economics for Foresters.* John Wiley and Sons, Inc., New York, 1987.

Henly, R. K., P. V. Ellefson, and M. J. Baughman. *Minnesota's Private Forestry Assistance Program: An Economic Evaluation.* Misc. Publication 58-1988. Minnesota Agricultural Experiment Station, University of Minnesota, St. Paul, 1988.

Meyer, R. D., and D. W. Klemperer. Current Status of Long-Term Leasing and Cutting Contracts in the South, in *Proceedings, 1984 Southern Forest Economics Workshop.* School of Forest Resources, North Carolina State University, Raleigh, pp. 125–130.

Minnesota Department of Natural Resources. *Water Quality in Forest Management: Best Management Practices in Minnesota.* St. Paul, 1989.

USDA-Extension Service. *Renewable Resources Extension Program: Five Year Plan, 1986 through 1990.* U.S. Department of Agriculture, Washington, 1986.

USDA-Forest Service. *State and Private Forestry Learning System: Quick Reference Guide*. U.S. Department of Agriculture, Washington, 1983.

______. *Forest Service Manual*. U.S. Department of Agriculture, Washington, 1985.

______. *Report of the Forest Service: 1988*. U.S. Department of Agriculture, Washington, 1989a.

______. *A Description of Forest Service Programs and Responsibilities*. Technical Document Supporting the 1989 RPA Assessment. U.S. Department of Agriculture, Washington, 1989b.

______. *Report of the Forest Service: 1989*. U.S. Department of Agriculture, Washington, 1990.

18

FISCAL AND TAX PROGRAM INITIATIVES

Even though private owners of forestland may be the focus of well-conceived public programs that supply important information and technical assistance, they may encounter conditions which make active management of their forests exceedingly difficult or even impossible to accomplish. Some form of publicly supplied financial assistance may be required. Although the assistance can take many forms, two primary types stand out: publicly sponsored financial payments made directly to landowners for forestry purposes and public payments made indirectly via reductions in the tax rates imposed on an owner's forest property or on the income from products produced by the property. Both approaches are predicated on the assumption that government efforts to support societywide interests in private forests are necessary for reasons such as the following (U.S. Department of Agriculture 1978).

- *Market systems are operating poorly* Pricing mechanisms may fail to encourage timber investments by forest owners.
- *Certain socially desirable benefits are not being produced* For example, clean water, nongame wildlife, or pleasing forest landscape.
- *Socially unacceptable distributions of income (or wealth) are occurring among forest owners* For example, regional inequities in forest owners' income.

The presumption is that society in general obtains benefits from the management of privately owned forests and that the production of such benefits should accordingly be supported by society in general—not only by the private landowners.

FISCAL INCENTIVES

Fiscal payments paid directly to private owners of forestland by government are designed to share all or a portion of the owners' costs of undertaking certain forestry practices. These payments are popularly identified by a variety of terms, including *subsidies, fiscal incentives,* and *cost-share payments.* There are many federal and state cost-share programs. In this section we shall consider a few examples.

Forestry Incentives Program

The Forestry Incentives Program (FIP) is a federally sponsored cost-share program authorized by the Agricultural and Consumer Protection Act of 1973 and subsequently made part of the Cooperative Forestry Assistance Act of 1978. As legitimized in the 1978 act, the purposes of the FIP are to:

> Encourage landowners to apply [forestry] practices that will provide for afforestation of suitable open lands, reforestation of cut-over or other nonstocked or understocked forest lands, [application of] timber stand improvement practices, and forest resources protection.

The USDA-Agricultural Stabilization and Conservation Service is the program's primary administrative agency; it handles eligibility, waiver procedures, and payments to participants. The USDA-Forest Service is the program's major source of technical forestry assistance (e.g., specification of forestry practices). State forestry agencies assist landowners in developing forest plans as required by the program. They also approve such plans and carry out inspections to ensure that installed forestry practices are in compliance with program guidelines and agreed-to plans.

The U.S. Department of Agriculture is responsible for distributing among states the funds which are appropriated for the program. As a part of the distribution process, the agency considers a state's timberland area, number of nonindustrial private forests, the need for reforestation, and the need for timber-stand improvement. Attention may also be given to the availability of vendors to carry out eligible forestry practices, the effective use of cost-share funding in the past, and the health and condition of forests in a particular location. Areas with an adverse timber growth-harvest ratio will be at an advantage.

Private individuals, groups, associations, and corporations whose stocks are not publicly traded are eligible for cost-share payments made available via the FIP. Cost-share payments can be made for up to 75 percent of the cost of tree-planting and timber-stand improvement practices. The exact rate varies by state, depending on maximums determined by state forestry officials and the USDA-Agricultural Stabilization and Conservation Service. To participate in the program, landowners must usually own less than 1,000 acres of forestland, but they may own up to 5,000 acres if significant public benefits will result from federal cost-share support. A minimum of 10 treatable forest acres is required for eligibility for cost-share payments; such land must be capable of growing at least 50 cubic feet of wood fiber per acre per year. The maximum annual cost-share per eligible forest owner is $10,000.

Appropriations for the FIP have ranged from $7 million to $18 million per year since 1974, the year the program was first implemented (Table 18-1). Program investments in 1990 totaled $10.6 million and resulted in reforestation of 151,000 acres of forestland and application of timber-stand improvement practices to over 33,000 forested acres. In 1988, over 90 percent of the former and 26 percent of the latter occurred in the South. In the early 1980s, the average tract size reforested nationwide was 41 acres, while the average area subject to timber-stand improvement was 31 acres. Over 70 percent of the forests treated in the same period were in the timber productivity class 85 to 120 cubic feet per acre per year (Risbrudt and Ellefson 1983).

The efficiency and effectiveness of the FIP has been evaluated a number of times (e.g., Mills 1976; Risbrudt and Ellefson 1983). An evaluation of the 1979 program, which had an appropriation of $14.5 million, determined that such

Table 18-1
Funding and accomplishments of the Forestry Incentives Program and the Agricultural Conservation Program, 1977–1990

	Forestry Incentive Program (FIP)			Agricultural Conservation Program (ACP)		
Year	Investment (million dollars)	Reforestation (thousand acres)	Timber stand improvement (thousand acres)	Investment (million dollars)	Reforestation (thousand acres)	Timber stand improvement (thousand acres)
1977	10.3	152.8	155.2	2.2	32.7	43.4
1978	12.0	168.8	154.1	2.3	32.1	43.3
1979	14.5	212.0	117.6	3.3	46.8	66.1
1980	16.8	219.0	123.0	4.1	48.7	61.7
1981	17.6	211.2	102.6	5.0	53.3	74.8
1982	12.1	155.2	74.3	4.8	55.2	54.3
1983	10.2	143.3	58.4	5.2	66.2	38.0
1984	8.9	144.7	36.4	3.9	54.6	24.7
1985	9.9	167.3	37.1	5.8	87.4	34.1
1986	11.2	190.0	35.8	6.4	98.1	26.6
1987	7.3	118.5	26.1	6.3	129.4	21.2
1988	10.4	152.1	28.7	9.4	145.1	37.0
1989	10.7	164.1	30.5	9.6	132.1	39.6
1990	10.6	150.7	33.2	10.2	124.9	43.1

Source: State and Private Forestry. USDA-Forest Service, Washington, 1991.

investments were yielding an average internal rate of return of 8.3 percent above inflation and a present net worth of nearly $70 million, at 7 percent interest (Risbrudt and Ellefson 1983). In the South, 61 percent of the 1979 investments were producing a 10 percent rate of return. As a result of 1979 program investments, 1.3 billion cubic feet of additional timber (compared to current management intensities) is expected over the first rotation. As for retention of installed practices over the years, an evaluation in 1981 of acres treated in 1974 found these installed practices (reforestation and timber-stand improvements) to be in existence on 94 percent of the acres treated. Retention rates were highest in the North and Rocky Mountain regions; no region had a retention rate below 92 percent (Risbrudt et al. 1983). In terms of income distribution, a significant proportion of 1981 program investments were paid to eligible landowners living in counties with below-average per capita incomes (Ellefson and Wheatcraft 1983).

Stewardship Incentives Program

The Stewardship Incentives Program is also a federally-sponsored cost-share program implemented in cooperation with state forestry agencies. Authorized by the Forest Stewardship Act of 1990, the program is designed to encourage nonindustrial private owners of forestland to more actively manage their land for a wide variety of purposes, including:

- Shelterbelts and windbreaks
- Sustained production of timber products
- Protection and restoration of forest wetlands
- Management of native forest vegetation to enhance water quality
- Trees grown for energy conservation purposes
- Recreation opportunities and fish and wildlife habitat

To be eligible for cost-share payments made available by the program, nonindustrial private owners of forestland must have an approved forest stewardship plan and must agree to implement practices specified in the plan for a period of not less than 10 years. Eligible landowners must generally own less than 1,000 acres of forestland. Cost-share payments to an eligible landowner cannot exceed 75 percent of the cost of preparing the plan and implementing the practices specified in the plan. Authorized for each of the fiscal years 1991 through 1995 is $100 million. The federal implementing agency, the USDA-Forest Service, is to distribute funds to each state according to:

- Area of nonindustrial private forest land
- Potential productivity of nonindustrial private forest land

- Number of nonindustrial private forestland owners that are eligible for cost sharing
- Need for reforestation
- Opportunities to enhance nontimber forest resources
- Anticipated demand for timber and nontimber resources

For 5 years, the Stewardship Incentives Program will provide cost-share assistance in addition to that provided by the Forestry Incentives Program. The latter expires on December 31, 1995.

Agricultural Conservation Program

The Agricultural Conservation Program is a federally sponsored cost-share program designed to encourage resource-conservation practices in farming communities. Authorized by the Soil Conservation and Domestic Allotment Act of 1936, the program is administered by the USDA-Agricultural Stabilization and Conservation Service through a system of state and county committees. Applicants file funding requests with their county Agricultural Stabilization Conservation Committee. State forestry agencies and the USDA-Forest Service provide technical forestry advice and assistance to program administrators and monitor the cost-share contracts that have been agreed to by the participating forest landowners. Although the program is focused on installment of agriculturally oriented soil conservation practices, it does promote tree planting, timber-stand improvement, and wildlife habitat management. Eligible landowners can receive up to 75 percent cost-share payments on approved practices. The maximum allowable annual payment to any one landowner is $3,500; there is no minimum or maximum acreage requirement. In 1990, $10.2 million of program funds was used to reforest nearly 125,000 acres and to perform timber-stand improvement practices on 43,000 acres of forestland (Table 18-1). Over the years, such practices have accounted for less than 3 percent of the program's total funding.

Agricultural Conservation Program investments in forestry practices have been periodically evaluated. In the late 1960s, for example, reforestation investments on medium to good sites located in Pennsylvania generated internal rates of return over 6 percent; returns were negative, however, for many timber-stand improvement practices on poorer forestlands (Manthy 1970). In Minnesota, program investments were generally found to have acceptable benefit-cost ratios for all forestry practices installed in 1972. However, 70 percent of the program participants in Minnesota were willing to plant the same amount of forestland with less program funding (Gregersen et al. 1979). It was found, 10 to 15 years later, that over 96 percent of the acreage planted with program funds in five eastern states in the 1960s had been retained for timber-production purposes (Kurtz et al. 1980).

Conservation Reserve Program

The Conservation Reserve Program is a federal cost-share program, the purposes of which are to reduce soil erosion on highly erodible cropland. Authorized by the Food Security Act of 1985, the program is focused on long-term cropland retirement. Participants (farm owners and operators) are paid an annual rent per acre plus half the cost of establishing permanent land cover (usually grass or trees), in exchange for retiring highly erodible land for 10 years. The maximum annual rental payment per participant is $50,000. The USDA-Agricultural Stabilization and Conservation Service is the primary administering agency; field implementation in each county is carried out by the state government and the county Agricutural Soil Conservation Committee.

The USDA-Soil Conservation Service is responsible for determining technical eligibiity requirements and for developing conservation plans, while the USDA-Forest Service provides technical assistance for tree-planting activities. Applicants must agree to implement a conservation plan which designates grasses, trees, or other acceptable vegetative cover to be planted on erodible cropland. Participants must also agree not to harvest, graze, or make other commercial use of the planted vegetation during the 10-year contract. The program is authorized for the years 1986 through 1995, and for a cumulative retirement acreage of not less than 40 million acres nor more than 45 million acres. Approximately 5 million (12.5 percent) of the specified acreage is to be planted in trees. This program could become one of the largest single tree-planting efforts in the nation.

The area planted with trees financed by the Conservation Reserve Program during the period 1986 through 1988 exceeded 1.4 million acres. In 1988, 499,000 acres was planted with trees; the total program cost in that year was $19.5 million. The 1988 plantings were concentrated in the South (91 percent), as follows: the Southeast, 59 percent; the Delta states, 24 percent; and Appalachia, 8 percent. The average cost-share per acre in those regions was $36. Over a 45-year period, an average Conservation Reserve Program acre planted to trees will probably produce 7,400 cubic feet of food fiber; a million such acres would produce 7.4 million cubic feet of wood over the same time span—equivalent to about half the current annual national consumption of softwood products. Other results of the 1988 program were 381,000 acres of land treated for permanent wildlife habitat, 1,300 acres planted to field windbreaks, and 5,500 acres treated as wildlife food plots (Osborn et al. 1989; U.S. Department of Agriculture 1989).

A variety of other federal programs provide financial support for the management of privately owned forests. For example, the Farmers Home Administration offers loans to farmers who are unable to obtain reasonable-priced credit elsewhere; to be eligible, landowners must earn a substantial portion of their income from family-size farms. Among the purposes for which the loans can be used are improvement of forest resources, harvesting and processing of timber, and conversion of portions of farms to income-producing recreational enterprises (often based on forest resources). Loans are also available for such purposes from the Federal Land Bank. Land Bank foresters appraise an applicant's timber and timberland and use the information to determine the amount and terms of a loan. Forest resources are often used as collateral for loans made by the bank.

State Programs

State governments have also established and implemented cost-share programs. For example, the Virginia Reforestation of Timberlands Program is administered by the Virginia Division of Forestry. This program authorizes 50 percent cost-share payments to private forest owners who are interested in furthering the production of pine. The maximum payment per acre is $60, and the maximum number of acres per year is 500. The program is funded by a tax on harvested pine timber and matching funds from state general revenue. Applicants must be willing to enter into a reforestation agreement with the state.

The Mississippi Forest Resource Development Act also provides cost-share payments (up to 50 percent) to landowners who want to establish or improve their forests for timber production and game management. Program funds are derived from a tax on harvested timber. Participation in a state-approved forest-management plan is a prerequisite. The North Carolina Forest Development Program and the South Carolina Forest Renewal Program are two more examples of state cost-share programs. Both are supported by taxes on primary or manufactured wood products and require participation in state-approved forest plans for eligibility. In both states, cost-share payments for reforestation are 40 percent and for timber-stand improvement are 50 percent. In Texas, the Texas Reforestation Foundation (privately funded by voluntary contributions from forest industry) pays nonindustrial private owners of forests on a matching basis for the completion of certain forest practices. The fund is administered by the Texas Forestry Association, and technical assistance is provided by the Texas Forest Service.

The California Forest Improvement Act cost-shares up to 90 percent of a private owner's cost of reforesting, improving timber stands, and enhancing wildlife habitat. The program is funded by receipts from the sale of timber from state-owned forests.

Other states that have cost-share programs focused on nonindustrial private owners of forests include Illinois and Minnesota (Cubbage and Haynes 1988; Meeks 1982). Minnesota also sponsors an erodible cropland retirement program—the Reinvestment in Minnesota Program (RIM)—which is similar to the federal Conservation Reserve Program. The extent of state interest in cost-share programs is highlighted by actions of the Council of State Governments, which has prepared for state use a model state forestry incentives act (Council of State Governments 1986).

Fiscal Incentive Challenges

Like public assistance programs in general, programs involving financial assistance are often troubled by administrative problems and concerns over program efficiency and effectiveness. Such problems include (Fedkiw 1983; Irland 1982):

- Conflicting intentions arising from multi-agency involvement in program administration
- Insufficient targeting of financial support on those landowners owning larger more productive forests
- Inadequate coordination with extension, technical assistance, and public regulation programs
- Meager program-wide support relative to the magnitude of private landowner acreage that economically deserves public cost-share investments
- Political bias against subsidy programs and associated problems in getting annual political support needed for their continuation
- Difficulties in convincing landowners of the need for a planned succession of treatments over the life of a forest stand

Some especially pointed criticisms that have been leveled at the FIP by the President's Office of Management and Budget are that (Office of Management and Budget 1981):

- Cost-shares are income transfers which register as windfalls to forest landowners at the expense of the general taxpayer.
- Cost-shares lead to unfair competition with unsubsidized private timber.
- Cost-shares stabilize timber prices at levels below those which would encourage market-induced private investments. (The question is whether timber prices should be allowed to increase.)
- There is only limited evidence showing that more work is accomplished via the program than would have been accomplished without federal cost-shares. The concern is that financially able landowners may delay or fail to perform needed forestry practices in hopes of obtaining federal cost-share payments, leading to substitution of public capital for private capital.

Many of these problems have been lessened in recent years as a result of recommendations originating from numerous evaluations of cost-share programs.

TAX INITIATIVES

Society can also influence the amount and type of benefits originating from privately owned forests by altering the timing, type, and amount of tax expected from a forest property or from the income produced by such property. Reducing forest taxes, for example, may put landowners in a better position to invest (by at least the amount of the reduction) in necessary forest-management practices.

From a social perspective, tax policies should be (Gregory 1987, pp. 159–161; Hickman 1982):

- *Neutral in effect* Tax policies should interfere as little as possible with the optimum allocation and use of resources. For example, a given tax policy should not encourage forest exploitation.
- *Equitable in application* Distribution of the tax load among citizens and producing entities should help in attainment of a desired pattern of income distribution. For example, tax policy should not encourage unequal tax treatment of equal parcels of forest property.
- *Efficient to collect and administer* Real costs of collecting a tax should be minimal, and procedures should not be inconvenient to taxpayers. For example, tax policy should not require forest landowners to maintain complex records over long periods of time.
- *Certain as to amount* Tax rates should be dependable over time. For example, tax policy should not be dramatically changed over the long periods of time often required to grow timber.

Within such a framework, tax policy accomplishes important social objectives, including furnishing government with revenue needed to provide public goods and services, redistributing income among disparate income classes, and encouraging private investment in favored enterprises and economic sectors. For the sake of this discussion, forest taxes can be categorized as income taxes and property taxes.

Income Taxes

Capital Gains Federal and state income taxes are many in type and complicated in application. What's more, they often vary from year to year (Hoover et al. 1989). Ordinary income from the sale of timber or timber products has traditionally been taxed at either personal or corporate rates. In contrast, capital gains income (gains from the sale or exchange of capital assets such as land, timber, buildings, and equipment) was treated in a special manner by federal tax law, until enactment of the 1986 Tax Reform Act. Such gains were taxed at rates below those applied to ordinary income—gain from wages, salary, or business profit. Timber was considered to be a capital asset. The rationale for this treatment was that timber is a long-term investment and should be subject to the same favorable tax rates afforded other long-term investment (such as stocks). Treating timber income as a capital gain, and taxing it accordingly, was expected to encourage investment in the long-term processes required to produce timber, while simultaneously discouraging the premature harvesting that might result from application of ordinary corporate income tax rates. The capital gains option for timber was seen as a means of increasing both cash flows and after-tax rates of return on forestry investments (Fedkiw 1983). Through 1987, opportunity to apply the favorable capital gains tax rate to timber income was available to corporate as well as nonindustrial private owners of forests. Most tax savings were accrued by forest industries (Sunley 1976).

The Tax Reform Act of 1986, which was implemented in 1988, repealed the tax rate differential between ordinary income and capital gains income, thus eliminating the favorable treatment of income from long-term capital gains. Income gains from the sale of timber were included in the effects of the act (Hoover et al. 1989; Siegel 1989). In addition to being affected by the frenzy of general tax reform, the demise of capital gains treatment of timber income was accelerated by the lack of empirical evaluations attesting to the ability of favored capital gains tax treatments to elicit investments in private timber-producing activities. The few evaluations that were accomplished presented mixed results (Chang 1983; Dennis 1985; Gorte and Taylor, 1985; Society of American Foresters 1986). What was lacking was, in the first place, empirical evidence demonstrating the responsiveness of the private sector to the federal tax subsidy, in the form of production of additional timber at specified prices. The federal tax subsidy was estimated to be as much as $360 million annually (Boyd and Daniels 1985). The second thing lacking was proof that capital gains tax rates were an efficient means of eliciting the desired response, as compared to other programmatic means of augmenting private timber supplies. Cost-shares and forced investment via legally mandated forestry activities are example of other programatic means. (General Accounting Office 1981; Klemperer 1989).

Investment Tax Credits The production of timber by private forest owners has also been facilitated by investment tax credits for reforestation. As authorized by the Recreational Boating and Facilities Improvement Act of 1980, private landowners can receive a 10 percent investment credit (not to exceed $1,000 per year) plus an amortized deduction for annual reforestation expenses up to $10,000 per year over an 8-year period. In 1982, the Tax Equity and Fiscal Responsibility Act altered rules for such credits and deductions generally. If landowners elect the full 10 percent credit, they can claim only 95 percent of previously allowed deductions over 8 years; alternately, they may claim 8 percent as an investment tax credit and receive full deductions over the 8-year amortiztion period (Dennis 1983). The intent of the investment tax credit is to encourage investments in the reforestation of privately owned forests.

Estate and Inheritance Taxes Tax policies are not always an incentive to the management of private forests. Estate and inheritance taxes are often considered deterrents to forestry investments. Federal tax law exempts estates valued up to $600,000 from estate taxes. Inheritance taxes, however, are imposed on the share of an estate left to an heir; in essence, an inheritance tax is equivalent to an income tax. Thus, an heir to a valuable forest property may be required to pay inheritance tax. To raise the money to pay the tax, the heir may have to liquidate all or a portion of the timber located on the property or may have to sell the entire property or small parcels of it. The result may be premature harvesting of timber or the breaking up of large tracks of forestland into smaller properties that are more difficult to manage (Tedder and Sutherland 1979). At the federal level, the problem has been partially resolved by rules which allow an heir to pay inheritance taxes in installments over a 10-year period.

Property Taxes

Property taxes (ad valorem taxes) are also applied to forests and the land they occupy, and are levied by state and local governments. Forests that are subject to a traditional property tax system are annually assessed and taxed on the basis of their fair-market value in highest and best use. The process involves determination of a property's value (assessment), review and appeal of assessed value, determination of tax rate, collection of taxes, and imposition of penalties for delinquency. Traditional property tax systems applied to forests have been strongly criticized over the years. The following criticisms are especially noteworthy (Hickman 1982, 1989):

- Property taxes are not convenient because they mandate annual collections even though most forest properties do not provide annual incomes.
- They are not equitable because they tend to take an excessively large share of tax revenue from properties such as forests that produce deferred income.
- They are not neutral because they encourage shorter rotations, lower stocking levels, and shift marginal forestlands into other uses.
- They are not predictable as to amount; uncertainty regarding future tax obligations discourages long-term investments in timber production.
- They are not efficient, because government often chooses to annually appraise the value of forests. Furthermore, forests are particularly difficult to assess because of the many variables that determine their value.
- They do not bear any relationship to a property's current income-producing potential. Thus they force transformation of rural lands (including forests) into more immediate income-producing uses, such as urban development.

Because of the real and perceived faults and deficiencies of property taxes, state and local governments have established special property tax policies to be applied in forestry situations. They can be summarized as follows (Hickman 1982, 1989; Clements et al. 1986):

- *Exemption laws* Forestland or timber may be excluded from the provision of property tax laws, either permanently or for a specified period of years.
- *Rebate laws* Landowners engaging in an approved forestry activity (such as tree planting) may subsequently apply for abatement (refund) of a portion of

the taxes levied on the value of their timber, timberland, or both. Rebates generally continue for a limited period of time, and may be given as direct cash payments or as reductions in taxes.

Deferred-payment laws Annual taxes on forestland and timber are determined as for other classes of property, but some portion of each year's tax is postponed until the time of timber harvest.

• *Modified-rate laws* Forestland and timber are assessed like other forms of property, but a different tax rate (lower than that used generally) is used to compute the tax.

• *Modified assessment laws* Forest properties are valued differently from other forms of taxable property. Forest assessments (valuations) are frozen or calculated using a reduced assessment ratio.

• *Productivity tax laws* A calculated productivity value which varies with the quality of the forestland (not with timber-stocking levels) is applied. The tax is figured on per-acre value, which varies with different levels of timberland productivity.

Yield tax laws Land and timber values are conceptually separated. Land values continue to be subject to some form of annual property tax, but timber values go untaxed until the time of harvest. When timber is harvested, it is generally taxed at a percentage of its estimated stumpage value.

Severance tax laws A tax is imposed on those exercising the privilege of cutting timber. Severance taxes are levied in addition to, not in place of, traditional ad valorem taxes. Severance taxes differ from yield taxes in that they are calculated as a fixed amount per unit of product (not as a percentage of stumpage value).

In the early 1980s, the distribution of states employing property tax laws such as the above was as follows (Hickman 1982):

- *Exemption and rebate laws* 11 states
- *Deferred-payment laws* no states
- *Modified-rate laws* five states
- *Modified assessment laws* 43 states
- *Productivity tax laws* 7 states
- *Yield tax laws* 17 states
- *Severance tax laws* 4 states

In all cases, the presumption is that when forests are subject to special tax treatment within a property tax system, private investment in forest holdings will be encouraged, conversion of forestland to other uses will be impeded, and forfeiture of forestland to government entities will be thwarted.

An additional type of tax potentially applicable to forests is the value-added tax (i.e., a tax on the value added in each stage of production). Although common in Europe, the tax is rarely applied in the United States.

Tax Program Challenges

Tax policies designed to encourage investment in private forests or to deter conversion of forestland to nonforest uses face challenges similar to those encountered by programs which provide direct fiscal subsidies. These challenges include appropriate targeting of landowners, the debate about substitution of public capital for private capital, and coordination with other assistance programs. From a property tax perspective, failure to separate taxes on timber from taxes on the land on which the timber resides is often argued as double taxation. Although yield taxes reduce tax incentives for premature harvesting and establish a close correspondence between receipt of income and payment of taxes, when broadly applied these taxes can disrupt the annual flow of tax revenues to local governments (Hickman 1982). Probably the most troubling problem in development and implementation of tax policies applied to forests is the lack of information concerning their efficiency and effectivness as means for encouraging investments in the management of private forests (Klemperer 1989). It is largely unknown whether modifications of (which generally means decreases in) income taxes and property taxes for forestry purposes result in production of additional timber supplies with values commensurate to or greater than the revenues foregone by federal, state, and local treasuries. In this respect, the General Accounting Office made the following observation regarding capital gains treatment of income from timber: ''None of the many sources GAO contacted could provide firm evidence to support generally claimed values for conservation and reforestation from capital gains tax treatment. . . . [Thus effect of capital gains tax expenditures] on site improvement or reforestation needs cannot be definitively assessed'' (General Accounting Office 1981, p. ii, 12).

REFERENCES

Boyd, R., and B. J. Daniels. Capital Gains Treatment of Timber Income: Incidence and Welfare Implications. *Land Economics* 61(4):354–362, 1985.

Chang, S. J. Reforestation by Nonindustrial Private Landowners: Does the Capital Gains Tax Matter? in *Nonindustrial Private Forests: A Review of Economics and Policy Studies,* by J. P. Royer and C. D. Risbrudt. School of Forestry and Environmental Studies, Duke University, Durham, N.C., 1983.

Clements, S. E., W. D. Klemperer, H. L. Haney, and W. C. Siegel. Current Status of Timber Yield and Severance Taxes in the United States. *Forest Products Journal* 36(6):31–35, 1986.

Council of State Governments. *Model State Forestry Incentives Program Act. Suggested State legislation: 1986.* Vol. 45. Committee on Suggested State Legislation, Lexington, Ky., 1986.

Cubbage, F. W., and R. W. Haynes. *Evaluation of the Effectiveness of Market Responses to Timber Scarcity Problems.* Marketing Research Report No. 1149. USDA-Forest Service, Washington, 1988.

Dennis, Donald F. Tax Incentives for Reforestation in Public law 96–451. *Journal of Forestry* 81(5):293–295, 1983.

———. Capital Gains Treatment of Timber Income: An Economic Assessment. Research Paper NE-556. Northeastern Forest Experiment Station, USDA-Forest Service, Washington, 1985.

Ellefson, Paul V., and Andrew M. Wheatcraft. Equity and the Allocation of Forestry Incentives Program Funds, in *Nonindustrial Private Forests,* by J. P. Royer and C. D. Risbrudt (eds.). School of Forestry and Environmental Studies, Duke University, Durham, D.C., 1983, pp. 204–215.

Fedkiw, John. *Background Paper on the Nonindustrial Private Forest Lands: Their Management and Related Public and Private Assistance.* USDA-Forest Service, Washington, 1983.

General Accounting Office. *New Means of Analysis Required for Policy Decisions Affecting Private Forestry Sector.* EMD–81–18. U.S. Congress, Washington, 1981.

Gorte, Ross W., and Jack H. Taylor. *Impact of the President's Tax Reform Proposal on Timberland Owners and the Forest Products Industry.* ENT 85–905. Congressional Research Service, Library of Congress, Washington, 1985.

Gregersen, H., T. Houghtaling, and A. Rubinstein. *Economics of Public Forestry Incentive Programs: A Case Study of Cost-Sharing in Minnesota.* Technical Bulletin 315–1979. Agricultural Experiment Station, University of Minnesota, St. Paul, 1979.

Gregory, G. Robinson. *Resource Economics for Foresters.* John Wiley and Sons, Inc., New York, 1987

Hickman, Clifford A. Emerging Patterns of Forest Property and Yield Taxes, in *Proceedings, Forest Taxation Symposium II,* by H. L. Haney and W. C. Siegel (eds.). FSW-4–82. School of Forestry and Wildlife Resources, Virginia Polytechnic Institute and State University, Blacksburg, 1982, pp. 52–69.

———. *Inconsistent Forest Property Tax Policies within Selected Southern States.* Research Paper SO-253. Southern Forest Experiment Station, USDA-Forest Service, New Orleans, La., 1989.

Hoover, W. L., W. C. Siegel, G. A. Myles, and H. L. Haney. *Forest Owners' Guide to Timber Investments, the Federal Income Tax, and Tax Recordkeeping.* Agriculture Handbook No. 681. USDA-Forest Service, Washington, 1989.

Irland, Lloyd C. Federal and State Assistance to Nonindustrial Private Forest Owners in New England and USA [in general]: Issues for the Year 2000. *The Consultant* 27(2):36–41, 1982.

Klemperer, W David. Taxation of Forest Products and Forest Resources, in *Forest Resource Economics and Policy Research: Strategic Directions for the Future,* by Paul V. Ellefson (ed.). Westview Press, Boulder, Colo., 1989, pp. 276–287.

Kurtz, W. B., R. J. Alig and T. J. Mills. Retention and Condition of Agricultural Conservation Program Conifer Plantings. *Journal of Forestry* 78(5):273–276, 1980.

Manthy, R. S. *An Investment Guide for Cooperative Forest Management in Pennsylvania.* Research Paper NE-156. Northeastern Forest Experiment Station, USDA-Forest Service, Upper Darby, Pa., 1970.

Meeks, Gordon. State Incentives for Nonindustrial Private Forestry. *Journal of Forestry* 80(1):18–22, 1982.

Mills, Thomas J. *Cost Effectiveness of the 1974 Forestry Incentives Program.* Research Paper RM-175. Rocky Mountain Forest and Range Experiment Station, USDA-Forest Service, Fort Collins, Colo., 1976.

Office of Management and Budget. *Budgetary Memo and Fact Sheet Concerning Con-*

servation Cost-Sharing (*Forestry Incentives Program*). Executive Office of the President, Washington, 1981.

Osborn, C. T., F. Llacuna, and M. Linsenbigler. *The Conservation Reserve Program: Enrollment Statistics for 1987–88. Statistical Bulletin Number 785. Economic Research Service, Washington, 1989.*

Risbrudt, Christopher D., and Paul V. Ellefson. *An Economic Evaluation of the 1979 Forestry Incentives Programs.* Bulletin 550–1983. Agricultural Experiment Station, University of Minnesota, St. Paul, 1983.

______, M. H. Goforth, A. Wheatcraft, and P. V. Ellefson. *1974 Forestry Incentive Investments: Retention as of 1981.* Bulletin 552–1983. Agricultural Experiment Station, University of Minnesota, St. Paul, 1983.

Siegel, William C. New Federal Tax Issues: Capital Gain Status for Timber Production, in *Proceedings, Appalachian Society of American Foresters* (*68th Annual Meeting*). Society of American Foresters, Bethesda, Md., 1989, pp. 86–92.

Society of American Foresters. *Timber Aspects of the Federal Income Tax.* Report of the Federal Forest Taxation Task Force, Bethesda, Md., 1986.

Sunley, E. M. Capital Gains Treatment of Timber: Present Law and Proposed Changes. *Journal of Forestry* 74(2):75–78, 1976.

Tedder, P., and C. Sutherland. The Federal Estate Tax: A Potential Problem for Private Nonindustrial Forest Owners. *Southern Journal of Applied Forestry* 3(3):109–111, 1979.

U.S. Department of Agriculture. *The Federal Role in the Conservation and Management of Private Nonindustrial Forest Lands.* Interagency Committee, Soil Conservation Service, Forest Service, and Extension Service, Washington, 1978.

______. *Conservation Reserve Program: Progress Report and Preliminary Evaluation of the First Two Years.* Washington, 1989.

USDA-Forest Service. *Report of the Forest Service: 1988.* U.S. Department of Agriculture, Washington, 1989.

19

GOVERNMENT REGULATORY INITIATIVES

Society may furnish private owners of forests with important technical advice and financial assistance, only to find out that the owners continue to use and manage their forests in ways that are inconsistent with the desires and wishes of the public generally. Pesticides may be used indiscriminately, wildlife may be endangered by the loss of forest habitat, timberlands may be left unforested after harvest, and forestland may be transferred to inappropriate nonforest uses. Circumstances of this nature are classic examples of market inefficiencies, specifically negative externalities. Such conditions may well persuade society (via government) to regulate the forestry activities of private forestland owners—to require them by law to meet socially prescribed forestland use and management standards. In such an environment, the landowners' incentive for meeting such standards becomes the fear of punishment and penalties, including fines, jail sentences, and legally imposed limits on forestry conduct.

The regulatory environment in which owners of private forestland conduct business has grown in scope, complexity, and severity since the early 1970s. State governments have been most active in this environment, although federal programs have provided significant impetus for state regulatory action. Examples are the land-use implications of pollution-control laws and constraints placed on grants-in-aid sought by state governments. The scope of regulations imposed on owners of forestland and on processors of forestland products is well illustrated by the various forest and wood-based manufacturing activities regulated in the state of Washington (Department of Natural Resources 1980):

- Industrial safety and labor regulations (state and federal)
- Payroll deductions, taxes, and insurance regulations (state and county)
- Interstate transportation regulations (federal)
- Timber-harvesting regulations (state)
- Land-use and zoning regulations (county and municipal)
- In-state transportation regulations (state and county)

In this chapter we shall discuss regulatory programs constraining landowner decisions regarding the use of forestland, the creation of environmental pollutants, and the application of forest practices to specific tracts of forestland.

LAND-USE REGULATION

Decisions regarding the use of forestland are among the most important determinants of the type and magnitude of benefits to flow from private forests. Not infrequently, state and federal governments have intervened to directly or indirectly determine socially appropriate uses of private forest property. Such action frequently focuses on:

- *Fragile lands, such as wetlands and shorelands* These lands maybe considered an especially important wildlife habitat or an important link in the food chain.
- *Natural hazards, such as flood plains or highly erodible lands* Improper use of management of these areas may have devastating economic and social consequences.
- *Natural resources, such as wild and scenic rivers* Such resources are considered valuable for cultural or recreational reasons.

Federal Regulations

Federal actions which can influence private land-use decisions include efforts to protect and cautiously develop natural resources (including forests), the use and management of which can significantly affect the nation's coastal waters. Via the Coastal Zone Management Act of 1972 (as amended), states are encouraged to define ''what shall constitute permissible land and water uses within the coastal zone'' and to specify ''means by which [a] state proposes to exert control over land and water uses'' (Environmental Law Institute 1974 p. 832). Similarly, federal administration of the Wild and Scenic Rivers System (protection of free-flowing rivers which possess outstanding scenic, geologic, recreational, historic, and wildlife values) may entail cooperative management agreements with state governments. Such agreements pay ''particular attention . . . to scheduled timber harvest, road construction, and similar activities which might be contrary to the purposes of the [Wild and Scenic Rivers Act of 1968 (as amended)].'' The administering agency is also authorized to acquire scenic easements (including land-use restrictions) from private landowners whose property falls within the boundary of a designated river. These examples reflect intentional federal involvement in private land-use decisions.

Federal involvement in land-use matters can also occur incidentally, through implementation of federal pollution-control laws. These laws, for example, may place limits on point or ambient sources of air and water pollution. Such standards indirectly limit the use of property on which a point source is located or limit the intensity of land use in a regional context. Regional ambient air and water-quality standards may imply land-use limitations to the extent that a forested region becomes less conscious of growth and development (Grad 1988; Mandelker 1988).

State Regulations

State governments have also proceeded to establish rules and regulations constraining the use of privately owned land, including forestland (Healy and Rosenberg 1979). At the local and municipal level, such restrictions are frequently imposed via zoning and subdivision ordinances. More broadly focused, some states have established comprehensive statewide land-use plans and planning processes. These states include Oregon, Colorado, Nevada, North Carolina, and Maine. Maine's Land Use Regulation Act created the Land Use Regulation Commission, which has wide-ranging control over development and land use within the state's unorganized lands. In Oregon, the Land Conservation and Development Commission has adopted goals for various types of land use (including agriculture, forest, recreation, and open space), with which all comprehensive land-use plans and regulations adopted by local governments and state agencies must comply. In California, the California Board of Forestry must act on proposed conversion of designated forestland to nonforest uses.

Most states have been reluctant to forcefully and rapidly impose statewide restrictions on land use, especially at the expense of the regulatory prerogatives of local governments. More common is planning for and regulation of land use within critical areas, especially when significant historical, environmental, and natural resources are affected. Florida's Land and Water Management Act of 1972, for example, focuses on planning and regulation of land use within the Big Cypress, Green Swamp, and Florida Keys area. Maryland's Chesapeake Bay Critical Area Act of 1985 controls development and use of land adjacent to sensitive shoreline around Chesapeake Bay. California's Tahoe Conservancy Act of 1985 and Coastal Act of 1976 impose significant restrictions on the use of land (including forestland) in two environmental sensitive regions of the state. Another example is New York's Adirondack Park Agency, which controls private land use in the Adirondack Mountain Region, so as to further goals and objectives developed for six types of land use (such as rural use areas and resource management areas), as specified in the region's detailed land-use plan.

Similarly, the New Jersey Pineland Commission divides the state's Pinelands Region into two distinct types of areas and establishes land-use regulations for each. The two types are preservation areas, in which all new development is virtually prohibited, and outer rings, which are subject to less stringent development regulations (Grad 1988).

POLLUTANT REGULATION

Forestland use is but one focal point for public regulation of the forestry activities of private forest owners. Government can also impose restrictions on a landowner's use of certain materials, such as chemicals, solid wastes, and hazardous substances, and on actions which cause pollution of air and water (Novick 1988; Schoenbaum 1982).

Pesticides

Regulations can be imposed on the use of pesticides for managing undesirable vegetation, insects, diseases, and the like. Federally, pesticides are regulated by the Environmental Protection Agency (EPA) via authority set forth in the Federal Environmental Pesticide Control Act of 1972 (as amended). The act prohibits the sale and use of pesticides that are not registered with the EPA and requires them to be classified as either general-use pesticides (the use of which will not cause unreasonable effects on the environment) or restricted-use pesticides (all other pesticides). Restricted-use pesticides may be handled only by certified applicators. States are given the opportunity to develop certification programs which must attest to individuals' competence in the use and handling of specific pesticides or pesticides generally. To apply a restricted-use pesticide forest landowners must either be certified or seek the services of a person who is certified. Private applicators who violate provisions of the law may be assessed a civil penalty of up to $1,000 (Grad 1988).

Water Pollutants

Water pollutants are also regulated in a major way at the federal level by the EPA. Such pollutants are addressed by amendments to the Federal Water Pollution Control Act of 1972 (as amended), which recognizes two sources of pollutants:

- *Point source pollutants* Water pollutants discharged from a readily identifiable source such as a pipe or ditch. Industrial wastes and municipal sewage are common point sources of pollutants.
- *Nonpoint source pollutants* Water pollutants originating from diffused sources such as the application of management practices to agricultural land and

forestland. Diffused pollutants that reach water over land or through related runoff.

Point sources are regulated via an elaborate permitting system. The National Pollution Discharge Elimination System is designed to keep pollutants below specified effluent standards. Because of the nature of point-source pollutants, the permitting system focuses on manufacturing enterprises (say, a pulp mill that releases effluent) and government operations (such as municipal sewerage treatment facilities).

Effluent standards for nonpoint source pollutants are difficult to establish. Even if they are established, their administration via a permitting system is difficult to implement. The 1972 act therefore merely authorizes the EPA to issue guidelines regarding methods and procedures to be used for the control of such pollutants, and requires states to identify and implement plans for the abatement of nonpoint sources of pollutants (including those related to silvicultural practices). State plans identify practices (called *best-management practices*) considered most appropriate for reducing nonpoint sources of water pollutants in forested areas (such as modified road construction and altered timber harvesting). State plans also set forth means of securing the implementation of such practices by private owners of forests. Implementation may occur via voluntary compliance with prescribed best-management practices or via involuntary compliance, or regulation—possibly state forest practices laws.

The Federal Water Pollution Control Act (as amended by the Clean Water Act of 1977) also acknowledges the discharge of dredge or fill material into waterways as a source of pollutants requiring regulation. Since the definition of waterways encompasses certain wetlands, including forested wetlands, vegetative management that accelerates the deposit of pollutants in wetlands is a subject for concern. (Normal forestry operations and temporary forest roads, however, are excluded.) The U.S. Army Corp of Engineers is responsible for administering the permit system focused on dredge-and-fill activities.

Air Pollutants

Regulation of private forestry actions can also occur via implementation of the Clean Air Act of 1970 (as amended), which mandates compliance with administrative air-quality standards established by the EPA. Subject to EPA approval, the law requires state formulation and implementation of plans which result in achievement of national ambient air-quality standards (those necessary to protect the public health). Amendments to the act (in 1977) addressed prevention of significant deterioration of air that is cleaner than ambient air standards. Areas in which little or no deterioration can take place (class I areas) include national parks larger than 6,000 acres and national wilderness areas exceeding 5,000 acres. In other designated areas, moderate to significant deterioration of air quality is allowed; however, deterioration cannot exceed national ambient standards. Regulation of air pollutants resulting from wood-based manufacturing activities and from land-management practices such as prescribed burning in forested environments are most commonly subject to the provisions of the Clean Air Act.

The forestry actions of private owners of forestland are also affected by federal programs that regulate the following sources of pollution:

- *Noise* Noise Control Act of 1972, as amended
- *Extraction of minerals* For example, the Surface Mining Control and Reclamation Act of 1977
- *Solid waste disposal* The Resource Recovery Act of 1970, as amended

Private forestry actions that affect historic or culturally significant sites are also regulated most often by state and local governments. In addition, private forestry activities that have environmental consequences may be subject to regulation if they involve, for example (Fisher and Phillips 1983):

- *Monopoly and price discrimination in the market place* Regulated by, for example, the Sherman Act of 1890 (as amended) and the Clayton Act of 1914 (as amended)
- *Employer-employee relations* Regulated by, for example, the Occupational Safety and Health Act of 1970 (as amended), and the Civil Rights Act of 1964 (as amended)
- *Consumer protection and product liability* Regulated by, for example, the Truth in Lending Act of 1968, the Fair Credit Reporting Act of 1970, and the Magnuson-Moss Warraty Act of 1975

FOREST PRACTICES REGULATION

Regulation of the forestry activities of private forestland owners can also be focused on forestry practices. Constraints can, for example, be placed on the types of silvicultural practices used, the methods by which timber is harvested, the manner in which reforestation is to take place, and the procedures to be followed in order to protect fish and wildlife habitat. The most common means of imposing such constraints on private forests are state forest practices laws. Between 1937 and 1955, 16 states enacted laws designed to guide the conduct of forest-management activities on privately owned forestland. Narrow in scope, these laws were primarily concerned with the adequacy of future timber supplies, and especially with achievement of a minimal level of reforestation after timber harvesting. Most of these laws, which are known as *seed-tree laws,* were never seriously enforced, and nearly all have long since been ignored or repealed.

Beginning in the early 1970s, however, state governments began to seriously consider—and to enact—stringent, comprehensive forest practices laws. Especially noteworthy were laws enacted in Alaska (1978), California (1973), Idaho (1974), Massachusetts (1983), Nevada (1971), Oregon (1971), and Washington (1974).

Arising from growing public concern over environmental protection and from federal laws mandating state programs to protect water quality (especially the 1972 Amendments to the Federal Water Pollution Control Act), such forest practices laws were established to protect a much wider range of forest resources than earlier laws, including water, soils, wildlife, recreation, fisheries, and natural beauty. The laws provide for rigorous administrative and enforcement structures that assure achievement of specified resource protection goals. Requirements are established for timber-harvesting permits, management plan preparation, resource protection, achievement of adequate reforestation levels, and state inspection for compliance (Henly and Ellefson 1986). In this section, we shall consider the administrative environment of forest practices laws, the number and types of laws currently in force, the nature of the costs such laws impose on government and private landowners, and the type and extent of benefits that can be attributed to forest practices laws.

Administrative Environment

State forest practices laws and the regulations which result from them have a significant legal foundation (Hickman and Hickman 1990). Although land ownership has traditionally been considered a bundle of rights regarding acquisition, disposal, and exclusive use, such rights have never been considered absolute. They have always been limited and conditioned by the overall interests of society (as administrated by government), including protection and promotion of public health, safety, morals, and the general welfare. State governments have customarily assumed police power to regulate the actions of private landowners.

Authority to regulate the forest practices of private landowners stems from common law and legal doctrine, especially the doctrine of private nuisance and, less frequently, the doctrine of waste (Carmichael 1975; Freeman 1975). The *doctrine of private nuisance* argues that individuals may not use their property in a manner that will injure the property of others. Accountability is established between owners of property in the present tense. If one owner's property use is immediately injurious to the use and enjoyment of another's, the injured party may seek both monetary damages and an injunction against the nuisance. Originally applying to actions between private parties, the nuisance doctrine has been expanded to protect the property rights of society in general. Forest practices laws fall within this expanded version of private nuisance (as do most land-use and zoning regulations).

The *doctrine of waste* declares that a property owner must not wantonly decrease the value of the property during the course of ownership. The property must be transferred to future owners (future generations) in a substantially unimpaired condition. The doctrine covers accountability for property interests over time. Needless destruction or overly consumptive use of property by a present owner is viewed as impermissible destruction of the patrimony of future generations. The first legal challenge to early state forest practices regulations was denied on the basis of the doctrine of waste (Pacific Reporter 1947).

Crucial to the application of doctrines of waste and private nuisance is whether laws based on such doctrines constitute a taking. The Fifth Amendment of the U.S. Constitution states, "nor shall private property be taken for public use without just compensation." If a regulation is judged to be so severe that a taking without compensation occurs, the regulation is considered unconstitutional. Judgments regarding taking involve a balancing test—a weighing of the public benefits of regulation against the extent of loss of property values. In recent years, increasing knowledge of the environmental damage caused by some patterns of land use and subsequent management makes the public's interest in private activities so important that it can seldom be outbalanced by an individual's loss of property rights or values (Bosselman et al. 1973).

State regulation of privately prescribed forest practices involves a number of interacting components, often including several state agencies and multiple regulatory systems. These components can best be viewed as a system (Table

TABLE 19-1
Major components of a state's private forest practices regulatory system

- *Forest practice law* A law establishing in broad terms the objectives of regulation, the activities to be regulated, the manner in which forest practices are to be regulated, the responsible administering agency, and penalties to be served for noncompliance.
- *Rule-making body* A body (board or individual) authorized to promulgate forest practice rules (standards) with which private forest landowners must comply.
- *Forest practices rules* Rules (such as reforestation standards) which are enforced by an administering agency and are to be complied with by a forest landowner or forest user.
- *Administering agency* A public agency responsible for carrying out the intent of a forest practices law. In most states, the state forestry agency is given primary responsibility; nonforestry agencies (concerned primarily with water quality, and fish and wildlife) may be given secondary administrative responsibility, including review of permit applications or inspection of operations.
- *Interested publics* Organized or unorganized publics with an interest in regulation of forest practices. Includes the regulated public (the forest landowner), the antiregulation public, and the proregulation public.
- *Related regulatory systems* Additional and separate regulatory systems with which private forestry activities must comply (such as air and water-quality standards). A state's forest practices law may be but one component of a more comprehensive state system used to further society's interest in environment quality.

Source: State Forest Practice Regulation in the U.S.: Administration, Cost and Accomplishments, by Russell K. Henly and Paul V. Ellefson. Bulletin AD-SB-3011. Agricultural Experiment Station, University of Minnesota, St. Paul, 1986.

19-1). A thorough understanding of a comprehensive state forest practices regulation system would entail detailed knowledge of:

- *Statutory resource protection goals* General forest resources of concern, specific resources to be protected, and degree of protection to be sought.
- *Administering agencies* Forestry and nonforestry agency responsibilities.
- *Rule-promulgating authority* Responsible board or individual, composition of board or background of individual, method of appointment and term of office, and rule promulgation process.
- *Advisory bodies* Regional basis for formation of multiple bodies, composition of bodies (education, experience, and interest-group representation), and responsibilities (initiation of new rules and rule modifications).
- *Ownership categories regulated* Private or public.
- *Activities regulated* Timber harvest, timber production, nontimber management activities, and conversion of forestland to nonforestland.
- *Activities exempt from regulation* Small timber harvests, Christmas tree harvest, and timber harvested for personal use.
- *Licensing requirements* Foresters and timber harvesters.
- *Procedural requirements* Before commencing activity, filing notice of operation with forestry agency (simple notice versus application with accompanying management plan which requires approval), notification of abutting property owners, and notification of general public. After completion of activity, filing completion reports and restocking reports.
- *Application or management plan approval process* Agencies involved in review and approval, inspection of site before approval, and appeal of application denials.
- *Enforcement procedures* Inspection for compliance, enforcement sanctions (such as stop-work orders, repair damage orders, fines, and imprisonment), appeal of enforcement actions (administrative, judicial).
- *Local government authority to regulate practices* None granted, as granted by forest practices board, or no restrictions on local regulations.

Modern forest practices laws are basically environmental protection laws applied to forestry activities. They often pertain to a wide range of forest ownerships, including industrial, nonindustrial private, state, county, and municipal. Federal agencies usually comply, at a minimum, with state forest practices standards. Landowners and timber harvesters may be required to file for and receive approval before commencing timber-harvesting activities. Inspection of harvesting operations by a state forestry agency may be required. Comprehensive forest practices laws enacted since the early 1970s usually require landowners or timber harvesters to reforest after harvest and to take appropriate steps to protect fish and wildlife habitat, soil productivity, aesthetic appearance, and water quality. For example, a landowner may be required to leave merchantable timber standing in streamside protection zones.

Contrary to popular impressions of regulatory activities, state programs regulating private forest practices are neither harshly nor ruthlessly administered. In general, rule compliance is sought through cooperative and consultative means. Educational programs are commonly employed to help landowners and timber harvesters to understand promulgated rules and their environmental soundness. Most states impose citations, stop-work orders, fines, or other enforcement mechanisms only as a last resort. For example, Washington's forest practices rules and regulations state, "It is the policy of the act and the [state forest practices] board to encourage informal, practical, and result-oriented resolution of alleged violations and actions needed to prevent damage to public resources" (Henly and Ellefson 1986, p. 3).

State forest practices laws commonly provide for technical advisory committees to assist in the development of forest practices rules. The rules often take the form of performance standards that specify a level of resource protection of reforestation (such as 300 seedlings per acre) with which a landowner or timber harvester must comply, using whatever procedures are deemed appropriate to a given forest site. The application of performance standards entails use of professional judgment and technical skill by foresters, timber harvesters, and landowners. Such qualities are necessary because of the highly variable forest conditions found within most states—conditions which make geographically uniform, highly prescriptive forestry standards rarely appropriate. However, to ensure some degree of administrative uniformity in the application of forest practices standards, some states have seen fit to license timber harvesters. Massachusetts and California, for example, require the licensing of all timber harvesters, and Massachusetts's harvesters must pay a fee and pass a written exam attesting to their understanding of forests generally and forest practices specifically. Some states, such as California, require foresters to meet educational and experience requirements and to pass a written exam before they are allowed to prepare timber-harvesting plans for private forest landowners.

Regulatory Landscape

Regulatory programs focused on the forestry practices of private forest landowners are common throughout the world (Eckerberg 1987; Hummel and Hilmi 1989). In the United States, states in all regions of the nation have forest practices laws (Table 17-1). Nearly half of the states have legal forestry regulations focused on protection of water quality; 14 employ regulation to ensure proper timber management and reforestation of private forestlands; and a like number use regulatory mechanisms to ensure use of forest practices conducive to quality wildlife habitat and aesthetic management. Prospects that additional states will elect to enact forest practice laws in order to impose social interests on private forest landowners are mixed. Four states (Connecticut, Maine, Maryland, and Montana) were considering forest practices legislation in early 1985, and Maine has since enacted such legislation. An additional eight states had considered such legislation in the recent past. In the same year, 16 states considered the proliferation of local ordinances restricting timber management and harvesting as a primary factor motivating them to consider forest practices regulation. In the early 1980s, for example, 100 of New Jersey's 567 municipalities had established restrictive local ordinances on timber harvesting. Similarly, in Connecticut, 25 of the state's 169 townships had enacted forest practices regulations (Henly and Ellefson 1986; Hickman and Martus 1991).

In 1985, concern over future timber supplies was not a primary factor in forcing consideration of forest practices regulation, except in four of eight southern states studied (Henly and Ellefson 1986). In 1988, knowledgeable forestry administrators in four southern states (Florida, Georgia, Virginia, and North Carolina) believed that most citizens, interest groups, and natural resource agencies would support greater regulation of forestry practices (Cubbage, Siegel, and Lickwar 1989). In 1985, the following obstacles to establishing a forest practices law in their state were listed (in order of importance) by state natural resource administrators (Henly and Ellefson 1986):

- Industrial landowner resistance to regulation
- Nonindustrial private forest landowner resistance to regulation
- Costly state administrative procedures
- Nonindustrial private forest landowner cost of compliance
- Cumbersome state administrative procedures
- Industrial landowner cost of compliance

The regulatory environment imposed by a state forest practices law can best be appreciated by example. We shall consider laws in two states, Massachusetts and Washington (Henly and Ellefson 1986).

Massachusetts The Massachusetts Forest Cutting Practices Act establishes broad multiple-use, multiple-resource protection of forests as a major focus. Rules and regulations for the act are promulgated by the director of the Division of Forests and Parks (the lead agency) and are subject to the approval of the commissioner for the Department of Environmental Management. The eight-member advisory State Foresty Committee, appointed by the governor, provides the director with recommended regulations. The director can appoint additional advisory committees to aid in the preparation of regulations. Such a committee helped to prepare the state's current forest practices regulations.

All forestland in Massachusetts (public and private) is subject to the Forest Cutting Practices Act, although the regulated activities are primarily those directly related to timber harvesting. In general, any commerical harvest of over 25,000 board feet or over 50 cords on any one parcel of land at any one time is subject to the act's provisions. Exempted harvesting activities include clearance for highway rights of way, noncommercial cutting for use by the landowner, small timber harvests (less than threshold size), and clearing land for building or cultivation.

Before harvesting timber, landowners (or their agents) must file a notice of intent to harvest and a harvesting plan with the Division of Forests and Parks and the local advisory conservation commission. They must also notify other landowners whose property is within 200 feet of the boundary of the proposed harvest area, of the pending operation. Harvesting activities may commence upon the division's approval of the cutting plan or 10 days after the filing date; the plan is valid for 2 years. The division may modify the cutting plan or attach additional requirements as conditions of approval.

The forest practices regulations which landowners are subject to entail both standards and guidelines. The standards must be followed; they include requirements for regeneration, retention of buffer strips along highways and around bodies of water, construction of roads and skid trails, and use of specified timber-harvesting systems. Guidelines are not mandatory (unless written into the cutting plan); they involve forest practices recommended for the benefit of fish and wildlife, for protection of residual trees, and for preservation of a stable road network.

The Massachusetts Forest Cutting Practices Act also requires all persons who harvest timber and other commerical forest products to be licensed. Timber-harvesting licenses must be renewed annually, at which time applicants must pass a test of their knowledge of the state's timber-harvesting regulations. The licenses may be revoked for noncompliance with timber-harvesting regulations. In addition, persons may be fined up to $500 per act of harvesting timber without a license.

The Division of Forests and Parks may inspect harvesting operations and can issue stop-work orders. Such an order, once issued, remains in effect until the operation is made to comply with the approved plan, or until the director holds hearings to consider revocation of the timber harvester's license.

When an operation is completed, the landowner is required to notify the division. Upon notification, the division inspects the completed operation and, if it is satisfactory, issues a certificate of compliance. If it is not satisfactory, the certificate is withheld until the deficiencies are corrected.

A landowner or timber harvester may appeal any decision (including the compliance decisions) of the Division of Forests and Parks. The Massachusetts superior court is empowered to enforce the provisions of the Forest Cutting Practices Act. Fines of up to $100 per acre may be levied for cutting that occurs in violation of the landowner's approved cutting plan or in violation of filing and approval procedures. Such fines may be levied against the landowner, the stumpage owner, or the independent contractor, whichever is responsible.

Washington Established in 1974, the Washington Forest Practices Act (as amended) recognizes a broad range of forest resources (including scenic beauty and recreation) but focuses primarily on the regulation of forest practices that have an impact upon timber productivity; forest soils; and related public resources such as water, fish, and wildlife. Forest practices are regulated on all state and private lands capable of growing merchantable stands of timber and not devoted to incompatible uses. The Division of Private Forestry and Natural Heritage (within the state's Department of Natural Resources) has primary responsibility for administration of the act.

The forest practices regulations called for by the Washington act are promulgated by the Forest Practices Board, although regulations related to water quality are established by both the board and the Department of Ecology. The 11-member Forest Practices Board is composed of the commissioner of public lands; the directors of the Department of Agriculture, the Department of Ecology, and the Department of Commerce and Economic Development; a member of a county legislative body; and six members of the public, one of whom must not own more than 500 acres of forestland and another of whom must be an independent logging contractor. The commissioner of public lands serves as chair of the board. The act also calls for the 11-member Forest Practices Advisory Committee to prepare regulations for the board's consideration.

Washington is divided into two regions (the western and eastern regions, separated by the crest of the Cascade Mountain Range) for purposes of applying promulgated forest practice regulations. Practices that fall within the scope of the act are grouped into one of four categories (one of which has two subcategories), as specified by the law and as influenced by potential impact on variously categorized bodies of water. These forest practice categories are:

- *Class I* Forest practices that have no direct potential for damaging the environment or a public resource, do not require the filing of notifications or applications, though compliance with forest practices regulations is mandated. Examples are the culture of Christmas trees, general road maintenance, construction of landings less than 1 acre in size, precommerical thinning and pruning, and removal of less than 5,000 board feet of timber in any 12-month period.
- *Class II* Forest practices that have less than ordinary potential to damage the environment and a public resource. These practices cannot be commenced until 5 days after the filing of a notification with the Department of Natural Resources. Examples, are construction of advance fire trails, salvage logging, construction of 600 or more feet of road, and harvesting of more than 40 acres of forest outside the streamside management zone of a specified category of water.
- *Class III* Forest practices that require the filing of an application; the Department of Natural Resources must either approve or deny the application within 14 days. Examples are replacement of bridges and culverts over specified categories of water, opening of new pits or extensions of existing pits over 1 acre in size, aerial application of insecticides, and application of chemicals on an area over 160 acres in size.

• *Class IV* Forest practices that have potential for substantial impact on the environment or a public resource. *Class IV General Practices* require submission of an application to the Department of Natural Resources and may be commenced within 30 days of filing. Examples of general practices are harvesting, site preparation, and road construction in areas known to contain especially significant wildlife, or on important historic or archeological sites. *Class IV Special Practices* require evaluation of whether a detailed environmental statement is necessary. Examples of special practices are harvesting, site preparation, and road construction in areas known as habitat for threatened or endangered species, or in areas landlocked within local, state, or national parks. Another example is the widespread use of persistent pesticides.

The promulgated forest practices rules which implement the Washington Forest Practices Act contain specific requirements for the conduct of forest practices falling within these classes. Depending on the class, they are often very detailed, for example:

> Stocking levels are acceptable if 300 well-distributed, vigorous seedlings of commerical trees species have survived on the site at least one growing season. . . .
>
> On all aerial applications of pesticides, operators shall maintain for three years daily records of spray operations. . . .
>
> Stream beds shall be cleared of slash and debris 50 feet upstream from a culvert inlet. . . .
>
> Leave all nonmerchantable vegetation which provides midsummer midday shade over the surface of specified categories of water; leave sufficient shade to retain seventy-five percent of the midsummer midday shade over the surface of specified categories of water, except where the ambient midsummer midday maximum water temperature would not normally exceed 60 degrees, the shade requirement shall be 50 percent.

Applications to carry out forest practices must include (Figure 19-1):

• The names and addresses of the landowner, the timber owner, and the operator
• A description of the proposed forest practices
• Beginning and expected ending dates of operation
• Maps showing the locations of all bodies of water in and adjacent to the proposed operation
• Proposed reforestation plan
• Soil, geological, and hydrological data with respect to the proposed forest practices
• Provisions for continuing maintenance of roads

The Department of Natural Resources is required to send review copies of all notifications and applications to the Department of Ecology, the Department of Fisheries, and the Department of Game; to the county in which the operation is to occur; and to appropriate park administrators, when proposed operations

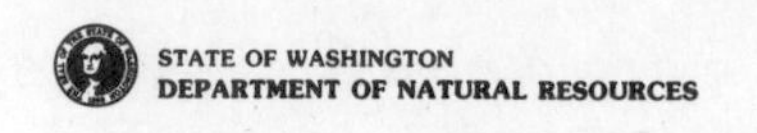

FOREST PRACTICES APPLICATION/NOTIFICATION

NOTE TO APPLICANT: Read but DO NOT write in shaded areas. Complete items 1-18. Use additional sheets as needed.

SHUTDOWN ZONE	BURN PERMIT ZONE	DATE RECEIVED	DATE DUE	APPLICATION/NOTIFICATION NUMBER

CHECK BOX FOR ANY REQUEST YOU HAVE. NOTE UPLAND MANAGEMENT AREAS AND RIPARIAN MANAGEMENT ZONES.

REQUESTS

DELAYED EFFECTIVE DATE ______ REQUESTED APPROVAL DATE
ALTERNATE RMZ PLAN ☐
PRE-FILING REVIEW ☐

REQUESTS

FP APPLICATION RENEWAL NO. ______
HYDRAULIC PROJECT APPROVAL ☐
FOREST PRACTICE ASSISTANCE ☐
LANDOWNER WANTS FOREST MANAGEMENT ASSISTANCE ☐

DATA UPLAND MANAGEMENT AREAS ☐ NOs. ACRES HARVESTING OLD GROWTH ☐

RIPARIAN MANAGEMENT ZONES (RMZ) BY:
WATER TYPE ____ NO. UNITS ____ LINEAL FEET ____ STREAM ☐ SIDE(S) 1 or 2 POND/LAKE/BAY ☐
WATER TYPE ____ NO. UNITS ____ LINEAL FEET ____ STREAM ☐ SIDE(S) 1 or 2 POND/LAKE/BAY ☐
WATER TYPE ____ NO. UNITS ____ LINEAL FEET ____ STREAM ☐ SIDE(S) 1 or 2 POND/LAKE/BAY ☐
(ALSO SHOW UPLAND MANAGEMENT AREAS AND RIPARIAN MANAGEMENT ZONES ON MAP)

NOTE: TYPE OR PRINT INFORMATION IN PERMANENT BLACK INK.

1. NAME OF OPERATOR	2. NAME OF LANDOWNER	3. NAME OF TIMBER TAXPAYER (TIMBER OWNER)
ADDRESS	ADDRESS	ADDRESS
CITY STATE ZIP	CITY STATE ZIP	CITY STATE ZIP
TELEPHONE NUMBER ( )	TELEPHONE NUMBER ( )	TELEPHONE NUMBER ( )

4. TIMBER TAXPAYER FULL LEGAL NAME — TIMBER TAX REGISTRATION NO.
TYPE OF LEGAL ENTITY — UNIFIED BUSINESS IDENTIFIER
SOLE OWNER ☐ PARTNERSHIP ☐ CORPORATION ☐ OTHER ☐
CALL 1-800-548-TTAX FOR TAX NUMBER OR TAX INFORMATION

5. COUNTY
6. WITHIN CITY LIMITS? NO ☐ YES ☐ NAME:
7. LAND PLATTED AFTER JAN. 1, 1960 (INCLUDES SHORT PLATS) (CONTACT CO. ASSESSOR FOR VERIFICATION) NO ☐ YES ☐

8. IN PUBLIC PARK OR WITHIN 500 FT (e.g. PORT, CITY COUNTY, STATE, FED. PARK) NO ☐ YES ☐ PARK NAME:

9. LEGAL SUBDIVISION OF OPERATION(S)—INCLUDE PARCEL OR TAX LOT NUMBER IF CONVERSION OR IF PLATTED	SEC.	TWP.	RGE. E/W
A.			
B.			
C.			

FED. THREAT. AND ENDANG. SPECIES OR CRITICAL WILDLIFE HABITAT: NAME ______ (SHOW ON MAP)

10. SLOPE CONDITION & PERCENT
UNSTABLE ☐
HIGHLY ERODIBLE ☐
STEEPEST 10% OR SIGNIF. PART OF OPERATION AREA ____ %
AVERAGE SLOPE ____ %

11. START DATE — END DATE

12. TYPE OF OPERATION (ALSO SHOW ON MAP)	13. METHOD OF OPERATION AND TYPE OF EQUIPMENT (ALSO SHOW ON MAP)	14A. ACREAGE OPERATION	14B. ROAD MILES	15. EST. VOLUME TO BE CUT A. TOTAL	B. PERCENT
A.					
B.					
C.					

ROAD ABANDONMENT PROPOSED/PLANNED FOLLOWING COMPLETION OF OPERATION NO ☐ YES ☐ (SHOW ON MAP)

16. LOCATION OF WATER OR WATER INTAKES ON OR WITHIN 1/4 MILE OF THE OPERATIONS

PLAN FOR ROAD CONSTRUCTION OR IMPROVEMENTS OR HARVEST OF 30% OR MORE OF VOLUME IS WITHIN 200 FT. OF SHORELINE OF THE STATE ☐ OR SHORELINE OF STATEWIDE SIGNIFICANCE ☐ (WATER TYPES 1 AND +1, CONTACT LOCAL GOVT. FOR INFORMATION)

APPLICANT IS AWARE OF 1 OR MORE INTAKES (DOMESTIC, AGRIC., INDUSTRY) ON OPERATION NO ☐ YES ☐ (SHOW ON MAP)

SHOW THE TYPE AND LOCATION OF ALL WATERS ON THE ATTACHED MAP (WATER TYPE MAPS ARE AVAILABLE AT REGIONAL OFFICES OF DEPT.) SHOW SIZE AND LOCATION OF ANY WATER CROSSING STRUCTURES. INDICATE ALL ACTIVITIES PLANNED WITHIN OR ABOVE ORDINARY HIGH-WATER MARKS OF ANY WATER OR CHANNEL AND INDICATE WATER TYPE AND OTHER DETAIL:

WATER TYPE ____ WATER TYPE ____ WATER TYPE ____ WATER TYPE ____ WATER TYPE ____
OPERATE EQUIPMENT (KIND OF) ____ INSTALL CULVERTS (SIZE OF) ____ BRIDGES (KIND OF) ____ FELL TIMBER ☐ SKID OR YARD TIMBER ☐

WATER TYPE ____ WATER TYPE ____ WATER TYPE ____ WATER TYPE ____
ROAD OR LANDING ☐ SIDECAST, WASTE ☐ DEBRIS DISPOSAL ☐ OTHER ____ NONE ☐

GIVE APPROXIMATE START AND END DATES FOR THIS WORK: START ____ END ____ (SHOW ACTIVITY & LOCATION ON MAP)

17. LANDOWNER OR PERPETUAL TIMBER OWNER REFORESTATION PLAN OR LANDOWNER CONVERSION PLAN

HARVESTED LANDS ARE TO BE REFORESTED EXCEPT WHEN CONVERTED TO AN INCOMPATIBLE NON-FOREST USE OR EXEMPT ACCORDING TO RCW 76.09.060(3) AND WAC 222-34-050. THE LANDOWNER DOES ☐ DOES NOT ☐ INTEND TO CLEAR, IMPROVE OR DEVELOP LAND WITHIN THREE YEARS ☐ SIX YEARS ☐. IF THE LANDOWNER INTENDS TO CONVERT OR DEVELOP LAND, ATTACH SEPARATE STATEMENT WITH DETAILS INDICATED IN INSTRUCTION SHEET. LOCAL GOVERNMENT MAY REQUIRE ADDITIONAL INFORMATION. AN ENVIRONMENTAL CHECKLIST MAY BE REQUIRED.

50% OR MORE OF THE MERCHANTABLE TIMBER VOLUME IS TO BE REMOVED NO ☐ YES ☐ (ADEQUATE STOCKING REQUIRED)
LANDOWNER INTENDS TO LEAVE ACCEPTABLE STOCKING NO ☐ YES ☐ SPECIES ____ AVERAGE NO. STEMS/ACRE ____
OR THE LANDOWNER INTENDS TO REFOREST YES ☐ * NO ☐ (ATTACH EXPLANATION OR CONVERSION STATEMENT)
OPERATOR OR NAMED ALTERNATE IS TO PREPARE SITE. (LANDOWNER OR PERPETUAL TIMBER OWNER IS RESPONSIBLE FOR REFORESTATION)
*REFORESTATION WILL BE COMPLETED BY ____ NAME ____ DATE
POST HARVEST SITE PREPARATION WILL BE COMPLETED BY ____ NAME ____ DATE
DESCRIBE SITE PREPARATION METHOD. (ATTACH SEPARATE STATEMENT IF NEEDED.)

METHOD OF REFORESTATION: PLANTING ☐ ARTIFICIAL SEEDING ☐ NATURAL SEEDING ☐ OTHER ☐ (Specify reforestation on additional sheet)
IF PLANTING OR SEEDING: SPECIES ____ STOCKING SIZE ____ PROPOSED STEMS/ACRE ____
IF NATURAL: SEED TREES ☐ SEED BLOCKS ☐ (SHOW ON MAP AND IF ON ADJACENT LANDOWNER, ATTACH AGREEMENT.)
METHOD TO CONTROL COMPETING VEGATATION:

18. WE AFFIRM THAT THE INFORMATION CONTAINED HEREIN IS TRUE TO THE BEST OF OUR KNOWLEDGE AND UNDERSTAND THAT THIS PROPOSED FOREST PRACTICE IS SUBJECT TO THE CURRENT RULES AND REGULATIONS AND THE FOREST PRACTICES ACT.

OPERATOR'S SIGNATURE	LANDOWNER'S SIGNATURE	TIMBEROWNER'S SIGNATURE

RES (12/87) -1352-

SEE FOLLOWING CONDITIONS AND ATTACHMENTS

FIGURE 19-1
Forest practices application-notification form required of forest landowners intending to harvest timber in the state of Washington, 1990. (*Source: Washington Department of Natural Resources. Olympia, 1990.*)

are within 500 feet of a public park. Such organizations may send advisory comments about a proposed operation to the Department of Natural Resources. All application approvals are subject to compliance with any conditions specified by the Department of Natural Resources on an approved application. In addition, the Department of Natural Resources may issue supplemental advisory directives.

In all cases in which rules governing the application of forest practices apply, an operation may commence as if approved if the Department of Natural Resources fails to act on the application within the required time span. Landowners are required to file a reforestation report with the department after planting a harvested area or at the end of a normal planting season. In turn, the department is required to inspect harvested sites for adequate stocking within 1 year after the owner has filed a reforestation report.

The Department of Natural Resources' primary enforcement tools are the notice to comply and the stop-work order. The notice to comply sets forth steps that must be taken to correct a violation of established regulations, while the stop-work order directs cessation of all work connected with any violation (in addition to specifying the same corrective measures as stated in the notice to comply). If a landowner or operator fails to comply, the department may use available funds to carry out a required action; the cost (if not repaid by the landowner) may be attached as a lien on the forestland involved. Landowners and operators may appeal departmental rulings to the governor-appointed Forest Practices Appeals Board. Both civil and criminal penalties can be levied for failure to comply with the law and its rules and regulations. Civil penalties up to $500 per violation may be assessed. Criminal gross misdemeanor penalties of up to $1,000 and/or 1 year in prison may be imposed for each violation. Each day of violation constitutes a separate offense.

The 1984 Department of Natural Resources budget for administration of the state's forest practices law and regulations exceeded $1.6 million (3.5 percent of its annual budget). Forest practices program staffing of the department's Division of Private Forestry and Natural Heritage totaled 43.2 full-time equivalents in the same year, 91 percent of whom were field personnel.

Costs and Accomplishments

Costs Forest practices regulations cannot be implemented without costs, both financial and other. In 1984, state government expenditures for administering the regulatory programs focused on private forest practices in seven states (comprehensive laws) totaled over $10 million (Table 19-2), California spent the most ($4.6 million) while Idaho invested the least ($93,000). Agencies concerned with water quality, fisheries, and wildlife participated to some extent in the administration of such laws; the bulk of the administrative effort, however, was borne by each state's forestry agency.

TABLE 19-2
Administrative and enforcement expenditures for state forest practices programs, by selected states and state agencies, 1984

State and agency	Expenditure
Massachusetts	
Division of Forests and Parks	$ 500,000
Metropolitan District Commission	1,000
Total	$ 501,000
Nevada	
Division of Forestry	$ 100,000
Alaska	
Division of Forestry	$ 317,000
Department of Fish and Game	450,000
Total	$ 767,000
Idaho	
Bureau of Private Forestry	$ 33,000
Bureau of Water Quality	45,000
Department of Water Resources	15,000
Total	$ 93,000
Oregon	
Department of Forestry	$ 1,600,000
Department of Fish and Wildlife	40,000
Department of Environmental Quality	6,000
Total	$ 1,646,000
Washington	
Division of Private Forestry and Natural Heritage	$ 1,635,000
Department of Game	395,000
Department of Fisheries	305,000
Department of Ecology	30,000
Department of Agriculture	5,000
Department of Commerce and Economic Development	5,000
Total	$ 2,375,000
California	
Department of Forestry	$ 4,377,000
Department of Fish and Game	120,000
Regional Water Quality Control Boards	120,000
Coastal Commission	12,000
Department of Parks and Recreation	6,000
Total	$ 4,635,000
Total Expenditures	$10,117,000

Source: Evaluating State Regulation of Private Forest Practices: What Accomplishments at What Costs? by R. K. Henly, P. V. Ellefson, and R. J. Moulton. Reprinted by permission of *Journal of Evaluation and Program Planning,* 11: 325–333, 1988. Pergamon Press.

Though the private sector's financial costs of compliance with state forest practices regulations (management costs beyond those which would normally occur in the absence of regulation) are not easy to determine, they also can be significant. The added direct costs of private-sector compliance in these same

TABLE 19-3
Public- and private-sector costs of administering and complying with state forest practices programs, by selected states, 1984

State	Public agencies	Private sector	Total
Massachusetts	$ 501,000	NA	$ 501,000
Nevada	100,000	$ 2,000	102,000
Alaska	767,000	1,010,000	1,77,000
Idaho	93,000	1,911,000	2,004,000
Oregon	1,646,000	36,936,000	38,582,000
Washington	2,375,000	28,360,000	30,735,000
California	4,635,000	52,325,000	56,960,000
Total	$10,117,000	$120,544,000	$130,661,000

Source: State Forest Practical Regulation in the U.S.: Administration, Cost, and Accomplishments, by R. K. Henly and P. V. Ellefson. Bulletin AD-SB-3011. Agricultural Experiment Station, University of Minnesota, St. Paul, Minn., 1986.

seven states exceeded $130 million in 1984 (Table 19-3); these costs were more than 10 times as much as the state administrative costs incurred in the same year. The combined public- and private-sector financial cost of forest practices regulation in the seven states in question was estimated to be about $131 million in 1984 (Henly, Ellefson, and Moulton 1988). The amount would have been larger if all programs had been funded at levels which program directors considered adequate to fully accomplish the intent of the regulatory legislation.

Other private-sector costs are the psychological traumas resulting from the complex and often frustrating paperwork which must be dealt with by landowners and from their involvement with state forestry officials who are carrying out their legally mandated oversight responsibilities.

Political costs involved in implementation of forest practices regulations may include enlarged bureaucracies and the loss of individual freedom to exercise private rights over forestland (Vaux 1983).

Accomplishments Forest practices regulation can be very costly; thus, questioning its effectiveness is only reasonable. Unfortunately, definitive quantitative estimates of the benefits resulting from the regulation of forest practices are lacking. Not only are there few production functions which relate specific forest practices to benefits generated (e.g., increased wildlife, additional scenic beauty, improved water quality), but also there are serious problems involved in attempts to place a monetary value on such outputs. Furthermore, it is extremely difficult to separate the effects of forest practices regulation from changing landowner practices precipitated by a growing public interest in quality forest environments generally and by increases in timber values (possibly leading to more intense application of forest practices) which have occurred since many state forest practice laws were enacted.

In the states that gather information on whether landowners meet reforestation requirements, the effectiveness of forest practices regulation is substantial. In Massachusetts, the requirements are met 98 percent of the time; in Idaho, the figure is 80 percent; in Washington, it is 80 percent; and in Oregon, it is 95 percent. In Oregon, an estimated 30 to 40 percent more area is reforested than otherwise would have been; in Washington, 10 percent more area (Henly and Ellefson 1986). By one estimate, regulation in California has resulted in an annual $2 million to $3 million reforestation investment which would not have otherwise occurred (Vaux 1983).

The rate at which harvest sites are inspected can be another indicator of the effectiveness of forest practices regulation. The assumption is that the more often an operation is inspected, the more likely it is to comply with state forest practices standards. There are some examples (Henley and Ellefson 1986):

- In 1983, the Oregon Department of Forestry conducted 14,268 inspections. On average, the department inspects operations more than once per year; 10,124 notifications were filed in that year.
- In Washington, reforestation inspection for the period 1983 through 1985 averaged 1,346 per year, substantially below the average 5,992 new operations for the same period.
- On average, 80 percent of timber-harvesting permit applications in California receive an inspection by the state's Department of Forestry. In 1984, when 1,187 new timber-harvesting operations were approved, the department made 6,793 inspections. In addition to preharvest inspections, each harvesting operation in California receives an average of 2 to 3 inspections over its lifetime.
- Inspections in Idaho have been very limited; only 6 percent of the estimated 3,000 operations were inspected in 1984. Low funding levels for administration of Idaho's regulatory program have prevented a more intense inspection program in the state.

The number and severity of enforcement actions required to secure private landowner compliance with forest practices standards can also be a measure of effectiveness. The fewer the number of enforcement actions taken by a state, the more successful a state's forest practices program, since formal enforcement actions represent failures in cooperative resolution of violations. Applying such logic, state forest practices regulation must be given high marks, because very few formal enforcement sanctions (e.g., fines, jail sentences, license revocations) are applied by states. Here are some examples (Henly, Ellefson, and Moulton 1988).

- In 1985, Massachusetts imposed no fines and revoked no timber-harvesting licenses.
- Alaska experiences an average of 3 to 4 violations per year; full prosecution has been sought in only two cases.
- In the fiscal years 1981–1984, Idaho issued only 11 citations for the 378 violations found to be unsatisfactory.

• In Oregon, only 706 citations were issued on a total of 43,213 operations during the period 1980–1983.

• Washington issued only 53 criminal citations for the 20,912 operations occurring during the period 1983–1985.

• In California, though an average of 1,117 violations per year were detected during the period 1980–1984, formal enforcement actions were initiated on an annual average of only 82.

Clientele satisfaction is another indicator of program effectiveness. If a high degree of satisfaction exists among those who benefit from a regulatory program, the program may well be effective. The same may be true if those whose behavior is being limited through regulation do not perceive the regulation as overly burdensome. Study of nine major clients or parties interested in forest practices regulation (e.g., public and private forestry professionals, industrial and nonindustrial private owners of forestland, timber harvester, local units of government, wildlife and environmental groups) in seven states found all such groups to be supportive of or at least neutral toward forest practices regulation. Even among the groups that carried the burden of regulation (landowners and timber harvesters), the attitude toward regulation was fairly positive. Such results indicate that, while regulation may not be painless, neither is it unreasonably onerous (Henly and Ellefson 1986).

Forest Practice Regulation Challenges

Modern comprehensive state forest practices laws represent a major departure in the means by which society's interests in the management of private (and sometimes public) forests are expressed. The general acceptance of such laws as a legitimate tool of forest policy is a major departure from the nationwide resistance to threatened federal regulation of private forest practices that prevailed in the 1930s and early 1940s. Although forest practices regulation is now considered an appropriate policy mechanism in many states, it continues to face a number of challenges (Ellefson 1988). For example, the cost of regulation to both the public and the private forestry sectors continues to be of major concern, as does the frequent lack of private compensation for public taking of private property. Also challenging are complications involved in specifying acceptable forest practices standards; great uncertainty often clouds the relationship between a proposed standard and its supposed ability to exert a positive influence upon the quality of natural environments.

Administration and organization of regulatory programs involving forest practices also pose challenges. Many questions arise. For example:

• Should forest practices standards be set forth in law, or should they be promulgated by a rule-making body?

• If rule-making and advisory bodies are appropriate, who should appoint their members? What combination of interests should be represented?

• Under what circumstances should states employ *notification systems* (in which a landowner notifies an agency of intent to harvest and then begins operation if no response has been received within a specified number of days) versus *inspection-permit systems* (in which a landowner notifies an agency of intent to harvest but proceeds with the operation only after receiving appropriate inspections and agency approval?

• How can forest practices regulations be properly coordinated with the myriad of related state and federal environmental regulations that are often imposed on owners of private forestland?

• To what extent should the ability of landowners, timber harvesters, and forestry professionals to comply with and carry out the standards established by forest practices regulation be certified?

REFERENCES

Bosselman, Fred, D. Callies, and J. Banata. *The Taking Issue*. Government Printing Office, Washington, 1973.

Carmichael, Donald M. Fee Simple Research as a Variable Research Concept. *Natural Resources Journal* 15 (4):739–764, 1975.

Cubbage, F. W., W. C. Siegel, and P. M. Lickwar. State Water Quality Laws and Programs to Control Nonpoint Source Pollution from Forest Lands in the South, in *Proceedings, Conference on Water—Laws and Management*. American Water Resources Association, Bethesda, Md., 1989, pp. 8A-29 to 37.

Department of Natural Resources. *Guide to Regulations Affecting Harvesting and Marketing Forest Products in Washington*. Division of Private Forestry, Olympia, Wash., 1980.

Eckerberg, Katarina. *Environment Protection in Swedish Forestry: A Study of the Implementation Process*. Research Report 1987:12. Department of Political Science, University of Umea, Sweden, 1987.

Ellefson, Paul V. Private Forest, Public Interest: Forest Practice Laws. *Habitat* (Journal of the Maine Audubon Society) 5(4):28–30, 1988.

Environmental Law Institute. *Federal Environmental Law*. West Publishing Company. St. Paul, Minn., 1974.

Fisher, Bruce D., and Michael J. Phillips. *The Legal Environment of Business*. West Publishing Company, St. Paul, Minn., 1983.

Freeman, Alan D. Historical Development of Public Restrictions on the Use of Private Property, in *Public Control of Privately Owned Land*. Center for Urban and Regional Affairs, University of Minnesota, Minneapolis, 1975, pp. 5–17.

Grad, Frank P. *Treatise on Environmental Law*. Vol. 3. Matthew Bender, New York, 1988.

Healy, Robert G., and John S. Rosenberg. *Land Use and the States*. The Johns Hopkins University Press, Baltimore, Md., 1979.

Henly, Russell K., and Paul V. Ellefson. *State Forest Practice Regulation in the U.S.: Administration, Cost and Accomplishments*. Station Bulletin AD-SB-3011. Agricultural Experiment Station, University of Minnesota, St. Paul, 1986.

———, and Robert J. Moulton. Evaluating State Regulation of Private Forest Practices: What Accomplishments at What Costs? *Journal of Evaluation and Program Planning.* 11:325–333, 1988.

Hickman, C. A., and M. R. Hickman. Legal Limitations on Governmental Regulation of Private Forestry in the United States, in Forestry Legislation, report of IUFRO Working Party S4.08.03. Zurich, 1990, pp. 118–136.

Hickman, C. A., and C. A. Martus. Local Regulation of Private Forestry in the Eastern United States. Southern Forest Experiment Station, USDA-Forest Service, New Orleans, 1991.

Hummel, F. C., and H. A. Hilmi. *Forest Policies in Europe; An Analysis.* FAO Forestry Paper 92. Forestry Department, Food and Agriculture Organization of the United Nations, Rome, Italy, 1989.

Mandelker, Daniel R. *Land Use Law.* Michie Company, Charlottesville, Va., 1988.

Novick, S. M. (ed.). *Law of Environmental Protection.* Vol. I and II. Environmental Law Institute, Clark Boardman Company, New York, 1988.

Pacific Reporter. *State versus Dexter.* Vol. 202. 1947, p. 906.

Schoenbau, Thomas J. *Environmental Policy Law.* The Foundation Press, Inc., Mineola, N.Y., 1982.

Vaux, Henry J. State Interventions on Private Forests in California, in *Government Interventions, Social Needs and the Management of U.S. Forests,* by R. A. Sedjo (ed.). Resources for the Future, Washington, D.C., 1983, pp. 124–168.

20

GOVERNMENT FOREST OWNERSHIP

Public ownership of forestland is a general expression of society's concern over the consequences of private ownership of forests. Technical assistance, financial assistance, and regulatory programs aside, society by design or by chance has often preferred to retain exclusive and direct control over a significant portion of the nation's forests. The value of public goods and the likely occurrence of extensive externalities become so important that total government control is deemed appropriate. Exclusive public authority over a large portion of the nation's forestland is significant; historical events leading to such authority often make for exciting reading (Coggins and Wilkinson 1987; Dana and Fairfax 1980; Gates 1968). Of concern in this chapter, however, is the extent of public forestland ownership, along with the goods and services provided by such land and the policies and programs which public agencies have chosen to focus on it.

The reality of public ownership of forestland is reflected by the fact that over 280 million acres of unreserved forestland, or 38 percent of the nation's forests, are controlled by a variety of federal, state, and county governments. (*Unreserved forestland* is land not withdrawn from timber use by statute or administrative regulation.) In addition, an important portion of the 75 million acres of reserved land managed by the USDI-National Park Service is forested, as are large parts of reserved public land controlled by state parks and wildlife refuges and by the USDI-Fish and Wildlife Service, the U.S. Department of Defense, the Tennessee Valley Authority, and the USDI-Bureau of Indian Affairs (which holds land in trust for Native Americans). Of the nation's 483 million acres of timberland, society has retained 28 percent in public ownership; over 71 percent of this land is federally managed.

EXTENT AND CHARACTER

The federal government owns over 727 million acres of land, approximately one third of the nation's total surface area. Six federal agencies are responsible for the management of nearly 720 million of these acres, namely:

- U.S. Department of the Interior
  - National Park Service—75 million acres
  - Fish and Wildlife Service—90 million acres
  - Bureau of Land Management (BLM)—334 million acres
  - Bureau of Indian Affairs—over 6 million acres
- U.S. Department of Agriculture
  - Forest Service—191 million acres
- Other federal agencies
  - Department of Defense—23 million acres
  - Other agencies—14 million acres

Although the land administered by such agencies is not all forested, some federal agencies control significant amounts of forestland. For example, the BLM is responsible for managing 5.8 million acres of timberland, while the USDA-Forest Service oversees 119 million acres of unreserved forestland, of which 85.2 million acres is classified as timberland (Table 16-1). Federal agencies are not the only public owners of forests. In 1987, a total of 25 percent of the nation's publicly owned timberland was controlled by state governments (26.7 million acres) and by county and municipal governments (7.0 million acres).

Public ownership of forests is not evenly distributed throughout the United States (Table 16-1). Nearly 61 percent of publicly owned timberland, for example, or 84.5 million acres, is located in the western portion of the nation. Nearly three-quarters of the latter is in the National Forest System. State-owned timberland is concentrated in the North Central (7.5 million acres), Pacific Northwest (7.5 million acres), and Northeast (5.8 million acres) regions. County and municipal ownership is especially prevalent in the North Central region (4.9 million acres), notably Minnesota and Wisconsin. In some states, public ownership of forests is especially pervasive. Of Idaho's 14.5 million acres of timberland, 79 percent is owned by public agencies. Similar patterns exist in Arizona (which has 99 percent public ownership), Washington (51 percent), and Colorado (73 percent). In contrast, public ownership of timberland in the South is very modest, constituting only 10 percent of the regional total.

PRODUCTS AND SERVICES

The goods and services provided by publicly owned forests are as diverse as the forest ecosystems located within such ownerships. Most provide the range of benefits forests are capable of producing, namely water, recreation, timber, wilderness, range, and fish and wildlife. In some cases, management policies emphasize one good or service at the expense of others. For example, the management of designated forest wilderness precludes the production of timber. Forest-based recreation is an extremely important benefit of public forests. In 1986, for example, the USDI-National Park Service's National Parks and National Monuments were the focal point for nearly 94 million recreation visits, many of which were predicated on the existence of forests within a park or a monument. Similarly, the BLM in 1986 provided opportunity for an estimated 20 million land-based recreational visits (often forest-oriented) of various types, including hunting, camping, and motorized recreation (USDI-Bureau of Land Management 1988a). The National Forest System in 1988 created opportunities for the occurrence of 242.3 million recreational visitor-days, of which 29 percent (70.2 million visitor-days) involved forest campgrounds, picnic areas, and swimming sites (USDA-Forest Service 1989). State parks and related recreation areas—many of which have forested environments—received a total of nearly 694 million visits in 1987 (Bureau of the Census 1989).

Public forests are also major reservoirs of timber and are likely to continue to be so in the future. As such, they are often critical to the wood-based industries of local and regional economies. The National Forest System, for example, was the source of 12.6 billion board feet of harvested timber in 1988—13 percent of the timber harvested from all lands in the United States (USDA-Forest Service 1989). This timber was valued at over $1.2 billion, and 60 percent of it was harvested from National Forests located in Pacific Coast states. In 1987, the USDI-Bureau of Land Management was the source of 1.3 billion board feet of timber, valued at more than $142 million. Most of the timber (98 percent of the sale value) offered by the agency originated from agency-administered Oregon and California Revested Lands located in western Oregon. Other federal public agencies (for example, the U.S. Department of Defense and the USDI-Fish and Wildlife Service) provide timber in lesser amounts. State agencies responsible for the management of public forests are also significant sources of timber.

Wilderness is a unique resource associated with publicly owned forestland. Via the National Wilderness Preservation System, for example, 89 million acres (in 1985) of land is made available for recreational experiences involving solitude in areas offering opportunities for appreciation of the unaffected evolvement of various natural ecosystems. A significant portion of the land in the federal system (10 million acres in 1981) is forested. The USDI-National Park Service is responsible for 42 percent of the system's area (36.8 million acres), the USDI-Fish and Wildlife Service for 22 percent (19.3 million acres), and the

USDA-Forest Service for 36 percent (32.2 million acres). The BLM accounted for less than 1 percent in 1985. About 5 percent (11.8 million recreation visitor-days) of the USDA-Forest Service's recreation activity occurred on lands that are part of the National Wilderness Preservation System (USDA-Forest Service 1989). Education, scientific study, habitat, and ecosystem preservation are growing uses of designated forest wilderness.

Recreation, timber, and wilderness are but three of many forest-based benefits provided by publicly owned forestlands, which are also significant sources of forage for domestic animals and wildlife, as well as major suppliers of water from forested watersheds. In association with forests, they are also major sources of important minerals. The range of benefits provided by public forests is large and diverse. These forests are an extremely important part of the nation's social and economic fabric, and many of them play significant roles in the economic and social lives of nearby communities. To ignore or downplay their importance in the course of developing natural resource policies for the nation would be a serious mistake.

POLICIES AND PROGRAMS

The benefits which flow from forested public lands are the result of an enormous array of policies, programs, and management strategies that have been developed over the years. Many of them have been the result of the intense public debate which invariably surrounds the search for the public interest in public lands. Always sought but never actually found, the ambiguity of *public interest* is made more tangible when debates occur over specific public land-use and management issues, focused on questions such as:

- Should formal designation of a wilderness area override state water laws?
- Should off-road vehicles be banned from certain public forests?
- Should a road be constructed through the forested habitat of a potentially endangered species?
- Should forested environments be open to all-terrain vehicles, at the expense of hikers and cross-country skiers?
- Should public forestland be designated for timber production? Or should forestland be opened for the extraction of valuable hard-rock minerals?

In all such debates, the question that is eventually focused on is: What is the public interest? Some idea of the public interest can be gained by briefly reviewing selected policies which have been developed to guide the management of some public forests.

USDA-Forest Service Policies

A significant participant in management of federal public forestland is the USDA-Forest Service, which is responsible for the 191 million acres of the

National Forest System. The agency's forest-management responsibilities date back to the late 1890s and early 1900s. It was during this period that withdrawals from the public domain (largely in the West) were made to create the Forest Reserves which, in 1905, were transferred to the U.S. Department of Agriculture to be managed as National Forests. The system was enlarged during the period 1912–1930, when the USDA-Forest Service purchased several million acres of agriculturally depressed or abandoned land east of the Mississippi.

The Forest Reserves that eventually became the National Forests were established "to improve and protect the forest . . . [and] for the purpose of securing favorable conditions of water flows, and to furnish continuous supply of timber." For decades, the management of the National Forests was primarily custodial; they supplied less than 5 percent of the nation's annual timber harvest prior to 1940. With the postwar housing boom and a burgeoning interest in the recreational out-of-doors, National Forest management approaches became more intense. Today, there is considerable public interest in the goods and services the National Forests are capable of providing. Among the more important policies guiding their management are (USDA-Forest Service 1983):

- *Multiple use* Requires the National Forests to be managed for recreation, range, timber, water, wildlife, and fish in a "combination that will best meet the needs of the American people . . . with consideration being given to the relative values of the various resources and not necessarily that combination that will give the greatest dollar return or greatest unit output" (Multiple-Use Sustained Yield Act of 1960).
- *Sustained yield* Requires the National Forests to be managed for the sustained yield of several products and services. They are to be managed so as to achieve and maintain in perpetuity a high-level annual or periodic output of renewable resources without impairment of the productivity of National Forest land (Multiple-Use Sustained Yield Act of 1960).
- *Wilderness* Requires certain areas in the National Forests to be managed so as to retain their primeval character, without permanent improvements or human habitation. They are to appear as having been affected primarily by the forces of nature; to be outstanding opportunities for solitude and for primitive and unconfined types of recreation; and to be of at least 5,000 acres in size. They may contain especially important ecological, educational, or historical features. Such lands are part of the National Wilderness Preservation System (Wilderness Act of 1964).
- *Timber management* Requires the National Forests (when managed for timber purposes) to be managed so that harvest rates are nondeclining and evenflow in nature; biological diversity of plant and animal communities is maintained; clear-cutting is limited to prescribed sizes and configurations; rotation ages are set at culmination of mean annual increment; and timber management

is practiced on only those lands which are physically and economically suitable for timber-production purposes (National Forest Management Act of 1976).

• *Land-management planning* Requires that for each National Forest a land-management plan be prepared and updated every 10 years (National Forest Management Act of 1976).

• *Timber sale receipts* Requires that a certain portion of National Forest timber sale receipts be retained for use in reforesting and applying silvicultural practices to harvested areas (Knutson-Vandenberg Act of 1930). Also requires a proportion (25 percent) of revenues collected by National Forests (e.g., timber, grazing, recreation, minerals) to be returned to state and local governments in the areas in which the National Forests are located.

• *Wild and scenic rivers* Requires certain areas in National Forests to receive special management status because of their proximity to environmentally and historically exceptional free-flowing rivers (Wild and Scenic Rivers Act of 1968).

• *Endangered Wildlife Species* Requires certain habitats within the national Forests to receive special management if threatened or endangered species of wildlife are involved (Endangered Species Act of 1973).

USDI-Bureau of Land Management Policies

The BLM was established in 1946 with the merger of the U.S. Department of the Interior's Grazing Service and General Land office. It has responsibility for a large portion of the nation's publicly owned land—a total of 334 million acres, of which 5.8 million is timberland. Total acreage includes virtually every type of ecosystem to be found west of the Mississippi, including deserts, glaciers, arctic plains, and forests. BLM holdings are primarily public-domain lands that are not set aside as National Forests or National Parks. The exception is the 24 million acres of revested railroad and wagon-road lands in western Oregon that contain some of the nation's most valuable forests and timberland (Barton 1987; Muhn and Stuart 1988).

The agency administers and supervises mining operations on an additional 732 million acres of federal mineral estates underlying other federal, state, and private ownerships. Among the many policies guiding the management of land administered by the BLM are:

• *Land ownership* Requires that "public lands [BLM] be retained in Federal ownership, unless as a result of the land use planning procedures provided for . . . that disposal of a particular parcel will serve the national interest" (Federal Land Policy and Management Act of 1976).

• *Multiple-use sustained yield* Requires BLM lands to be managed for multiple uses, namely "a combination of balanced and diverse resource uses . . .

including, but not limited to, recreation, range, timber, minerals, watershed, wildlife and fish, and natural scenic, scientific and historical values'' and that such uses be managed to achieve in perpetuity ''a high-level annual or regular periodic output of the various renewable resources . . . consistent with multiple use'' (Federal Land Policy and Management Act of 1976).

• *Land-management planning* Requires planning for the use and management of BLM lands. ''The national interest in such lands will best be realized if the public lands and their resources are periodically and systematically inventoried and their present and future use projected through a land use planning process.'' Furthermore, the agency is to ''. . . with public involvement, . . . develop, maintain, and, when appropriate, revise land use plans which provide by tracts or areas for the use of the public lands'' (Federal Land Policy and Management Act of 1976).

• *Local government compensation* Requires payment to state and local governments compensation for burdens created as a result of the immunity of BLM lands from state and local taxation (Federal Land Policy and Management Act of 1976).

• *Wilderness* Requires intensive inventory of BLM lands of 5,000 acres or more which are roadless and which have wilderness characteristics as described in the Wilderness Act of 1964. Recommendations are to be made as to the suitability of such areas for inclusion in the National Wilderness Preservation System (Federal Land Policy and Management Act of 1976; Wilderness Act of 1964).

• *Oregon and California lands* Requires sustained-yield and multiple-use management of revested Oregon and California Railroad Grants Lands and Coos Bay Wagon Road Grant Lands (Oregon and California Act of 1937).

• *Range* Requires issuance of permits to private individuals to graze livestock on certain BLM lands and the charging of fees for the privilege of doing so. A portion (50 percent) of collected grazing fees are to be retained for range improvements, half of which must be spent in the district in which the fees were collected (Taylor Grazing Act of 1934; Federal Land Policy and Management Act of 1976).

• *Recreation* Requires provision of a broad spectrum of recreation in the amounts needed to meet the needs and desires of public-land visitors. The BLM must plan for all outdoor recreation activities, limit recreation use where resources are in danger, use special permits to ensure equity in recreation opportunity, and charge appropriate fees for certain recreational use of public lands (USDI-Bureau of Land Management 1988b).

• *Wildlife* Requires appropriate species and habitat management of wildlife, including endangered species, bald eagles, and wild horses and burros (e.g., Wild Free Horse Roaming and Burro Act of 1971; Bald Eagle Protection Act of 1972; Endangered Species Act of 1973).

USDI-National Park Service Policies

The USDI-National Park Service is also an important manager of forested land. An orientation toward preservation has characterized the agency's mission since its inception in 1916, although recent years have seen a broadening of responsibilities to include management of recreation areas, urban parks, cultural areas, and the like. The USDI-National Park Service has always been torn by mandates to preserve certain public lands and the resources found thereon, while at the same time making such lands available for the benefit and enjoyment of the public. Among the many policies affecting the use and management of forests within the agency's jurisdiction are:

- *Mission* Requires the agency "to conserve the scenery and the natural historic objects and the wildlife . . . and to provide for the enjoyment of the same in such manner and by such means as will leave them unimpaired for the enjoyment of future generations" (National Park Service Act of 1916).
- *Land-use and land-management planning* Requires each park to establish a general management plan which sets forth a park-management concept; a role for the park in the context of regional trends (e.g., transportation, economic development); and a series of management strategies appropriate to achievement of park-management objectives (USDI-National Park Service 1988).
- *Ecosystem management* Requires management of natural systems in accord with objectives set for natural zones, cultural zones, and development zones within a park. Landscape conditions caused by natural phenomena in natural zones, for example, will not be modified unless required for public safety (USDI-National Park Service 1988).
- *Interpretation and education* Requires conduct of interpretive programs designed to instill in park visitors an understanding of unique natural and cultural phenomena (USDI-Fish and Wildlife Service; USDI-National Park Service 1988).

USDI-Fish and Wildlife Service Policies

The USDI-Fish and Wildlife Service, created in 1966, is responsible for the management of over 90 million acres of public land within the National Wildlife Refuge System. Its forests often provide important habitat required for the propagation of various species of wildlife. Policies guiding the agency's actions (including the use and management of forests) are set forth in federal law as well as in national plans (e.g., the Service Management Plan, which sets forth overall directions for 5- to 10-year periods), regional plans (5-year plans defining regionwide strategies and objectives), and field plans (e.g., policies for management of wildlife refuges, hatcheries, and laboratories) (Chandler 1985). Among policies guiding the use and management of the National Wildlife Refuge System are:

• *Mission* Requires the "conservation, protection, restoration, and propagation of selected species of native fish and wildlife," to be accomplished in part by the newly created National Wildlife Refuge System. Power lines, telephone lines, roads, and ditches (and related uses) which are determined to be consistent with the uses for which the System was established are permitted (National Wildlife Refuge Administration Act of 1966).

• *Recreation* Requires curtailment of "public recreation use generally or certain types of recreation uses within individual [refuges] . . . in order to avoid adverse effects upon fish and wildlife populations and management operations." Public recreational use of refuges is appropriate when judged incidental or secondary.

Public forests administered by the federal agencies discussed are also subject to a number of policies which are not unique to any one agency. For example, general policies concerning the quality of the environment are contained in the National Environmental Policy Act of 1970. Federal policies concerning water and air quality are contained in the Clean Air Act of 1970 (as amended) and the 1972 Amendments to the Federal Water Pollution Control Act (as amended). Similarly, the Land and Water Conservation Fund Act of 1964 (as amended) provides policy direction for federal land-management agencies by making funding available for the acquisition of additional public land. Also uniformly appropriate are policies concerning management of endangered species and wild and scenic rivers, as well as those concerning provision for wilderness resources.

State and County Policies

State and county governments are responsible for the management of significant parcels of forestland, including 33.7 million acres of timberland. State and county forests are usually managed for multiple uses, with a focus on providing a sustained yield of forest benefits. In recent years, state and county forests have assumed important roles in facilitating state and regional interests in economic development (Nothdurft 1984). Frequently defined economic development goals for state-owned forests include the following (Council of State Governments 1982):

• *Provide for net additions to state gross product.*

• *Facilitate stabilization of state and regional economies* Reduce swings and cyclical variations in economies generally.

• *Enhance economic diversification* Reduced dependence on a single industry.

• *Ensure fiscal stability for government programs* Enhance private-sector growth as needed for uniform flow of tax revenue.

• *Provide for balanced economic growth, particularly in areas of high unemployment.*

State interests in economic development via the forestry sector are often accomplished via approaches such as guaranteeing that specific companies will have access to state-owned timber resources, and offering timber at reduced prices to favored wood-based industries or companies. States which have especially aggressive economic development programs based in part on state-owned forests include Michigan, Minnesota, Oregon, Washington, and Wisconsin. Michigan's target industry program attempts to improve the business climate and marketing opportunities for forest products industries, to assure stable supplies of progressively more valuable timber in the future, and to more fully coordinate and focus public and private forestry activities (including those concerning state-owned forests) on economic development goals (Webster 1985). In addition to economic development goals, some states have constitutional or legislative provisions which designate the use of revenue obtained from the sale of goods and services produced by state forestlands. Revenue from Washington's state forests, for example, is deposited in school trust funds designated for general and construction use.

Government Ownership Challenges

Publicly owned and managed forests embody a variety of challenges which are faced on a continuing basis by policymakers and public land administrators alike (Foss 1987; Brubaker 1984). The intensity of public debate over the appropriate use of public forests is a disturbing challenge. This is especially true when such debate concerns the allocation of forestland to special land-use categories, including forested habitat for rare and endangered wildlife and designated forest wilderness for recreational use. Participants in public-land-use battles usually have strongly held ideologies and a propensity to shun necessary political compromise; this makes such issues visible, intense, and politically divisive. Of similar concern are the often annual clashes over appropriate levels of investment in public forests, as well as the ability of public treasuries to sustain such levels over long periods of time. The short-term planning horizons of legislative systems often mean that the long-term financial needs of public forests are neglected or postponed. Also challenging, though subtle, is continuing apprehension over the appropriate role of public forests in a free-market system. Such apprehension frequently rises to the level of public attention and debate under the banner of privatization or so-called sagebrush rebellions involving transfer of public land to private ownership.

The challenges to public land management are many. Their number and intensity will in all likelihood increase as the myriad of public interests begin to fully realize the enormous value of forests currently being managed by public agencies.

REFERENCES

Barton, Katherine. Bureau of Land Management, in *Audubon Wildlife Report: 1987,* by R. L. Di Silvestro (ed.). Academic Press, New York, 1987.

Brubaker, S. *Rethinking the Federal Lands.* Resources for the Future, Inc., Washington, 1984.

Bureau of the Census. *Statistical Abstract of the United States.* U.S. Department of Commerce, Washington, 1989.

Chandler, William J. The U.S. Fish and Wildlife Service, in *Audubon Wildlife Report: 1985,* by R. L. Di Silvestro (ed.). National Audubon Society, New York, 1985.

Coggins, G. C., and C. F. Wilkinson, *Federal Public Land and Resources Law.* The Foundation Press, Inc., Mineola, N.Y., 1987.

Council of State Governments. *Forest Resource Management: Meeting the Challenge in the States.* Lexington, Ky., 1982.

Dana, Samuel T., and Sally K. Fairfax. *Forest and Range Policy.* McGraw-Hill Book Company, New York, 1980.

Foss, Phillip O. (ed.). *Federal Lands Policy.* Greenwood Press, Inc., Westport, Conn., 1987.

Gates, Paul W. *History of Public Land Law Development.* Government Printing Office, Washington, 1968.

Muhn, James, and Hanson R. Stuart. *Opportunity and Challenge: The Story of the BLM.* Government Printing Office, Washington, 1988.

Nothdurft, William E. *Renewing America: Natural Resource Assets and State Economic Development.* The Council of State Planning Agencies, Washington, 1984.

USDA-Forest Service. *The Principal Laws Relating to Forest Service Activities.* Agriculture Handbook No. 453. U.S. Department of Agriculture, Washington, 1983.

———. *Report of the Forest Service: Fiscal Year 1988.* U.S. Department of Agriculture, Washington, 1989.

USDI-Bureau of Land Management. *Public Land Statistics: 1987.* U.S. Department of the Interior, Washington, 1988a.

———. *Recreation 2000: a strategic plan.* U.S. Department of the Interior, Washington, 1988b.

USDI-National Park Service. *Management Policies.* U.S. Department of the Interior, Washington, 1988.

Webster, Henry H. Forestry: A Target Industry for Economics Development in Michigan, in *Proceedings of Forum on Forest Resources in Regional Economics Development.* Publication 420–002. Virginia Cooperative Extension Service, Blacksburg, Va., 1985.

FIVE

SUMMARY AND OBSERVATIONS

Forest resource policies are the product of a seven-stage policy process involving agenda setting, formulation, selection, legitimizing, implementation, valuation, and termination. Among the important political actors that engage the process are legislatures, judicial systems, bureaucracies, interest groups and certain segments of the media, the general public, and national political parties. The outcome of interactions between actors and the policy process is a vast array of forest resource policies and programs. Science has an important role to play in policy development as does personal and professional ethics. Research on policy development is important to a properly functioning policy process.

SUMMARY AND OBSERVATIONS

The importance of forests as contributors to the social, economic, and political fabric of nations throughout the world is seldom questioned. Rather, the public and private sectors of many societies focus extensive attention on the use and management of forest resources. Thus, there is a certain logic to people's pervasive curiosity about events which lead to the establishment of forest resource policies, who participates in the development of such policies, and the policy consequences of interactions between policy events and policy participants. The motives for such curiosity can range from a selfish interest in manipulating the policy development processes for one's own betterment to an interest in seeking the development of policies that lead to more positive societywide effects. For practicing foresters and for students who aspire to be forestry professionals, an understanding of how public forest resource policies are established is an important aspect of professional development.

Forest resource policies are developed at various levels within government, and a multitude of participants in both public and private service are involved in the process. Furthermore, these policies have varying degrees of impact on many people, organizations, and geographic locations. To generalize about complicated policy development processes is risky. Even so, making educated generalizations about policy development can be a logical start toward developing a better understanding of how forest resource policies come into being. In this book, policies are treated as the product of a seven-stage process; agenda setting, formulation, selection, legitimizing, implementation, evaluation, and termination. This process, however, should be understood as an abstraction of reality; the policy events that make up the process are seldom discrete events, and they are not necessarily accomplished in the indicated order. However, treating policy as a process does provide a logical framework for development of an appreciation of how forest resource issues come into existence, how they are processed, and how they are ultimately disposed of.

The process by which forest resource policies are developed cannot be divorced from the political actors who give it life. To attempt to understand the process without an appreciation of the participants would be a mistake. From the perspective taken in this book, the major participants in policy development are legislatures, judicial systems, bureaucracies, interest groups, and (to varying degrees) certain segments of the general public, the media and national political parties. None of these participants can be viewed as a distinctly separate entity in the policy process. Legislatures and interest groups often interact for the common good, as do judicial systems and large bureaucracies. The vitality of the policy process and the merits of the resulting forest resource policies reflect the energies expended by a variety of participants in the policy process.

The substance of forest resource policies is incredibly complex in both subject and application. Attempting to categorize and summarize forest resource policies is as risky as generalizing about policy processes. Yet such categorizing and summarizing are necessary in order to gain an appreciation of the nature of public forest resource policies.

For purposes of this discussion, policies and programs are viewed as being developed over an extended period of time, and as involving ever-greater intrusions of government into the forestry business of the private sector. At one extreme, forests are privately owned, and industrial and nonindustrial forest landowners make their own policies. Greater societal involvement in the private sector occurs when government undertakes information and service initiatives—and these initiatives are often closely followed by fiscal subsidies and tax incentives. Government policy involvement in the private forestry sector becomes even more intense when public regulatory policies constraining the activities of private landowners are established. The ultimate in government involvement is government ownership of forest resources. Again, neatly described categories belie the complexity of the policy products which have actually been produced by the policy process.

FOREST SCIENCE IN THE POLICY PROCESS

The events of the policy process are extremely demanding of various types of information. In a political context, the information required often takes the following forms:

- Who opposes and who supports a proposed policy?
- What rewards or sanctions are necessary in order to secure an opponent's support for a proposed policy?

• Will there be a greater probability of agreement on a proposed policy if the policy is suggested at a later date, when the composition of participants in the policy process is different?

As important as such information is to policy development, equally important is ready access to scientific and technical information concerning the substance of existing or proposed forest policies. Scientific and technical information can provide a much needed perspective on the consequences of existing policies—or the consequences of their lack of existence. Such a perspective can facilitate agenda setting and the government's eventual recognition of the need to establish a new policy or modify an existing one. Similarly, scientific information can greatly facilitate the formulation of forest resource policies by clearly defining the consequences of proposed policies, a necessity if the linkages between proposed policies and desired forestry benefits are to be accurately determined (for example, Burns 1989). The infusion of scientific information into the policy process can also encourage the proposers of forest policies to define the substance of their goals and objectives more carefully. More careful definition helps to focus debate on differences in values—the area in which difficult political judgments are most needed (Schultz 1968).

Scientists who are interested in the use and management of forest resources can play a number of roles in policy development and implementation. For example, they are commonly called upon to serve as advisers to government agencies. In this capacity they provide technical counsel to administrators, legislators, and related policymakers. Recently constituted scientific advisory committees have addressed a number of forestry subjects including the role of fire in natural forest ecosystems managed by federal agencies, the development of regulations to implement the National Forest Management Act of 1976, and the determination of endangered or threatened status for certain species of forest wildlife. What government wants from scientists is their keen concern for facts, their objectivity, and their logical problem-solving abilities. However, scientists values and biases often come to the surface and affect their work. In addition, scientists working as advisers in a political setting may underestimate the complexity of the policy choices faced by government, may become frustrated by an inability to secure sufficient information on which to base policy advice, and may become annoyed and impatient with the political bargaining that is necessary in order to secure compromise and agreement on a particular policy option (Lakoff 1977).

Scientists play an important role in the development of policies by publicly debating questions of fact concerning the use and management of forests. All too often, the consequences of an existing policy or a pending issue are plagued with uncertainties; standards for action on a particular issue have not yet been scientifically developed and established. Through debate in public meetings, such as legislative hearings, and by publishing their finding in scientific journals, scientists confront uncertainties and clarify the substance of important public issues or redefine the merits of policies suggested to address a particular issue. Classic examples include scientific debates over the effects of pesticides and the consequences of tropical deforestation on global warming. Widespread debate in the scientific community usually results in better policies (Lakoff 1977; Weiss 1978).

The forestry topics subjected to scientific inquiry vary considerably according to the geographic scope of an issue—whether it is regional, national, or worldwide—and according to the mission and responsibilities of a particular research organization. Certain conditions predicted for the future have been judged likely to precipitate issues that will need economic and policy research. The information obtained will in turn influence the development of forest resource policies. Example conditions are as follows (Ellefson 1989):

- *Forestry will experience growing linkages to the world economy,* necessitating research on international trade in forest and related products, and on social and economic growth of developing nations.
- *Forest resource policymaking and program administration will increase in complexity,* necessitating research on policy development and program administration, on forest resources law and legal processes, on taxation of forest products and forest resources, and on institutional arrangements for directing use and management of forests.
- *Forest-based industries will continue significant restructuring within domestic and worldwide economies,* necessitating research on the economic structure and performance of wood-based industries.
- *Rural needs for social and economic development will increase as a focus of forestry interests,* necessitating research on community and regional economic growth and development.
- *Scientific and technical developments will increasingly influence resource use and productivity,* necessitating research on development, dissemination, and adoption of new technology.
- *Forest management and production process will grow increasingly complex,* necessitating research on wood-fiber production, timber harvesting, and timber marketing; on production and evaluation of forest and wildland recreation; on management of fire, insects, and diseases in forested environments; on production and valuation of water from forested environments; and on structure and performance of nonindustrial private forests.
- *Information requirements and information management will increase as a concern of forestry interests,* necessitating research on resource assessment, information management, and communication technology; and on forecasting the demand and supply of forests resources, products, and services.
- *Forest use and management decisions will increasingly reflect environmental quality interests,* necessitating research on forestry-sector environmental effects.

An example of suggested forestry research is the 1980–1990 national program of research for forests and associated rangelands which was developed as a joint effort by 60 forestry schools and universities and the USDA-Forest Service (1982). Seven important program areas were identified as in need of scientific investigation, namely:

- *Multiresource inventory, appraisal, and evaluation* Alternative uses of forestland, and multiple-use potential and evaluation
- *Timber management* Genetics and breeding of forest trees; economics of timber production; and biology, culture, and management of timber crops
- *Forest protection* Control of insects and diseases, and prevention and control of forest and range fires
- *Harvesting, processing, and marketing of wood products* Forest engineering systems; properties, processing, and protection of wood; and economics and marketing of wood products
- *Forest watersheds, soils, and pollution* Watershed protection and management; soil, plant, water and nutrient relationships; and alleviation of soil, water, and air pollution
- *Forest wildlife and fisheries habitat improvement*
- *Forest recreation and environmental values*

In 1990 the Committee on Forestry Research of the National Academy of Science's National Research Council further identified needs for forestry research (National Research Council 1990). Mentioning concerns over loss of biological diversity, effect of global deforestation on climate, ability to sustain wood production, and increasing demands for preservation of pristine forested areas, the committee recommended that research be strengthened in five areas:

- *Biology of forest organisms* Genetic bases for forest health and productivity, long-term site productivity, and pest management
- *Ecosystem function and management* Landscape ecology, global change, biological diversity, intensive timber production, and alternative silvicultural systems
- *Human-forest interactions* Sociology of humans and forests
- *Wood as a raw material* Structure and properties of wood, recycling of wood materials, cost-control mechanisms, and timber harvesting
- *International trade, competition, and cooperation* Supply and demand conditions, and trade and development centers

In sum, ample supplies of scientific and technical forestry information are critical ingredients for a well-functioning policy process. As stated by Lakoff (1977, p. 383):

> Insofar as most political decisions reflect the interplay of interest groups, scientists and technologists can serve a particularly useful role in articulating policy alternatives

> and contributing to public debate. . . . They can contribute to the strengthening of the heart of the parliamentary system of government. There can be no guarantee that even informed and enlightened political leaders will make wise decisions, but it is surely better for society to have the best and fullest advice from experts in all fields.

ETHICS AND POLICY DEVELOPMENT

The development and implementation of forest resource policies can often place forestry professionals in difficult ethical and moral situations. The values and moral standards that individuals use to guide their relations with others should also be used in developing the substance of a proposed forest resource policy. Such standards—ethics—involve principles of right and wrong, or good and bad behavior. In general, forestry professionals involved in any of the many events of the policy process should follow the principles of basic honesty and conformity to law, avoidance of conflicts of interest, an orientation to public service, and attention to procedural fairness. Public managers should consider competing interests in the development of public policies and should exercise informed moral judgment regarding the balance of such interests (Rosenbloom 1986).

Ethical principles have application in at least two broad policy circumstances, namely, the ethical conduct of forestry professionals in the development of policies, and the application by professionals of ethical norms or values in selecting the substance of a particular forest resource policy.

The ethical conduct of professionals engaged in the development of policies and programs can be tested by many circumstances (Callahan and Jennings 1983). Here are some examples:

- Should a program manager withhold information that casts doubt on the usefulness of a policy proposal that the manager favors?
- Should a public land manager actively support an interest group (by serving on its committees and testifying at its hearings) if the interest group vigorously opposes forest-management policies that are being implemented by the manager's agency?
- Should a forestry professional actively engage in logrolls and sidepayments in order to resolve a bitterly contested matter of forest policy?
- Should a policymaker select a policy which yields substantial forestry benefits for society as a whole even if the policy inflicts severe costs on a small group of forest users?
- Should a policy analyst yield to pressure to alter evaluations so that they will better support an administrator's position in an intensely political situation?
- After estimating the cost range of a proposed program to be $5 million to $15 million, should an analyst at the request of an administrator publicly distribute only the lowest cost estimate, in order to greatly improve chances of legitimizing the proposed program?

• Should an analyst respond favorably to a request to use more favorable pricing estimates when forecasting an expected demand for timber, thereby assuring continued public investments in timber programs?

• Should an analyst protest publicly when an administrator distorts or misrepresents the results of an evaluation?

• Should analysts slant their analyses so as to counter the slanted analyses of others?

Ethical concerns also arise when professionals face policy and program proposals which conflict in substance with their own ethical or political standards. Again, here are some examples:

• A professional who cherishes efficiency and effectiveness in program operation faces a certain moral dilemma when asked to continue the operation of a program that is clearly ineffective and inefficient. The problem becomes even more acute when the program's benefits are distributed to a select and wealthy few—a potentially disquieting situation for an ethical-minded professional.

• A forestry professional who is asked to implement forest resource policies that have been selected without the benefit of public participation, adherence to procedural due process, or full disclosure of the policy consequences for those likely to be impacted may be greatly disturbed.

Equally sensitive to a professional may be policies that conflict with personally held liberal or conservative values. Consider the following examples:

• Should a conservative-minded forestry professional be asked to actively implement government forestry programs which intervene in the forestry activities of private individuals?

• How can a professional be comfortable participating actively in forestry programs that are clearly designed to distribute income more equally among members of society, when the professional's value scheme justifies such inequities on the basis of ability and (possibly) ancestry?

• What ethical obligation does a professional have to implement programs with goals which may conflict with his or her personal values (such as nondiscrimination along ethnic or gender lines)?

Such ethical and moral dilemmas are often very real for forestry professionals.

Then too, as society struggles with concepts of ecological and environmental ethics, professionals face additional ethical difficulties. Of fundamental concern is the ethical responsibilities of humans toward trees, wildlife, nature, and the biosphere in general. For example, a question that often arises is: Do trees and animals have rights? Of more explicit concern is the proper role of forestry professionals in activating such responsibilities (Rolston 1988; Stone 1988).

The search for solutions to such problems leads to consideration of a number of tenets regarding natural systems, the elements which make up such systems, and the human role in such systems. A forerunner in this respect was Aldo Leopold (1970), who argued that humans are members of an interdependent community composed of soils, water, plants, and animals. Leopold held that the stability, integrity, and beauty of this community should be the basis for the ethical dimensions, or land ethic, used to guide the use and management of such resources. Stability of forest ecosystems, for example, may be adversely affected by decades of fire suppression. The integrity of wild and undeveloped forests may be lost when roads are constructed through them, and the natural beauty of forests may be decimated when indiscriminate timber harvesting is permited (McQuillan 1990).

Among more recently suggested liberal and reformist concepts of environmental ethics are the principles of "deep ecology," which focuses on tenets such as protection of the nonhuman world, maintenance of biological diversity, reduction of human population, and consumption only for vital human needs (Devall and Sessions 1985).

Although environmental ethics continues to evolve, there is a growing awareness that it is right to protect natural systems and wrong to abuse them, and that natural systems have an intrinsic value and consequently have at least the right to exist, as reflected in such laws as the Endangered Species Act of 1973 (Nash 1989). The forestry professional who adheres to such principles may confront extremely discomforting ethical dilemmas when asked to pass judgment on the substance of particular forest resource policies.

A professional's response to a serious ethical conflict depends on individual moral standards, sense of analytical and professional integrity, and conception of what makes for a good society. One possible response is to abide by ethical codes which professions have developed for their members, some of which have application to policy development. For example, the American Society of Public Administration specifically addresses the ethics and values to which public program administrators should adhere (Table S-1), while similarly the Society of American Foresters has established the Code of Ethics to guide its members in matters concerning forestry (Table S-2). Both codes of ethics provide guidelines for dealing with common ethical predicaments faced by managers; they represent a consensus of beliefs held by the members of the respective professional organization.

Professionals may also respond to ethical and value dilemmas in other ways. Consider the professional who is assigned the task of implementing an agency program which the professional believes to be morally wrong—which is in sharp conflict with the professional's ethical standards for resource management. A number of options are available, including remaining with the agency and taking action to eliminate or change the program; leaving the agency in hopes that such a dramatic act will force the agency to rethink the merits of the program; and engaging in some sort of disloyal action that will lead to the program's demise (Weimer and Vining 1989) (Figure S-1).

TABLE S-1
Code of ethics: American Society for Public Administration, 1985

1. Demonstrate the highest standards of personal integrity, truthfulness, honesty, and fortitude in all public activities, in order to inspire public confidence and trust in public institutions.
2. Serve in such a way as to not realize undue personal gain from the performance of official duties.
3. Avoid any interest or activity which is in conflict with the conduct of official duties.
4. Support, implement, and promote merit employment and programs of affirmative action to ensure equal employment opportunity via recruitment, selection, and placement of qualified persons from all elements of society.
5. Eliminate all forms of illegal discrimination, fraud, and mismanagement of public funds, and support colleagues if they are in difficulty because of responsible efforts to correct such discrimination, fraud, mismanagement, or abuse.
6. Serve the public with respect, concern, courtesy, and responsiveness, recognizing that service to the public is service beyond service to oneself.
7. Strive for personal professional excellence and encourage the professional development of associates and those seeking to enter the field of public administration.
8. Approach organizational and operational duties with a positive attitude. Constructively support open communication, creativity, dedication, and compassion.
9. Respect and protect the privileged information which is available during the course of official duties.
10. Exercise whatever discretionary authority is granted under law to promote public interest.
11. Accept as a personal duty the responsibility to keep up to date on emerging issues and to administer the public's business with professional competence, fairness, impartiality, efficiency, and effectiveness.
12. Respect, support, study, and when necessary, work to improve federal and state constitutions and other laws which define the relationships among public agencies, employees, clients, and all citizens.

Source: Code of Ethics and Implementation Guidelines, by American Society for Public Administration, Washington, adopted Mar. 25, 1985. Reprinted by permission of the American Society of Public Administration.

Working to change the objectionable program from within the organization might entail *protest* by holding informal discussions with supervisors, by submitting a formal memorandum to higher-ranking authorities within the agency, or by speaking out against the program at staff meetings. The professional might request that the program be assigned to someone else, which would emphasize the intensity of the professional's objections to the program.

If such avenues prove fruitless, the person's ethical objections to the program may justify *resignation,* the personal costs of which may be high and may include unemployment and family strife. If the professional's skills are necessary to implementation, the program may come to a halt. If, however, the agency is

TABLE S-2
Code of ethics: Society of American Forsters, 1989

1. A member's knowledge and skills will be utilized for the benefit of society. A member will strive for accurate, current, and increasing knowledge of forestry, will communicate such knowledge when not confidential, and will challenge and correct untrue statements about forestry.
2. A member will advertise only in a dignified and truthful manner, stating the services the member is qualified and prepared to perform. Such advertisement may include references to fees charged.
3. A member will base public comment on forestry matters on accurate knowledge and will not distort or withhold pertinent information to substantiate a point of view. Prior to making public statements on forest policies and practices, a member will indicate on whose behalf the statements are made.
4. A member will perform services consistent with the highest standard of quality and with loyalty to the employer.
5. A member will perform only those services for which the member is qualified by education or experience.
6. A member who is asked to participate in forestry operations which deviate from accepted professional standards must advise the employer in advance of the consequences of such deviation.
7. A member will not voluntarily disclose information concerning the affairs of the member's employer without the employer's express permission.
8. A member must avoid conflicts of interest or even the appearance of such conflicts. If, despite such precautions, a conflict of interest is discovered, it should be promptly and fully disclosed to the member's employer, and the member must be prepared to act immediately to resolve the conflict.
9. A member will not accept compensation or expenses from more than one employer for the same service, unless the parties involved are informed and consent.
10. A member will engage, or advise the member's employer to engage, other experts and specialists in forestry or related fields whenever the employer's interest would be best served by such action, and member will work cooperatively with other professionals.
11. A member will not by false statement or dishonest action injure the reputation or professional associations of another member.
12. A member will give credit for the methods, ideas, or assistance obtained from others.
13. A member in competition for supplying forestry services will encourage the prospective employer to base selection on comparison of qualifications and negotiation of fee or salary.
14. Information submitted by a member about a candidate for a prospective position, award, or elected office will be accurate, factual, and objective.
15. A member having evidence of violation of these canons by another member will present the information and charges to the [Society of American Forester's] Council in accordance with the Bylaws.

Source: Ethics Guide, by Society of American Foresters, Bethesda, Md., 1989. Reprinted by permission of the Society of American Foresters.

Figure S-1
Possible responses to ethical and value conflicts. (*Source:* Policy Analysis: Concepts and Practices, *by David L. Weimer and Aidan R. Vining. Reprinted by permission of Prentice Hall, Inc., Englewood Cliffs, N.J., 1989.*)

able to continue the program, the ethical value of the professional's resignation becomes questionable. In addition, the professional will have forfeited the possibility of direct influence upon the future of the program. Protest and resignation can be combined by *issuing an ultimatum* threatening to resign if the program is not terminated. In this instance, the professional must be willing to carry out the threat.

Disloyalty is also an option. The professional may *leak* the agency's proposed plans for the program to persons (such as journalists or legislators) of organizations (perhaps interest groups) that can disrupt the plans. Whenever professionals are tempted to act covertly and dishonestly, however, they should closely examine the morality of their actions, remembering that dishonesty involves betrayal of the trust of their superiors (Weimer and Vining 1989). To avoid dishonesty, a professional may *resign and disclose* the agency's plans to potential opponents—thus acting honestly but disloyally.

Staying with the agency and *speaking out until silenced*—a course of action that is known as *whistle-blowing*—is another option. In such a case, the agency may exclude the professional from access to important information, exile the employee to a remote location, or attempt to discharge the professional. Under what conditions might whistle-blowing be justified? First, all channels of protest within the agency should be exhausted; second, the professional should determine beyond a doubt that certain moral or legal bounds have been exceeded; third, the whistle-blower should be convinced that the proposed program would have harmful immediate effects on humans or the natural environment; and fourth, he or she should have clear evidence to support the alleged claims (French 1983).

If all else fails, consideration may be given to *sabotage.* A subtle flaw that would ensure the program's demise could be added to the plans for implementing the program. However, it is difficult to imagine a situation so dire as to justify such an action.

Forestry professionals have faced and will continue to face ethical and moral dilemmas as they engage in various segments of the policy process. The following advice, directed primarily but not exclusively to policy analysts, is valuable: ''As teachers and practitioners of policy sciences, we should explicitly recognize our obligations to protect the basic rights of others, to support our democratic processes as expressed in our constitutions, and to promote analytical and personal integrity'' (Weimer and Vining 1989, pp. 27–28). These values should dominate professional thinking. Professionals who are charged with resolving difficult value conflicts involving the use and management of forest resources should also be given considerable understanding.

INDUSTRIAL-SECTOR POLICY DEVELOPMENT

Forest resource policies implemented by public agencies are often the focal point of public interest in forests. The development and implementation of these policies are often severely scrutinized and criticized by various forestry interests. Although public forest resource policies are of major concern, recognition must be given also to the development and implementation of forest resource policies by private establishments, especially the wood-based firms that own nearly 71 billion acres of timberland in the United States. How do such firms accommodate complex internal and external business environments? Through what sequence of events do they develop strategic plans and the policies embodied therein? Who is responsible for the development of strategic corporate plans? What problems do wood-based firms face as they attempt to develop policies considered necessary for steering the corporate ship?

The complexity and sophistication of business decision making requires firms to employ a systematic process for determining their goals and the means by which they will be achieved. Among the products of such processes are *strategies,* or large-scale, future-oriented plans that depict how a firm will interact with its competitive environment and thus achieve its business objectives. Strategies contain no detailed operational directions; instead, they are a company's game plan. They reflect the company's awareness of how to compete against whom, when, where, and for what.

The process by which strategies are developed and implemented entails *strategic management,* or sets of decisions and actions that result in the formulation and implementation of strategies (Pearce and Robinson 1985). The formality of the process is dependent upon factors such as firm size, management style of executives, complexity of external environments, and the production processes carried out by a firm. Small single-product firms (often led by one or two individuals) tend to use intuition and instinct as cornerstones for strategic management, whereas large multinational wood-based firms may employ numerous planners and managers who use sophisticated review processes and complex analytical tools.

It has been suggested that strategic management, as employed in the development and implementation of corporate strategies, involves eight critical subject areas (Ackoff 1981; Pearce and Robinson 1985):

- *Company mission* Identifies the firm's fundamental purpose, philosophy and goals. A general, enduring statement of a company's basic purpose, principal business, and geographic scope of operation.
- *Company profile* Identifies the firm's capabilities based on existing or accessible resources; depicts available financial, human, and physical resources and the strengths and weaknesses of the firm's management and organizational structure.
- *External company environment* Assesses external conditions affecting the company, but typically beyond the firm's control. Operating environment assessment would identify potential impacts of actions taken by competitors (new entrants), consumers (greater price consciousness), suppliers (timber price increase), and creditors (tightening of credit), whereas remote environment assessment would identify affects of changes in general economic conditions (inflation), political conditions (export restrictions), social conditions (demographic changes), and technological conditions (obsolete production process) (Sonnenfeld 1981).
- *Identification of possible and desirable strategies* Names possible avenues for investments and screens such avenues through the firm's mission statement, so as to determine desirable strategies.
- *Selection of strategy* Chooses a strategy from among desirable strategies. The choice based on top-level managers' attitude toward risk, managerial flexibility, interest in growth, concern over profitability, and apprehension over access to needed resources.
- *Development of long-term and annual objectives* Within the context of a selected strategy, develops a 4- to 5-year horizon, long-term objectives (expected profitability, return on investment, competitive position, public image, employee relations), and annual objectives (expected product growth in 1 year).
- *Implementation objectives* Carries out objectives via budget allocations and assignment of tasks to appropriate people and organizational units.
- *Review and evaluation* Monitors implementing actions to determine whether objectives are being accomplished and whether there are signs of marketplace response to strategies and objectives. Makes necessary corrections, based on review and analysis.

Strategic management is typically carried out by a company management team which reflects three levels of company operations: the *corporate level* (the board of directors, the chief executive, and administrative officers), the *business level* (business and corporate managers), and the *functional level* (managers of product and geographic areas). Corporate-level actors are primarily responsible for identifying and approving corporatewide missions and strategies. Their strategy decisions tend to be value-oriented and very conceptual in nature, with priorities such as growth, finding sources of long-term financing, and exploring possibilities for business acquisitions. Business-level team members work to flesh out long-term objectives and are responsible for bridging corporate- and-functional-level objectives and targets. Annual objectives are typically developed and implemented by functional-level managers; their decisions lead directly to actions that will accomplish corporate-level strategies and business-level objectives. A strategic management team often relies also on corporate-level planning departments and business-level staff planning groups. Corporate-level planning departments assist in the development of mission and profile statements, and work to integrate the objectives of various company business departments (Pearce and Robinson 1985).

Strategic management is designed to facilitate decision making when a company is faced with certain strategic-type issues, such as when consideration is being given to the allocation of large amounts of company resources and when concern is focused on actions that could have a significant effect on the long-term prosperity of the firm—say, 4 to 5 years into the future. In addition, such management can assist corporate managers who are faced with multifunctional or multibusiness issues within a company (such as product mix or business structure) or who must deal with external conditions that are of special concern and that are beyond a company's control (such as government, labor, or competitors). Strategic management processes can enhance the problem-prevention capabilities of a firm—can, for example, predict future raw material shortages. Because of the interactions and screenings by specialists that take place at various corporate levels, these processes also enhance the range of strategic options available for choice and can make the technical merits of such options more dependable. In addition, strategic management is capable of identifying duplication among diverse groups and individuals within complex companies. Finally, it can play a positive role in enhancing employee motivation, because it allows employees to participate in development of strategies and objectives (Pearce and Robinson 1985).

RESEARCH ON POLICY DEVELOPMENT

The process of developing policies and programs focused on forest resources throughout the world has grown more complex in recent years. Likewise, the number of disciplines and the number and type of organizations expressing an interest in the use and management of forests has also expanded. Indeed, the conceptual foundation for policy development has broadened considerably. No longer is economic efficiency the most pervasive criterion for judging the merits of forest resource policies. Institutional arrangements, distribution of benefits, and consensually determined standards are often of paramount concern. Furthermore, policy and program development is no longer the task of the few. Interest groups, the media, the general public, and judicial and legislative systems are actively involved in the development of policies. The social science disciplines which contribute the conceptual foundations of policy have also expanded; law, sociology, and political science are offering important insights into policy development.

Unfortunately, the process by which policies are developed, the role of participants in the process, and the incorporation of broader conceptual frameworks into policy development are not always models of efficiency and effectiveness. What they lack in virtue is frequently reflected by the products they produce: poorly designed and poorly implemented forest resource policies. In part, such ineffectiveness stems from the forestry community's misunderstanding of how policies are developed. Notable problems, however, exist because of the limited amount of information which is available for designing more effective means of establishing and implementing policies. Needed is research dealing specifically with policy development and administration, including research focused on persons and organizations engaged in such activities. One possible research agenda is described below. It focuses on the policy development process, the participants involved in policy development, budgetary and fiscal processes, the planning of use and management of forests, and the design of new and more effective program initiatives (Ellefson and Lyons 1989).

A. *Develop an improved understanding of the process by which forest resource policies are developed.* The process by which forest resource policies are developed and implemented includes numerous political events which range from agenda setting to termination. To date, comprehensive research designed to improve the process in the forestry community has been limited. Researchers have been inclined to focus on the product of the process—a specific forest policy. Few have researched opportunities for making the process itself more effective and, subsequently, the policy product more appropriate. The areas listed below deserve research attention.

Evaluate the effectivness of agenda-setting processes germane to the development of forest resource policies.

- How effective are existing agenda-setting procedures?
- What shapes state, national, and worldwide forestry agendas?
- What role do various individuals and organizations play in agenda-setting processes?
- How important are general political atmosphere and political windows of opportunity to agenda setting?

• Why are certain issues diverted, deferred, or displaced?
• Under what circumstances are issues ignored and agenda status obstructed?
• How can less immediate problems that need long-term attention be assured of agenda status?

Assess the effectiveness of processes by which forest resource policies are formulated.

• How effective are existing formulation processes and how might they be improved?
• Who is (and who is not) involved in formulation activities?
• What circumstances promote an especially creative atmosphere for policy formulation?
• How might such conditions be institutionalized?

Evaluate alternative means by which forest resource policies are selected.

• How appropriate are current selection models—the rational comprehensive approach, rational incrementalism, mixed scanning, and organized anarchy—to forestry decision environments?
• What role do personality and various information sources play in the selection process? Under what circumstances is group decision making more appropriate than individual decision making?
• What should be the nature and application of strategies and tactics involving bargaining, negotiation, and mediation?
• Under what circumstances should criteria of various sorts be employed to make judgments about competing policy alternatives?

Evaluate means by which forest resource policy choices are legitimized.

• By what means are forest resource policies legitimized?
• What forms of legitimacy are more effective than others (law versus agency rule making, for example)?
• What role does political feasibility play in the legitimizing of forest resource policies?
• What type of information about political feasibility would be most useful under what circumstances?
• What is the relative efficiency of various strategies and tactics used to secure legitimacy of a forest policy?

Assess means by which forest resource policies are implemented.

• Why do policies and programs often fail in the implementation stage?
• How might the process be made to function more efficiently?
• How common are problems—such as limited financial resources and limited professional talent—in organizing a bureaucracy for implementation purposes?
• How might such problems best be resolved?
• How are laws and regulations translated into feasible directives that will achieve widely agreed-to objectives?

• What about follow-through (the routine provision of services, payments, and obligations) as an important segment of implementation?

Assess approaches to evaluating forest resource policies and programs.

• Why do certain policies and programs continue to be implemented, even though weighty evidence demonstrates their inefficiencies?
• What mechanisms (such as legislative and interest-group oversight, and internal auditing and accounting) play a role in evaluation and control of forest resource programs?
• How might these mechanisms be made to operate more effectively?
• What motivates interest in evaluation of forest resource programs?
• Why is there often such fierce resistance to evaluation among program administrators?
• Who are the clients of evaluation processes?
• What informational products do these clients desire?
• For systematic evaluations of program efficiency, what procedures, criteria, and analytical tools are most appropriate?
• What role should policy analysts play in the evaluation process?

Evaluate means by which forest resource policies and programs are terminated or succeeded.

• What should be the nature of efficient processes for terminating policies and programs that are judged to be lacking?
• How can policies and programs be designed for eventual termination, especially when the policy issue being addressed has a well-defined time dimension or when termination is expected because of pending changes in economic and political conditions?

B. *Acquire a more discerning understanding of participants involved in the development of forest resource policies.* The process by which forest resource policies are developed and implemented is an important focus for research. It cannot, however, be separated from the actors or participants who make the process operate. For example, organized interest groups may resist adoption of a policy, but the nature of such resistance will not be appreciated until the objectives, structure, and operation of interest groups are adequately understood. The areas listed below deserve research attention.

Evaluate the role of legislative and parliamentary systems in the development and implementation of forest resource policies. Legislative systems are a major source of forest resource policies; hence, the manner in which they operate in a forestry context is especially important.

• What legislative structures have an especially important impact on matters of forest resource policy?
• How might such structures be made more effective?

• What is the nature of legislative decision making regarding forest resource topics?

• How might strategies toward consensual decision making be enhanced when forest resource issues are being addressed?

• What is the nature of the forest resource information that flows to legislative systems? Especially what types of information are required, what are the sources of such information, and what are the means by which it is provided (e.g., legislative hearings, special commissions, bureaucratic or legislative staff)?

• What role do legislative systems play in each of the major steps involved in the development and implementation of forest resource policies (i.e., agenda-setting, formulation, selection, legitimizing, implementation, evaluation, and termination)?

Assess the role of judicial systems in the development and implementation of forest resource policies. The judiciary has had substantial impact on the course of forestry in recent years. The laws that have been interpreted and the rulings that have been issued have often been nothing short of revolutionary. Ironically, relatively little systematic research has been focused on the judiciary from a forestry perspective.

• What is the fundamental structure of state, federal, and worldwide judicial systems?

• What makes such systems so influential on matters of forestry?

• Over what subjects does the judiciary have jurisdiction?

• What is the nature of the forest resource issues that are likely to appear on judicial agendas?

• Under what conditions do forestry interests have standing in the eyes of the courts?

• What circumstances foster class-action suits?

• What is the nature of the forestry information utilized by the judiciary (kind, source, quality)?

• To what extent do courts make policy?

• How do they make it—by reinterpreting laws or constitutions, by designing administrative remedies, or by extending the reach of existing law to subjects not previously covered?

• What role does the judiciary play in the various steps leading to the development and implementation of forest resource policies?

Evaluate the role of organized special-interest groups in the development and implementation of forest resource policies. Organized interest groups active in matters of forest resource policy have proliferated in recent years. Likewise, their influence on the nature of forest policies has flourished to the point of becoming an advocacy explosion. Though the forestry community has not been totally comfortable with interest groups, understanding interest-group operations and facilitating their involvement in policy development has substantial merit.

- What is the nature of the interest-group landscape that is relevant to the forestry community? For example, what groups exist, and how large are they both financially and in terms of membership?
- What forest resource issues (e.g., subjects of concern, or geographic limits on activities) are likely to elicit the involvement of interest groups?
- What is the nature of interest-group governance and operation?
- How might the governance and operation of interest groups be made more efficient?
- What is the nature of interest-group influence?
- How can interest-group influence be reorganized for the sake of improving concensual decision making on matters of forestry?
- What is the nature of interest-group lobbing activities?
- How can such activities be made more effective within the forestry community?
- How can the financial and staffing difficulties which often face interest groups be overcome?
- How can organized interest groups become more effective in each of the major steps involved in the development and implementation of forest policies?

Evaluate the role of bureaucratic systems in the development and implementation of forest resource policies. Bureaucracies are the means by which most forest resource policies are implemented. They are the cohesive force which leads to specific accomplishment of a desired forestry objective. The number and complexity of bureaucracies have grown substantially in recent years. Ironically, very little research has been focused on how forestry and related organizations should be structured and staffed in order to achieve legislative or judicial mandates.

- How should forestry organizations be structured in terms of labor, authority, and departmentalization?
- What structures would facilitate communication, decision making, and performance evaluation?
- How can administrators facilitate intergroup behavior and manage conflict more effectively?
- What is (and should be) the relationship between politics and administration? Between ethics and public service?
- What structural and managerial actions will facilitate the relationship between public and private management of forests and forest resources?
- How can bureaucracies become more effective in each of the major steps involved in the development and implementation of forest policies?

Evaluate the role of the media, political parties, and related organizations in the development and implementation of forest resource policies. Journalistic interest in the use and management of forests is keen, yet it is often overwhelming to forest resource professionals. Media interpretation of forest resource issues can have a substantial effect on public perceptions of forestry and the forestry profession. Most certainly there is a need and an opportunity for research that would lead to a better understanding of journalistic interest in and effects upon forest policies and programs. Likewise political parties and similar organizations deserve research attention from a forestry perspective.

C. *Acquire improved understanding of budgetary and fiscal investment processes germane to forestry programs.* The production of goods and services by forests often requires the investment of land, labor, and capital over unusually long periods of time. Rotation lengths of 100 or more years are not uncommon for the production of timber. Similar time periods are often required to produce the aesthetically pleasing landscapes demanded by forest recreationists. The development of wildlife habitat may also require extended periods of time, as do many long-term forestry research activities. Because of forestry's unique requirements for extended periods of time, the need for long-term, sustained levels of investment in forestry and related programs is essential. Financial and professional commitment cannot be erratic if the flow of forest outputs is to have long-term uniformity and sustainability. The problem becomes especially acute during times of austere public budgets.

Much neglected by the research community, however, are budgeting and fiscal investment processes. A concerted effort is needed to understand existing budgetary processes and to develop processes that can accommodate the unique conditions which result from the management of forests. For example:

- Exactly what difficulties do forest resource programs face because of the manner in which budgeting cycles operate?
- How do budgets and fiscal demands influence forest policies and programs?
- Are alternative revenue sources (e.g., user fees, bonding programs, and private loans) more accommodating of the public's interest in forests?
- If so, how effective have they been?
- What budgetary problems arise because the benefits of the forest are often nonmarket in nature?
- How might such problems be addressed in the process of allocating scarce public funds?
- Are forest planning processes linked to budgeting processes?
- If not, how can such a linkage be effectively accomplished?
- What impact do managerial sciences have on budgetary processes?
- If an increase in their impact is deemed desirable, how can they be made to be more influential?
- Are there cumbersome administrative reviews of proposed budgets that could be streamlined to encourage greater efficiency?

D. *Develop more effective means of planning the use and management of forest resources.* Planning the use and management of forest resources has become an important activity of government forestry organizations. Fostered by an abundance of recently established laws in countries throughout the world, planning programs have enabled forestry interests to more clearly define agreed-to forestry objectives. Such planning programs have also enabled forestry professionals to develop programs that can accomplish forestry objectives more effectively. Relatively new comprehensive program and land-management planning activities have brought to the surface a number of planning issues that deserve research attention. Such issues do not involve the substance of the plans that are produced; rather, they are concerned with the efficiency and effectiveness of the processes used to produce plans and the administrative uncertainties implied therein. The planning topics listed below deserve research attention.

Evaluate planning objectives and commitment.

- What fundamental purposes are served by planning activities?
- If benefits are expected, what is their nature, and what client groups are likely to receive them?
- If planning is deemed important, what mechanisms (e.g., legislative directive, agency regulation, professional interest, nongovernmental pressure) can be used to secure agency commitment to planning activities?
- In order to maintain their relevance to the forestry community, how should the effectiveness of planning programs be judged?

Assess alternative planning processes. Planning can employ any one of many procedures.

- Should the process rely on issue-driven mechanisms, goal-driven mechanisms, or some other approach?
- How can planning processes be designed so as to accommodate unforeseen events, such as a change in political leadership or a severe economic downturn?
- What is the most effective means of integrating planning processes with other governmental processes, such as budgeting and program implementation?
- What forest resources (single versus multiple uses), functions (transportation system versus fire management), and ownerships (public versus private) should planning processes be designed to accommodate?

Evaluate resources required to plan. Financial and related resources are consumed by planning activities.

- Under varying circumstances what financial commitments are required to carry out planning activities?
- Are there economies of scale to be captured in the design of planning programs?
- From what source will financial support originate (from the planning unit, for example, or from fees from serviced unit)?
- What type of and how much professional talent should be invested in planning activities?

• Are the investments made in planning worth the benefits obtained?

Assess administrative structure for planning.

• What unit or units of government should be responsible for planning regional versus national units? Should planning be done within functional units? Should a planning unit be operationally integrated or independent?
• What agencies (natural resource, economic development, pollution control, universities) should provide input to planning activities?
• At what level should such agencies be involved (chief executive staff, department level, division level, field level)?
• How can coordination of planning within an agency and among agencies be most effectively accomplished?
• Who at what level, is to decide whether a plan is acceptable and by what criteria?

Evaluate public involvement processes. Planning public forestry programs often implies accommodation of public desires for the products and services of such programs.

• What techniques are most effective for satisfying client and public perspectives on the use and management of forests?
• What portion of the total planning budget should such activities consume?
• If the public is to be involved in the planning process, how should such involvement take place? (Should the public be asked for help, briefed on progress, asked to comment on results?)
• What is the most effective means of integrating public input into the preparation of a plan?

Assess information-management procedures. Planning activities require the use of much information.

• What type of information is necessary for the development of plans?
• How can such information be most effectively generated?
• Who should supply it?
• When should it be made available in the planning process?
• What type of commitment to information management will be required?
• Should an information-management system be unique to a planning effort, or should it be designed to serve other purposes (such as program budgeting)?

Evaluate implementation and monitoring.

• What forms of commitment are most effective for securing the implementation of forestry plans (formal policy statements)?
• Who should impose the requirement that a plan be implemented. (Should implementation begin in response to a legislative directive, for instance? Administrative rules? Interest-group pressure?)

- How can plans be effectively integrated with budgeting processes?
- What flexibility should be built into plans in order to accommodate unforeseen events, such as new technologies or new, more reliable information?
- How may added budget constraints resulting from economic downtowns be addressed?
- How can changing program emphasis due to changes in political leadership be accommodated?

E. *Develop more effective program initiatives focused on the use and management of forest resources.* Public forestry programs focusing on private forestry activities are substantial in number and in importance. They are subject to continuing analysis and evaluation, a healthy process that fosters efficiency and effectiveness. Such assessments should occur on a continuing basis and should accommodate technological advances in evaluation procedures. The topics described below deserve continuing evaluation.

Evaluate information and service program initiatives. A variety of information and service programs exist at various levels of government—for instance, extension of information, and both direct and indirect technical assistance. The effectiveness and efficiency of such programs have been assessed in the past, often resulting in positive administrative responses to improve program design and focus. Evaluations of this sort should continue.

Assess fiscal subsidies and tax program initiatives. Fiscal subsidies (cost-share programs) designed to encourage the management of private forests have been a significant worldwide activity. Likewise, tax policies and programs designed to encourage the production of forest products (such as property tax initiatives and capital gains treatment of income from the sale of timber) have met with considerable favor in the forestry community. Tax policies need rigorous evaluation; the aim should be to design new tax policies that will efficiently achieve public objectives in private timber management.

Evaluate regulatory actions focused on the use and mangement of forests. Public regulatory programs focused on the management of forests and the products and services they produce have grown significantly in recent years. In the United States, for example, many states have developed programs for legal regulation the forest-management activities of private landowners. Presumed is that regulatory programs are more effective than other programmatic alternatives, such as extension of information, technical assistance, and fiscal and tax subsidies. Unfortunately, however, there have been few economic evaluations of regulatory programs. These programs certainly deserve research attention, with the goal of designing more efficient programs to meet a variety of public interests.

In sum, the development of forest resource policies and the administration of programs designed to implement such policies have increasingly been acknowledged as important areas of research. In recent years, however, the scope of the policy development arena has been significantly broadened, requiring the talents of a variety of social, managerial and institutional sciences. To be most effective in the years ahead, policy-type research should strategically focus on improving several aspects of policy development and implementation, including:

- The forestry community's understanding of the processes by which policies and programs are developed
- The participants involved in the development of forest resource policies
- The means by which the use and management of forests are planned
- The budgeting and fiscal processes important to forestry
- The innumerable public programs which are focused on the use and management of forests

In addition, researchers should be actively involved in the evaluation and analysis of appropriate forest resource issues. All such activities should be undertaken in the spirit of developing more effective policies and programs focused on the world's public and private forests.

REFERENCES

Ackoff, Russell L. *Creating the Corporate Future: Plan or Be Planned For.* John Wiley and Sons, Inc., New York, 1981.

Burns, Russell M. (technical compiler). *The Scientific Basis for Silvicultural and Management Decisions in the National Forest System.* General Technical Report WO-55. USDA-Forest Service, Washington, 1989.

Callahan, D., and B. Jennings (eds.). *Ethics, The Social Sciences, and Policy Sciences.* Plenum Press, New York, 1983.

Devall, Bill and George Sessions. *Deep Ecology.* Peregrine Smith Publishers, Salt Lake City, Utah, 1985.

Ellefson, Paul V. *Forest Resource Economies and Policy Research: Strategic Directions for the Future.* Westview Press, Boulder, Colo., 1989.

Ellefson, Paul V., and James R. Lyons. Policy and Program Development Research, in *Forest Resource Economics and Policy Research: Strategic Directions for the Future,* by Paul V. Ellefson (ed.). Westview Press, Boulder, Colo., 1989, pp. 228–240.

French, Peter A. *Ethics in Government.* Prentice Hall, Inc., Englewood Cliffs, N.J., 1983.

Lakoff, Sanford A. Scientists, Technologies and Political Power, in *Science, Technology and Society: A Cross-Disciplinary Perspective,* by Ina Spiegel-Rosing and Derek de Solla Price (eds.). Sage Publications, Beverly Hills, Calif., 1977.

Leopold, Aldo. *Sand County Almanac.* Ballantine Publisher, New York, 1970.

McQuillan, Alan G. Is National Forest Planning Incompatible with a Land Ethic? *Journal of Forestry* 88(5):31–37, 1990.

Nash, Roderick F. *The Rights of Nature: A History of Environmental Ethics.* The University of Wisconsin Press, Madison, 1989.

National Research Council. *Forestry Research: A Mandate for Change.* Committee on

Forestry Research, National Academy of Sciences, National Academy Press, Washington, 1990.

Pearce, John A., and Richard B. Robinson. *Strategic Management: Strategy Formulation and Management.* Richard D. Irwin, Inc., Homewood, Ill., 1985.

Rolston, H. *Environmental Ethics: Duties to and Values in the Natural World.* Temple University Press, Philadelphia, 1988.

Rosenbloom, David H. *Public Administration: Understanding Management, Politics, and Law in the Public Sector.* Random House, Inc., New York, 1986.

Schultz, Charles L. *The Politics and Economics of Public Spending.* The Brookings Institute, Washington, 1968.

Sonnenfeld, Jeffrey A. *Corporate Views of the Public Interest: Perceptions of the Forest Products Industry.* Auburn House Publishing Company, Boston, 1981.

Stone, Christopher D. Moral Pluralism and the Course of Environmental Ethics. *Environmental Ethics* 10(2):139–154, 1988.

USDA-Forest Service. 1980–1990 National Program of Research for Forests and Forested Rangelands. General Technical Report WO-32. U.S. Department of Agriculture, Washington, 1982.

Weimer, David L., and Aidan R. Vining. *Policy Analysis: Concepts and Practices.* Prentice Hall, Inc., Englewood Cliffs, N.J., 1989.

Weiss, Carol H. Improving the Linkage between Social Research and Public Policy, in *Knowledge and Policy: The Uncertain Question* by L. L. Lynn Jr. (ed.). National Academy of Sciences, Washington, 1978, pp. 23–81.

SIX

GLOSSARY AND SELECTED FEDERAL LAWS

Development and implementation of forest resource policies involves a number of concepts and principles that can be identified by various terms and phrases. Some of these terms and phrases are defined and presented in a glossary. In addition, many important forest resource policies are legitimized in federal laws. Abstracts of selected major laws are also presented.

GLOSSARY

Students and practitioners of forestry and policy sciences commonly encounter a variety of terms used to describe similar concepts and principles. In the spirit of encouraging consistency, a list of relevant terms and their definitions is offered below. The definitions are based on a variety of sources, including *The Public Policy Dictionary,* by E. R. Kruschke and B. M. Jackson (ABC-Clio, Inc., Santa Barbara, 1987); *American Public Policy: Process and Performance,* by B. G. Peters (Franklin Watts Publishers, New York, 1982); and *Law Dictionary* by S. H. Gifis (Barron's Educational Series Publishers, New York, 1984). Some terms have been constructed and some definitions modified to accommodate unique forest resource circumstances.

activist An individual who is extensively and vigorously involved in political activities, especially policy advocacy.

administration The art and science of managing organizations or implementing policies of programs. Public administration is primarily concerned with the activities of the executive branch of government.

agency A unit of government that has major policy and program responsibilities. Examples are the Environmental Protection Agency, the USDI-National Park Service, the USDA-Forest Service, and state divisions of forestry.

agenda An enumeration of issues of concern to a person or some organization. Agendas may be systemic (including the total range of issues perceive by a community as meriting pubic attention) or institutional (unique to a specific institution, such as a legislature or a bureaucracy). *See also* institutional agenda; systemic agenda.

agenda setting The process of securing a place for an issue on an institutional agenda.

Agenda setting can be viewed as a functional activity, an opportunity activity, a power and influence activity, or a strategic activity. *See also* functional agenda setting; opportunity agenda setting; power and influence agenda setting; strategic agenda setting.

amicus-curiae brief Written statement presented to a court on behalf of a person or organization that is not party to a lawsuit. Means of providing the court with information necessary to make a proper decision, or means of urging a particular outcome on behalf of the public or a private interest of third parties who will be affected by the resolution of the lawsuit. *See also* civil law; common law; criminal law; law.

analogous formulation Design of policy options on the basis of experiences with previously formulated policies that addressed similar issues. An example would be formulating state policies on the basis of federal policy experiences. *See also* creative formulation; formulation; routine formulation.

appropriation (budget) Legislatively granted money which can be spent on behalf of a program that has an authorized budget level. *See also* authorization.

attention group A group of individuals or organizations that become involved in an issue because of its substance and the policies that have been suggested to resolve it. The focus of an attention group is on the issue, not on the parties involved. *See also* identification group; interest group.

attentive public A segment of the unorganized public that is composed of well-informed individuals who consistently express interest in and concern over government and public affairs.

authorization (budget) Legislatively granted authority for a program (or an agency) to exist and to spend money. Although authority to spend money may be granted, the money may not be spent until it has been legislatively appropriated. *See also* appropriation.

bargaining A process in which two or more persons or organizations with conflicting positions seek agreement on how they will behave one to another. Bargaining is facilitated by compromise, logrolling, and side payments. Its product is a bargain or a negotiated settlement. *See also* compromise; logroll; mediation; side payment.

base budget Current or lowest level of funding that an agency expects for a coming year. Having a program included in a base budget is usually a sign that the program will continue, will be funded, and will not be intensively scrutinized during ensuing steps of the budgetary process.

best management practices Combination of forest management practices that result in the least water pollutants from forestlands. *See also* nonpoint source pollutants; point source pollutants.

bureaucracy A formal administrative organization that provides the necessary structure and procedures for efficiently accomplishing its own goals. Spans of control, specialization of work, lines of authority, formal rules, and neutral and impersonal decision making are relied upon extensively.

business and manufacturing interest group Interest groups composed primarily of companies and related groups that have an interest in the position of a product (such as wood) in the marketplace and in the availability of natural resources to sustain the manufacture of that product. Membership is generally predicated upon a business interest in the specific product and upon the payment of dues or fees. *See also* interest group.

civil law A body of law establishing standards of conduct (including rights and duties) among private individuals. Civil law is generally compensatory, taking the form of monetary damages or equity relief. Equity relief can consist of a temporary restraining order, a preliminary injunction, or a permanent injunction. *See also* law; permanent injunction preliminary injunction; temporary restraining order.

clientele A specific group targeted for service by a government agency. *See also* agency.

committee (legislative) A segment of a legislature empowered to either advance or dispose of proposed policies, and to oversee the implementation of such policies by the executive branch of government. Committee types include standing committees (which are permanent and have responsibility for a defined subject area), select committees (which are temporary and have a specific purpose), joint committees (which have members from both legislative chambers), and conference committees (which are temporary, have members from both chambers, and are empowered to resolve differing versions of a law that have been proposed by the two chambers).

common law Law based on judicial precedent rather than on legislatively enacted laws. *See also* civil law; criminal law; law.

compromise A bargain of the form ''You want X, I want Z, let's settle on Y.'' Position Y is somewhere between positions X and Z. *See also* bargaining; logroll; side payment.

conservation and environmental interest groups An interest group composed primarily of lay citizens having an interest in the quality and quantity of goods and services provided by forest and related environments. The only requirement for membership is payment of dues. *See also* interest group.

conservatism The attitude of a person or group preferring stability in society and government and discouraging disruptive change or reform. Represents ideologies that accept and defend the values of the status quo. *See also* liberalism.

constituency Supporters or clients of a legislator or a government agency or program. Legislators have geographic constituencies (persons within a fixed geographic boundary), reelection constituencies (supporting voters), primary constituencies (voting supporters actively engaged in campaign activities), and personal constituencies (the legislator's closest supporters, confidants, and advisers). *See also* legislature.

cost sharing *See* fiscal incentive.

creative formulation The process of designing policy options for which there is little past experience; used when the options needed to address an issue are unprecedented, representing significant breaks with options used in the past. *See also* analogous formulation; formulation; routine formulation.

criminal law A body of law establishing standards of conduct designed to protect society's interest in individuals (for example, law prohibiting theft and bodily harm). Criminal law is generally punitive. *See also* law; civil law; common law.

criteria Standards against which policies and programs are compared and subsequently judged. Categories of criteria are technological and ecological, efficiency and effectiveness, equity and ethical, values and ideologies, and procedural.

delphi technique Forecasting technique that emphasizes iterative processes to refine opinions about future occurrences. Seeks from experts and policy makers repeated adjustments in opinions until consensus is reached.

doctrine of private nuisance Legal principle that property owners may not use their property in a manner that will injure the property of others. Accountability is established between owners of property in the present tense. *See also* doctrine of waste.

doctrine of waste Legal principle that property owners must not wantonly decrease the value of their property during the course of ownership. Property must be transferred to future owners (future generations) in substantially unimpaired condition. Accountability is established for property interests over time. *See also* doctrine of private nuisance.

due process Guarantee that procedures used by government are fair and reasonable. May also denote guarantees that government actions are not arbitrary and capricious nor beyond the scope of government authority.

elitism A concept of society as composed of powerful individuals who control the policy process. This concept is in contrast to pluralism. *See also* pluralism.

equity relief *See* civil law.

ethics The principles of right and wrong, or good and bad behavior. An individual's ethics may pose implications for policy development and implementation (for example, efficiency and due process) and for professional and related relationships to other persons (for example, truthfulness and provision of high-quality services). A personal ethical system may include environmental ethical standards, with a focus on responsibilities toward trees, wildlife, nature, and the biosphere in general.

evaluation The process of determining whether implemented policies and programs are accomplishing agreed- to goals and targets in an acceptable manner. Involves selection of the policy or program to be evaluated, measurement and analysis of policy and program outcomes, and judgment of policy or program consequences. *See also* experimental design evaluation; with-and-without evaluation design.

executive branch The branch of government composed of agencies and bureaus commonly responsible for day-to-day provision of government services. Contrasted with government's legislative and judicial branches. *See also* agency; bureaucracy.

experimental design evaluation Evaluation involving application of a policy or program to an experimental group but not to a control group. Measurable differences in outcome between the two groups are attributable to the policy or program. *See also* evaluation.

extension program A public or private program that transfers information about forest resources to groups of interested individuals or to the public in general. Extension programs are a major means by which forest technologies (especially the products of research) are transferred to large numbers of individuals.

externalities Goods or services that are produced and subsequently imposed on others without their permission or that are produced and subsequently consumed by others without payment.

facts Representations of reality. Void of personal preferences and political commitment. For example, the statement "This forest can produce timber" is a statement of fact. *See also* values.

fiscal incentive Public or private financial payment to private owners of forestland for purposes of encouraging certain land uses and management practices. Public payments are usually only a portion of the total cost of the practice undertaken by the private landowner, and the arrangement is one of cost sharing.

forest practice regulation Enforcement of a government (usually state) program of legal standards which must be met when forestry practices are carried out on private forestland. Among the standards commonly established are those concerning reforestation, road construction, and protection of water quality. *See also* regulation.

forest resources Goods and services that society demands from forests (or the production of which is influenced by forests). Examples are recreation, timber, forage, natural beauty, wilderness, water, wildlife, and fisheries.

forestland Land that is at least 10 percent covered by trees of any size, including land that formerly had such a tree cover and for which natural or artificial regeneration is planned.

formulation Design of appropriate policy responses to an issue. Involves clarification and definition of an issue, identification of potential policy options, and analysis of pros and cons of identified policy options. Formulation may be analogous, creative, or routine. *See also* analogous formulation; creative formulation; routine formulation.

functional agenda setting Agenda setting in which individuals or organizations perceive and subsequently define an issue, discover that others have similar interests in the issue, conclude that satisfaction on the issue can best be secured via a formal organizational structure (e.g., an interest group), and seek to secure institutional agenda status via a representative or spokesperson. *See also* agenda setting.

gatekeeper An individual (for example, a legislative committee chair) or organization (for example, an organized interest group) that has the authority or influence to determine which issues will appear on an institutional agenda. Gatekeepers can operate as promoters (who strive to get issues on an agenda they oversee), vetoers (who deny issues status on an agenda), bargainers (who negotiate the appearance of issues on an agenda), and neutralists (who decline to actively propose or oppose agenda status for an issue).

general public The segment of the unorganized public that is the least active, the least informed, and the least interested in government and related public affairs.

grass-roots lobbying Influencing policy decisions by organizing and encouraging citizens to contact and persuade policy makers. May involve telephone and letter-writing campaigns and political protests. *See also* lobbying.

identification groups An individual or organization that becomes involved in an issue because peers or counterparts are actively involved in it. The focus of an identification group is on the parties involved, not on the issue. *See also* attention group; interest group.

ideology A set of political, economic, and social views concerning the form or activities of government.

implementation The process of translating a legitimized policy into an operational program. The steps involved are interpretation (translation of broad statements of policy into specific targets and workable directives), organization (designation of agencies and departments to carry out programs and projects), and application (actual provision of goods and services).

industrial forestland Forestland owned by companies interested primarily in the production of timber for manufacture into wood products.

initiator *See* issue entrepreneur.

inspection-permit system Forest practice regulatory system in which a landowner notifies a government agency of intent to carry-out a forest practice (such as harvesting), but proceeds with the practice only after receiving appropriate inspections and government approval. *See also* notification system.

institutional agenda An agenda composed of explicit issues that are listed for active and serious consideration by government policymakers who have authority to deal with them. *See also* agenda; systemic agenda.

interest group A private, formally organized collection of individuals (or other private organizations) established for purposes of promoting the common interests of group members. *See also* attention group; identification group.

iron triangle A power and influence relationship between interest groups, an executive agency, and a legislative committee, each of which has a common interest in a policy, a program, or a budget level.

issue A conflict between groups of individuals or organizations over policies of procedure or substance concerning the distribution of power or resources.

issue character The traits of an issue, including its ambiguity, complexity, and social significance, as well as its precedence and related implications. Important to strategic agenda setting. *See also* strategic agenda setting.

issue entrepreneur An individual who manages or exploits issue agenda-setting opportunities; sometimes called an ''initiator.'' Typically, an issue entrepreneur has special technical expertise, is in a position to speak for others (is an interest-group leader), possesses special political connections and superb communication skills, and has a history of persistance in agenda-setting processes.

law Rules of conduct established and enforced by authority or custom of a community, a region, or a nation. Major categories are private law (which governs relationships between private citizens, such as torts and contract law) and public law (which governs relationships between citizens and government, such as constitutional law, statutory law, administrative law). *See also* civil law; common law; criminal law.

legislative oversight The process by which a legislature exercises control over the executive branch's implementation of policies and programs. Oversight involves questioning, reviewing, assessing, modifying, and rejecting the actions of executive agencies. It is accomplished via legislative hearings, special studies, and personal inquiries.

legislature An elected body of persons responsible for and having the power to select and legitimize public policies. Functions include representing constituencies, informing and educating various publics, legitimizing policies and programs, and carrying out legislative oversight. *See also* constituency; legislative oversight; legitimizing.

legitimizing The process of giving selected policies official status. A policy is embellished in an official form, such as a statute, a judicial ruling, or an administrative regulation.

liberalism The attitude of a person or group preferring changes in the status quo, generally in the direction of greater social, economic, and political equality of individuals. Represents ideologies advocating government involvement as a positive force in ensuring human rights and social and economic equity. *See also* conservatism.

litigation Process of engaging in legal proceedings wherein legal rights and obligations are determined and enforced.

lobbying The process of attempting to influence policy decisions through direct contact with persons responsible for such decisions (for example, legislators and agency heads). Grass-roots lobbying involves efforts by individual citizens (or members of interest groups) to influence policy decisions.

logroll A bargain of the form ''You give me what I want and I'll give you what you want.'' *See also* bargaining; compromise; mediation; side payment.

management by objective Process of establishing specific policy or program goals, obtaining commitment to goals, and periodically assessing performance toward the accomplishment of goals. Employees participate in the establishment of goals. *See also* agency; bureaucracy, evaluation; formulation; implementation; policy; program.

mediation Bargaining in which a mediator helps disputing individuals or organizations to reach an agreeable position. The mediator is independent of the parties involved in the dispute and uses a variety of negotiating skills to help them to reach agreement. *See also* bargaining; compromise; side payment.

mixed scanning selection Policy selection via a search for and a cursory review of a wide range of available policy options, followed by a detailed examination of options considered worthy of additional attention. Combines portions of rational comprehensive selection and rational incremental selection. *See also* organized anarchy selection; programmed selection; rational comprehensive selection; rational incremental selection; selection; unconventional selection.

negotiating *See* bargaining.

nonindustrial private forest A forest owned by an individual, who is often interested in the production of a variety of goods and services, including recreation, wildlife, timber, forage, water, and natural beauty. The typical nonindustrial private forest is small (less than 500 acres).

nonpoint source pollutants Water pollutants originating from diffused sources such as the application of management practices to agricultural land and forestland. Diffused pollutants that reach water over land or through related runoff. *See also* best management practices; point sources pollutants.

notification system Forest practice regulatory system in which a landowner notifies a government agency of intent to carry-out a forest practice (such as harvesting) and then begins the practice if no response has been received from the government within a specified number of days. *See also* inspection-permit system.

opportunity agenda setting Agenda setting resulting from the occurrence of a unique combination of social and political conditions that facilitates placement of an issue on an institutional agenda. Usually the mood of the citizenry is favorable, a window of opportunity exists, an issue entrepreneur is present, and there is general agreement that the issue must be dealt with. *See also* agenda setting; issue entrepreneur; window of opportunity.

organization *See* bureaucracy.

organized anarchy selection Policy selection which is dependent upon the mix of issues confronting an organization, the pool of policies available to address an issue at a particular time, and the inclination of available policymakers to become involved in policy-making activities. Preferences for goals, objectives, and policies are not well defined; the structure and activities of complex organizations are not well understood by the policymakers; and policymakers drift in and out of policy-making roles. *See also* mixed scanning selection; programmed selection; rational comprehensive selection; national incremental selection; selection; unconventional selection.

oversight *See* legislative oversight.

permanent injunction A court-issued order, issued after a full trial, prohibiting certain conduct, either indefinitely or until the results of an appeal are known. *See also* civil law; preliminary injunction; temporary restraining order.

pluralism Principle of a society composed of a large number of diverse ethnic, cultural, political, social, and economic groups which compete for government attention and support. Political power is widely dispersed among such groups. *See also* agenda setting; political influence; power.

point source pollutants Water pollutants discharged from a readily identifiable source such as a pipe or ditch. Industrial wastes and municipal sewage are common point sources of pollutants. *See also* best management practices; nonpoint source pollutants.

policy A generally agreed-to, purposeful course of action that has important consequences for a large number of people and for a significant number and magnitude of resources. *See also* program.

policy advocacy Use of analyses and persuasive techniques to promote the establishment of a desired policy. The focus is on what policies government ought to pursue, and often there is a definite linkage to preferred political ideologies. *See also* policy analysis.

policy analysis Examination of a policy for the purpose of explaining its consequences. The focus is on determining the anticipated effects of current or proposed government policies. *See also* policy advocacy.

policy analyst (evaluator) A person who evaluates policies and programs. May be an objective technician (who conducts evaluations for their own sake), a client advocate (who does evaluations to support the cause of a client or clients), or an issue advocate (whose purpose in making evaluations is to support his or her own cause or interests).

policy process A sequence of political events—agenda setting, formulation, selection, legitimizing, implementation, evaluation, and termination—leading to achievement of policy outcomes.

policy statement A formal (written or spoken) expression or articulation of policy (for example, a statute, a speech, or a judicial ruling.)

policymaker (decision maker) An individual, group, or organization that is responsible for choices at various stages in the policy process.

political Concerned with the business and activities of government.

political action committee A group of individuals organized for purposes of influencing the development of public policies by soliciting money and using it to provide, for example, financial support to candidates for elected public office. The conduct of PACs is regulated by the Federal Election Commission.

political feasibility The likelihood that a selected policy will receive sufficient political support to be legitimized and subsequently implemented.

political influence The ability of an individual or a group to shape activities of government by various means, including suasion, rhetoric, and oratory abilities.

political mood Common thoughts of the citizenry on general social or political subjects (for example, people may be generally antigovernment, environmentally sensitive, pro-development, anti-big business). Political mood is important in opportunity agenda setting. *See also* opportunity agenda setting.

political system A group of interrelated government activities or institutions which together form a whole.

power Capacity to control or change the behavior of others. Ability to dominate others with respect to values, goals, and resources.

power and influence agenda setting Agenda setting involving powerful and well-known elites that determine which issues are to appear on an institutional agenda. *See also* agenda setting.

preliminary injunction A court-issued order designed to prevent the occurrence of irreparable damages until a full trial has been conducted. The defendant is given an opportunity to appear and present contrary information. *See also* civil law; permanent injunction; temporary restraining order.

professional interest group An interest group composed primarily of professionals. In general, membership is predicated on attainment of a degree from a recognized university or college in a professional field and on the periodic payment of dues. Membership is independent of a professional's employer.

program A defined and relatively specific sphere of government activity. A particular package of laws, organizations, projects, and resources. *See also* policy.

programmed selection Routine policy selection that relies on experiences with previously defined policies, criteria, and procedures. For example, a modest increase in an agency's budget would be programmed selection. *See also* mixed scanning selection; organized anarchy selection; rational comprehensive selection; rational incremental selection; selection; unconventional selection.

public interest The concept of the overriding collective good of a community or nation, as opposed to the good of any one narrow or specialized interest.

public policy A policy generated or at least processed within the framework of government procedures and organizations.

rangeland Land on which the potential natural vegetation is predominantly grasses, grasslike plants, forbs, or shrubs, including land revegetated naturally or artificially that is managed like native vegetation. Rangelands include natural grasslands, savannas, shrublands, most deserts, tundra, alpine communities, coastal marshes, and wetlands that are less than 10 percent stocked with forest trees of any size.

rational comprehensive selection Policy selection via a series of logical choices focused on achieving or optimizing a goal, an objective, or an outcome. Involves definition and ranking of values, specification of objectives consistent with values, formulation of policy options capable of achieving objectives, identification of consequences (pro and con) associated with formulated options, specification of a decision rule for selecting an option, and application of the decision rule. *See also* mixed scanning selection; organized anarchy selection; programmed selection; rational incremental selection; selection; unconventional selection.

rational incremental selection Policy selection via a series of minimal adjustments in existing policies as a result of bargaining over values, objectives, and formulated policy options. Involves minimal debate about values and goals, consideration of relatively few policy options that are vigorously advocated by many interests, continuous comparison of existing and proposed options, search for satisfactory or acceptable policy options, reliance on bargaining and negotiation, and use of agreement as a criterion denoting the selection of a good policy. *See also* mixed scanning selection; organized anarchy selection; programmed selection; rational comprehensive selection; selection; unconventional selection.

regulation Enforcement of a government program of legal standards which must be met when forest and related resources are used and managed. Regulation typically focuses on private forests, especially their use, various pollutants, and forest practices. *See also* forest practice regulation.

routine formulation Design of policy options to address an issue that appears on an institutional agenda at predictable intervals (for example, an annual agency budget). *See also* analogous formulation; creative formulation; formulation.

selection Choosing from among the many (or few) policy options that have been formulated. *See also* mixed scanning selection; organized anarchy selection; programmed selection; rational comprehensive selection; rational incremental selection; unconventional selection.

side payment A bargain of the form "You give me what I want, and I will reward you," or conversely, "If you do not give me what I want, I will punish you." *See also* bargaining; compromise; logroll.

staff (legislative) Non-elected employees of a legislature. The categories of professional staff are personal staff (responsible to an individual legislator) and committee staff (responsible to a committee chair).

strategic agenda setting Agenda setting undertaken as a purposefully planned activity involving the use of triggering events, political symbols, sympathetic gatekeepers, and unique issue definitions. *See also* agenda setting.

strategic program planning A process designed to produce fundamental decisions that guide what an organization is, what it does, and why it does it. Requires broad-scale information gathering, exploration of far-reaching alternatives, emphasis on future implications of present decisions, and ability to accommodate divergent values and interests.

systemic agenda An agenda composed of all issues commonly perceived by members of a community, region, or nation as meriting public attention and as being within the legitimate jurisdiction of government. *See also* agenda; institutional agenda.

tax incentive A public program designed to alter the timing, type, or amount of tax expected from a private forest property or the income produced by the property. The purpose of offering tax incentives is to encourage certain land uses and management practices. Income taxes and property taxes are major categories of tax programs by which incentives are commonly offered.

technical assistance programs Public or private programs that provide information on the use and management of forest resources. In direct programs, trained professionals provide technical advice and counsel to individual owners of forestland. In indirect programs, advice and service are directed to the combined needs of many landowners (for example, research programs and forest protection programs).

temporary restraining order A court-issued order designed to preserve the status quo among disputants until a detailed examination of the dispute can be made by the court. Commonly issued for a period of 10 days. The defendant need not be present at the time the order is issued. *See also* civil law; permanent injunction; preliminary injunction.

termination The deliberate conclusion or cessation of a policy, a program, or an organization.

timberland Forestland which produces or is capable of producing crops of industrial wood, and which has not been withdrawn from timber utilization by statute or administrative regulation. In a technical sense, timberland from natural stands must have the capability of producing annually more than 20 cubic feet of industrial wood per acre.

triggering event An unannounced administrative, political, or environmental event that generates significant public attention, often culminating in placement of an issue on an institutional agenda.

unconventional selection Selection from among formulated policy options that are unique and unstructured, and that involve substantial uncertainty as to their outcome. Substantial reliance upon intuition and considerable tolerance for uncertainty on the part of policymakers are required. *See also* mixed scanning selection; organized anarchy selection; programmed selection; rational comprehensive selection; rational incremental selection; selection.

values Beliefs that certain conditions are inherently good while others are inherently bad. Values may involve an individual's political commitment, personal preferences, and policy orientation. For example, the statement ''This forest ought to be used for timber production'' is a value statement. *See also* fact.

window of opportunity A unique occurrence in social and political conditions, involving the political mood of the citizenry, perhaps a change in leadership, and the existence of an issue entrepreneur. Facilitates placement of an issue on an institutional agenda and is important in opportunity agenda setting. *See also* opportunity agenda setting.

with-and-without evaluation design Evaluation design in which conditions before and after implementation of a policy are compared. *See also* evaluation; experimental design evaluation.

SELECTED MAJOR FEDERAL LAWS ADDRESSING THE USE AND MANAGEMENT OF FOREST RESOURCES

Federal law legitimizes an enormous range of policies which direct the use and management of forest resources. Abstracted below are a few highly selected such laws. More detailed information about a particular law can be found in the statutes (for example, *United States Code Annotated,* West Publishing Company, 1990) or in any of various complications of federal forest and natural resource laws—for example, *The Principal Laws Relating to Forest Service Activities* (USDA-Forest Service, Agriculture Handbook No. 453, 1983), *Wildlands Management Law, Volumes I and II* (compiled by R. E. Shannon, Montana Forest and Conservation Experiment Station, 1983), *The Evolution of National Wildlife Law* (by M. Beam, Praeger Publishers, 1983), *Digest of Public Land Laws* (Public Land Law Review Commission, 1968), *Laws Relating to National Park Service, Volumes I through IV* (U.S. Government Printing Office, 1980), *Federal Public Land and Resources Law* (by G. C. Coggins and C. F. Wilkinson, Foundation Press, 1987), and *Public Natural Resources Law* (by George C. Coggins, Clark Boardman Company, 1990). Readers should be mindful that an equally far-ranging set of forest resource policies have been legitimized in the laws of state governments (*State Environmental Law* by D. P. Selmi and K. A. Manaster, Clark Boardman Company, 1990).

Organic Administration Act of 1897 (*as amended*) Authorizes the establishment of the National Forest System and provides direction on matters concerning uses of, access to and surveys of the system.

Transfer of National Forest Receipts to States Provision of 1908 Authorizes the transfer to states of 25 percent of the receipts received from the sale of goods and service produced by National Forests. The transferred receipts are to be used to benefit public schools and public roads located within the county in which a National Forest is situated.

Weeks Law of 1911 Authorizes the Secretary of Agriculture to purchase for inclusion in the National Forest System certain forested, cut-over, or denuded lands that must be wisely managed for the production of timber or the regulation of water flows into navigable streams.

National Park Service Act of 1916 Authorizes estabishment of the National Park Service within the Department of the Interior and requires the director of the service to promulgate rules and regulations considered necessary for the management of National Parks.

Knutson-Vandenberg Act of 1930 Authorizes the Secretary of Agriculture to establish such forest tree nurseries as may be required to supply seedlings for National Forests. The secretary is also authorized to require purchasers of National Forest timber to deposit with the secretary sufficient funds to cover the cost of undertaking certain timber management (e.g., reforestation) and related forestry activities (e.g., wildlife habitat management) on harvested areas in National Forests.

Federal Aid in Wildlife Restoration Act of 1937 Authorizes establishment of the Federal Aid to Wildlife Restoration Fund in the Department of Treasury. Revenues for the fund are derived from a federal excise tax on the sale of firearms, shells, and cartridges. A major portion of the revenue from the fund is apportioned to states for purposes of research, management, and land acquisition concerning the use and management of wildlife. (Commonly known as the *Pittman-Robertson Act.*)

Multiple-Use Sustained-Yield Act of 1960 Directs the Secretary of Agriculture to administer the National Forests for multiple uses (in the combination that best meets the needs of the American people) in a sustained-yield fashion (i.e., in a way that will yield high levels of outputs in perpetuity without impairing land productivity). Directs that National Forests "be administered for outdoor recreation, range, timber, watershed, and wildlife and fish purposes." Establishment and maintenance of designated wilderness areas are considered to be consistent with the provisions of the act.

McIntire-Stennis Act of 1962 Directs the Secretary of Agriculture to cooperate with state colleges and universities for purposes of carrying out programs of forestry research, including the training of research workers. Authorizes funds for doing so.

Wilderness Act of 1964 Establishes the National Wildnerness Preservation System, composed of federally owned areas designated by Congress as wilderness areas. The latter are areas "where the earth and its community of life are untrammelled by man, where man himself is a visitor who does not remain." Such areas generally appear to have been affected primarily by the forces of nature; offer outstanding opportunities for solitude or primitive-type recreation; are at least 5,000 acres in size; and often contain ecological or geological features with special scientific, educational, scenic, or historic value. Establishes a planning process for consideration of areas to be potentially included in the system.

Land and Water Conservation Fund Act of 1965 (as amended) Authorizes the Secretary of the Interior to provide funds and assistance to states for purposes of planning, acquisition, and development of land and water areas needed for recreation purposes, and authorizes provision of funds for federal acquisition and development of certain lands for recreation purposes, including inholdings of the National Parks, National Forests, and National Wildlife Refuges. Authorizes admission fees for certain federally administered recreation areas.

Historic Preservation Act of 1966 Authorizes the Secretary of the Interior to develop and maintain a national register of districts, sites, buildings, structures, and related objects of significance to American history. Authorizes the granting of funds to states for preparing comprehensive statewide historic survey plans.

Wildlife Refuge System Administration Act of 1966 Authorizes the consolidation of various federal wildlife refuges into the unified Wildlife Refuge System. Places restrictions on the exchange and disposal of lands within the system and authorizes the establishment of regulations concerning the use and management of units which make up the system.

Wild and Scenic Rivers Act of 1968 Establishes the National Wild and Scenic Rivers System, composed of certain free-flowing rivers with outstanding scenic, geologic, recreational, or cultural conditions, including related surroundings. The Secretary of the Interior and the Secretary of Agriculture are given major responsibility for management of rivers within the system. Cooperative procedures with state governments are established.

National Trails System Act of 1968 Establishes the National Trails System for purposes of promoting the preservation of public access to, travel within, and enjoyment of the open-air, outdoor areas and historic resources of the nation. The Secretary of the Interior and the Secretary of Agriculture may establish (with consent of the federal, state, or local government having jurisdiction over the lands involved) national recreation trails, national historic trails, and national scenic trails.

National Environmental Policy Act of 1970 Declares as national policy the encouragement of productive harmony between people and the environment, the promotion of efforts to eliminate damage to the environment, and the enrichment of citizens' understanding of ecological systems and the nation's natural resources. Requires all federal agencies to prepare reports of environmental impacts of certain proposed actions and to suggest alternative actions. Establishes the Council on Environmental Quality.

Youth Conservation Corps Act of 1970 Authorizes the Secretary of the Interior and the Secretary of Agriculture to carry out a program of gainful employment of American youth during summer months. Employment occurs in the National Forest System, the National Park System, the National Wildlife Refuge System, and other federal systems as appropriate.

Mining and Minerals Act of 1970 Declares as in the national interest the encouragement of private enterprise in the development of economically sound and stable domestic mining and reclamation activities and the orderly and economic development of domestic mineral resources. Promotes mineral resource research and the study and development of methods for the disposal, control, and reclamation of mineral wastes.

Volunteers in the National Forests Act of 1972 Authorizes the Secretary of Agriculture to recruit and train individuals for work without compensation in the National Forest System, in capacities such as visitor services, interpretative functions, and conservation measures. Volunteers may also serve in related programs administered by the USDA-Forest Service.

Federal Water Pollution Control Amendments of 1972 (as amended) Authorizes the Environmental Protection Agency (EPA) to undertake programs to achieve specified goals for the quality and use of the nation's waters. The focus is on point and nonpoint sources of water pollutants and mechanisms for curtailing such pollutants. States are to identify nonpoint sources of water pollutants (such as silvicultural practices) and prepare plans for curtailing such pollutants. The Secretary of the Army (through the Chief of Engineers) is to establish and implement a permit system for controlling the discharge of dredged or fill material into certain navigable waters and wetlands.

Federal Insecticide, Rodenticide, and Fungicide Act of 1972 (as amended) Authorizes the EPA to establish a system for registering pesticides and certifying the users of certain pesticides. Pesticide uses classified as restricted pose potentially harmfully affects to health and the environment. Such uses must be carried out under the auspicious of certified (trained) pesticide applicators.

Endangered Species Act of 1973 (as amended) Authorizes the Secretary of the Interior (and other secretaries as appropriate) to undertake a program for conserving the ecosystems upon which endangered and threatened species depend. Endangered species are those in danger of extinction throughout all or a significant portion of their range, while threatened species are those likely to become endangered in the foreseeable future. The secretary establishes detailed rules for determining endangered or threatened status and sets forth protective regulations and recovery plans for identified species. Civil and criminal penalties apply when provisions of the act are violated.

Forest and Rangeland Renewable Resources Planning Act of 1974 (as amended) Authorizes the Secretary of Agriculture to periodically assess the nation's renewable forest resources and to prepare for congressional approval a 40-year program for the management of such resources. The assessment (updated every 10 years) and the program (updated every 5 years) are prepared by the USDA-Forest Service. The program is implemented primarily by USDA-Forest Service activities concerning the National Forests, Research, and State and Private Forestry.

Eastern Wilderness Act of 1975 Declares as in the national interest the inclusion in the National Wilderness Preservation System of certain federal lands in the eastern United States.

National Forest Management Act of 1976 Authorizes the Secretary of Agriculture to prepare and periodically revise (every 10 years) comprehensive land-management plans for forests and related areas within the National Forest System. Establishes sustained-yield, even flow as a principle guiding harvesting rates in National Forests, and requires consideration of the suitability of forestland for timber-production purposes. Requires extensive public participation in the development of National Forest plans. (The act is an amendment to the Forest and Rangeland Renewable Resources Planning Act of 1974.)

Federal Land Policy and Management Act of 1976 Declares that public lands adminstered by the Secretary of the Interior through the USDI-Bureau of Land Management shall be retained in federal ownership, that such lands shall be periodically inventoried, and that plans for their use and development shall be prepared. Establishes procedures and authorities regarding land planning, acquistion, and disposition; regarding functions and programs of the USDI-Bureau of Land Management; regarding planning the use and management of rangeland; and regarding the granting of rights-of-way across public lands. Provides for a planning process to identify potential public land to be included in the National Wilderness Preservation System. Specifies multiple-use, sustained-yield approaches to management of certain lands. Requiries public involvement in decisions concerning public lands.

Payments in Lieu of Taxes Act of 1976 Authorizes the Secretary of the Interior to make direct payments (at a specified rate per acre) to local governments for certain federal land located within a local government's jurisdiction. The federal land (entitlement land) of concern is that found within the National Park System and the National Forest System.

Surface Mining and Reclamation Act of 1977 Authorizes cooperation between the Secretary of the Interior and state governments with regard to the regulation of surface coal mining and the acquisition and reclamation of abandoned mines. Prohibits (after enactment of this law) coal mining within the boundaries of the National Park System, the National Wildlife Refuge System, the National Trails System, the National Wilderness Preservation System, and the National Wild and Scenic Rivers System. Also prohibits coal mining within the National Forest System, except when no significant recreational, timber, economic, or other values are determined to be incompatible with surface mining.

Clean Air Act Amendments of 1970 (*as amended*) Authorizes the EPA to establish a system for managing and controlling pollutants emitted to the nation's air. Requires prevention of significant deterioration in air quality (including visibility) in certain areas with specified characteristics (for example, minimum size)—National Parks, National Wilderness Areas, National Wildlife Refuges, and the like.

Renewable Resources Extension Act of 1976 Authorizes the Secretary of Agriculture to establish (in cooperation with state governments) extension programs for disseminating the results of research on renewable resources; for developing educational activities for nonindustrial private forest owners; for providing educational opportunities concerning range, fish, and wildlife management; and for providing continuing education opportunities for renewable resource professionals. National and state renewable resource extension plans are to be developed. Authorizes funds for accomplishing the provisions of the act.

Soil and Water Resources Conservation Act of 1977 Authorizes the Secretary of Agriculture to promote the conservation and wise use of soil and water resources via the periodic appraisal (every 5 years) of the nation's soil, water, and related resources, and the development and periodic updating (every 5 years) of a program for conserving and protecting such resources. Presidential statements of policy and proposed budgets are to be prepared on the basis of the appraisal and the program. The USDA-Soil Conservation Service is the primary implementing agency.

Forest and Rangeland Renewable Research Act of 1978 Authorizes the Secretary of Agriculture to implement a comprehensive program of forest and rangeland research and to disseminate the findings of such research. Authorizes the maintenance of forest experiment stations, experimental forests and related research facilities. Establishes a competitive forest research grants program available to public and private research organizations. (Repeals the McSweeney-McNary Act of 1928.)

Cooperative Forestry Assistance Act of 1978 Authorizes the Secretary of Agriculture to establish and implement programs concerning rural forestry assistance, forestry cost-share incentives, insect and disease control, urban forestry assistance, rural fire prevention and control, and general management and planning assistance. State governments are designated as major cooperators in implementing the provisions of the act. (Repeals all or major portions of the Forest Pest Control Act, the Clark-McNary Act, the Cooperative Forest Management Act, the Forestry Incentives Program, and the White Pine Blister Rust Protection Act.)

Public Rangelands Improvements Act of 1978 Authorizes the Secretary of the Interior and the Secretary of the Interior and the Secretary of Agriculture to develop and maintain an inventory of federal rangeland conditions. Establishes procedures for the development of allotment plans and the establishment of grazing leases, permits, and fees. Authorizes funding for the accomplishment of range-improvement practices.

Recreational Boating Safety and Facilities Improvement Act of 1980 Authorizes amortized deduction of up to $10,000 annual reforestation expenses for federal income tax purposes and allows a 10 percent investment credit for reforestation expenses. Establishes a Department of Treasury trust fund, the proceeds of which are to be used for reforestation purposes as specified in certain portions of the Forest and Rangeland Renewable Resources Planning Act of 1974. Revenues for the fund are tariff receipts on wood and wood products. (Known as the *Reforestation Tax Incentives and Trust Fund Program.)*

Alaska National Interest Land Conservation Act of 1980 Authorizes the establishment in Alaska of over 103 million acres of new (or additions to) National Forests, National Parks, National Wildlife Refuges, and National Wild and Scenic Rivers. Also authorizes an annual $40 million fund to maintain timber-management and timber-harvesting activities in the Tongass National Forest.

Food Security Act of 1985 Authorizes the Secretary of Agriculture to establish the Conservation Reserve Program (CRP), wherein the Secretary enters into contractual agreements with farm owners to retire highly erodible land from agriculture. Contracts are for a 10-year period, during which landowners receive annual payments for participation in the program. Land accepted for inclusion in the program must be protected from erosion via the establishment of stands of trees or permanent vegetative cover (which cannot be harvested or grazed during the contract period). Also limits farm owner eligibility for Department of Agriculture programs generally if highly erodible lands are brought into cultivation without an approved soil and water conservation plan (''sodbuster provision''). Similar denial of general program access occurs if wetlands are altered or drained (''swampbuster provision'').

Forest Ecosystem and Atmospheric Pollution Act of 1988 Authorizes a major federal research effort focused on the effects of atmospheric pollutants on forests and forest productivity.

Forest Stewardship Act of 1990 Authorizes the Secretary of Agriculture to establish a rural forestry assistance program designed to provide financial, technical, and educational assistance to the forestry community via state foresters and state extension directors. With the objective of enhancement of multiple forest benefits, cost-share assistance focused on nonindustrial private forests is provided by a stewardship incentive program. Establishes a forest legacy program for protection of environmentally important forested areas. Expands forestry research authority, especially competitive grants programs and numerous special research centers, and enlarges the forestry responsibilities of the USDA-Extension Service. Also initiates the America the Beautiful program (including a tree-planting foundation) and the Rural Revitalization through Forestry program. (Amends portions of the Cooperative Assistance Program of 1978 and the Renewable Resources Extension Act of 1976.)

Log Export Restrictions of the Customs and Trade Act of 1990 Authorizes a permanent ban on the export of unprocessed timber from federal lands west of the 100th meridian in the contiguous 48 states. In the same region, prohibits persons from substituting (for export purposes) timber purchased from federal agencies for unprocessed timber obtained from private forestlands.

INDEX

Ad Valorem taxes (*see* Property taxes)
Agencies:
accountability, 156, 157
budgets and employees, 272
challenges to
accountability, 291, 292
ideologies, 289
policy selection, 290, 291
political-technical skills, 291
public-private interests, 289
decision-making in, 269–272
descriptions (examples) of, 272–289
evaluation of
change fostered by, 157
destructiveness of, 159
operating staff and specialized staff, role of, 182–185
federal agencies, 275–282
interaction with legislatures, 238–240
international examples of, 286–289
iron triangle involvement, 238, 239 (*See also* Interest groups, Legislative committees)
leadership of, 272
legislative oversight and review of, 214–217
legitimizing policies, 272, 273
organization of, 265–269, 286, 287
legal perspective of, 267
Agencies (*Cont.*)
managerial perspective of, 266, 267
political perspective of, 267
participative management of, 268, 269
policy implementation by, 148
power, 270
procedural characteristics of, 265, 266
state agencies, 282–286
termination of
design for, 196, 197
strategies and tactics for, 194–199
Agenda setting:
chief executive officer role in, 54
civil servant role in, 54
definitions
agenda, 32
institutional agenda, 32, 33
systemic agenda, 32
examples, 63, 64
forestry professional role in, 56–58
functional activity, 34–36
gatekeepers involved in, 40–41
issues
defined, 31, 32
characteristics for expansion, 44–46
obstruction of agenda status, 51–53
mass media role, 55–56, 333–336
opportunity activity, 35–38

Agenda setting (*Cont.*)
  participants in
    government, 54–55
    nongovernment, 55–56
  power and influence activity, 35
  public opinion role in, 56
  publics involved in, 41–44
  relation to other policy process events, 65
  strategic activity, 38–47
  symbols, 46–47
  triggering events, 38–39
Agendas (*see* Agenda setting)
Agricultural Conservation Program, 401
Air pollutant regulation, 416
Analogous formulation, 69
Analysts (*see* Evaluators)
Attention groups, 42
Attorneys (*see* Judicial systems)

Bargaining:
  attitudes that facilitate, 106, 107
  defined, 104
  mediation, 113–114
  role in policy selection, 98, 103–114
  strategies called for by
    compromise, 108, 109
    logrolling, 109–111
    side payments, 111
  tactics for, 112
Best land-management practices, 394, 395
Budget and Accounting Act, 360
Budget and Impoundment Control Act, 360
Bureaucracies (*see* Agencies)
Business and manufacturing interest groups, 299 (*See also* Interest groups)

Capital gains tax policy, 405, 406
Charitable and scientific organizations (*see* Interest Groups)
Coastal Zone Management Act, 413 (*See also* Regulatory initiatives)
Code of Federal Regulations, 127–128, 361
Committee hearings (*see* Legislative committees, Legislative systems)
Committee on Forest Development in the Tropics, 61, 75, 78, 180
Committee on Forestry Research (National Academy of Sciences), 180
Compromise, 108, 109 (*See also* Selection)
Congressional Budget Office, 238
Congressional Research Service, 179, 238
Conservation and environmental interest groups, 299 (*See also* Interest groups)
Conservation Reserve Program, 402, 403
Cooperative forestry assistance programs (*see* Direct technical assistance)
Cooperative Forestry Assistance Act, 398
Cooperative fire protection programs, 186, 393
Cooperative pest management programs, 393
Cost-share programs (*see* Fiscal incentives)
Council on Environmental Quality, 61, 128, 274
County and municipal forests, 442, 443
Courts (*see* Judicial systems)
Creative formulation, 69
Criteria:
  for evaluation, 163–165
  for judging usefulness of evaluations, 169, 170
Criteria for selection:
  categories of
    efficiency and effectiveness, 116
    equity and ethical, 116–118
    procedural, 120
    technical and ecological, 115–116
    values and ideologies, 119–120
  character of useful, 114, 120
  defined, 114
  examples, 121–122

Decision makers (*see* Policymakers)
Democracy:
  framework for policy development, 20
Direct technical assistance: (*See also* Extension, Indirect technical assistance)
  budgets, 383
  challenges to, 389, 390
  effectiveness of, 385
  objectives of, 384
  privately provided
    consultants, 388
    industrial forestry programs, 388, 389
    Tree Farm Program, 388
  services provided by, 383, 385

Elites:
  framework for policy development, 21
Endangered Species Act, 439, 454
Environmental Protection Agency:
  activities, 279, 280, 394
  air pollutant regulation by, 416
  budget, 274
  employees, 274, 279
  organization, 279–282
  pesticide regulation, 415
  water pollutant regulation, 415, 416

Estate and inheritance taxes, 406, 407
Ethics:
 defined, 452
 example
  situations involving, 452, 453
  codes (professional) of, 454
 judges, 257
 response to dilemmas involving, 454–458
Evaluation:
 approaches to, 161–165
  formal, 161
  informal, 161
 defined, 154, 155
 examples, 185, 186
 experimental design used for, 167, 168
 linkages between clients and evaluators involved in
  clients (administrators), 168–170
  evaluators (analysts), 170–174
 major events in, 155, 156
 obstacles to, 174–178
 participants in
  legislatures, commissioners, citizens, 178–182
  operating staff, specialized staff, 182–185
 problems with, 159, 160
 purpose and intention of, 156–159
 quantitative tools for, 168
 relation to other policy process events, 186
 results of, 160, 161
 scope of, 156
 standards (criteria) for judging policies, 163–165
 standards for judging usefulness of, 169, 170
 symbolic (ritual) use of, 158
 with-and-without design used for, 165–167
Evaluators:
 choosing policies (programs) to evaluate, 173, 174
 interaction with clients (administrators), 172
 operating environment of, 172
 role of, 170, 172
Extension: (*See also* Direct technical assistance, Indirect technical assistance)
 activities, 390, 391
 administration of, 391
 budget, 391
 challenges to, 392
 employees, 391, 392
Experimental evaluation design, 167, 168 (*See also* Evaluation)
Externalities, 382, 397
FAO Forestry Department: (*See also* International forest resource activities)
 budget, 288
 employees, 288
 forestry responsibilities of, 288
 organization, 288, 289
Federal Water Pollution Control Act (amendments to), 417, 418, 442
Fire Management Policy Review Team, 184
Fiscal incentives:
 challenges to, 404
 examples
  Agricultural Conservation Program, 401
  Conservation Reserve Program, 402, 403
  Forest Stewardship Incentives Program, 400, 401
  Forestry Incentives Program, 398–400
 rationale for, 397
 state implementation of, 403
Forest and Rangeland Renewable Resources Planning Act (*see* Resources Planning Act)
Forest ownership (*see* Government forest ownership, Industrial
 private forests, Nonindustrial private forests)
Forest practices regulation:
 accomplishments of, 429–431
 challenges to, 431, 432
 cost of implementation, 427–429
 examples
  of legitimized policies, 130, 131
  Massachusetts, 422, 423
  Washington, 423–427
 challenges to, 431, 432
 cost of implementing, 427–429
 frequency of regulation laws, 421, 422
 historical perspective, 417, 418
 legal foundation for, 418, 419
 system components, 419–421
Forest products industry (*see* Wood-based industry)
Forest recreation:
 importance of, 7
 provided by nonindustrial private forests, 371, 372
Forest Stewardship Incentives Program, 400, 401
Foresters (*see* Forestry professionals)
Forestry Incentives Program, 185, 398–400
Forestry professionals:
 response to ethical dilemmas, 454–458

Forestry professionals (*Cont.*)
role in
agenda setting, 56–58
policy evaluation, 184–185
types of
consultants, 388
extension foresters, 391, 392
industrial foresters, 379
Forests:
area, 3–4
ownership of, 370, 371, 375, 376, 435, 437–439, 441, 442
potential outputs from, 4
social and economic importance of, 6–8
political and institutional prominence of, 8–11
Formulation:
case examples, 78
communities of formulators
examples of, 75–76, 78
characteristics of, 76–77
cost, benefit, and risk estimation, 71
defined, 67
examples
cases, 78
of policies resulting from, 74–75
forms of
routine, 68–69
analogous, 69
creative, 69
issue definition importance to, 69, 78
policy options to be avoided during, 72–73
political context of, 70
relation to other policy events, 78–79
relationship to evaluation, 71
source of ideas for
knowledgeable authorities, 73
deductive reasoning, 73
institutional sources, 73–75

General Accounting Office, 238
Governor's Task Force on Northern Forest Lands, 74, 180
Government forest ownership:
challenges to, 443
land area, 434, 435
policies and programs, 437–443
regional distribution of, 435
Gramm-Rudman-Hollings Act, 360
Group competition: (*See also* Interest groups)
framework for policy development, 20, 21
Group decision-making:
advantages, 90
Group decision-making (*Cont.*)
characteristics, 88–90
disadvantages, 90

Hearings (*see* Legislative committees, Legislative systems, Oversight and review: legislative)
Historical significance of forests, 11–13

Identification groups, 41–42
Implementation:
defined, 139
examples, 152
guidelines for
development of, 141–142
successful, 143–148
major events involved in
interpretation, 140
organizational assignment, 141
application, 141
participants in
agencies, 148
courts, 150–151
general public, 148
interest groups, 151
legislatures, 149
relation to other policy process events, 65
Indirect technical assistance: (*See also* Extension, Direct technical assistance)
examples, 392
fire protection, 393
pest management, 393
research, 393, 394
Industrial private forests:
area, 375
major companies owning, 375, 376
policy development process for, 458–460
polices for ownership and management of, 377, 378
problems resulting form ownership of, 378, 379
professional forestry staff involved with, 379
public interest in, 379
timber productivity of, 376, 377
timber harvest from, 377
Inequities (distributional), 116–120, 382
Information:
legislative system requirements for, 226, 227
needed for legitimizing, 133–135
scientific requirements for policy development, 448–452

Interagency Task Force on Tropical Forests, 184
Interest groups:
administration (effective) of, 327, 328
agenda-setting role of, 55
attention groups, 42
attraction (public) to, 294, 295
categories (types) of, 298–303
challenges to, 325–328
defined, 294
examples, 303–307
function of, 295, 296
governance of, 308
group competition model of, 20–21, 297
identification groups, 41–42
international examples of, 299–303
involvement in
policy evaluation, 181
policy implementation, 151
iron triangles, 238, 239 (*See also* Agencies, Legislative staff)
issues of concern to, 61–62
leadership of, 307, 308
lobbying by, 316, 317, 325
number of, 299–303
organization of, 308, 323
policy advocacy by
issue selection, 311–314
policy products, 314–316
strategies and tactics, 316–321
political action committees of, 319–321
regulation of
ethics, 309
lobbying, 309
tax status, 309, 310
staff, 308, 309
success in influencing policy process, 322–325
International forest resource activities (*See also* FAO Forestry Department)
Committee on Forest Development in the Tropics, 61, 75, 78, 180
forestry agencies involved in
general, 286, 287
Norway, 287
Spain, 287
Sweden, 286
Union of Soviet Socialist Republics, 286
United Kingdom, 287
United Nation's FAO Forestry Department, 287–289
Interagency Task Force on Tropical Forests, 184
International forest resource activities (*Cont.*)
interest group involvement in
Earthwatch, 303
Greenpeace International, 303
International Society of Tropical Foresters, 303
International Union of Forestry Research Organizations, 303
issues of concern to, 61
Tropical Forestry Action Plan, 78
World Commission on Environment and Development, 61
Investment tax credits, 406
Iron triangles, 238–239
Issues:
characteristics required for expansion of, 44–46
defined, 31, 32
disposition of, 48–53
entrepreneurs involvement in, 37
examples, 58–63, 344
initiators of, 39–40
obstructing agenda status for, 51–53
publics relevant to, 41–44
redefined by evaluation, 160
triggering events for, 38–39

Judicial systems: (*See also* Law)
attorneys involved in, 256
adversary process followed by, 250, 251
challenges to
judicial activism, 261
policy making, 261
technical forestry decisions, 262, 263
courts: type of
appellate courts, 225
trial courts, 254, 255
example rulings by, 127, 245–248
frequency of rulings by, 245
illegal conduct addressed by, 251
interest group involvement in, 318
involvement in policy implementation, 150–151
judges participation in
decision making, 256, 257
responsibilities, 256, 261
organization, 252, 253
policy making by
appropriate scope, 260
characteristics, 258
relief granted by
monetary relief, 254
permanent injunction, 255

Judicial systems (*Cont.*)
preliminary injunction, 255
temporary restraining order, 254, 255
strengths of, 251, 251

Land and Water Conservation Fund Act, 442
Land-grant universities:
extension forestry programs, 391
Law:
categories of
civil, 249
common, 249
criminal, 249
public and private, 248, 249
defined, 248
Laws, enactment of:
calendar, 212
committee action, 211, 212
concurrence, 213, 214
drafting, 211
executive approval (veto), 214
floor action, 212, 213
introduction, 211
referral, 211
Leadership:
characteristics of, 272
influence on policy (program) termination, 196
interest group, 307, 308
judicial, 261
Legislative committees: (*See also* Legislative systems, Oversight and review: Legislative)
Congressional examples of, 232, 233
hearings
conducted by, 354, 355
interest group participation in, 317, 318, 325
iron triangle involvement, 238, 239 (*See also* Agencies, Interest groups)
role of, 233–235
types, 232
Legislative staff:
committee
functions, 236, 237
norms of conduct, 237
number of, 235
personal, 236
of state legislatures, 237
Legislative systems:
challenges to
decision-making process, 241
innovation, 241

Legislative systems (*Cont.*)
parochialism, 241
power, 242
responsiveness, 240
committees, 232–235
functions and responsibilities of
constituent representation, 208
enactment of laws, 209–214
information and education, 208, 209
oversight and review, 214–217
summary of, 240
hearings conducted by, 324, 354, 355
information flows to
political relevance, 226
biased character, 226
forest resource, 227
interaction with agencies (bureaucracies), 238–240
involvement in policy implementation, 149
leadership, 227–231
legislators, 219–277 (*See also* Legislators)
organization of, 217–218
policy evaluation role of, 178, 179
staff, 235–237 (*See also* Legislative staff)
supporting agencies, 237, 238
rules of procedure, 218, 219
Legislators:
agency deference to, 239
agenda-setting role, 54–55
constituencies, 223
decision making, 224–226
issues addressed (types) by, 221, 222
limits on discretion of, 226
motivation for
oversight and review, 214, 215
serving in legislatures, 219
norms of conduct, 221
relationships with other legislators, 220, 221
representation styles
delegate, 223
trustee, 224
work loads, 219, 220
Legitimizing:
defined, 126
examples of, 137, 272
forms of, 126–131
political feasibility of
defined, 131
criteria for judging, 132–133
information requirements, 133–135
strategies and tactics, 135, 136
relation to other policy process events, 137, 138

Library of Congress (*see* Congressional Research Service)
Lobbying:
  interest group practice of, 316, 317
  grass roots, 318, 319
  government regulation of, 309
  requirements for effective, 325
Logrolling, 109–111

Market inefficiencies, 382
Mass media:
  agenda-setting role, 55,56
  characteristics of news, 334, 335
  communication process of, 334
  influence over policy substance, 336
  involvement in policy process, 335, 336
  types of, 333, 334
McIntire-Stennis Act (Cooperative Forestry Assistance Program Act), 394
Mediation, 113
Media (*see* Mass media)
Minerals, 8
Mixed scanning selection model, 99–100

National Environmental Policy Act, 128, 442
National Forest Management Act: (*See also* National Forests, Planning)
  challenges resulting from, 359–360
  cost of planning required by, 359
  example forest plan (Clearwater National Forest), 357–359
  planning process and procedures required by, 354–359
National Forest Products Association:
  budget, 301, 306
  employees, 301
  Forest Industries Political Action Committee, 321
  issues of concern to, 306
  membership, 301,
  objectives, 305
  organization, 305, 306
  publications, 306
National Forests:
  area, 435, 437, 438
  legitimized policies for, 130
  planning use and management of, 353–360, 439
  policies and programs, 438, 439
  recreation visits, 436
  timber harvest, 436
  wilderness recreation, 76, 436, 437, 438
National Wilderness Preservation System:
  agencies responsible for, 436, 437
  area, 7, 436, 437
  recreation visits to, 7, 437
National Wildlife Federation:
  issues of concern to, 62
Negotiation (*see* Bargaining)
Nonindustrial private forests:
  area, 370, 371
  benefits produced by, 371–373
  management problems of, 373
  policy options available to, 373–375
  timber harvest, 371
  timber productivity, 371

Office of Management and Budget, 272
Office of Technology Assessment, 238
Organized anarchy selection model:
  characteristics of, 101–102
  operation of, 102–103
Oversight and review: legislative: (*See also* Legislative committees)
  motivation for, 214, 215
  problems with, 217
  types of, 215–217

Pesticide regulation, 415
Planning: (*See also* National Forest Management Act, Planning: Fiscal and budgetary, Resources Planning Act)
  benefits of, 353
  fiscal and budgetary, 360–367
  for land use and management, 353, 354
  National Forest Management Act, 353–360
  principles for
    land use and management planning, 360
    strategic planning, 352, 353
  Resources Planning Act, 340–353
  strategic program directions, 339, 340
Planning: fiscal and budgetary:
  appropriation, 364
  authorization, 364
  base budget, 364
  challenges to, 366, 367
  character and process of, 361–364
  phases involved in, 361
  strategies for budget development, 365, 366
Policy:
  characteristics of, 17–20
  clarification caused by evaluation, 157
  defined
    policy, 15–16
    forest resource policy, 15–16

Policy (*Cont.*)
development of, 20–23
industrial development of, 458–460
judging success of, 163–165 (*See also* Evaluation)
Policy participants (*see* Bureaucratic systems, Interest groups, Judicial systems, Legislative systems, Mass media, Political parties, Public: general)
Policymakers:
difficulty with evaluations
conflicting expectations, 174, 175
mistrust of results, 175, 176
resistance, 175
involvement in selection, 84–88
judges as, 258–260
legislators as, 219–227
personality of, 88
Policy process: (*See also* Agenda setting, Evaluation, Formulation, Legitimizing, Selection, Implementation, Termination)
major events of, 21–23
operation of, 23–28
research into, 460–470
summary of, 447, 448
Political action committees, 319–321
Political moods, 36
Political elites:
framework for policy development, 21
Political parties:
purpose of, 332
environmental and forestry platforms of, 332, 333
President's Commission on Americans Outdoors, 74–75, 180
Presidential Commission on State and Private Forests, 180
Private forest-management assistance, 185, 186, 383 (*See also* Direct technical assistance, Extension, Indirect technical assistance)
Private forestry (*see* Industrial Private Forests, Nonindustrial Private Forests)
Professional interest groups, 299 (*See also* Interest groups)
Professionals (*see* Forestry professionals)
Property taxes, 407, 408
Public: general
activism, 330, 331
acknowledgement of forests, 8–9
involvement in policy process
agenda setting, 56, 331, 332
evaluation, 182
Public (*Cont.*)
implementation, 148
mass public, 42–44
attentive public, 42–43
general public, 44
Public goods, 381
Public interest:
interest group influence over, 296
in industrial private forests, 379
in public lands, 437
Public Land Law Review Commission, 121

Rational incremental selection model:
characteristics of, 93–98
difficulties with application of, 98–99
role of agreement in, 98
Recreation user fees, 128
Regulatory initiatives:
business structure and activities as focus of, 417
land-use, 413–415
pollutant, 415–417
forest practices, 417–432 (*See also* Forest practices regulation)
Research: (*See also* Indirect technical assistance)
policy process, 460–470
programs of, 393, 394
Resources Planning Act: (*See also* Planning)
challenges to, 352, 353
example of formulation, 78
legislative oversight and review required by, 216
objectives of, 351
planning process authorized by
accomplishment statements, 345
assessment, 340, 341
policy statement, 345
program, 341–345
Routine formulation, 68–69
Rural forestry assistance programs (*see* Direct technical assistance)

Scientific information:
role in policy process, 448–452
scientists role in providing, 449
Selection:
bargaining and negotiation in, 103–114
criteria used in, 114–120
defined, 81
examples, 122, 123
group versus individual involvement in, 88–91

Selection (*Cont.*)
mediation role in, 113
mixed scanning model of, 99–100
organized anarchy model of, 100–103
policymaker considerations in, 84–88
programmed versus unconventional, 82–83
rational comprehensive model of, 91–93
rational incremental model of, 93–99
relation to other policy process events, 123
Service foresters (*see* Direct technical assistance)
Side payments, 111
Sierra Club:
budget, 300, 304
employees, 300, 304
issues of concern to, 62, 303–305
membership, 300, 304
organization, 303, 304
publications, 304
Sierra Club Committee on Political Education, 320, 321
Society of American Foresters:
budget, 302, 307
employees, 302, 307
issues of concern to, 62, 307
membership, 302, 307
organization, 306, 307
policy development by
procedures, 315
issue selection criteria, 313
publications, 307
State forest practices laws (*see* Forest practices laws)
State forestry:
direct technical assistance programs administered by, 383–388
example agencies involved in
Georgia Forestry Commission, 283–285
Minnesota Division of Forestry, 285,286
Oregon Department of Forestry, 282, 283
extension service programs, 390–392
fiscal incentive programs, 403
forest practices regulation programs, 417–432
indirect technical assistance
fire protection, 393
pest management, 393
research, 393, 394
land area owned by, 435
landownership policies of, 442, 443
land-use regulations, 414, 415
organization of agencies involved in, 274, 275
State forestry (*Cont.*)
policies legitimized in forest plans, 129–130
property taxes focused on, 407, 408
public investment in, 10
recreation visits to state forests, 436
Statutes (*see* Laws, enactment of, Law)
Sunset laws, 197 (*See also* Termination)
Subgovernments (*see* Iron triangles)
Subsidies (*see* Fiscal incentives)
Symbols:
agenda-setting role of, 46–47
defined, 46
used during legislative oversight, 216, 217

Tax incentives:
challenges to, 409
income
capital gains, 405, 406
investment tax credits, 406
estate and inheritance tax credits, 406, 407
principles for, 404, 405
property, 407, 408
Technical assistance (*see* Direct technical assistance, Extension, Indirect technical assistance)
Tennessee Valley Authority, 274
Termination:
defined, 190, 191
examples, 202
obstacles to
ideology of permanence, 200
legal deterrents, 201
moral resistance, 199, 200
political opposition, 200, 201
strategies for, 194–197
sunset laws requiring, 197
reasons for, 191, 192
relation to other policy process events, 202, 203
tactics for, 197–199
types of, 192–194
zero-based budgeting requiring, 197
Triggering events, 38–39

United Nations' FAO Forestry Department (*see* FAO Forestry Department)
U.S. Congress:
example policy evaluations, 178, 179
enactment of laws, 209–214
U.S. Congress Office of Technology Assessment, 179
U.S. General Accounting Office, 179

USDA-Agricultural Stabilization and Conservation Service, 398, 401, 402
USDA-Extension Service:
budget, 272
employees, 272
relation to state extension services, 391
USDA-Forest Service:
budget, 10, 274
budget preparation time schedule, 362, 363
direct technical assistance programs, 383–388
employees, 274, 279
evaluation staffs, 183, 184
example functions, 278
example policy selection criteria, 121, 122
fiscal incentive programs, 398–401
indirect technical assistance programs
pest management, 393
fire protection, 393
research, 393, 394
land ownership, 435 (*See also* National Forests)
legitimized policies in manuals and handbooks, 128, 129, 273
objectives, 275–276
organization, 276–278
planning
land use and management planning, 353–360
strategic program planning, 340–353
policy formulators, 76
policy responsibilities, 279
public acknowledgement of, 9

USDA-Soil Conservation Service:
direct technical assistance programs, 385
small watershed programs, 394
strategic planning activities, 340
USDI-Bureau of Indian Affairs:
land area managed by, 435
recreation visits, 436
timber harvest, 436
USDI-Bureau of Land Management:
budget, 10, 274
criteria used for policy selection, 121
employees, 274
land ownership, 435, 439
land-management planning, 440
Oregon and California Lands, 436, 440
policies and programs, 439, 440
policy evaluation staffs, 183
recreation visits, 436
wilderness, 436, 437, 440
USDI-Fish and Wildlife Service:
budget, 10, 274
employees, 274
land ownership, 435, 441
planning activities, 441
policies and programs, 442
policy evaluation staffs, 183
wilderness, 436
USDI-National Park Service:
budget, 10
land ownership, 435
policy evaluation staffs, 183
recreation visits, 436, 437
wilderness, 436, 437

Water pollutant regulation, 415, 416 (*See also* Forest practices regulation)
Wild and Scenic Rivers Act, 413, 414, 439 (*See also* Regulatory initiatives)
Wildlife and fish:
importance of, 7
Window of Opportunity, 36
(*See also* Agenda setting)
With-and-without evaluation design, 165–167
(*See also* Evaluation)
Wood products:
importance of, 6
Water and Watersheds:
importance of, 6
Wilderness (*see* National Wilderness Preservation System)
Wood-based industry:
issues of concern to, 62–63
policy development process, 458–460
World Commission on Environment and Development, 61

Zero-based budgeting, 197 (*See also* Termination)